Innovation, Industrial Dynamics and Structural Transformation

Uwe Cantner · Franco Malerba
Editors

Innovation, Industrial Dynamics and Structural Transformation

Schumpeterian Legacies

With 69 Figures and 45 Tables

Springer

Prof. Dr. Uwe Cantner
Friedrich-Schiller-University of Jena
Faculty of Economics and Business Administration
Chair for Microeconomics
Carl-Zeiss-Str. 3
07743 Jena
Germany
uwe.cantner@wiwi.uni-jena.de

Prof. Franco Malerba
CESPRI-Bocconi University
Via Sarfatti 25
20136 Milan
Italy
franco.malerba@uni-bocconi.it

Some of the contributions have been published in Journal of Evolutionary Economics, Volume 16, 1-2, 2006

Library of Congress Control Number: 2006938716

ISBN 978-3-540-49464-5 Springer Berlin Heidelberg New York

Springer is part of Springer Science+Business Media

springer.com

Production: LE-TeX Jelonek, Schmidt & Vöckler GbR, Leipzig
Cover-design: Erich Kirchner, Heidelberg

SPIN 11935131 42/3100YL - 5 4 3 2 1 0 Printed on acid-free paper

Table of Contents

Part 6: Patents – patenting strategies and impacts of patents

Part 7: Public Policy – competition policy and the patent system

Franco Malerba · Uwe Cantner

Introduction

The general theme of the 10th International Joseph A. Schumpeter Society Conference, held during June 9th–12th, 2004 at Università Luigi Bocconi, Milan, was the exploration of the dynamics of industries driven by the highly interrelated processes of innovation and of structural transformation. The phenomena addressed are at the core of Schumpeter's work and the discussion of these endogenously created modes of change constitutes one of the major Schumpeterian legacies of today. Indeed, the influence of the Schumpeterian approach to economic dynamics is far reaching and covers evolutionary as well as neoclassical theories.

This book provides an account of work in the Schumpeterian and evolutionary tradition of industrial dynamics and the evolution of industries. Here, the main analytical concern is that, over time, industries evolve and change their structure, and that, in this dynamic process, knowledge and technologies, the capabilities and incentives of actors, new products and processes (as well as variants of existing ones), and institutions affect and constrain change, sometimes smoothly and sometimes in a radical way. Thus, what is meant here by the term structure and structural change is not the traditional market structure view (as reflected, for example, in the classical discussion of the Schumpeterian Hypotheses), but all those elements and relations between actors, knowledge and technologies which drive innovative activities and greatly affect economic performance in an industry. Investigations into these complex phenomena show a deep interdependence between empirical work, delivering a rich account of regularities and stylized facts in the structure of industries and their change alteration, and theoretical analyses, ranging from appreciative theorizing to more formal modeling.

The various contributions in this book reflect upon this general theme in a number of ways. The book is organized in the following way.

F. Malerba
CESPRI, Bocconi University, Via Sarfatti 25, 20136 Milan, Italy
E-mail: franco.malerba@unibocconi.it

U. Cantner
Friedrich-Schiller-University Jena, Carl-Zeiss-Str. 3, 07743 Jena, Germany
E-mail: uwe.cantner@wiwi.uni-jena.de

Part 1 is conceptual and theoretical, and deals broadly with the main themes tackled by the book: the relationship between innovation and the evolution of industries, entrepreneurship and economic evolution, and the role of knowledge in the theory of the firm. Franco Malerba's Presidential Address reflects on innovation and the evolution of industries. After demonstrating the state of the art in the realm of industrial dynamics and its roots in the Schumpeterian work, Malerba identifies the following issues at the top of tomorrow's research agenda: demand, knowledge, networks and coevolution. Brian Loasby relates entrepreneurship to a theory of human mind and places them in an evolutionary perspective. Markus Becker, Patrick Cohendet and Patrick Llerena examine the role of, and the relationship between, the division of knowledge and the division of labor, and discuses their relevance for evolutionary theories of the firm that are knowledge based.

The next part of the book considers two key elements in industrial dynamics and structural transformation: new firms and networks. *Part 2* investigates entry and the role of small firms in industrial dynamics. Massimo Colombo and Luca Grilli analyze the relationship between the characteristics of the human capital of the founding team and the growth of new technology based firms. Roberto Fontana and Lionel Nesta examine the market positioning of entrants and incumbents in a fast growing industry – the LAN switch market. *Part 3* investigates networks in two complementary ways, by focusing on the demand side (and on network externalities) and on the development of innovation networks. Daniel Birke and Peter Swann examine the role of network effects in consumers' choice of mobile phone operators in the UK, while Robin Cowan, Nicholas Jonard and Jean Benoit Zimmerman develop a model of formation of R&D alliances, where firms combine their knowledge for innovation. Paul Muller and Julien Pénin propose a simulation model of the emergence and dynamics of innovation networks, with particular emphasis on open knowledge disclosure. Alicia Giacchero, Nora Donnini and Fanny Martin evaluate the capacity of a local SME productive system in Argentina to incorporate scientific knowledge and technological innovation by identifying principle external and internal conditions of enterprises required to form a critical mass.

Part 4 moves the analysis to the sectoral level. Stefano Brusoni and Giorgia Sgalari show that creative destruction through new combinations may affect the structure of a mature industry – tire manufacturing – over the long wave of the process of diffusion. Simone Ferriani, Gino Cattani and Charles Baden-Fuller examine how human capital and network resources interact with the structure of the industry and influence different dimensions of fitness in the Hollywood filmmaking industry. José Vitor Bomtempo, Edmar Luiz Fagundes de Almeida and Ronaldo Goulard Bicalho analyze the way the selection environment of gas to liquid technology is challenging its natural trajectory and the consequent firms R&D strategies. Finally, for 22 sectors in 10 European countries Tommaso Antonucci focuses on testing the hypothesis of the contrasting effects of product and process innovation strategies on employment at the sectoral level.

The next part of the book discusses several related aspects of industrial dynamics and transformation. *Part 5* looks at the dynamics of knowledge diffusion, absorption and spillovers and firm strategy. Andrea Morone, Piergiuseppe Morone and Richard Taylor investigate interactive learning in the dynamics of knowledge diffusion by means of a laboratory experiment and a simulation model. Aykut

Lenger and Erol Taymaz study spillover and technology transfer of domestic and foreign firms in Turkish manufacturing industries, and reiterate the importance of tacit knowledge. Gianbattista Dagnino and Marcello Mariani claim that firm strategy is fundamentally a dynamic process of gap-bridging between capability space and opportunity space and investigate the strategic role of the firm in coevolution. *Part 6* discusses the role of patents by looking at the strategic aspects of knowledge protection and the impact on corporate growth. Carine Peeters, Bruno van Pottelsberghe de la Potterie examine innovation strategies, in terms of R&D, product or process orientation and collaborative research, and their effect on patenting behavior. Michaela Niefert examines econometrically the relationship between patenting activity and employment growth in the German start-up firms.

Finally, the last part (*Part 7*) investigates two policy issues: competition policy and its ability to foster innovative activitities and the effectiveness of the patents system. Jean-Luc Gaffard and Michel Quéré discuss the aim of competition policies in terms of the dilemma between optimizing market structure and favoring innovative behavior through appreciative and discretionary policies. Tomas Valée and Murat Yıldızoglu develop an evolutionary model of industry dynamics in order to examine the consequence of a stronger patent system.

All these contributions confirm the richness of a Schumpeterian view of industrial dynamics and structural change and point to the need to look at all the aspects and dimensions examined in this book as different but interrelated elements that characterize the evolution and transformation of firms and industries.

Part 1:

Innovation, entrepreneurship, the firm and industry evolution

Franco Malerba

Innovation and the evolution of industries

Abstract The analysis of innovation and the evolution of industries has witnessed major progress in several areas. In the last years, several contributions at the empirical, appreciative, econometric and modelling levels have greatly advanced our understanding of innovation, industrial dynamics and the evolution of industries. This paper reviews these contributions. A discussion follows on four key challenges that are required for a better understanding of the relationship between innovation and the evolution of industries: the analyses of demand, knowledge, networks and coevolution.

Keywords Technological change · Market structure · Firm strategy · Market performance · Industry studies

JEL Classification O3 · L1 · L6

1 Introduction

The relationship between innovation and the evolution of industries is at the core of Schumpeter's work and is one of the major Schumpeterian legacies. It is a central theme in the Schumpeterian approach to economic dynamics, as well in evolutionary and neoclassical theories.

In this address, I will focus on the progress that has been made in the last years and on the challenges that lie ahead in understanding the relationship between innovation and the evolution of industries. In the first part of the paper, I will discuss the progress, while in the second I will focus on four big research challenges that I think lie ahead.

This is the Presidential Address delivered at the International Schumpeter Society Meeting held in Milan in June 2004.

F. Malerba
CESPRI, Bocconi University, Via Sarfatti 25, 20136 Milan, Italy
E-mail: franco.malerba@unibocconi.it

2 Innovation and the evolution of industry: a key Schumpeterian theme

The relationship between innovation and industrial change has always been central in Schumpeter's work in various ways and specifications. In *The Theory of Economic Development* (Schumpeter 1934), *Business Cycles* (Schumpeter 1939), and *Capitalism Socialism and Democracy* (Schumpeter 1950), Schumpeter was very much interested in innovation either as a process of creative destruction or as a process of creative accumulation (as Keith Pavitt would call it). He placed innovation in the evolution of industries and within the process of economic transformation: for him, innovation was very closely linked to the emergence, growth and decline of industries, which historically marked the development of capitalism. In Schumpeter's work, one may identify several lines of research that have been developed in the following decades at the empirical or at the theoretical level: innovations as clustered historically in specific industries, industrial development and transformation as associated with the emergence and growth of different sectors, dynamic competition as the struggle between new firms that introduce new technologies and products, and incumbents that focus on existing technologies.

After Schumpeter's death, many of the central messages of Schumpeter regarding innovation, industrial development and dynamic competition were put at the margin of mainstream economic research. There was a shift of attention away from transformation and industrial dynamics towards the relationship between innovation and firm size, on the one hand, and innovation and market structure, on the other. The mainstream literature referred to this field of research as the test of the two Schumpeterian hypotheses. But this testing did not have much in common with the original spirit of Schumpeter's work on this issue. The research that emerged and prospered during the 1950s and 1960s had a static flavor, as one may find in the tradition of the Structure–Conduct–Performance Paradigm, and did not pay any attention to industry evolution and transformation. With the advent of game theory, the focus moved to firms' strategies in R–D and licensing. But also in this case, the original message of Schumpeter that had been focused on innovation, industry evolution and transformation got lost.

Since the late 1970s and the early 1980s, however, a new focus of empirical and theoretical research on innovation and the evolution of industries has emerged. Most of this research has developed the original message of Schumpeter on innovation, the evolution of industries and structural transformation in various directions. Of course, not all of this research fits in a truly Schumpeterian framework: indeed the conceptual and theoretical approaches of several of these contributions have been quite different and eclectic. Most of them, however, share two common elements: the recognition of the key role of innovation in economic processes, and the need to use a dynamic analysis.

3 The progress

The progress obtained in the analysis of innovation and the evolution of industries has been on several fronts. First of all, many case studies of innovation and industry evolution have been developed for several industries. The so-called "SPRU tradition" has greatly contributed to the recognition of the major differences that exist across different industries, technologies and countries in the relationship between innovation and the evolution of industries (Freeman 1974; Pavitt 1984).

Second, our appreciative understanding on several aspects of the relationship between innovation and the evolution of industries has greatly increased.

(a) Progress has been obtained in examining the extent and effects of heterogeneity of firms in terms of different knowledge, competences and learning processes. As Richard Nelson, Sidney Winter, Wesley Cohen, Dan Levinthal, Richard Langlois and Giovanni Dosi among others have shown, different agents know how to do different things in different ways. Here the contributions of evolutionary and of organizational theory have been great, by placing emphasis on cognitive aspects such as routines, beliefs, objectives and expectations, in turn affected by previous learning and experience and by the environment in which agents act. Relatedly, heterogeneity and its dynamics have been related to processes of variety generation and selection, which take place in various ways and forms during the evolution of an industry (Nelson 1995; Cohen and Levinthal 1989; Gavetti and Levinthal 2000; Levinthal 1993).
(b) There is now enormous evidence on the contributions of universities, public research organizations, the military, other public actors and financial organizations (such as venture capital) in the generation and diffusion of technological advance in industries. Their roles, however, has been shown to be quite different in different industries (Levin et al. 1987; Cohen et al. 2002).
(c) The role of different institutions—some of them national, other sectoral—has been recognized as relevant. The major point here is that innovative activities in industries are shaped by institutions, which include standards, regulations, norms, routines, common habits, established practices, rules, and so on (Edquist 1997).
(d) Industries have been interpreted as systems, in which actors are related and interact in various ways (formal as well as informal relationships, market and non market interactions, and so on) and are strongly influenced by their competences, learning processes, the knowledge base of sectors and the institutions. In this frame, the notion of sectoral systems of innovation (Malerba 2004) is a useful tool for examining innovation in a sector.
(e) Finally, we know that industries follow specific dynamics of innovation, firm entry and growth and market structure, as the industry life cycle tradition (Abernathy and Utterback 1978; Utterback 1994; Malerba and Orsenigo 1996) shows. We also know that these dynamic sequences are different from one industry to another (Klepper 1997).

Third, econometric work on innovation and industrial dynamics has tremendously progressed in the last thirty years. With the availability of advanced computer technology and new firm level data, econometric analyses have moved from cross sections work during the 1960s and 1970s to panel data and longitudinal analyses since the early 1990s. Great progress has been obtained in identifying, measuring and understanding stylized facts and statistical regularities, and the factors explaining them. This has begun to shed light also on the statistical properties of change in terms of industrial demography, entry and innovation, firm growth, stability of firm size distributions, and persistence in asymmetric firm performance (see the work by Audretsch 1995; Baldwin 1995; Geroski 1994; Bottazzi et al. 2002 among others). All this work has identified robust inter-industry differences in concentration, firm age distribution and the characteristics of innovations.

Fourth, at the modelling level, different strands have developed, in various ways and directions.

(a) At one extreme, one can find models of industry dynamics with rational actors and technological learning by incumbents or entrants or both, and the competitive process weeding out the heterogeneity in firm populations (see for example, Jovanovic 1982; Ericson and Pakes 1995), with not much consistency with some stylized facts such as inertial asymmetric performance, irrational entry processes and so on.
(b) In some models such as Sutton (1998), technological and demand related factors set bounds on industrial structures via no arbitrage conditions, entailing corresponding Nash equilibrium on industry specific entry processes. No attention, however, is paid to the learning processes of firms.
(c) The tradition of population ecology has modelled industrial dynamics as the growth and decline of various groups of firms, each of which with different characteristics (Carroll and Hannan 1999).
(d) More focused on the learning processes of firms and on the general features of industrial dynamics are the evolutionary models à la Nelson–Winter, which have boundedly rational actors, learning and processes of experimentation and imperfect trial and error. Selection processes take place on a heterogeneous population of firms (Nelson and Winter 1982; Dosi et al. 1995), These models have a destrategizing conjecture, in that differences in structures and processes of change are understood as independent from firm micro strategies (Winter et al. 2000).
(e) Then one can find formal models of industry life cycle, analyzing together product and process innovations; rate and type of entrants; selection; firm size and growth; market concentration and market niches (Klepper 1996, 2002; Klepper and Simons 2000). Here there are both a strong link between stylized facts, empirical analyses, econometric analysis and formal theory, and an explanation of different types of industry life cycles.
(f) More attention to the specificities and histories of various industries is paid by history friendly models, which fall into the evolutionary tradition (Malerba et al. 1999; Malerba and Orsenigo 2002). These models pay attention to the evidence and the specific dynamics of the evolution of industries, by modelling the appreciative explanations and historical events that have shaped it.
(g) Finally, progress has been made at a more macro level, by linking innovation and industry evolution to structural change and the changing sectoral composition of the economy, as one can see from the from the early work of Dahmen (1989), Kuznets (1965) and Pasinetti (1981) to the more recent ones by Metcalfe (1998), Dopfer et al. (2004), Dosi (2001), Saviotti (1996), and Montobbio (2002).

4 The challenges

The point to be made in this paper is that, while a lot of ground has been covered in understanding innovation, industrial dynamics and the industry life cycle from an empirical and a theoretical point of view, some key challenges still lie ahead if we want to understand fully the relationship between innovation and the evolution of industries. Some sketchy observations of how industries are structured and how they change over time provide us some questions and hints in this respect. In *telecommunication equipment and services*, a convergence of different technologies, demand and industries has taken place, with processes of knowledge integration. This convergence has been associated with the creation of a wide variety of different specialized and integrated actors, ranging from large equipment producers to new service firms. In this broad sector, innovation is very much affected by standards, the institutional setting and the processes of privatization and liberalization. *In machine tools,* the evolution of the industry has been related to an application-specific knowledge base and has been associated with firm specialization. Here, user-producer interaction, local networks of innovators and in-house experienced human capital are key factors for innovation. However, products are increasingly being modularized and standardized, and suppliers of components are increasingly involved in innovation. In *pharmaceuticals and biotechnology*, a wide variety of science and engineering fields are playing important roles in renewing the search space. Universities, venture capital and national health systems play a major role in the innovative process. Several are the relevant actors—large firms, small firms and new biotech firms (NBFs)—and an extensive division of labor through networks exists. NBFs have entered the sector, competing as well as cooperating with (or being bought up by) the established large pharmaceutical firms. In this sector, demand and institutions (such as regulation, IPR and national health systems) affect the innovation process. In *software,* a highly differentiated knowledge base in which the context of application is relevant has created several different and distinctive product groups. The role of large computer suppliers in developing integrated hardware and software systems has been displaced by many specialized software companies innovating either in package software or in customized software. User–producer interaction, global and local networks and highly mobile skilled human capital are present. The role of the university has become important in the open source domain. IPR, standards and standard setting alliances play a major role in innovation, diffusion and competition. The emergence of *new clusters* that span over several sectors, such as internet–software–telecom, biotechnology–pharmaceuticals, and new materials, is one of the most relevant current transformation processes in sectoral systems. Here, a great role is played by the integration and fusion of previously separated knowledge and technologies and by the new relations involving users, consumers, firms with different specializations and competences, and non-firm organizations grounded in previously separated sectors.

This description points to some additional and quite relevant dimensions of the relationship between innovation and industry evolution. During its evolution, an industry undergoes a process of transformation that involves knowledge, technologies, learning, the features and competences of actors, types of products and processes, and institutions. An industry also changes its structure, where the term "structure" here means not market structure, but rather the network of relationships (competitive and cooperative, market and non market, formal and informal) among actors that affect innovation and performance in an industry.

In this frame, innovation and industry evolution may be seen as:

- the outcome of learning processes by firms and by individuals;
- based on a specific knowledge base which characterizes the industry;
- the result of the competitive and cooperative, market and non market, formal and informal interactions of several actors with different knowledge and competences;
- taking place in specific institutional settings, some of which are national while other are specific to the sector;
- bringing change and transformation not only to products and processes, but also to the actors, links, institutions and knowledge itself.

This view of industries characterized by a specific knowledge base, by firms with different competences, knowledge, learning processes, networks of interactions and a variety of institutions bring to the forefront at least four challenges for research in the analysis of innovation and the evolution of industries: *demand, knowledge, networks and coevolution*. Let me examine these in detail in the following pages.

5 Demand

The first challenge that I want to explore regards the role of demand in the evolution of industries.

As a way of introduction, let me first disagree with the usual complaint that demand has not been studied in its relationships with innovation and industries in the last decades. In the literature, we have various empirical and theoretical strands, from the old debate regarding demand pull vs technology push (Schmookler 1966; Meyers and Marquis 1969 and the SPRU-SAPPHO project), to the analysis of demand, market structure and innovation (from Kamien and Schwartz 1975 to Sutton 1991, 1998), imperfect information among consumers, user initiated innovation (Von Hippel 1986) user–producer interaction (Lundvall 1988) and value networks (Christensen and Rosenbloom 1995). In particular, advertising, bandwagon and networks have been shown to be important factors in influencing the magnitude and orientation of inventive effort and the degree of industry concentration. The presence of submarkets plays a role in affecting the growth and size distribution of firms within an industry, as Klepper and Thompson (2002) show for the laser industry. Demand has been key with respect to the emergence of disruptive technologies, as Christensen (1997) has documented in the case of hard disk drives, earthmoving equipment, retail stores and moto controls. Here, the early development of disruptive technologies serves niche segments that value highly their non standard performance attributes. Further developments in the performance and attributes of the disruptive technologies lead to a level sufficient to satisfy mainstream customers. Finally, when demand and innovation are examined, one has to mention the whole literature on diffusion: all the major empirical advancements and theoretical models regarding diffusion are nothing but contributions regarding the demand of innovation. The same holds for the literature on competing technologies, which pays attention to externalities and increasing returns. In sum, rather than saying that demand has been disregarded by economic studies on innovation, it would be more correct to say that it has been the Schumpeterian tradition that for a long time has kept demand rather marginal, because it has emphasized the supply side. I have to add that Schumpeter himself might have been responsible for that, given his emphasis on the passivity of the consumer in the innovation process.

We also need to recognize that some ground has been covered at the empirical, appreciative and modelling levels of demand with respect to the evolution of industries. At the empirical level, the role of demand during specific stages of the evolution of an industry has been shown to be relevant. In semiconductors and computers, public procurement has been important for innovation in the early stages of industries (Malerba 1985). In computers experimental customers have been major actors in the emergent phase of the industry (Bresnahan and Malerba 1999; Bresnahan and Greenstein 2001). In information technology, user involvement has been key for the development and modification of standards. In pharmaceuticals, demand channelled through agencies, physicians and the health system has played a significant role in the diffusion of new drugs. In instrumentation (Von Hippel 1988) and machine tools, lead users have played a major role in innovation and in shaping both the supplier industry and the user industry.

However, we need to ask analytically in which ways demand affects innovation in specific industries. Standard economic analysis claims that demand provides incentives to innovation during industry evolution. This is indeed a correct statement. The size, growth, structure and composition of demand, differentiation and market segmentation affect innovation in various ways in different stages of the evolution of an industry. One could also add that, in terms of incentives, demand is not homogeneous: it is highly heterogeneous in terms of segments, types of firms (private vs public) and individual customers.

But there is the need to move from this correct claim and to explore new conceptual grounds, as Loasby (1999, 2001), Metcalfe (1998), Witt (2003), Saviotti (2001), Teubal (1979), and others have started to do. In this respect,

(a) Consumer behavior plays a major role in affecting innovation. Consumer behavior may include the presence of information asymmetries and imperfect information with respect to new markets and submarkets, as well as routines, inertia, and habits on the part of consumers.
(b) The knowledge and mental frameworks of consumers and users greatly affect innovation and performance. There is a "knowing that" and knowing how" on the part of consumers. There is also learning and knowledge growth in consumption, much of which is local. Consumers are characterized by routines similar to firms, and deliberative decisions interact with habits when consumers confront new opportunities or new products. However, producers and consumers should not be treated as completely symmetrical because of their different stress on standardization and variety (Langlois 2001; Devetag 1999; Aversi et al. 1999).
(c) Consumer competences play a major role in influencing innovation. One could mention the absorptive capabilities of users, or the role of lead users and

experimental users. In addition, the distribution of competencies among users greatly affects the dynamics of industries.

(d) An analysis of the role of demand framed in the perspective discussed above can be linked to some strong empirical evidence on specific industries. In some industries, users participate intensively in the innovation process. User involvement is more than simple participation in the innovation process. Involvement implies a psychological dimension and a behavioral dimension. It is quite relevant, for example, in participatory design, IT and standard settings. Here the relationship between knowledge and the mental frameworks of producers and of users plays a major role in innovation.

(e) In other industries, coinvention is quite relevant: innovation by sellers and complementary investments and innovation by buyers (in terms of new products, services, applications and investments in human capital). As Timothy Bresnahan and Cristiano Antonelli have shown for IT, coinvention involves the technology of the user as well as that of the supplier. User coinvention is particularly important in explaining technological change in IT applications (package software, semi-custom IT solutions, turn-key solutions). Coinvention pulls technological change in a variety of directions and ways. This means that, in IT, there is not "one" standard type of adoption. Rather, coinventions in IT and its applications represent developments in tightly coupled interconnected technologies (Bresnahan and Greenstein 2001; Antonelli 2003).

(f) In other industries, such as instrumentation as well as IT, lead users play a great role (Von Hippel 1988). Lead users face needs that will be general in a market place but face them months or years before the bulk of the marketplace encounters them. They are also positioned to benefit significantly by obtaining a solution to their needs (Urban and Von Hippel 1988). One has to recognize, however, that the contribution of lead users comes from knowledge related to their experience. Therefore, they have a major role in periods of stability of uses and applications, but they may be less relevant when radical change or instability affects demand.

(g) Finally, in some sectors, as in software (open source software), community of practices are major sources of innovation and change. They act as facilitators of innovation, because members who innovate are able to share their ideas with other members, assist them and even obtain resources to develop their innovations. As Harhoff et al. (2003) show, for innovators it might be beneficial to reveal information inside a community because they may induce improvements by others; be helped in achieving a standard; face low rivalry conditions; and expect reciprocity. Franke and Shah (2003) adds to these four reasons a fifth: the fun and enjoyment that arise through engagement in the task and community. Historically, a similar process can be found in other sectors: for example in machinery (see the case of the development of the Cornish steam engine during the British industrial revolution discussed by Nuvolari 2004).

Such a variety of roles in so many different industries calls for the need to examine systematically consumers and users in different industrial settings, as well as to develop taxonomies of the different role of demand in innovation in industries.

At the modelling level, the challenge is to examine theoretically some of the processes presented above. Broadly, one would like to model the links between demand dynamics, firm dynamics and technology dynamics. In fact, on the one hand, the emergence and development of new technologies create new markets, submarkets and niches. On the other, the dynamics of demand in terms of consumer learning may stimulate technological change and the entry of new firms. This is indeed a challenging task.

In this perspective, one interesting model by Adner and Levinthal (2001) examines the way in which demand in terms of performance thresholds, types of preferences, changing utility and differences across market segments interacts with technological change to guide the evolution of technology and competition during the life cycle of an industry. Adner and Levinthal (2001) model a coevolutionay process in which there is a demand life cycle: early on product innovation increases performance, but then (to the extent to which some performance thresholds are met) process innovation takes over. Later on, a new phase starts in which firms, given a certain willingness to pay by demand, focus again on performance increases and product innovation. In a sense, mature consumers may demand performance, but their appreciation for performance improvements is not reflected in their willingness to pay for the improved product. In addition, demand is also modelled in terms of different market segments. In this was, Adner (2003) shows that technology disruption à la Christensen may be the result of the interplay between the preference overlap of different market segments and preference symmetry. This is a very interesting avenue of research because it takes into account the demand life cycle and demand segments. However, in these models, demand is still static in its basic structure. In fact, thresholds, preferences and decreasing marginal utility are fixed. In reality, contrary to the model, the value consumers derive from performance improvements can change in response to changes in the environment in which products are used (changes in complements and changes in standards and regulations) or to marketing techniques.

With another methodology, similar issues are tackled by models of industry evolution in a "history friendly" fashion (Malerba et al. 2003). Here, various types of customers are present: "standard" customers attracted by established products and guided by product characteristics such as price and performance; experimental customers who crave new technologies in existing products; consumers in new demand segments who look for completely new products. This history-friendly model is inspired by the case of the computer industry, in which experimental users and new demand have played a major role in affecting innovation, competition among technologies and the dynamics of market structure. Here the successful introduction of a radically new technology in an industry, in which a dominant design and a small collection of dominant firms are present, may be dependent upon a group of experimental customers, who are willing to experiment and buy the new products related to new technology. This allows new firms with the new technology to stay around long enough to be viable. A similar dynamics is played by potential customers with different preferences, when potential markets are not served by incumbent firms. Both cases of demand permit new technologies effectively to grow, either within established firms or through new firms. This can have a profound, long-run effect on industry structure. Within this framework, it is also possible to model a convergence among different demand segments and its effects on market structure.

The discussion on modelling brings the coupled dynamics of demand and technology into the forefront of the analysis of industry evolution. In this frame, the interaction between producers and users changes the capabilities and preferences of both producers and users, and sets in motion a coevolution of technology, knowledge, market structure and innovation. Some of these processes have started to be tackled at the conceptual and appreciative levels, by examining learning and specialization in consumption and the increasing variety of goods and services offered in the market (Witt 2001; Saviotti 2001). These processes are indeed evolutionary in the sense that they imply learning, routinized behavior and selection (Metcalfe 2001).

One of the most evident examples of coevolution is coinvention. As Bresnahan and Greenstein (2001) show for IT, coinventions generate new trajectories of improvements in the original technology, new organizational change and new institutions, which in turn generate new coinventions between users and suppliers. For example, the rise in demand for the world wide web has set in motion entirely new waves of coinventions, with new application developments, new business models and new institutions, which in turn feed back on demand, changing it in various ways and so on.

In conclusion, the progress in understanding the relationship between demand, innovation and industry evolution calls for new challenges in terms of richer and more detailed empirical analyses, deeper appreciative understanding and formal modelling. They have to include feedbacks, coevolution and processes of specialization in consumption, proliferation of niches and convergence in demand.

6 Knowledge

A different reasoning could be employed for the second challenge: knowledge. Here we face one of the key building blocks of the evolutionary approach: the key role of knowledge and learning by agents in innovation. The work done on knowledge by Richard Nelson, Paul David, Stan Metcalfe, Sidney Winter, Brian Loasby, Giovanni Dosi, Bengt Ake Lundvall and Dominique Foray, among others, in the last twenty years has greatly enriched our understanding of what knowledge is and of its role, properties and relevant dimensions in innovation and change. In particular, the evolutionary literature has proposed that sectors and technologies differ greatly in terms of the knowledge base and learning processes related to innovation. In some sectors, science is the force driving knowledge growth, while in others, learning by doing and cumulativeness of advancements are the major forces. We also know that knowledge differs across sectors in terms of sources (firms, universities, and so on), domains (i.e. the specific scientific and technological fields at the base of innovative activities in a sector), and applications (Cowan et al. 2000). Knowledge has also different degrees of accessibility (i.e. opportunities of gaining knowledge that are external to firms and may be internal or external to the sector) with major consequences on entry and concentration, and may be more or less cumulative (i.e. the degree by which the generation of new knowledge builds upon current knowledge). In addition, knowledge may flow more or less intentionally across individuals and organizations, as the huge literature on knowledge spillovers (see for example Jaffe et al. 2000) and the work on knowledge and mobility of inventors shows (Kogut 2000; Balconi et al. 2004; Breschi and Lissoni 2004).

Given this progress, the challenge here is to move from the broad identification of the main characteristics and effects of knowledge across sectors to a detailed understanding of the specific type and structure of knowledge, of its various effects on innovation and the organization of innovative activities, and on the two-way relationship between knowledge evolution and industry evolution in different sectors.

In this respect, several issues emerge clearly. First, finer grained dimensions of knowledge (such as the ones mentioned for example in Winter 1987) and links and complementarities have to be taken into account (Malerba 1992). Links and complementarities may refer to scientific, technological or application knowledge. These complementarities are often a major source of increasing returns. Related major areas of inquiry include knowledge transmission, flows and spillovers within industries and across industries, the coordination and integration of knowledge, and the relevance of modularity of knowledge, as in the work by Loasby (1999, 2001), Foray (2004), and Brusoni et al. (2001).

This calls also for the use of appropriate indicators of knowledge, knowledge flows and knowledge structure. Patents and patent citations are one indicator, and have been used to describe knowledge flows among people, organizations and sectors. Patent citations represent indeed relational data, and identify a paper trail left by the knowledge flowing from the inventor/applicant of the cited document to the inventor/applicant to the citing one. But criticism may be advanced for the use of patent citations as an exact measure of interpersonal or interorganizational knowledge flows. In fact, often the flow comes from the patent examiner rather than the inventor. When the inventor cites a patent, it is not necessarily the case that there is an interpersonal knowledge flow (but more simply the inventor may retrieve information on the cited patent directly from a database). However, I remain convinced of the importance of patent citations in providing some evidence of a paper trail about knowledge links, and in describing some features of knowledge and knowledge networks of an industry, as flows of knowledge can be captured by patent citations even when inventors are unaware of those citations. The analyses by Jaffe and Trajtenberg (2002) and by Thompson and Fox-Kean (2005) show that patent citations are a noisy signal of the presence of spillovers, and that aggregate citations flows can be used as a proxy for knowledge spillover intensity. In any case, the use of patent citations in order to examine knowledge flows and networks is a very fruitful research direction, provided that one is aware of their limitations and uses them jointly with other qualitative and quantitative indicators.

Second, a deeper understanding of the effects of the type of knowledge on the organization of innovative activities and the relationships among firms in various sectors is required. This may be accomplished in various ways. One is to link one specific dimension of knowledge with the organization of knowledge production. For example, one could link codified knowledge, the specialization in knowledge production and the division of innovative labor, as Arora et al. (2001) do. Another is to link the leaning and knowledge environment to the patterns of innovative activities in a sector. The identification of some key properties of knowledge such as accessibility, opportunity and cumulativeness can be related to the notion of technological and learning regimes (dating back to Nelson and Winter 1982), providing a description of the knowledge environment in which firms operate. Malerba and Orsenigo (1995, 1997) have proposed that a technological regime is composed by opportunity and appropriability conditions, degrees of cumulativeness of technological knowledge and characteristics of the relevant knowledge base. Then one could advance and test empirically some general propositions on the relationship between technological regimes and patterns of innovation in industries, related to a fundamental distinction between the Schumpeter Mark I and the Schumpeter Mark II models—the first is characterized by "creative destruction" with technological ease of entry and a major role played by entrepreneurs and new firms in innovative activities, the second by "creative accumulation" with the prevalence of a stable core of few large firms and the presence of relevant barriers to entry for new innovators (Breschi et al. 2000). Although rather archetypical, these analyses show that knowledge and learning regimes indeed affect the way innovative activities take place in industries (Marsili and Verspagen 2002). Finally, one could go into much more depth and relate the structure of knowledge in a sector to the type of actors and their relationships. As an example, let me mention here the work that Orsenigo et al. (1997) have done in pharmaceuticals and biotechnology. They have found that there is an isomorphism between the cognitive structure underlying the dynamics of knowledge and the structure of the network. The impact of science has been the proliferation of more specialized and new hypotheses that have generated new subdisciplines, requiring new sets of search techniques, testing procedures and skills. Over time, entrants tend to be more specialized in terms of the scientific hypotheses they are trying to test and the search techniques they are employing. The intrinsic characteristics of the search techniques and the patterns of learning in pharmaceutical R–D explain simultaneously why the network expands over time, why it remains relatively stable in its core-periphery profile and why entrants make agreements with incumbents or older NBF, rather than with firms of the same generation.

A third issue refers to the boundaries of knowledge in an industry and their effects on innovation and industry evolution. A focus on boundaries means highlighting all the interdependencies and complementarities outside the industry in related sectors or scientific fields. Think for example of multimedia, in which the convergence of different types of technologies and demand has originated a new sector with continuously expanding knowledge boundaries. A focus on boundaries also means that there are some bounds to knowledge growth which are related to the specificity of the technologies and the sector. This is in the notion of technological regimes discussed above, and is present in the evolutionary literature and, from a different perspective, also in the notion of bounds, as seen in the work of Sutton (1998). If we take this view of knowledge, the specificities of technological regimes and the knowledge base of an industry provide a powerful restriction on the patterns of firms' learning, competencies, behavior and organization of innovative and production activities. In general, a given knowledge base, technological environment or demand defines the nature of the problems firms have to solve in their innovative and production activities and the types of incentives and constraints to particular behavior and organizations. Within these constraints, however, great and persistent heterogeneity in firm innovative and productive behavior and organization is possible.

One last issue concerns the processes of change in knowledge and the knowledge base during the evolution of an industry. This goes to the heart of the evolution of industries and of the factors affecting the changes in industrial structure. This is a very difficult task to accomplish, even for a single case study, let alone the identification of regularities and laws of knowledge change in sectors.

7 Networks

Let me move to the next challenge: networks. This challenge starts from the recognition that innovation and industry evolution are highly affected by the interaction of heterogeneous actors with different knowledge, competences and specialization, with relationships that may range from competitive to cooperative, from formal to informal, from market to non-market.

Recently, the literature on networks among economic actors has boomed, and networks have been studied in a variety of ways. Tremendous progress has been made on various fronts by social networks analysis. Major developments have been achieved in the organizational and strategic studies of networks, the examination of the specific role and strategy of actors within networks, the analysis of firms and industries as networks, and the approach of social embeddedness. In a similar vein, great progress has been obtained in modelling networks in various ways. One may go from static models regarding the effects of different network architectures on performance, to dynamics of networks (in which the structure influences individual actions and performance and attention is paid to the efficiency and stability of networks and the feedback mechanisms), to network evolution in which the focus is on processes and rules. Similarly, the types of models are quite different, from small world models, to evolutionary game theory, percolation theory, neural networks and so on.

Within this burgeoning interest on networks, one major area of research has been that related to networks and innovation in an evolutionary perspective and with a system view. The innovation system literature has put the role of links and relationships among various actors at the center of the analysis (see Lundvall 1993; Edquist 1997; Teubal et al. 1991). In a similar vein, evolutionary theory has stressed that, in uncertain and changing environments, networks emerge because agents are different, thus integrating complementarities in knowledge, capabilities and specialization (see Nelson 1995). Along these lines, progress has been made in the analysis of the characteristics and structure of networks in several industries: biotechnology (Powell et al. 1996; Arora and Gambardella 1998; McKelvey et al. 2004; Orsenigo et al. 2001; Nesta and Mangematin 2002); ICT (Saxenian 1994; Langlois and Robertson 1995); automobiles (Dyer 1996); aircraft (Bonaccorsi and Giuri 2001); flight simulation (Rosenkopt and Tushman 1998); steel (Rowley et al. 2000); semiconductors (Stuart 1998).

Progress has been obtained by the recognition that the emergence of certain types of networks is a function of specific knowledge, industrial settings, demand and institutions, and their evolution is the result of the interplay between firms' internal capabilities and technological, social and institutional factors (Kogut

2000). Therefore, the types and structures of relationships and networks differ from industry to industry. If we go along this line, it will be seen as relevant to develop taxonomies of network structures for groups of industries (as the useful one by Kogut 2000).[1]

However, the empirical enquiry on networks has to be accompanied by answers to some basic questions. What is a network for the purpose of the analysis of innovation and industry evolution? How might we define it in such a way that is understandable and useful for research on innovation and industry evolution? This was the question that Freeman (1991) posed some years ago and which has not been fully resolved yet. Freeman wrote about networks of innovators as driven by technological complementarities, and pointed to the relevance of formal as well as informal networks. Once the domain of a network is clarified for the purpose of the analysis of innovation and industrial dynamics, the identification of the relevant interactions among actors, the indicators of these interactions and the measures of their structure and evolution can be developed (avoiding accepting uncritically the measures of social network analysis, often created for another purpose of analysis).

A related issue is how and why the specific features and characteristics of networks affect innovation, profitability and growth in an industry. In this respect, we are still at the beginning of the research agenda. From exploratory empirical analyses it seems that strong ties favor exploitation and weak ties favor exploration (as it has been found in longitudinal analyses for the chemical, semiconductor and steel industries). But additional robust evidence and deep appreciative theorizing on this and other connected issues are needed.

And a final, broader question regards the role of networks in different stages of industry evolution, and the related coevolutionary processes. At the general level, we know that networks show stability and change over the evolution of an industry, as longitudinal data for some industries show. Stable networks are often formed early in the industry life cycle. The evidence also shows that major industry specific events shape indeed the structure of networks (see for example Rosenkopf and Thusman 1998). Again, a lot of ground needs to be covered by moving from general statements to the analyses of specific network evolutions and coevolutionary processes. In this respect, the work by Lundgren (1995) on digital image processing and by Bonaccorsi and Giuri (2001) for aircraft are good examples of the coupled dynamics of networks and technology.

At the formal theoretical level, only few models explore the relationship between the dynamics of networks and the dynamics of industries. Among them, Cowan et al. (2004) show that, in industries in which tacit knowledge is relevant and technological opportunities are high, regular structures generate higher knowledge growth, while in industries in which knowledge is codified and technological opportunities are lower, communication without any structure performs better. Although very promising, this line of research is in its infancy, and needs to be developed further. For example, under which conditions early on in the life cycle of a sector do certain types of collaborations (for example to explore knowledge) emerge? And under what conditions in industry maturity do other

[1] Kogut relates the type of network to factors such as technology, resource bottleneck, competing and regulatory rules and strength of property rights, and does it for broad industries such as microprocessors, information technology, software operating systems, pharmaceuticals and biotechnology, automobile and financial markets.

types of collaborations (for example to exploit knowledge) gain in importance? Also, when and why at certain stages of industry evolution do large informal networks rather than formal ones become major sources of knowledge generation (see for example Pyka 2000). Finally, what is the relationship between different types of industry life cycles and different types of network dynamics?

These remarks bring us directly to the last point of the paper: the need to understand the coevolutionary processes that take place during industry evolution.

8 Coevolution

Coevolution goes to the very heart of the dynamic analysis of innovation and the evolution of industries, and more broadly addresses the issue of transformation and structural change. In a broad sense, coevolutionary processes involve knowledge, technology, actors, demand and institutions, and are often path-dependent (David 2000). Moreover, local learning, interactions among agents, and networks may generate increasing returns and irreversibilities that lock sectoral systems into inferior technologies, as Richard Nelson, Stan Metcalfe, Paul David and Richard Langlois among others have shown.

These coevolutionary processes are indeed sector-specific. And discussions of on the joint change of several variables during industry evolution have been proposed (Nelson 1994). For example, just looking at three elements such as technology, demand and firms, one could claim that, in sectors characterized by a system product and consumers with a rather homogeneous demand, coevolution leads to the emergence of a dominant design and industrial concentration (Klepper 1996). However in sectors with either a heterogeneous demand or competing technologies with lock-ins, specialized products and a more fragmented market structure may emerge. In general, one could say that changes in the specific knowledge base of an industry or in the features of demand may affect the specific characteristics of the actors, the type of organization of R–D, the features of the innovative process and of networks, the structure of the market and the specific role of the institutions. All these changes may in turn lead to further modifications in the technology, the knowledge base, demand, and so on. The cases of specific industries provide interesting examples. In chemicals, Arora and Gambardella (1998) have discussed the long run coevolution of technology, organization of innovative activities and market structure, and Murman (2003) has examined the joint interrelated evolution of the dye technology, the population of firms and market structure, national organizations (such as universities and firms), and the international leadership and decline of specific countries. In computers, coevolutionary processes involving technology, demand, market structure, institutions and firm organization and strategies have differed greatly in mainframes, minicomputers, personal computers and computer networks, involving different actors, mechanisms, entry processes and producer–customers relationships (Bresnahan and Malerba 1999). In pharmaceuticals and biotechnology, the interactions between knowledge, technology, institutions and country-specific factors have shaped the evolution of the industry. Changes in the knowledge base and in the relevant learning processes of firms have induced deep transformations in the behavior, structure and interaction of agents. These, in turn, have changed the knowledge and learning processes, leading to new products and so on (as McKelvey 1997; McKelvey et al. 2004). Another case refers to telecom equipment and services. The convergence within ICT and between ICT and broadcasting-audio-visual and the emergence of the internet produced a more fluid market structure with a lot of different actors with different specializations and capabilities, and new types of users. This, in turn, greatly expanded the boundaries of the sector by creating new segments and new opportunities, and also by creating national differences in the organization of innovation. Moreover, the emergence of the internet has generated more pressure in favor of open standards and has led to the rise of new actors (such as ISP and content providers) (Dalum and Villumsen 2003). In software, since the early 1980s, the spread of networked computing, embedded software, the internet, the development of open system architecture and open source, and the growth of web-based network computing has led to the decline of large computer producers as developers of integrated hardware and software systems and to the emergence of many specialized software companies, and to changes in software distribution from licensing agreements in the early days, to the rise of independent software vendors, to price discounts for package software, and, with the diffusion of the CD-ROM and the internet, to shareware and freeware (D'Adderio 2004).

The challenge for research here is to go to a much finer analysis at both empirical and theoretical levels, and to move from the statement that everything is coevolving with everything else to the identification of what is coevolving with what, how intense is this process and whether indeed there is a bi-direction of causality.

A very interesting case of coevolution can be found in vertically related industries. The empirical work by Bonaccorsi and Giuri (2001) for the aircraft industry has already been mentioned. At a more theoretical level, one could find a first discussion of these vertical interdependent processes in the contribution by Young (1928), in which the size of the market depends on prices, which in turn depend upon the cost of production, which depends on the extent of the division of labor. In other words, ultimately the division of labor is limited not by the extent of the market but by the division of labor itself.

In terms of appreciative theorizing and modelling, such coevolutionary processes between vertically related industries can be tackled in various ways. Arora and Bokhari (2000) examine vertical integration and specialization as driven by the interplay of the Babbage effect, the division of labor effect and coordination advantages. Jacobides and Winter (2005) explain the vertical scope of firms in terms of the coevolution of firms' capabilities and transaction costs, and focus their analysis on four factors - knowledge accumulation, capability differences, selection processes and endogenous transaction costs. Finally, in Malerba et al. (2005), vertical integration and specialization is driven by firm capabilities, the rate of technological change and the size and structure of markets, and it is the result of the coevolutionary processes of these variables in the upstream and downstream industries.

9 Conclusions

In this paper, I have suggested that the analysis of innovation and industry evolution has progressed in several areas. Contributions at the empirical, appreciative, econometric and modelling levels have greater advanced our understanding of innovation, industrial dynamics and the different evolution of industries. However, the main part of the paper is centered around the point that a full understanding of the relationship between innovation and the evolution of industries has also to cope with a finer grained analysis of demand, knowledge, networks and coevolution.

I have implicitly stressed that the challenge has to be faced by using a methodology that is quite common to researchers in the Schumpeterian and evolutionary tradition: identify some empirical regularities, stylized facts or puzzles that need to be explained, develop appreciative theorizing, do quantitative analyses and then formal modelling. Consistency between cases-appreciative theorizing-econometrics and modelling has to be present. In a sense, research should not be guided by techniques, but theory should be driven by empirical questions and facts.

In addition, in the realm of innovation and the evolution of industries, research needs to be interdisciplinary. This means that the full understanding of topics such as innovation and the evolution of industries requires the integration of economics, history, sociology, technology, management and organization. Interdisciplinarity means eclecticism and openness to new contributions coming from different field of research. Schumpeter might, I think, have approved, as he was after all as much a sociologist as he was an economist.

Being myself an economist, I may add that a major task is also to have very fruitful confrontation between different approaches within economics, as Schumpeter did with his fellow mainstream economists. I think that a fruitful confrontation is possible because the Schumpeterian approach and the evolutionary research paradigm have developed rich enough contributions to be shared with others and discussed openly.

In conclusion, I am convinced that the strength of a scientific society does not lie only in the discussion of the achievements of the past but in the intellectual challenges it is willing to take. It has to look ahead and not behind. Be bold in the challenges it takes. I am sure many, particularly the new generations, will enthusiastically take up some of the challenge that I have discussed in this paper.

References

Abernathy WJ, Utterback JM (1978) Patterns of innovation in technology. Technol Rev 80 (7):0–47

Adner R (2003) When are technologies disruptive: a demand-based view of the emergence of competition. Strateg Manage J 23(8):667–688

Adner R, Levinthal D (2001) Demand heterogeneity and technology evolution: implications for product and process innovation. Manage Sci 47:611–628

Antonelli C (2003) The digital divide: understanding the economics of new information and communication technology in the global economy. Inf Econ Policy 15:173–199

Arora A, Bokhari F (2000) Vertical integration and dynamics and industry evolution. Carnegie Mellon University, mimeo

Arora A, Gambardella A (1998) Evolution of industry structure in the chemical industry. In: Arora A, Landau R, Rosenberg N (eds) Chemicals and long-term economic growth. Wiley, New York

Arora A, Fosfuri A, Gambardella A (2001) Markets for technology: economics of innovation and corporate strategy. MIT Press, Cambridge, MA

Audretsch DB (1995) Innovation and industry evolution. MIT Press, Boston

Aversi R, Dosi G, Fagiolo G, Meacci M, Olivetti C (1999) Demand dynamics with socially evolving preferences. Ind Corp Change 8(2):353–408
Balconi M, Breschi S, Lissoni F (2004) Networks of inventors and the role of academia: an exploration of Italian patent data. Res Policy 33(1):127–145
Baldwin J (1995) The dynamics of industrial competition. Cambridge University Press, Cambridge, MA
Bonaccorsi A, Giuri P (2001) Network structure and industrial dynamics. The long term evolution of the aircraft-engine industry. Struct Chang Econ Dyn 12:201–223
Bottazzi G, Cefis E, Dosi G (2002) Corporate growth and industrial structures: some evidence from the italian manufacturing industry. Ind Corp Change 11:705–723
Breschi S, Lissoni F (2004) Knowledge networks from patent data: methodological issues and research targets. In: Moed H, Glänzel W, Schmoch U (eds) Handbook of quantitative science and technology research: the use of publication and patent statistics in studies of S&T systems. Springer, Berlin Heidelberg New York, pp 613–643
Breschi S, Malerba F, Orsenigo L (2000) Technological regimes and Schumpeterian patterns of innovation. Econ J 110(463):388–410
Bresnahan T, Greenstein S (2001) The economic contribution of information technology: towards comparative and user studies. J Evol Econ 11(1):95–118
Bresnahan T, Malerba F (1999) Industrial dynamics and the evolution of firms and nations competitive capabilities in the world computer industry. In: Mowery D, Nelson R (eds) The sources of industrial leadership. Cambridge University Press, Cambridge, MA
Brusoni S, Prencipe A, Pavitt K (2001) Knowledge specialization, organizational coupling and the boundaries of the firm: why firms know more than they make? Adm Sci Q 46(4):597–621
Carroll GR, Hannan MT (1999) The demography of corporations and industries. Princeton University Press, Princeton
Christensen CM (1997) The innovators dilemma: when new technologies cause great firms to fail. Harvard Business School Press, Boston, MA
Christensen CM, Rosenbloom R (1995) Explaining the attacker's advantage: technological paradigms, organisational dynamics and the value network. Res Policy 24:233–257
Cohen WM, Levinthal D (1989) Innovation and learning: the two faces of R&D. Econ J 99 (397):569–596
Cohen WM, Goto A, Nagata A, Nelson RR, Walsh P (2002) R&D spillovers, patents and the incentives to innovate in Japan and the United States. Res Policy 31(8–9):1349–1367
Cowan R, David PA, Foray D (2000) The explicit economics of knowledge: codification and tacitness. Ind Corp Change 9(2):211–253
Cowan R, Jonard N, Ozman M (2004) Knowledge dynamics in a network industry. Technol Forecast Soc Change 71(5):469–484
D'Adderio L (2004) Inside the virtual product: how organizations create knowledge through software. Elgar, Cheltenham
Dahmen E (1989) Development blocks in industrial economics. In: Carlsson B (ed) Industrial dynamics. Kluwer, Boston Dordrecht
Dalum B, Villumsen G (2003) Fixed data communications: challenges for Europe. In: Edquist C (ed) The internet and mobile telecommunications system of innovation. Edward Elgar, Cheltenham Northampton
David PA (2000) Path dependence and varieties of learning in the evolution of technological practice. In: Ziman J (ed) Technological innovation as an evolutionary process. Cambridge University Press, Cambridge
Devetag G (1999) From utilities to mental models: a critical survey on decision rules and cognition in consumer choice. Ind Corp Change 8(2):289–351
Dopfer K, Foster J, Potts J (2004) Micro-Meso-Macro. Schumpeter Conference, Milan
Dosi G (2001) Innovation, organization and economic dynamics: selected essays. Edward Elgar, Cheltenham Northampton
Dosi G, Marsili O, Orsenigo L, Salvatore R (1995) Learning, market selection and the evolution of industrial structures. Small Bus Econ 7(6):411–436
Dyer JH (1996) Specialized supplier networks as a source of competitive advantage: evidence from the auto industry. Strateg Manage J 17:271–291
Edquist C (ed) (1997) Systems of innovation; technologies, institutions and organizations. Science, technology and the international political economy series. Pinter, London
Ericson R, Pakes A (1995) Markov perfect industry dynamics: a framework for empirical work. Rev Econ Stud 62

Foray D (2004) The economics of knowledge. MIT Press, Boston
Franke N, Shah SK (2003) How communities support innovative activities: an exploration of assistance and sharing among end-users. Res Policy 32:157–178
Freeman C (1974) The economics of industrial innovation. Penguin, Harmondsworth, UK
Freeman C (1991) Networks of innovators: a synthesis. Res Policy 20:499–514
Gavetti G, Levinthal D (2000) Looking forward and looking backward: cognitive and experiential search. Adm Sci Q 45(1)
Geroski P (1994) Market Structure, corporate performance and innovative activity. Oxford University Press, Oxford
Harhoff D, Henkel J, von Hippel E (2003) Profiting from voluntary information spillovers: how users benefit by freely revealing their innovations. Res Policy 32(10):1753–1769
Jacobides MG, Winter SG (2005) The co-evolution of capabilities and transaction costs: explaining the institutional structure of production. Strateg Manage J 26(5):395–413
Jaffe AB, Trajtenberg M (2002) Patents, citations, and innovations: a window on the knowledge economy. MIT Press, Cambridge, MA
Jaffe AB, Trajtenberg M, Fogarty M (2000) Knowledge spillovers and patent citations: evidence from a survey of inventors. Am Econ Rev 90(2):215–218
Jovanovic B (1982) Selection and the evolution of industry. Econometrica 50(3):649–670
Kamien MI, Schwartz NL (1975) Market structure and innovation: a survey. J Econ Lit 13:1–37
Klepper S (1996) Entry, exit, growth and innovation over the product life cycle. Am Econ Rev 86:562–583
Klepper S (1997) Industry life cycles. Ind Corp Change 6:145–181
Klepper S (2002) Firm survival and the evolution of oligopoly. RAND J Econ 33(1):37–61
Klepper S, Simons KL (2000) The making of an oligopoly: firm survival and technological change in the evolution of the U.S. tire industry. J Polit Econ 108(4):728–760
Klepper S, Thompson P (2002) Submarkets and the evolution of market structure. Carnegie Mellon University
Kogut B (2000) The network as knowledge: generative rules and the emergence of structure. Strateg Manage J 21:405–425
Kuznets S (1965) Economic growth and structure. Norton, New York
Langlois RN (2001) Knowledge, consumption, and endogenous growth. J Evol Econ 11(1):77–93
Langlois RN, Robertson PL (1995) Innovation, networks, and vertical integration. Res Policy 24:543–562
Levin RC, Klevorick AK, Nelson RR, Winter SG (1987) Appropriating the returns from industrial research and development. Brookings Pap Econ Act 3:783–820
Levinthal D (1993) The myopia of learning. Strateg Manage J
Loasby B (1999) Knowledge, institutions and evolution in economics. Routledge, London
Loasby B (2001) Cognition, imagination and institutions in demand creation. J Evol Econ 11:7–21
Lundgren A (1995) Technological innovation and network evolution. Routledge, London
Lundvall B-Å (1988) Innovation as an interactive process: from user–producer interaction to the national system of innovation. In: Dosi G, Freeman C, Nelson R, Silverberg G, Soete L (eds) Technical change and economic theory. Pinter, London, pp 349–369
Lundvall B-Å (1993) User–producer relationships, national systems of innovation and internationalisation. In: Foray D, Freeman C (eds) Technology and the wealth of nations. Pinter, London
Malerba F (1985) The Semiconductor business: the economics of rapid growth and decline. Pinter, London
Malerba F (1992) Learning by firms and incremental technical change. Econ J 102(413):845–859
Malerba F (ed) (2004) Sectoral systems of innovation. Cambridge University Press, Cambridge, MA
Malerba F, Orsenigo L (1995) Schumpeterian patterns of innovation. Camb J Econ 19:47–65
Malerba F, Orsenigo L (1996) The dynamics and evolution of industries. Ind Corp Change 5 (1):51–87
Malerba F, Orsenigo L (1997) Technological regimes and sectoral patterns of innovative activities. Ind Corp Change 6:83–117
Malerba F, Orsenigo L (2002) Innovation and market structure in the dynamics of the pharmaceutical industry and biotechnology: towards a history friendly model. Ind Corp Change 11(4):667–703

Malerba F, Nelson R, Orsenigo L, Winter S (1999) History friendly models of industrial evolution: the computer industry. Ind Corp Change 1:3–41
Malerba F, Nelson R, Orsenigo L, Winter S (2003) Demand, innovation and the dynamics of market structure: the role of experimental users and diverse preferences. CESPRI WP No. 135
Malerba F, Nelson R, Orsenigo L, Winter S (2005) The dynamics of the vertical scope of firms in related industries, the coevolution of competences, technical change and the size and structure of markets. CESPR, Bocconi University, Milan
Marsili O, Verspagen B (2002) Technology and the dynamics of industrial structures: an empirical mapping of Dutch manufacturing. Camb J Econ v. 11:791–815
McKelvey M (1997) Coevolution in commercial genetic engineering. Ind Corp Change 6 (3):503–532
McKelvey M, Orsenigo L, Pammolli F (2004) Pharmaceuticals analysed through the lens of a sectoral innovation system. In: Malerba F (ed) Sectoral systems of innovation. Cambridge University Press, Cambridge, MA
Metcalfe S (1998) Evolutionary economics and creative destruction. Routledge, London
Metcalfe S (2001) Consumption, preferences, and the evolutionary agenda. J Evol Econ 11 (1):37–58
Meyers S, Marquis DG (1969) Successful industrial innovations. National Science Foundations, Washington
Montobbio F (2002) An evolutionary model of industrial growth and structural change. Struct Chang Econ Dyn 13(4):387–414
Murman JP (2003) Knowledge and competitive advantage, the coevolution of firms, technology, and national institutions. Cambridge University Press, Cambridge, MA
Nelson R (1994) The co-evolution of technology, industrial structure and supporting institutions. Ind Corp Change 3(1):47–63
Nelson R (1995) Recent evolutionary theorizing about economic change. J Econ Lit 33
Nelson R, Winter S (1982) An evolutionary theory of economic change. Harvard University Press, Cambridge, MA
Nesta L, Mangematin V (2002) Industry life cycle, knowledge generation and technological networks. In: Gault F, de La Mothe J (eds) Alliances, networks, and partnerships in the innovation process. Kluwer, Boston
Nuvolari A (2004) Collective invention during the British industrial revolution: the case of the Cornish pumping engine Camb J Econ 28:347–363
Orsenigo L, Pammolli F, Riccaboni M, Bonaccorsi A, Turchetti G (1997) The evolution of knowledge and the dynamics of an industry network. J Manage Gov 1(2):147–175
Orsenigo L, Pammolli F, Riccaboni M (2001) Technological change and network dynamics: lessons from the pharmaceutical industry. Res Policy 30(3):485–508
Pasinetti LL (1981) Structural change and economic growth: a theoretical essay on the dynamics of the wealth of nations. Cambridge University Press, Cambridge, MA
Pavitt K (1984) Sectoral patterns of innovation: towards a taxonomy and a theory. Research Policy 13:343–374
Powell WW, Koput KW, Smith-Doerr L (1996) Interorganizational collaboration and the locus of innovation: networks of learning in biotechnology. Adm Sci Q 41:116–145
Pyka A (2000) Informal networking and industrial life cycles. Technovation 20(11):25–35
Rosenkopf L, Tushman M (1998) The coevolution of community networks and technology: lessons from the flight simulation industry. Ind Corp Change 7:311–346
Rowley T, Behrens D, Krackhardt D (2000) Redundant governance structures: an analysis of structural and relational embeddedness in the steel and semiconductor industries. Strateg Manage J 21:369–386
Saviotti P (1996) Technicological evolution, variety and the economy. Edward Elgar, Cheltenham
Saviotti P (2001) Variety, growth and demand. J Evol Econ 11(1):119–142
Saxenian AL (1994) Regional advantage: culture and competition in Silicon Valley and Route 128. Harvard University Press, Cambridge, MA
Schmookler J (1966) Invention and economic growth. Harvard University Press, Cambridge, MA
Schumpeter J (1934) The theory of economic development. Harvard University Press, Cambridge, MA
Schumpeter J (1939) Business cycles. MacGraw Hill, New York
Schumpeter J (1950) Capitalism, socialism and democracy. Harper Brothers, New York
Stuart TE (1998) Network positions and the propensity to collaborate: an investigation of strategic alliance formation in a high-technology industry. Adm Sci Q 43(3):668–698

Sutton J (1991) Sunk costs and market structure. MIT Press, Cambridge, MA
Sutton J (1998) Technology and market structure: theory and history. MIT, Cambridge, MA
Teubal M (1979) On user needs and need determination: aspects of the theory of technological innovation. In: Baker M (ed) Industrial innovation: technology, policy and diffusion. Macmillan, London
Teubal M, Yinnon T, Zuscovitch E (1991) Networks and market creation. Res Policy 20:381
Thompson P, Fox-Kean M (2005) Patent citations and the geography of knowledge spillovers: a reassessment. Am Econ Rev 95(1):450–460
Urban G, von Hippel E (1988) Lead user analyses for the development of new industrial products. Manage Sci 34(5):569–582
Utterback JM (1994) Mastering the dynamics of innovation. Harvard Business School Press, Boston, MA
Von Hippel E (1986) Lead users: a source of novel product concepts. Manage Sci 32(7):791–805
Von Hippel E (1988) The sources of innovation. Oxford University Press, New York
Winter SG (1987) Knowledge and competence as strategic assets. In: Teece DJ (ed) The competitive challenge. Ballinger, Cambridge, MA
Winter SG, Kaniovski YM, Dosi G (2000) Modeling industrial dynamics with innovative entrants. Struct Chang Econ Dyn 11(3):255–293
Witt U (2001) Learning to consume—a theory of wants and the growth of demand. J Evol Econ 11(1):23–36
Witt U (2003) Evolving economy: essays on the evolutionary approach to economics. Edward Elgar, Cheltenham, UK
Young AA (1928) Increasing returns and economic progress. Econ J 38:527–542

Brian J. Loasby

Entrepreneurship, evolution and the human mind

Abstract Schumpeterian 'development from within' requires imagination, skill and motivation; so does Cattaneo's 'psychology of wealth'. Neither can be encompassed by models that rely on deductive rationality, but are twin products of Knightian uncertainty, where the absence of demonstrably correct procedures allows individuals to create domain-limited mental structures. The human mind (as studied by Smith, Marshall and Hayek), is a product of biological evolution which supports the evolution of knowledge and of economic systems. These are non-biological processes; both require (fallible) bounds to uncertainty, which are provided by (evolving) formal and informal organisation, including institutions.

Keywords Intelligence · Uncertainty · Imagination · Domain-limitation · Organisation
JEL Classification B52 · D83 · O12 · O31

1 Foreword

The objective of this paper is to relate entrepreneurship to a theory of the human mind, setting both within an evolutionary perspective. The first requirement, therefore, is to indicate what this perspective implies. As a generic concept, I take evolution to be the self-transformation of a system through the internal production and diffusion of novelty (Witt 2003, 280). The production of novelty may be a response to external events, but this response, and indeed the perception of the external events, is internally generated; this distinction is made explicit in Schumpeter's account of 'development from within' (Schumpeter 1934, 63). Changes in technology and consumer preferences are treated as changes in the data to which the economy adapts (Schumpeter 1934, 65); such adaptations are to be explained

B. J. Loasby
Department of Economics University of Stirling, Stirling FK9 4LA
Tel +44 1786 467470, Fax +44 1786 467469
E-mail: b.j.loasby@stir.ac.uk

by co-ordination theory, and are contrasted with entrepreneurial action to induce changes in preferences and (especially when entrepreneurship is located in large firms) changes in technology.

This evolutionary self-transformation requires a combination of three processes: the emergence of variety, selection among this variety according to some reasonably systematic criteria, and retention of the selected variants, at least over a time-scale that is significant for the particular phenomena being studied. Neo-Darwinian biology relies on a particular subset of these criteria, which prescribes, for example, absolute separation between the selection environment and the kinds of variation that are produced; it therefore excludes anything corresponding to Schumpeterian enterpreneurship, or indeed any kind of intentionality. Penrose (1952) rejected attempts to justify reliance on economic models that assumed perfectly-informed rationality by invoking biological analogies which excluded human purpose along with any traces of rationality; however there can, conceptually, be many evolutionary processes, operating at different levels and in a variety of ways. Some of these may be connected; and in the fourth section of this paper I shall suggest that the biologically-evolved characteristics of the human brain (the neoDarwinian explanation of which is not here at issue) have created the conditions for the significantly different processes of economic evolution. There may also be interdependence, or co-evolution, between similar processes, for example between the evolution of capabilities and of the firms which combine them; this particular relationship will not be explored here, but should be an important topic in evolutionary economics.

Though evolution provides the perspective and not the theme, it is important to recognise that this notion of evolution necessarily excludes the possibility that all the phenomena being studied – such as economic outcomes – are to be explained as rational choice equilibria, because these equilibria result from correct ex-ante deduction, whereas evolution operates through ex-post selection among alternatives that are based on conjectures. There is a methodological sub-theme to this paper, and it is explicitly Schumpeterian. Schumpeter was interested in ideas – for many reasons, but above all because ideas influence events. In particular, ideas about the economic system could change the way that the system worked, through their effects on individual actions and economic policy. This was the message of *Capitalism, Socialism and Democracy*. The criteria by which economic theories (and especially the models which purport to represent them) are appraised may have major impacts on the admissibility of ideas: the appraisal criteria of standard economics, such as internal coherence, tractability, and consistency with particular conceptions of rationality and equilibrium, make many topics undiscussable, and may force practitioners into perverse interpretations and lead to policy conclusions that are inappropriate or even disastrous. In *Capitalism, Socialism and Democracy* Schumpeter (1943) denounced the errors of both Marx and neoclassical economists. Much earlier he had set out the basis of his own ideas.

2 Schumpeter in Milan

Joseph Schumpeter took care to distinguish his theory of economic development from the theory of a fully co-ordinated economy. Development 'from within' was the result of human initiative; by contrast, and despite his admiration for Walras's analysis, Schumpeter's picture of the fully co-ordinated economy gives little credit to human agency. Although an economy which can be represented by a general equilibrium system is observationally equivalent to a system in which everyone is a completely rational optimiser, Schumpeter declares this to be an illusion. 'In so far ... as it is a question of adapting himself to the conditions and of simply complying with the objective necessities of the economic system without wishing to change them, one and only one particular way of acting commends itself to the individual' (Schumpeter 1934, 40), and this particular way of acting, although undeniably efficient – within this context – in both outcome and method, cannot be achieved by rational choice but only 'if things have time to hammer logic into men' (Schumpeter 1934, 80). The failure of economists to provide any account of equilibration that is consistent with the fundamental principles of general equilibrium models reinforces Schumpeter's position. His reliance on an implicit selection process within a stable environment to achieve co-ordination permits us to consider Schumpeter as an evolutionary economist in terms of our definition, while allowing him to preserve his distance from concepts of biological evolution, as he was clearly anxious to do.

As soon as people encounter conditions in which optimal behaviour requires a change of routine, the fiction of rationality is exposed; 'the choice of new methods is not simply an element in the concept of rational economic action, nor a matter of course, but a distinct process which stands in need of special explanation' (Schumpeter 1934, 80, fn. 2). The special explanation that Schumpeter offers requires the creation of 'new combinations', which cannot be derived by logical manipulation or by the application of experience-based routine. (Why routines are likely to have limited applicability will be explained in the fourth section.) It also requires motivations of the kind that do not fit easily into conventional preference functions or modern analyses of incentives; indeed they imply a distinctive psychological basis, which supports the capability, the confidence and the determination to achieve major change. This combination of imagination, skill and motivation defines the Schumpeterian entrepreneur.

Now I wish to suggest that a Schumpeter Conference in the city of Milan was the perfect occasion to explore the psychological basis of entrepreneurship; for it was here in 1861 that Carlo Cattaneo published his essay 'Del pensiero come principio d'economia publica', which has recently been published with an English translation under the title of 'Intelligence as principle of public economy' (Cattaneo 2001). Cattaneo traces the development of economics as the study of the relationships between land, labour and capital and the flow of outputs, which he calls 'the physics of wealth'; and he notes – as has been repeatedly noted since – that differences in these inputs often seem inadequate to explain observed differences in outputs, both between nations and over time. To explain these differences he urges economists to turn to what he calls 'the psychology of wealth', and specifically to the effects of intelligence and will. As we have just observed, in Schumpeter's theory of growth intelligence and will are both essential entrepreneurial qualities; we may therefore claim that his theory rests on a conception of the human mind which has much greater potential, both for economic analysis and the human agents whose actions are to be studied, than that of economic agents as rational optimisers. (That concept is clearly restricted, in Cattaneo's classification, to the physics of wealth.) The first claim of this paper is that entrepreneurship, intelligence, and evolutionary economics all require a psychological basis which rational choice cannot provide. Schumpeter should feel at home in Milan; so should members of the Schumpeter Society.

We may approach the relationship between entrepreneurship and intelligence by introducing a third perspective; a particularly appropriate link between Schumpeter and Cattaneo is supplied by Frank Knight's analysis of uncertainty. In his systematic attempt to clarify the suppositions and scope of the economic theory of perfect competition that was currently being refined into a precise model, Knight (1921) investigated the underlying knowledge requirements of this theory. He argued that precise information about future events was not necessary; risk could be readily incorporated, provided that there were procedures available for distributing probabilities over closed sets of possible outcomes. These procedures might be based on a priori calculation or appropriate evidence; what was essential was that they should be known to be correct. Any situation in which correct probabilities could not be supplied – including those in which probabilities were invented in order to make calculation possible, a practice that Knight (1933, xiv) criticised in the preface to the LSE reprint of *Risk, Uncertainty and Profit*, but which is now standard practice – he defined as one of uncertainty.

In Knight's view, uncertainty was not unusual, and its consequences were not of minor significance for economists. It substantially restricted the applicability of perfect competition theory; but in so doing it substantially expanded the scope of economics. Knight, like Schumpeter, believed that there were major economic phenomena that could not be explained by what we now call rational choice models. First, if publicly available procedures, which were known to be correct, could be applied to any decision, then no decision-maker could gain any advantage over rivals, except through the possession of some distinctive and non-imitable resource. However, such a resource would yield rent, not profit, and since all risk could be insured profit would be an empty category. Second, in such circumstances all activities could be co-ordinated by contracts for the exchange of goods and services; there would thus be no justification for contracts of employment, in which one person agrees to accept direction, however limited, from another. Third, because all opportunities could be fully specified, and would be open to all, there would be no entrepreneurship. Profit, firms, and entrepreneurship, Knight argued, all depended on uncertainty, defined as the absence of correct procedures for dealing with a range of possibilities. We may add to this the absence of correct procedures for defining the range of possibilities, so eloquently emphasised by George Shackle. Though less emphatically than Shackle, Knight clearly recognised that, despite its unfortunate implications for the applicability of models of rational choice equilibria, uncertainty created opportunities for economic development. Indeed Knight (1921, 348) regarded it as of fundamental importance for the human species, noting that 'a life with uncertainty eliminated or perhaps even very greatly reduced would not appeal to us'. For Shackle (1966, 133) it implied 'the reverse of hope, the opposite of freedom'. It certainly would not appeal to a Schumpeterian entrepreneur.

Knight (1921, 268) went on to observe that in the absence of uncertainty all organic adjustments would be mechanical and all organisms would be automata. This corresponds closely to Schumpeter's claim that the circular flow of economic activity in a Walrasian equilibrium is maintained by a precisely-defined structure of mutually compatible routines. However, Knight draws a conclusion that Schumpeter does not make explicit: 'it is doubtful whether intelligence itself would exist in such a situation' (Knight 1921, 268). Intelligence and entrepreneurship are both responses to uncertainty, to situations in which there is no correct procedure for deciding what to do. In fact they may be considered as aspects of a single response, because crucial to both is the consequence that in such situations people must create their own structures for interpretation and decision, or find some ready-made structure that they are prepared to adapt; and these activities, rather than the rational choices that can be produced by automata, are the distinguishing characteristics of human intelligence. Rational choice theory excludes both entrepreneurship and intelligence: the appraisal criteria that we apply to theory choice are important, because, as Schumpeter insisted, ideas influence events.

Those who accept rationality as the ideal might reflect on David Hume's (1978, 164) observation that 'no kind of reasoning can give rise to a new idea'; they might further reflect that whereas, judged from the perspective of rational choice equilibrium, new ideas, which invalidate the data from which individual optima and system equilibria are derived, are a threat to efficiency, the possibility of new ideas exposes the inadequacy of this concept of efficiency. Though Knight does not explore in any detail the processes by which we create interpretative structures, he offers some very helpful suggestions, notably the proposition that we form categories on the basis of similarities that we deem significant, while ignoring differences that we believe to be irrelevant, where significance and relevance are conditional on 'the purpose or problem in view' (Knight 1921, 206). If we link this proposition to his observation that 'the existence of a problem of knowledge depends on the future being different from the past, while the possibility of the solution of the problem depends on the future being like the past' (Knight 1921, 313), we can conclude that all decisions involve a selective mapping from past to future, where the principles of selection are themselves conjectural.

Intelligent behaviour depends on making sense by making selective connections, thus creating new combinations; this, of course, is Schumpeter's own summary of the entrepreneurial vision. Knight (1921, 241) also notes that people differ in their capacity to create appropriate structures, and that for each individual this capacity varies substantially between domains; both are natural consequences of the conditionality of relevance to which Knight had already drawn attention. The effectiveness of individual entrepreneurs will therefore be limited, though these limits may not correspond with conventional industrial categories, and may be misjudged, not least by the entrepreneur. We should not therefore be surprised that intelligence, combined with will, produces outcomes that cannot be predicted by any analysis which is restricted to what Cattaneo called the physics of wealth. Intelligence and entrepreneurship produce new ideas, most of which do not survive. Entrepreneurship, like any kind of evolution – even that driven by the search for success, entails widespread failure. The psychology of wealth is the psychology of behaviour in an uncertain environment, though not, as we shall see, in an environment of unbounded uncertainty. In the remainder of this paper we shall attempt to explore entrepreneurship as a particular application of human intelligence, and therefore of the human mind.

3 Economists and the human mind

A notable feature of twentieth-century theoretical development in economics, recently analysed by Nicola Giocoli (2003), was the multi-faceted endeavour to 'escape from psychology', notably by successive respecification of preferences, the increasing resort to game theory to circumvent difficulties in closing models, and generally by reducing all choice to pure logic – and therefore reducing human beings to Knight's automata. This development was prompted by the fear that the human mind is an alien topic for economists; it is therefore salutary (and for some of us encouraging) to recall that the operation of the human mind attracted the attention, early in their careers, of three people who are generally regarded as among the greatest of economists. The first was Adam Smith (1980 [1795]), who was probably responding to David Hume's demonstration that there could be no way of establishing the ultimate truth of any empirical proposition, either by deduction or experiment, and to Hume's proposal that we should therefore seek to understand how people come to accept certain empirical propositions as true. Schumpeter's proposal to move from the formal deduction of equilibria from data sets to the internal generation and diffusion of novelty is a cognate idea.

Smith developed a psychological theory of the emotional and aesthetic motivations and imaginative processes by which phenomena are gathered into categories, and causally linked to other categories, by the invention of 'connecting principles'. He argued that the gradual concentration of attention on developing connecting principles that could be applied to major categories of phenomena, such as astronomy, led to the emergence of science as an identifiable activity; and thereafter increasing specialisation between fields of science accelerated the growth of knowledge. The effects of a knowledge-generating division of labour within the economy subsequently became his fundamental explanation of the wealth of nations (Smith 1976 [1776]). Thus Smith substantially anticipated Cattaneo's call for a focus on the psychology of wealth, although Cattaneo was correct to observe that much of Smith's economic analysis is not thus oriented. The reasons for this change of orientation are beyond the scope of this paper; but we cannot fail to note that it had substantial implications for the subsequent history of economics.

Imagination, we should remember, is almost as central to Smith's thought as to George Shackle's. It is also essential to Schumpeter's conception of the entrepreneur, though Schumpeter uses the term 'vision', with the explicit implication, contrary to Smith's, that what is envisaged is likely to prove true. (Acknowledging the possibility of false visions might have diverted attention from his theory of successful development, and this was easy to avoid in an account that eschewed any mention of evolution.) It may seem remarkable that Schumpeter failed to recognise the similarities between Smith's explanation of the growth of knowledge and his own theory, especially as he drew attention to Smith's 'History of Astronomy' as the prime exemplar of the quality of Smith's thinking (Schumpeter 1954, 182), and because the role of Smith's (1976 [1776], 21 'philosophers and men of speculation' is precisely to envisage 'new combinations'. This apparent failure is a notable (but far from unique) illustration of the difficulty of selecting appropriate contexts of similarity as the basis for intelligent action, a difficulty that, as we shall see, is inherent in the characteristics of the human mind, and which supports the preference for evolution rather than rational choice as an analytical perspective.

A few years after Cattaneo's essay, Alfred Marshall was confronted with the problem of human knowledge as a young Fellow of St John's College, Cambridge (Butler 1991); and he too responded with a process theory, in which he combined Darwin's evolutionary principle with physiologically-based psychology, which had been developed by Alexander Bain (1864, 1865), and mechanical systems, as envisaged by Babbage. Marshall (1994) postulated a 'machine' that was equipped with an operating system (a 'body'), which could receive impressions and perform various actions, and a control system (a 'brain'), which communicated only with the 'body' and worked with 'ideas' of impressions and actions. This distinction between body and brain immediately excluded any direct perception of external data; 'ideas' are representations, even when they are ideas about facts, and (like Smith's 'connecting principles') they may be inaccurate.

Marshall went on to consider how these representations are formed, and how and why they may be modified. The machine's brain consisted of nodes and an array of potential connections between them, and in the course of the machine's operations a combination of positive and negative feedback would gradually build up strong linkages between 'ideas' of repeated impressions that the 'body' received from its environment and 'ideas' of those repeated actions in response which led to acceptable consequences. It thus gradually developed a cluster of self-reinforcing routines that, together with the routines developed by similar firms according to their varied circumstances, could support what Schumpeter was to call a circular flow. The resemblance to Marshall's subsequent account of the emergence of partial equilibria, which is set in a context of continuing development, may now seem as obvious as it was previously obscure; there are clues in the *Principles* which can be recognised only by those already equipped with appropriate connecting principles (Raffaelli 2003). Marshall suggested that, once a satisfactory repertoire of routines had been established, and therefore required no further explicit attention – but not before – then the development of a superior level of control within the 'brain' of a modified species of machine allowed that 'brain' to respond to novel situations by mental experiments in which ideas of novel actions might be generated and their possible consequences envisaged. Successful novelties would then be incorporated into the repertoire of routines.

Unlike most subsequent economists, Marshall, though as yet unschooled in economics, recognised that cognition was a scarce resource. Low-cost operation was therefore a precondition for mental experiments, and the occasions for such experiments must be carefully chosen. (Compare the philosopher Whitehead (1948 [1911], 42) on the need for operations of thought to be reserved for 'decisive moments'.) In this mechanical system innovation is triggered by a failure of routine, but only if this failure is limited to a small part of the repertoire; and any success is preserved by routinisation. This interdependence between routine and innovation is a feature to which we shall return – as Raffaelli (2003) has demonstrated, it is the core of Marshall's theory of growth through organisation; for the present we will note that in examining the potential of such psychological machinery Marshall postulated results that have some obvious resemblance to Smith's theory of the growth of knowledge (of which Marshall seems not to have been aware). We may also recognise the conceptual similarities with Nelson and Winter's (1982) theory, in which growth results from the stimulus to develop and install new routines. It should be noted that the development and publication of that theory preceded any recognition of the significance (and barely of the existence) of Marshall's model; the primary inspiration was a combination of Schumpeterian innovation and 'Carnegie' analyses of organisational behaviour, and the explicit analogy was between routines and genes. Subsequent claims for the role of neoDarwinian evolutionary genetics has raised associations which both Nelson and Winter believe are inappropriate to economic development; what once appeared a convenient context of similarity for persuading economists is now a context in which differences between evolutionary processes have become prominent. They now argue that the evolution of connections within purposeful organisations, but still through a process of problem-stimulated trial and error (which is the operating principle of the 1982 theory), does not need a specifically genetic analogy.

It was the third of our trio of people who went on to become economists, Friedrich Hayek, who produced much the most elaborate theory of the human mind. Like his two predecessors, the stimulus was an early encounter with problems of knowledge; in this instance what puzzled him were the substantial discrepancies between our sensory perceptions and the scientific formulations that were subsequently developed to represent the same phenomena. These formulations have progressively discarded sensory qualities in favour of the relations between objects (Hayek 1952, 2-3), and in the process have created physical categories that do not match our sensory categories. Hayek's solution to this problem entails the concept of the mind as a structure of connections, each cluster resulting from interaction with particular environments. These interactions might occur during the development of the species or of the individual; Hayek carefully avoided any attempt to delimit the domain of each kind of development, in order to emphasise his fundamental argument that the sensory and physical orders are built up by similar but separate evolutionary processes which result in differentiated systems of connections.

Sensory and physical orders – and by extension, all kinds of human knowledge, including 'knowledge that', 'knowledge why', 'knowledge who' and 'knowledge how' – are represented by physical connections within the brain and exist as sets of relations that are imposed on events. Therefore '*all* we know about the world is of the nature of theories and all 'experience' can do is to change these theories' (Hayek 1952, 143). Note that this conclusion may also be derived from Smith's psychological theory of knowledge – and indeed from Hume's argument against demonstrable truth. It applies both to knowledge which results from the development of the species and knowledge that is developed within the individual, despite the differences between these processes. The supersession through 'experience' of those theories which are genetically embedded in the structure of the brain occurs by the natural selection of genetic instructions at the level of the species, whereas the supersession of those theories which have been constructed in the course of the individual development of a genetically-endowed potential occurs by a reordering of connections at the level of the individual. The time-periods required by these two sequences differ by many orders of magnitude; hence, we shall argue shortly, the advantage of distinctive evolutionary processes.

Hayek's theory may be interpreted as combining a development of Marshall's ideas about mechanism with Smith's account of mental processes; the key concept shared by all three is the formation of selected connections, which are corroborated or refuted by the perceived consequences of their application to understanding or to action. Because individuals differ in their orientations, and because they are nevertheless motivated to observe the behaviour and study the ideas of others, and capable, within limits, of adapting some of their practices and ideas – characteristics to be considered in the following section, human communities exhibit the three basic evolutionary elements of variation, selection, and retention. This is the cognitive basis of 'development from within', and the foundation of the psychology of wealth that Cattaneo advocated.

As Herbert Simon insisted, human rationality is bounded. Three kinds of bounds may be identified. First, human beings are not good natural logicians, and consequently not good natural statisticians either; second, the premises for logical operations are often doubtful, and even more likely to be incomplete; and third, cognition is a scarce resource, and so rationality has to be applied very selectively. Within conventional econpomics bounded rationality is usually treated (if it is treated at all) as a kind of cognitive failure; but this perspective diverts attention from the remarkable human capability to create and use patterns – which is the common theme of Smith, Marshall and Hayek. Given the second and third bounds on rationality (which correspond to Knight's definition of uncertainty), this pattern-making capability seems to be much more useful than a high level of logical skills. (Chester Barnard (1938) presented a similar argument for the importance of non-logical processes, against a background of business experience.) With Marshall's and Hayek's accounts as examples, it is also easier to see how this capability might develop; and Smith shows how it might lead to the growth of both scientific and economic knowledge. We can therefore acknowledge Knight's insight that uncertainty is the precondition of intelligence – and intelligence of the kind that is particularly relevant to entrepreneurship.

4 The human mind and human capabilities

Hume's demonstration that there can be no way of proving any general empirical proposition may be augmented by Hayek's (1952, 185) argument that 'the capacity of any explaining agent must be limited to objects with a structure possessing a degree of complexity lower than its own'. An immediate consequence is that 'the human brain can never fully explain its own operations'; any psychological theory of the individual must always be incomplete. The theories by which the human brain attempts to explain any system that includes many other human brains (and usually many non-human elements) are inevitably much less complete. Even theories that are intended to match real structures are themselves confined to the space of representations, and their relationship to the real-world phenomena that they are intended to interpret is inescapably subject to Knightian uncertainty – a theme explored by Shackle (1972). Neither biological evolution nor human reason, backed by scrupulous experimentation, can avoid this.

In practice, both natural selection and the mind's selection among human ideas, artefacts, organisations and institutions are based on characteristics that are successful within particular environments (in accordance with Knight's principle of relevant similarity); and because all representations are incomplete, the opportunity cost of success is likely – indeed, one may say virtually certain – to include the absence of characteristics that would be essential for success in some other possible environments. Every successful system has its characteristic way of failing; and even systems that survive may exhibit persistent pathologies, which may be fatal to some members of the relevant population. The population of entrepreneurs is a notable example; although, as noted earlier, Schumpeter does not discuss entrepreneurial failure, it is a natural conclusion from this conception of the mind that the great majority of new combinations that can be imagined will turn out to be poor representations. If, as Popper argued, this is true in science, why should it not be true in the economy?

This coexistence of systematic advantage and systematic deficiency is a common finding of experimental economics and psychology; if such results seem surprising, that is because they are interpreted by inappropriate theories and because economists have forgotten that opportunity costs (revealed by systematic deficiencies) are inherent in their own professional activities. It is a notable merit of Gigerenzer and Selten's (2001) investigation of 'the adaptive toolbox' that they explicitly link the systematic advantages of 'fast and frugal heuristics' within a particular domain to systematic errors outside that domain, even though the boundaries of the domain are often difficult to recognise by those using a particular heuristic (as Hayek's comment on the brain's inherent limitations implies). Schumpeter's claim that the response to major innovations is not smooth adjustment but co-ordination failure – a claim that is the foundation of his business cycle theory – is soundly based in human psychology.

In this paper I shall follow Hayek's example in avoiding any discussion of the relative importance of species and individual development, which (as many readers will know) is a major topic of contemporary discussion among biologists and psychologists; a more extensive, but still limited, discussion will be found in another paper (Loasby 2004). For our present purposes it seems reasonable to assume that development at the level of the individual must draw on that individual's genetic endowment, which is the outcome of species development and may impose many programmes of behaviour, or elements of such programmes, but that the form of development for each person may be substantially influenced by interaction with particular environments, and may be difficult to predict in detail.

In the course of biological evolution, the relatively very small proportion of mutations that happened to confer some relative advantage led to the differentiation of genetically-programmed physical form and behaviour across species; that the diversity of species could be explained by Smith's principle of the division of labour as a source of differentiated knowledge and skills was noted by the Belgian biologist Milne-Edwards (1827), and this attracted the attention of Charles Darwin. Because the content of biological mutations does not respond to the environment, and because the effectiveness of natural selection in matching development to environment depends on the stability of that environment, biological evolution must be very slow (and relatively rapid environmental change tends to be followed by multiple extinction).

In principle, therefore, if some means of accelerating adaptation should appear it might well be favoured by natural selection; and this is what seems to have happened with the emergence of *homo sapiens*. A substantial enlargement of the human brain, unaccompanied by changes in the human genome which were sufficient to programme it, created a relatively extensive scope for programmes to be constructed within each brain in the process of individual development, in addition to those programmes that remain genetically controlled; and as much of this development now necessarily occurred after birth (because of the constriction of the birth canal as a consequence of upright posture) it could be influenced by interaction with the local environment of each individual. Thus a particular example of environmental selection among random biological mutations made possible a new evolutionary process that incorporated directed variation: intelligence was guided by will towards the search for solutions of perceived problems (as in Smith's, Marshall's and Hayek's theories).

This development at the level of the species greatly enhances the possibilities of distinctive development at the level of the individual, leading to the differential emergence of domain-relevant knowledge and skills, which are much less demanding of cognitive capacity and brain energy than general-purpose logical processing, against a continuing low-cost background of programmed bodily functions and brain operations. The conversion of novelty to routine releases capacity for creating further novelty. Hayek's analysis is an appropriate illustration of this sequence: though the physical order originated from sensory perception, it has led to innovations that could not have been produced without evading the constraints of the sensory order; but the sensory order is still essential to normal human activity. The evolutionary process has itself evolved; but it is nevertheless an evolution and not a revolution. This, we shall argue, is true of all innovation; discontinuities are never absolute.

This process cannot escape the context of uncertainty and so it is still governed, although in a different form, by the evolutionary principles of variation, selection and retention. Smith, like Hume, rejected the idea that there was any correct procedure for achieving scientific truth; the successive cosmological systems had all eventually been rejected, and in reminding us that even Newton's system was not a perception of reality but a product of Newton's imagination he allowed for the possibility of its eventual supersession. Evolution does not optimise; it rejects alternatives which are, or have become, insufficient within a particular environment. This makes the 'Carnegie' basis for Nelson and Winter's (1982) theory highly appropriate in explaining the stimulus to seek new routines, the sequence of search and the termination of search according to some criteria of sufficiency – until the next failure. The advantages for the human species of this genetically-endowed capacity for far more rapid adaptation than genetic evolution can permit has so far outweighed the substantial costs of this capacity. The personal costs, however great, of individual failure to adapt, or of membership of a group or organisation that has failed to adapt (costs which may be readily observed and which Schumpeter made no attempt to belittle) are, of course, irrelevant at the genetic level; their relevance for public policy is beyond the scope of this paper.

The conception of the human mind as an extensive cluster of quasi-decomposable and selective connections corresponds with Jason Potts' (2000) general proposition that the crucial fact about systems is the incompleteness of their connections. If connections are incomplete, then the performance of a system depends not only on what elements are included but also on the links between these elements and the specific pattern of connections (and absence of connections) to other systems. Performance may then be changed either by modifying the set of elements or by a rearrangement of connections, internal or external; and it is important to note that the breaking of established connections, both in ways of thinking and in organisational relationships, may be no less important, for good or ill, than the formation of novel cognitive and organisational relationships. Such changes are characteristic of intelligence and entrepreneurship. They cannot be achieved by purely logical processes, though logical processes may subsequently be invoked to check for consistency or to trace some of the implications.

It is not possible to get outside our own minds in order to reconfigure them, and the possibilities of reconfiguration open to any one of us are quite severely limited. These limits result partly from the programming installed by our biological inheritance and partly from the need to economise on cognition by relying at any moment on the great majority of connections which exist at that moment. Therefore it seems reasonable to conclude, with Potts, that most changes will consist of movement to some 'adjacent state'. We may, however, include within this category not only limited rearrangements within an existing system or modest adjustments to its boundaries, but also the transfer of an existing pattern of thought or action across domains which are thought to exhibit relevant similarities. Of course, what states are adjacent, in this range of senses, will vary greatly between individuals, according to the development of their cognitive systems in relation to the particular environments that they have encountered. There will also be some variation across individuals within similar environments because of the particular patterns that they have developed for interpreting and responding to them, as Marshall (1920, 355-6) noted. When many people are following different paths, these differences promote both differentiated knowledge and skills across domains and 'the tendency to variation' (Marshall 1920, 355) within each domain which together fuel economic evolution. In these conditions path dependency (which is not to be identified with path determination) is a source of novelty.

We may therefore conclude that what Cohen and Levinthal (1989) called 'absorptive capacity' will differ substantially across individuals, not only in degree, but also in dimension; what kinds of ideas or skills each of us can absorb is heavily influenced by the ideas and skills that we already possess. (An important corollary of this principle is that people who have recently absorbed a substantial change – of the kind that is sometimes called a 'paradigm shift' – are likely to be particularly resistant to further substantial changes.) Absorptive capacity helps to shape entrepreneurship. However, the distinctive entrepreneurial contribution identified by Schumpeter is the complement of this: the imagination of new connections from existing patterns to elements that lie outside these patterns. This we may call creative capacity; it generates the variety on which selection can work. Recognising the potential relevance of other people's knowledge, skills and motivation may also require an awareness of other people as intentional agents, which appears to be at best primitive even among other primates (Tomasello 1999); and this awareness, which is the foundation of Smith's (1976 [1759]) *Theory of Moral Sentiments*, is not only a major aid to cognition, including the spread of ideas and practical skills, but also, as Smith recognised, the basis of trade. These together allow new knowledge to diffuse incomparably faster than genetic evolution could permit. In turn it is this diffusion that makes entrepreneurship so powerful an instrument of economic development.

5 Problems and decisions

Hayek (1945, 523) declared that 'economic problems arise always and only in consequence of change', defining this as a departure from anticipated patterns. Such patterns might include considerable but familiar variations, which could be handled by standard procedures. Knight (1921, 313) had already identified unanticipated change as a prerequisite of uncertainty, and therefore as 'a condition of the existence of any problem whatever in connection with life or conduct'. But as Shackle above all emphasised, if external change creates uncertainty, uncertainty makes internally-generated change possible: problems – perceived inadequacies of existing patterns of interpretation and action – are also opportunities for imagining new patterns of interpretation and action which may be better fitted to the new circumstances. Innovation which is prompted by the perception of problems is essential to Smith's, Marshall's and Hayek's theories of the mind, to Smith's and Marshall's theories of economic development, and to Schumpeter's theory of entrepreneurship. Cattaneo argues that the psychology of wealth is necessary to explain how will that is directed towards wealth fosters the development of intelligence which generates what we now call entrepreneurship.

We may consider Kirzner's version of entrepreneurship in this perspective. In Kirzner's (1973) initial analysis, the entrepreneur is presented as the agent of equilibration, who responds to a change that has already happened. This allows Kirzner to deal solely with the polar cases in the spectrum of uncertainty: 'sheer ignorance' and certain knowledge. An arbitrage opportunity is inherent in any economic change: until it is perceived we have sheer ignorance and no action, but once it is perceived, the action required is obvious and unimpeded – in Kirzner's own phrase, the ten-dollar bill is discovered to be already in the entrepreneur's hand. This instantaneous transition between cognitive states is achieved by alertness, which we may interpret as the absorption of a new fact into a familiar framework. It does not require the creation of new patterns; indeed it relies on a fundamental assumption of well-defined markets for clearly-differentiated products.

The first-mover profit on which Kirzner's theory depends requires alertness to be domain-limited for each entrepreneur, because, as Richardson (1960, 57) had already pointed out, a 'profit opportunity, which is both known to everyone, and equally capable of being exploited by everyone, is, in an important sense, a profit opportunity for no one in particular'. It is therefore essential that each pair of locations within a specific market is observed by a limited number of people; and Kirzner's justification for this is that people will tend to notice what they are interested in. This differentiated interest appears to be a compound of motivation and specific knowledge; each entrepreneur is receptive to an opportunity of moving to profitable adjacent states, but what states are adjacent varies widely between entrepreneurs. Kirzner takes this as an obvious fact of life; we have suggested a cognitive explanation. The variation of interpretative frameworks naturally limits the number of people who are capable of recognising any particular opportunity, while allowing the entrepreneur's actions to attract sufficient competition to bring this particular market into profitless equilibrium.

Kirzner's initial model, which is illustrated by price differences between locations, is therefore compatible with our cognitive theory. However that theory suggests that later versions of Kirzner's analysis, in which the relevant arbitrage is between resources and the outputs that they might produce, or requires intertemporal transfers, should be modified by recognising that potential competitors may have difficulty in interpreting what is happening. Indeed as the complexity of the profit opportunity increases, the consequences of Kirznerian entrepreneurship may increasingly resemble those of Schumpeterian entrepreneurship, in disrupting established routines without indicating adjacent cognitive states to which those deprived of their routines might move. The cognitive bases of Kirznerian and Schumpeterian entrepreneurship also become increasingly similar.

Schumpeter's explicit exclusion of invention from his theory of entrepreneurship suggests that his entrepreneurs, like Kirzner's, benefit from changes that have already happened and which imply opportunities that are ready for exploitation. However, these are not presented as arbitrage opportunities which are immediately obvious to anyone who knows about the change and is equipped with the appropriate domain-specific cognitive apparatus; instead the opportunity can be revealed only by the construction of a new cognitive apparatus. This feature has prompted some comparisons between Schumpeter's theory of economic development and Kuhn's (1962, 1970) theory of scientific revolution; both kinds of progress depend on the creation of new ways of thinking. Neither author, it may be said, gives adequate consideration to the elements of continuity which are necessary to carry even radical transformation; Smith (1980 [1795]) is a better guide.

The creation of a new combination of economic significance is an act of intelligence, which as Cattaneo (2001, 101) tells us, can be stimulated by a will that is directed towards wealth. This relationship is surely a natural reading of Schumpeter's theory, especially in his discussion of the psychology of the entrepreneur (Schumpeter 1934, 90-94). From the perspective of this paper, what is particularly interesting about this theory (apart from its relationship with Cattaneo) is that external change is not sufficient, and not even necessary. The crucial element is a change within the individual that creates a new vision; what is required is not domain-specific alertness but domain-linking imagination. This is perhaps the most fundamental of Schumpeter's challenges to standard economics, because it is a challenge to the standard conception of human agency as closed-system optimisation.

Problems, and indeed all occasions for initiating any decision process, are defined by differences (Pounds 1969). Responding only to differences (and not to all of them) is an efficient way of allocating the scarce resource of attention, without the need to allocate resources to the process of allocating attention; and this is a pattern of behaviour that seems to have evolved long before the primates. Indeed the self-regulating systems of the body seem to be governed by the identification of differences that are large enough to trigger a response. We should not therefore be surprised by the abundant experimental evidence (Kahneman 2003) that people rarely follow the prescriptions of decision theorists to concentrate on the levels of outcomes but persist in looking for differences. Anyone designing a control system, whether mechanical or for members of any organisation, will focus on differences, and decide what differences should generate signals for action; and major disasters have often resulted from a failure to notice differences that have not been specified in advance. The perception of an external change defines a difference between our established interpretation of past events and some new phenomenon; and this perception itself depends on an established interpretation. Without a background of relevant stability change cannot be recognised as change (a principle that I learnt in conversation with Andy Van de Ven).

More generally, the perception of an occasion for decision requires a reference standard by which a difference can be defined. Our interpretation of a particular supposedly relevant segment of history, limited in time and scope, is an obvious and important example, and already indicates the importance of interpretative systems (such as those suggested by Smith, Marshall and Hayek) in influencing human thought and action. However, there are other bases of comparison, and in a development of Pounds' classification (Loasby 1976, 96-103; Loasby 2000, 723) which was intended for application to economic development I suggested that we should consider additional reference standards, including the performance of relevant comparators, our own intentions or specific plans, and imaginative standards – what Shackle (1979, 26) called 'the imagined, deemed possible'.

Even more clearly than interpretations of the past, all three of these reference standards are partly subjective, and they may themselves be the outcome of decision processes that are initiated in response to other differences: people may be motivated to search for comparators, to review plans, and to devise or import procedures that might generate novelty. It is therefore natural to find great diversity in the definition of problems, and consequently great diversity in ideas and actions. Moreover, since there are no guaranteed procedures for discovering the best responses to differences that are defined in any of these ways, it is not unusual for the consequences of decisions to generate new differences, leading to new decisions and so on in sequences that may extend over many years. Indeed, this is the characteristic pattern of major innovations. The 'new combinations' envisaged by Schumpeterian entrepreneurs are very rarely visions that simply turn out to be true; and the research departments of the large businesses which take over much of the entrepreneurial role in Schumpeter's later exposition do not find it easy 'to turn out what is required and make it work in predictable ways' (Schumpeter 1943, 132), however extensive and knowledgeable their members may be. This is what we should expect from our understanding of the human mind and its relationship to the universe.

The great majority of substantial innovation processes fail, and those that succeed often require a great deal of cognitive reconstruction of the initial understanding of the project, and may deliver outcomes which do not correspond at all closely with initial intentions. These processes, like so many managerial activities, cannot be controlled by logical procedures (Barnard 1938), and rational choice models of innovation are correspondingly misleading. Instead we may observe evolutionary processes, conspicuously in research and development but more generally in dealing with complex problems, in which selection leads directly and repeatedly to the generation of new variants over a time-scale that is extremely brief in relation to biological evolution; and each new variant is a conjecture that is offered for selection. (Decision cycles are discussed in Loasby 1976 and 2000.) Within these processes, differences continue to define the next issue to be tackled, and – not surprisingly – also the criteria for choice.

6 Organisation, intelligence and evolution

Everything in the universe appears to depend on selective connections (or, in Herbert Simon's language, quasi-decomposability). Because our ability to comprehend phenomena seems to be strictly dependent on the formation of selective connections, thus restricting the forms of representation that we are able to construct, this appearance may be deceptive; there may be other phenomena, of which we are not aware, which do not depend on selective connections. However, because selectivity appears to be inherent in both the structure and the functioning of all the phenomena that we encounter, it may be reasonable to assume that selectivity is not simply a property of our understanding but inherent in what we are trying to understand. Even if this assumption is not reasonable, it seems to be inevitable. It is nevertheless important to remember that the particular set of elements and connections by which we represent a phenomenon is unlikely to correspond precisely to those that constitute the phenomenon itself, and may be adequate for only a limited range of applications, as Knight implies; moreover, these limits may be discoverable only by exceeding them. We should also remember that different people may develop, or adopt, different representations of what they consider to be the same phenomenon – consider, for example, the variety of representations to which economists have applied the labels 'competition' and 'innovation'. Such differences sometimes cause misunderstandings and co-ordination failures; but they may also help to solve problems, improve knowledge, and stimulate enterprise. This is a principal means by which the tendency to variation – a crucial feature of intelligence as understood by Cattaneo and Knight – aids progress. All these possibilities are natural consequences of uncertainty, combined with human skills in pattern-making. They are elements of an evolutionary process: all knowledge is constructed, by individuals and within communities – though the proportion of these constructions that survive application is very small.

Systems of selective connections may be called organisations; thus organisation is a central issue in thinking about the human mind and the economy. Knowledge itself is organisation, imposed on phenomena, and it is the result of organising activity. In the realm of ideas, this organising activity is intelligence; in the realm that we call economic, it is entrepreneurship. (The concept of entrepreneurship, and its connection with Knightian principles of intelligence, may be readily extended to other realms.) However, both kinds of organising activity are dependent on the prior existence of structures which set bounds to uncertainty; for, as Shackle (1969, 224) observed, 'the boundedness of uncertainty is essential to the possibility of decision'. A rather similar notion seems to underlie Alfred Marshall's (1920, 138-9) principle that 'Organization aids knowledge', especially because of his insistence that it 'has many forms', which are needed, together with variety within each form, in order to provide the differentiated bounds to uncertainty which will permit the development of locally-appropriate knowledge, leading to locally-appropriate decisions, which is essential for both Kirzner's and Schumpeter's theories of entrepreneurship. Marshall's (1920, 318) definition of increasing return as operating through changes in organisation was expanded by Allyn Young (1928) into a cumulative process in which new knowledge stimulated extensive rearrangement of organisational boundaries, which redefined the context for generating further knowledge, leading to further restructuring.

However, in the context of this paper – and only in this context – I would like to conclude by emphasising three kinds of non-formal organisation that aid knowledge. The first of these is the foundational principle of this paper: the internal organisation of the human mind as a partly-adjustable cluster of selectively-connected systems. Two features of this organisation, discussed earlier, deserve specific attention, and they may both be considered to be economising principles. The first is the complementarity between automaticity and new combinations in allocating cognitive resources; this emphasis on complementarity is necessary to balance Schumpeter's emphasis on the contrast between them, and it suggests that Schumpeter's assignment of co-ordination and growth to incompatible but independent theoretical systems needs reconsideration. Marshall's attempt to combine them led to trouble, but his instinct was surely sound; and a good deal of his analytical apparatus may still be serviceable, because of its consistency with our current understanding of human cognition. The second feature is the contrast between the potential of the human mind for developing knowledge and skills within many possible domains and the limited range within which this potential can actually be developed by any single individual. It is this particular contrast between individual and population that underlies the significance of Adam Smith's great principle of the division of labour, which allows a community of differentiated specialists to develop a range of knowledge and skills far greater than even the greatest philosopher or scientist could accomplish.

It is also this feature that justifies economists' focus on co-ordination – for which, in modern societies, prices are indispensable but insufficient, even for many market transactions. The internal organisation of the individual human mind provides a balance – not always adequate – of differentiation and integration; but a different perspective on co-ordination results from thinking of the population of minds on which each particular individual can draw. In order to do this effectively people must develop a second form of organisation; for minds, like firms, flourish best when they have effective internal and external organisations. That is an important part of the message of Smith's (1976 [1759]) *Theory of Moral Sentiments*, and it allows us to bypass part of the problem of altruism; even if we have no direct concern for the welfare of others we have an interest in their success in generating knowledge because, in addition to the exchange opportunities that may result, we may be able to draw on their new knowledge to improve our own, taking advantage both of the speciation of knowledge and of variations within each species. For exactly this reason, firms have an interest in the success, not only of firms with complementary capabilities, but also of their rivals.

The third kind of non-formal organisation comprises the set of institutions that supply the procedures and premises which help us to economise on cognition. We rely on such premises and procedures even when no interaction with others is involved; indeed much of education, formal and informal, supplies us with rules to guide our thought and action. Because we share so many institutions for our private purposes, we are predisposed to look for institutions that will facilitate interactions and are likely to find many of them already part-formed in the similarity of patterns of thought and action within groups where interaction is most likely to occur (Choi 1993). Markets are the most obvious class of institutions within modern economies, though often not recognised as such (Ménard 1995); the creation of a new business is much easier if appropriate institutions are already in place, and the performance of any business is critically dependent on the institutions that develop within it, as Barnard (1938) well knew. The creation or modification of institutions is a necessary element in any entrepreneurial project, as Casson (1982, 2003) has shown.

Innovation processes are evolutionary. They also appear to be channelled by some of the consequences of biological evolution that are manifest in the human mind, where channelling involves a mixture of constraint and facilitation. The co-ordination of the growth of knowledge is supported by the ability to rely, most of the time, on the regularity of other people's behaviour, which is a product of automaticity, while having some expectation that they will occasionally produce ideas which are relevant to particular problem domains. The generative potential of an economic system depends on an adequate variety of organisational forms and interorganisational relationships (Richardson 1972), each with its domain-limited advantages embedded in combinations of cognitive proximity and cognitive distance, and also adequate variety within each form. The realisation of this potential depends on uncertainty, which provides scope for, though it does not ensure, the exercise of imagination and will, leading to intelligence and entrepreneurship.

References

Bain A (1864) The Senses and the Intellect. Longmans Green and Co., London, 2nd edn.
Bain A (1865) The Emotions and the Will. Longmans Green and Co., London, 2nd edn.
Barnard C (1938) The Functions of the Executive. Harvard University Press, Cambridge MA
Butler R (1991) The historical content of the early Marshallian work. Quaderni di Storia dell'Economia Politica, 9, 2-3: 269–88
Casson M (1982, 2003) The Entrepreneur: An Economic Theory. (1982) Martin Robertson , Oxford; (2003) Edward Elgar, Cheltenham UK and Northampton MA
Cattaneo C (2001) Del Pensiero Come Principio D'Economia Publica: Intelligence as Principle of Public Economy. Libri Scheiwiller, Milano
Choi YB (1993) Paradigms and Conventions: Uncertainty, Decision Making and Entrepreneurship. University of Michigan Press, Ann Arbor, MI
Cohen WM, Levinthal DA (1989) Innovation and learning: The two faces of R & D. Econ J, 99: 569–96
Gigerenzer G, Selten R (2001) Bounded Rationality: The Adaptive Toolbox. MIT Press, Cambridge MA
Giocoli N (2003) Modeling Rational Agents: From Interwar Economics to Early Modern Game Theory. Edward Elgar, Cheltenham UK and Northampton MA
Hayek FA (1945) The use of knowledge in society. Amer Econ Rev, 35: 519–30
Hayek FA (1952) The Sensory Order. University of Chicago Press, Chicago
Hume D (1978) A Treatise on Human Nature. Clarendon Press, Oxford, ed. Selby-Bigge, L. A., 2nd edn. revised by Nidditch, P. H.
Kahneman D (2003) Maps of bounded rationality: Psychology for behavioural economics. Amer Econ Rev, 93, 5: 1449–75
Kirzner IM (1973) Competition and Entrepreneurship. University of Chicago Press, Chicago
Knight FH (1921) Risk, Uncertainty and Profit. Houghton Mifflin, Boston
Knight FH (1933) Preface to the re-issue. In: Risk, Uncertainty and Profit, vol. 16 of *Reprints of Scarce Tracts in Economics and Political Science*, xi–xxxvi, London School of Economics, London
Kuhn TS (1962, 1970) The Structure of Scientific Revolutions. University of Chicago Press, Chicago
Loasby BJ (1976) Choice, Complexity and Ignorance: An Inquiry into Economic Theory and the Practice of Decision Making. Cambridge University Press, Cambridge
Loasby BJ (2000) Decision premises, decision cycles and decomposition. Ind Corp Change, 9, 4: 709–31

Loasby BJ (2004) Hayek's theory of the mind. In: Koppl R (ed.) Advances in Austrian Economics, 101–34, Elsevier, Oxford
Marshall A (1920) Principles of Economics. Macmillan, London, 8th edn.
Marshall A (1994) Ye machine. In: Raffaelli T (ed.) Research in the History of Economic Thought and Methodology, Archival Supplement 4, JAI Press, Greenwich CT
Ménard C (1995) Markets as institutions versus organizations as markets? Disentangling some fundamental concepts. J Econ Behav Organ, 28: 161–82
Milne-Edwards H (1827) Nerf. In: Bory de Saint-Vincent M (ed.) Dictionnaire Classique de l'Histoire Naturelle, Rey et Gravier, Paris
Nelson RR, Winter SG (1982) An Evolutionary Theory of Economic Change. Belknap Press, Cambridge MA and London
Penrose ET (1952) Biological analogies in the theory of the firm. Amer Econ Rev, 42: 804–19
Potts J (2000) The New Evolutionary Microeconomics: Complexity, Competence and Adaptive Behaviour. Edward Elgar, Cheltenham UK and Northampton MA
Pounds WF (1969) The process of problem finding. Ind Manage Rev, 11: Jan.19
Raffaelli T (2003) Marshall's Evolutionary Economics. Routledge, London and New York
Richardson GB (1960) Information and Investment. Oxford University Press, Oxford
Richardson GB (1972) The organisation of industry. Econ J, 82: 883–96
Schumpeter JA (1934) The Theory of Economic Development. Harvard University Press, Cambridge MA
Schumpeter JA (1943) Capitalism, Socialism and Democracy. Allen and Unwin, London
Schumpeter JA (1954) History of Economic Analysis. Allen and Unwin, London
Shackle GLS (1966) The Nature of Economic Thought: Selected Papers 1955-64. Cambridge University Press, Cambridge
Shackle GLS (1969) Decision, Order and Time in Human Affairs. Cambridge University Press, Cambridge, 2nd edn.
Shackle GLS (1972) Epistemics and Economics. Cambridge University Press, Cambridge
Shackle GLS (1979) Imagination and the Nature of Choice. Edinburgh University Press, Edinburgh
Smith A (1976 [1759]) The Theory of Moral Sentiments. Oxford University Press, Oxford, eds. Raphael DD, Macfie AL
Smith A (1976 [1776]) An Inquiry into the Nature and Causes of the Wealth of Nations. Oxford University Press, Oxford, eds. Campbell RH, Skinner AS, Todd WB
Smith A (1980 [1795]) The principles which lead and direct philosophical enquiries: Illustrated by the history of astronomy. In: Wightman WPD (ed.) Essays on Philosophical Subjects, Oxford University Press, Oxford
Tomasello M (1999) The human adaptation for culture. Annu Rev Anthropol, 28: 509–29
Whitehead AN (1948 [1911]) An Introduction to Mathematics. Oxford University Press, Oxford
Witt U (2003) Generic features of evolution and its continuity: A transdisciplinary perspective. Theoria, 18, 48: 273–88
Young AA (1928) Increasing returns and economic progress. Econ J, 38: 527–42

Markus C. Becker · Patrick Cohendet · Patrick Llerena

Division of labor and division of knowledge: Why the nature of the causality matters for the evolutionary theory of the firm

Abstract This article considers the role and interrelation of the division of labor and the division of knowledge, and its relevance for knowledge-based and evolutionary theories of the firm. As is well known, Adam Smith focused on the effect of the division of labor, while Charles Babbage focused on the effect of the division of knowledge. We are not the first to argue that both are connected in a loop, the division of labor guiding the division of knowledge via learning by doing, and the division of knowledge, in turn, guiding the development of competences for accomplishing steps of the work process. In this article, our main argument is that even knowledge-based and evolutionary approaches to the theory of the firm have focused more on the Smithian than the Babbagian part of the loop. We argue that the impact of the division of knowledge should not be glossed over and perhaps be considered first. The reason is that in the division of knowledge cognition enters the picture and makes a difference, amongst other things on the division of labor. Taking the division of knowledge into account also helps cast light on such important issues as understanding the emergence of routines.

Keywords Division of labor · Division of knowledge · Charles Babbage · Adam Smith

JEL Classification B52 · D8

Introduction

The classical vision of the division of labor presented by Adam Smith has cast a long shadow on economic theory. Any first grade student knows the famous pin

A first version of this paper has been presented at the 10th International Schumpeter Society Conference, Milano, June 9-12, 2004. We thank in particular B. Loasby for his comments during the conference and the referee for his/her remarks.

M. C. Becker · P. Cohendet · P. Llerena
BETA UMR 7522, Université Louis Pasteur, Strasbourg
E-mail: pllerena@cournot.u-strasbg.fr

factory example. Adam Smith's hand – in more or less visible ways – has also guided many an economist's vision of things, and many theoretical developments in economics. The Smithian argument on the importance of the division of labor and specialization has been perpetuated in the economics (and management) literature in the following classical form: The division of labor entails a process of learning by doing that contributes to increasing skills and expertise, and thus to enhancing the accumulation of specialized knowledge. From this accumulation, new gains of productivity can then be obtained in society that lead to a virtuous extension of markets, which in turn will trigger a new cycle of further division of labor.

Though it has been widely diffused and has left its stamp on economic theory, we consider this classical Smithian interpretation a *narrow* one. Our view is that, while it explains *why* labor is divided and the effects of this division on the formation of learning and knowledge in society, it says nothing on *how* labor is divided. The underlying hypothesis is just that a new division of labor that emerges from a new cycle of growth in the economy is supposed to be a finer division of labor than the previous one. As a result, this vision casts a veil on the fundamental cognitive process that may influence the choice of activities (and thus the division of labor) within firms. The classical Smithian interpretation restricts the relationship between the division of labor (which distributes tasks amongst agents) and the division of knowledge (which distributes interpretative capabilities between agents) to a mere causal relationship: the division of work *causes* the division of specialized knowledge. Once, and only once, the division of labor has taken place, the cognitive mechanisms, reduced to the sole functioning of learning by doing effects, shape the nature of competences and routines in the economy. The division of labor not only defines the problems to be solved, but also defines the direction, potential, and participants of the learning that can take place.

Thus, such a representation, driven by short term efficiency motivations, makes no mention of the complex cognitive process that may precede the choice of the tasks to undertake. The possibility that existing learning, routines, competences and representation could influence the choice of the division of labor (and drive the economy towards a new division of work which is not merely a finer decomposition than the previous one) is excluded. The aim of our contribution is precisely to introduce such a possibility in the Smithian scheme. We consider that, generally speaking, there is a co-evolution between the division of labor and the formation of knowledge. In other words, in the Smithian cycles, we introduce the possibility of a loop between the division of work and the division of knowledge. Our view is that the implications of such an assumption contribute to an enrichment of the classical representation, and in particular allow for new perspectives for the evolutionary theory of the firm.

Interestingly, an alternative to the classical Smithian interpretation exists: Babbage (1832/1989) argued the division of labor must itself be founded on differences of skill. Rather than understanding the division of knowledge as a consequence of the division of labor, he considered it its cause, or what should determine it.

Our main argument is that the Babbagian vision invites a shift in the focus of the analysis of routines (Nelson and Winter 1982) from an explanation of the functioning of existing routines towards the examination of the mode and the history of emergence of the routines. To a large extent, the evolutionary literature considers that routines result from a given existing division of work which is not part of the analysis. A Babbagian perspective is more nuanced in focusing on the cognitive process between agents that contribute to the emergence of the routine. The generation of routines is thus heavily dependent of the cognitive processes at stake.[1]

The article is structured as follows: In a first part, we recall the Babbagian vision of the division of labor and emphasize the way this vision could enlarge the classical Smithean interpretation. Having introduced the possibility of co-evolution between the division of labor and the division of knowledge, we develop in a second part the knowledge based vision of the firm, to identify the mechanisms by which the cognitive dimension may be introduced in the Smithian interpretation. A third part is devoted to the analysis of these issues at the microlevel of routines in order to open the discussion on the consequences for the evolutionary theory of the firm.

Part 1 : The Babbagian perspective of the division of labor: how it contributes to enlarge the classical Smithian interpretation

As we have seen, in the classical Smithian interpretation, the progressive specialization of work induces the creation of capabilities and a progressive increase of specialized knowledge (by mechanisms such as learning by doing) (Loasby 1999; Pagano 1985). The task one is assigned to carry out determines what one can learn by doing this task. For our argument, what is decisive about this idea is that even though specialization decides what capabilities are created, the criterion that decides on how labor is divided is efficiency (i.e., thanks to higher dexterity of who is specialized). It is here that considering the 'second step' from a Babbagian perspective has something important to add.

As did Smith, Babbage analyzed England's transition from an agricultural to an industrial economy. Babbage's work is considered a crown jewel of management literature and arguably one of the first management texts ever written, and the beginning of what would later be called operations research (Hyman 1982). In his analysis of the transition to an industrial economy, Babbage applied Smith's idea of the division of labor not only to mechanical but also to mental operations (Stigler 1991). He paid attention to mechanization and the development of machine-based production in the factory, but at the same time considered the social relations in industry (Hyman 1982). The reason was that, according to Babbage, while the progress of technology had an important part in the emergence of the industrial economy, manufacturers had to pay attention to principles other than merely technological ones. Manufacturers also had to take care of what he called "the domestic economy of the factory" – in modern terms, organization and management issues. To this end, Babbage considered problems such as workers' opposition to technological improvements, and even developed a system of profit sharing and co-operative production to ensure that new technology was adopted. Babbage's

[1] Moreover, many characteristics of a given routine (its strength, degree of resistance, consistency, power of replication, etc.) depend on the cognitive process that takes place at the phase of the emergence of the routine. To give an example, a routine that has naturally emerged within a community of economic agents sharing strong common social norms will have probably a much stronger power of replication than a routine which results from the functioning of a temporary team project constituted from heterogeneous agents who never met before.

contribution that is most relevant in the present context, however, is the so-called "Babbage principle" that he added onto the Smithian foundations of his analysis:

> 'That the master manufacturer, by dividing the work to be performed into different processes each requiring different degrees of skill and force, can purchase exactly that precise quantity necessary for each process; whereas, if the entire work is executed by one workman, that person must possess sufficient skill to perform the most difficult, and sufficient strength to carry out the most laborious of the operations into which the art is divided' (Babbage 1832/1989, 137-8).

Babbage's contribution to Smith's analysis of the division of labor was to distinguish different dimensions of work tasks, such as skill and physical force (Stigler 1991).

What matters for our argument is that, although not inconsistent with Smith, Babbage puts the priority on skills in the sequence of steps in the cycle of division of labor – learning – division of labor based on the higher degree of specialization and higher knowledge (driven by increased complexity in product markets and technology). He also puts the priority on skills in terms of the criterion that should decide how labor is divided. In fact, one could say that with his argument he *establishes* knowledge and skills as the criterion that should decide how labor is divided.

Possibly, this observation was a result of Babbage's many factory visits in Britain and continental Europe. Having adopted such a more fine-grained analytical lens, an additional problem in the Smithian endeavor of decomposing work in sub-tasks in order to maximize economies scale and gains from of specialization becomes visible. The problem is: if a work task has two dimensions, which of the two dimensions should be 'optimized', and thus be used for the decomposition scheme? The solution Babbage proposed is to divide labor *along its different dimensions*: if you separate out those tasks that require a high level of physical force (and at the same time a low level of skill, which can be found in any average person), then you can 'optimize' the provision of physical force in the execution of the subtask by allocating the task to the physically strongest person. As compared to the division of labor by a different logic, Babbage's system increases the economics of scale of specialization further. What is important for the present discussion is that Babbage thus presumed that the division of labor must itself be founded on differences of skill, a point also emphasized by Marshall (cf. Hodgson 1993, 412). By affording logical primacy to skills (or more broadly speaking, characteristics of human resources), Babbage therefore turned the relationship of causation between the division of labor and of knowledge on its head: the division of labor is determined (caused) by the skill characteristics (the characteristics of the human resources), rather than the other way around.

In fact, tracing the reception of Smith's and Babbage's perspectives in economics provides a clear picture of the bias to be found in much of the discussion. As Martens (1999) has outlined, three different 'streams' of the concept of division of labor can be discerned in the economics literature:

(i) the Classical stream, starting with Adam Smith (1776), who first used the concept of division of labor but without much reference to knowledge or information interpretations. In fact, there appears to be considerable confusion in the Classical school on how the division of labor is supposed to work and its impact on society

(ii) the Neo-Classical or modern mainstream economics, that has basically adopted the perfect information paradigm of general equilibrium theory and welfare economics, thereby excluding the possibility of knowledge differentiation among agents. There are, however, several attempts to reintroduce it, mostly under the guise of individual learning models

(iii) the imperfect or asymmetric information economic stream, which steps outside the perfect information paradigm: von Hayek, the bounded rationality school, Coase, Williamson (Martens 1999, 4).

From a classical Smithian perspective, the division of labor causes learning and skill development, and determines the division of knowledge by giving access to learning-by-doing. The implication of the development described by Martens, however, is that considerations of knowledge have been sidelined in the literature that builds on Smith. Only one stream of literature along the line of Penrose – Arrow – Loasby gives attention to knowledge in considering the division of labor (Penrose 1959/1995; Arrow 1962; Loasby 1999, 2000). In particular, Loasby has pointed out the consequences of a Smithian vision, as interpreted through a Penrosian lens: 'even within a single field of knowledge or activity, differences between individuals' environments, interacting with differences in their responses to these environments, lead to some differences between their knowledge and skills' (Loasby 2000, 7).

We now turn to an analysis of the implications of the debate between the Babbagian vision and the classical Smithian interpretation of the division of labor in the theory of the firm.

Part 2: The knowledge based view of the firm

As emphasized above, the debate between the Smithian and the Babbagian views found strong echoes in the theory of the firm. Our opinion is that, in particular, the existing controversy between competences and transactions, which is the hottest issue right now in the literature on the firm, is inherently related to the question of the causality between division of labor and division of knowledge. The literature dealing with transactions and competences does not in general recognize this underlying debate between division of labor and division of knowledge. Lacking awareness of this fact is, according to us, a source of misinterpretation and misunderstanding between the different theoretical approaches.

Division of labor, division of knowledge and the strategy of the firm

To paraphrase Porter, the issue in any debate on strategy is the underlying vision of the theory of the firm. To show the intensity and the importance of the debate between division of labor and division of knowledge, we propose thus to start by exposing the strategic version of it, then we will turn in examining its underlying theoretical foundations.

The debate on strategy from the 90's has been dominated by the dispute between the 'classical' vision focused on *activities* promoted by Porter, and the challenging *competence-based* vision inspired by the influential article by Prahalad and Hamel published in HBR in 1990.

- On the one hand, as is well known, the basic argument of the Porterian approach to strategy is that superior performance of a firm is explained by its competitive advantage derived by its unique positioning in the industry whether it is on costs or on differentiation. Such positioning is based on activities performed by the firm and systematized in a value chain[2]. It is also based on efficiency as criterion, not on competence development.
- On the other hand, the competence based approach, which renewed the perspective of the resource based approach due to Wernerfelt (1984), was clearly prescriptive as to the best way to set winning strategies for the firm, especially as to diversification and the excessive use of SBUs (Strategic Business Units) in highly decentralized profit centers. "In the 1990s, top executives will be judged on their ability to identify, cultivate, and exploit the core competencies that make growth possible".

According to Porter, the competence-based approach cannot be a substitute for the activity-based approach of strategy; it is at best a complement to Porter's vision[3]. In order to arrive to such conclusions, Porter first makes a distinction in its theory of strategy between the cross-sectional problem of strategy ("linking firm characteristics to market outcomes", 1994, 424) and the longitudinal problem of strategy ("the dynamic process by which positions are created", 1994, 424). For him, a comprehensive theory of strategy should address both the cross-sectional and longitudinal dimensions. The activity-based approach clearly does, the competence-based approach does not[4]. The competence approach just addresses the longitudinal perspective and sheds light on some kind of learning and accumulated capabilities over time that could explain why a firm is better able to perform some activities[5]. In the 1998 revised introduction to *Competitive Advantage*, Porter stresses: "Is a

[2] As Porter stressed in the new introduction to his 1985 book, "the book's core is an activity-based theory of the firm" (1998, xv).

[3] M.E. Porter is, among others (Spanos and Lioukas 2001), the one who has seen important limitations to the resource-based view and core competence approach, and, at best, considers it as a complement to the basic I/O framework. "To date, the most attention paid to the integration of the two perspectives has been by Michael E. Porter in Competitive Advantage: Creating and Sustaining Superior Performance, (1985) and, in the dynamic context, in his article "Towards a Dynamic Theory of Strategy" (1991)" Collis and Montgomery (1995, 61-62). We could add to those two basic contributions the new introduction to the book Competitive Advantage: Creating and Sustaining Superior Performance in 1998 where Porter clearly reformulates his thought on the fortune of I/O and competence approach.

[4] Looking to the resource based view, he comments "the promise of the resource based view for the strategy field is its effort to address the longitudinal problem, or the conditions that allow firms to achieve and sustain favourable competitive positions over time [...]. The resource based view cannot be an alternative theory of strategy [...] it cannot be separated from the cross-sectional determinants of competitive advantage or from the conception of a firm as a collection of activities" (1994, 445-446).

[5] "Activities involve human resources, purchased inputs and a technology for performing them broadly define to include organizational routine [...] Performing an activity, or a group of linked activities also creates assets in the form of skills, organizational routines and knowledge" (1994, 435-436). We could find almost the same reasoning in Williamson (1999) as to learning. In fact, Porter will come to conclude that resources are intermediate between activities and competitive advantage.

firm a collection of activities or a set of resources and capabilities? Clearly, a firm is both. But activities are what firms do and they define the resources and capabilities that are relevant. Activities provide the connections between factor markets and product markets positions" (1998, xix). In the Porterian view, the firm is a nexus of both activities and resources, but activities come first. They define competences (as a by-product, through a type of learning by doing) and shape strategy. We are thus clearly in a Smithian perspective, where the division of work (activities in the Porterian wordings) shapes the division of knowledge.

The firm as a "processor of information", or why dynamic considerations (building competences) should have priority over efficiency considerations

Consider Porter's strategic analysis, which relies on the traditional contractual approaches of the firm – transaction costs theory in particular – that are not designed to accommodate the fundamental characteristics of knowledge. As underlined by Fransmann (1998), these traditional approaches consider the firm to be a 'processor of information', the behavior of which can be understood as an optimal reaction to external signs and factors which are detected[6]. In the Porterian framework, the main result of the processing of information by firms is precisely those *activities* that emerge from the positioning of an end product within an industry structure. For Porter, once activities are shaped and defined by processing information, *then* these activities may drive *competencies* (accumulation of collective knowledge and learning in the firm). Processing information is thus the core of the economic decisions of the traditional firms, while creating and exchanging knowledge is just a by-product of the current activities.

There is, however, a condition for the above hypotheses (first focus on information processing, then consolidate knowledge by-products) to be valid: the economic environment is supposed to be *stable* (in particular, the industry structure is considered as given). In innovative environments, characterized by the development of hyper-competition, globalization, and knowledge based economy, the hypothesis is severely questioned. As Bierly and Chakrabarti (1996) underline: "We believe the development of a dynamic knowledge strategy typology or taxonomy will offer more insight than the basic static strategy typologies developed to date. Other generic typologies such as Porter's (1980) rely on the basic assumptions that are not valid for many industries to-day. Specifically; they assume that a) the primacy

[6] The behavior of the firm as a rational information processing machine signifies that the same signals will give rise through time to the same pattern of action, provided the technical conditions (as expressed by the production function) remain unchanged. The neo-classical theory of the firm, in particular principal/agent theory, has basically reduced the co-ordination principles to a bundle of bilateral contracts which are meant to achieve co-ordination by devising appropriate incentive schemes in order to align self interested individual action with the common organisational goals. The transaction cost approach, despite its different angle and its specific focus on the boundaries of the firm, comes to a similar fundamental conclusion: the firm could be seen as a 'nexus' of contracts. Its very reason of existence is to correct market failures, when the functioning of market mechanisms in terms of information processing is too costly. Transaction cost theory agrees with the principal/agent vision that information is imperfect and that the existence of potential asymmetries of information authorises unproductive rent-seeking behavior. The firm is thus conceived as an institutional mechanism creating a governance structure to solve the problem of misaligned incentives attendant on imperfect information. The focus is thus on the process of allocation of resources needed to cope with such adaptation.

focus of strategy is about the positioning of an end-product within an identifiable industry structure, and b) the industry structure is relatively stable and changes to the environment are mostly incremental, linear changes that do not redefine the product or industry. However, other researchers have observed the preponderance of boundary-less industry structures, hyper-competition, increasing globalization, an increasing rate of technological change and diffusion and a tremendous increase in access to information through the advance in computer and communications (Bettis and Hitt 1995; D'Aveni 1994; Hamel and Prahalad 1994). The "static" generic strategy typologies offer little practical value to top managers in determining how to develop a competitive advantage, which is the primary purpose of strategy" (Bierly and Chakrabarti 1996, 133).

For these reasons, following Fransman, we consider that the actual innovative economic environment invites a reconsideration of the foundations of the theory of the firm. In such environments, firms should be viewed *in primis* as processors of knowledge. As Nonaka and Takeuchi (1995, 58) note, "when organizations innovate, they do not simply process information from outside in, in order to solve existing problems and adapt to a changing environment. They actually create new knowledge and information, from the inside out, in order to redefine both problems and solutions and in the process, to re-create their environment". We consider that the competence-based approaches of the firm are the challenging theories in such a perspective that try to bring forwards the vision of the firm as a processor of knowledge. This perspective implies an in-depth reconsideration of the strategy of firms, by suggesting that *first* firms select their domain of competencies, and *subsequently* they manage their activities. Competencies are thus driving activities, reversing the Porterian order of priorities.

The firm as a processor of knowledge: A programmatic way to implement the perspective we have argued for

One of the main characteristic of the competence-based approach is that, within this approach, the firm is conceived as 'a processor of knowledge', as a locus of setting up, construction, selection, usage and development of knowledge. The governance of the firm is not focused on the resolution of informational asymmetries, but on the co-ordination of distributed pieces of knowledge and distributed learning processes. Cognitive mechanisms are essential in this endeavor, and routines play a major role in keeping the internal coherence of the organization. The focus of the theory thus falls clearly on the process of creation of resources. This perspective on the firm has been taken forward, among others, by Cyert and March (1963/1992); Cohen et al. (1972); Cohen (1991); Loasby (1976, 1983); Dosi and Marengo (1994); Marengo (1996, 1994).

In a perspective focused on knowledge, the competence-based approach sees the firm *in primis* as a processor of knowledge, for which the focus of attention is the key limiting factor[7]:

The theoretical approach supposes that firm fixes its attention, first, on the *core competence zone*, which expresses the *domain of knowledge* for which the firm aims to be ahead of the competition (for example, Prahalad and Hamel pointed out

[7] Following Simon (1982), it is attention rather than information that is becoming a rare resource as screening and selection of information become important functions.

that Sony has its core domain on "miniaturisation"). This domain of knowledge, with the highest focus of attention, encompasses different abilities to do things (the *activities* belonging to this domain), but also the capabilities to articulate them together. Moreover, in line with the focusing attention of knowledge in this zone, activities belonging to the core domain can be ranked along a decreasing index of importance (from the most important activity to the less important in this core domain)

Then, in a decreasing order of attention, the firm chooses its *domain of competences*, which encompasses what the firm "knows well how to do" (but not necessarily better than the others). In the case of Sony, this could be the domain of "mastering quality". Here again, once this second domain of knowledge has been delineated, then activities that belong to this domain can be classified along a decreasing order of attention.

Finally, the remaining domain is the *periphery* of the firm. This domain encompasses activities that are necessary to support the functioning of the firm, but which are not under a high focus of attention, so that they do not require by definition a strong commitment in terms of knowledge management. The periphery is a domain of knowledge, where the firm has no ambition to do better than others, and would agree to buy from the outside, if necessary[8]. As for the other domains of knowledge, the activities that belong to the periphery (which generally correspond to the larger number of activities and employment positions in the firm) can still be ranked along a decreasing order of attention.

As a result, the "division of knowledge" (as expressed by the delineation of the domain of core competences, the domain of competences and the periphery) "precedes" the division of labor (expressed by activities that are ranked along a decreasing order from the core activity). In the co-evolution of the division of labor and the division of knowledge which is suggested above, it is clear that, to some extent, the choice of the different domains of knowledge is shaped by the pre-existing division of labor. The choice of competences is partly guided by the activities that the firm has experienced. However, the influence of the pre-existing division of work is only partial. The cognitive deliberation that leads to the future division of work could focus the attention on developing fields of knowledge of which the firm has no experience. Some of the activities required for the selected domains of knowledge could be acquired by the firm, and some are not even known in advance and could be progressively built.[9]

---

[8] For further developments on the definition of the core domain and the periphery, see Amesse and Cohendet (2001), Cohendet and Llerena (2001) and Cohendet and Llerena (2005).

[9] As an example, when the Canadian telecom company Nortel decided at the beginning of the 1990s to abandon Nortel's traditional specialized field (digital switches) to build a core competence in a completely new domain (Internet protocol systems), Nortel headquarters did not know what would be the precise tasks to prescribe to the different departments of the company in order to achieve this goal. They were not in a position to distribute tasks between the different actors. What Nortel did was to "distribute interpretative capabilities" in order to accumulate strategic knowledge in the targeted domain. The mastering of the new domain of knowledge was achieved through different cognitive processes: intense R&D programs (in house as well as through contracts with external research centers), progressive learning through loose alliances and networks, absorption of new pieces of strategic knowledge through some key acquisitions, etc. It was only when sufficient strategic knowledge was accumulated by Nortel (about 1996), that the precise tasks and activities to be performed were revealed. The new division of labor could then take place (see Amesse and Cohendet 2001)

Moreover, such an order implies that the firm manages competencies and transactions simultaneously, but that it does so according to a specific lexicographic order of priorities.

First, within its sets of "high focus of attention" (the domains of competencies and domain of competencies), the firm functions as a knowledge processor giving full priority to the creation of resources. The activities belonging to this key domain of knowledge are not considered to be tradable on the market: They are 'disconnected' from the make-or-buy trade-off suggested by transaction-cost theory. Second, once the set of activities that belong to the set of competencies has been chosen, the other activities that do not belong to the core (the 'periphery' or 'non core activities') are then managed under traditional methods which may rely on the transaction cost approach[10].

A "dual" structure of the firms emerges, as a consequence of a clarified view on the respective role of division of knowledge and of labor.

Part 3: Implications of the Babbagian perspective for the evolutionary theory of the firm

The discussion in part 2 has emphasized that, in a knowledge based context, economic units tend to give priority to the division of knowledge first. The division of labor is then driven by the division of knowledge. In terms of firm strategy, the concept of the "dual firm" can bring clarity to the compromise between competences and transactions. However, what can be analyzed at the global strategic level of the firm still needs microeconomic foundations that are lacking for the moment. Our view is that this lack of micro-foundations is due to the fact that the evolutionary theory of the firm in its present state of development is still embedded in a Smithian vision, for which the division of work precedes the division of knowledge. This statement particularly applies to the central concept of the evolutionary theory, the notion of routine.

As emphasized by Nelson and Winter (1982), the notion of "routine" is a key concept in the foundation of their evolutionary theory. Routines are the "genes" of the organization, and constitute the element of heredity of the evolutionary theory. Evolutionary approaches to economics are grounded on an explicit dynamic account of the interaction between mechanisms of variation (which constantly introduce variety, novelty and heterogeneity amongst routines) and mechanisms of selection (which tend to reduce heterogeneity amongst routines). Indeed, there is a cognitive dimension of routines, when considering that routines encompass the organization's knowledge base and constitute the organizational memory ("organizations remember by doing", ibid pp 99). The notion of routine is used in an ambitious attempt to restore a theory of the firm based on the production side, as

[10] In other words, in the domain of competences, low-opportunism conditions prevail, and allow more valuable knowledge to be applied to the firm's activities. In the periphery domain, meanwhile, the risk of opportunism is high and leads to the situation described by the classical transactional approach. As Langlois and Foss (1996) claim, "as firms move increasingly from their core businesses, they confront increasing adverse selection and moral hazard, since management becomes increasingly unable to efficiently monitor employees or to evaluate their human capital. Agency costs rise correspondingly, producing the net profitability disadvantage associated with further integration"....

the elementary building block of the competence-based approach, and as alternative to the dominant transaction-cost approach (Langlois and Foss 1996; Teece and Pisano 1994; Dosi and Marengo 1994; Prahalad and Hamel 1990). As stated by Nelson and Winter (1982, 128), "the behavior of firms can be explained by the routines that they employ. Knowledge of the routines is the heart of understanding behavior. Modeling the firm means modeling the routines and how they change over time".

It seems to us that the problem with routines is that they are still considered as a stock of knowledge ("a repertoire") inherited from a preexisting division of work. The analysis of routine does not in general pay attention to the active and dynamic cognitive process, often involved at the origin of the emergence of routine. In the words of Cook and Brown (1999), the current analyses of routines are associated with an 'epistemology of possession of knowledge' in that both tend to see knowledge as something people possess, while for an approach fundamentally looking at the formation of knowledge, what matters is the 'knowing' that emerges from the pragmatics of individual and group practices. The emphasis should fall on knowledge as a process and practice, rather than a possession, on the pragmatics of everyday learning in situated contexts of embodied and encultured practice. A reading of knowledge as the fruit of practice rejects an ontology that places knowledge in a separate realm of possession (of mental faculty, memory, text, competence, skill), to be somehow activated by another set of impulses such as learning, social interaction, application, technological instruments. Instead, the emphasis falls on knowing in the process of enactment, in which all these 'actors' are merged into one and the same ontological plane, such that no differentiation between knowledge stimulus and knowledge agent can be found.

> "Much current work on organizational knowledge, intellectual capital, knowledge-creating organizations, knowledge work, and the like rests on a single, traditional understanding of the nature of knowledge. We called this understanding the 'epistemology of possession', since it treats knowledge as something people possess. Yet this epistemology cannot account for the knowing found in individual and group practice. Knowing as action calls for an 'epistemology of practice'.... We hold that knowledge is a tool for knowing, that knowing is an aspect of our interaction with the social and physical world, and that the interplay of knowledge and knowing can generate new knowledge and new ways of knowing. We believe this generative dance between knowledge and knowing is a powerful source of organizational innovation". (Cook and Brown 1999, 381)

The current vision of routine tends to consider routine as a result of a division of work that is "given", exogenous to the evolutionary process that is analyzed. The preexisting division of work shapes the cognitive process that is taking place, and shapes in particular the nature of the learning processes at stake[11]. The evolution of routines is viewed as resulting from two main learning processes: a trial and

---

[11] The fact that routines are analyzed as the cognitive result of a preexisting division of labor is reinforced when looking at the second dimension of routine (besides the cognitive dimension), the motivational dimension associated with the control of intra-organizational conflict – routines are "truces" amongst conflicts as Nelson and Winter (1982) put it. Routines may be considered "*as being a locus of conflict, governance, and a way of codifying micro-economic incentives and constraints*" (Coriat and Dosi 1998). When looking in this way at routine, one can deduce what is

error process in a learning-by-doing perspective, which determines the exploitation dimension, and a process of "search", the foundations of which are not clearly explained, and which determines the exploration dimension. Both these learning processes are in line with the Smithian vision. On the one side, the *learning by doing process* is the core learning mechanism that results from the division of labor. It is a passive learning mechanism which considers the cognitive capabilities of agents as given. In this process, agents are not supposed to change their representation of the world through time, they do not differ in their perception of the environment, or they do not pay attention to common sets of rules, codes and languages within the organization. On the other side, the "search" mechanism appears as a kind of probabilistic event determined in the realm of routine. The creative forces, negotiations and compromises that have led to this type of learning process are not the focus of attention.

What we are arguing for is an analytical representation where the focus is on the cognitive forces that influence the emergence of routines (and not on the emergence of cognitive forces out of a given organizational framework). It seems to us the very nature of the evolutionary process should be based on the analysis of these fundamental cognitive mechanisms – the manner in which agents through their current daily practices and interactions come to structure cognitive arrangements from which routines emerge. These cognitive buildings are the mere loci of the generation of diversity from which routines emerge in an evolutionary vision. In the same vein, when looking at the selection process at stake in the evolutionary approach, it appears that the selection mechanisms do not concern only the sets of existing routines. The selection mechanisms bear on the cognitive arrangements that are at the base of the formation of routines.

In such a perspective, many features and issues of the evolutionary vision of the firm could be revisited. When considering, for instance, the problem of the replication of routines, a preliminary analysis of the cognitive forces that contribute to create the routine will significantly improve the examination of the nature, the language, and even the need for the replication. Routines experienced in a functional group, in a project team, in a network of partners, in a community of different nature, are all different in terms of power of replication, of degree of inertia, of potential of search.

The message resulting from the Babbagian vision is thus that the cognitive process that precedes the formation of routine matters for the evolutionary theory.

Conclusion

The paper advocates a Babbagian perspective as a more relevant foundation to the evolutionary / knowledge based theory of the firm. It considers that division of knowledge precedes division of labor, supporting the arguments that competences come first and drive activities. This view allows locating the emergence, diffusion and duplication of routines as context specific (with their characteristics), specific to social groups (like communities). An evolutionary – knowledge based theory of the firm is define by a theory of why and how the division of knowledge first and

the kind of organizational structure and division of work (and associated structure of hierarchy) that preexists the routine.

the division of labor second organize the different communities in order to achieve its objectives.

This analysis can also help untangle the threads of the discussion between the evolutionary and knowledge-based theories of the firm. Following the seminal impulse given by Nelson and Winter (1982), eminent scholars in the field have developed the foundations of an evolutionary approach to the theory of the firm (Teece 1998; Dosi and Marengo 1994; Teece et al. 1997). One of their main hypotheses is that routines hold the key to understanding the organization of firms, their diversity, and their persistence. At the same time, building on the resource-based view in strategy (Wernerfelt 1984; Barney 1991; Peteraf 1993), a partly overlapping set of scholars have developed a knowledge-based approach to the theory of the firm (Grant 1996; Spender 1996; Teece et al. 1997; Winter 1987, 1988, 1995). In the following, we will use the term 'evolutionary/knowledge-based approach of the theory of the firm' to refer to this set of theories.

While much progress has been made in the evolutionary/knowledge-based approach of the theory of the firm, many open questions still prevail. The hypothesis we develop in this article is the following: a central cause of trouble for the evolutionary/knowledge-based approach to the theory of the firm is, that it is built on a Smithian foundation. The Smithian perspective, considering the division of knowledge as a *consequence* of the division of labor, works as a blinder that blocks out a good and important part of precisely what is the object of the evolutionary/knowledge-based approach to the theory of the firm – important characteristics of knowledge in firms.

We claim that reconsidering the issue of the division of knowledge, and placing it at the heart of the analysis following the principles of a Babbagian perspective, will contribute to a large extent to narrowing the gap between the evolutionary approach and the knowledge based approach of the firm, and to giving much more strength and coherence to the evolutionary theory of the firm.

References

Amesse F, Cohendet P (2001) Technology transfer revisited in the perspective of the knowledge based economy. Res Pol, 30, 9: 1459–79

Arrow K (1962) The economic implications of learning by doing. Rev Econ Stud, 29, 3: 155–173

Babbage C (1832/1989) The economy of machinery and manufacturers. In: Campbell-Kelly M (ed.) The Works of Charles Babbage Vol. 8., William Pickering, London

Barney J (1991) Firm resources and sustained competitive advantage. J Manage, 17, 1: 99–120

Bettis RA, Hitt MA (1995) The new competitive landscape. Strategic Manage J, 16, Summer Special: 7–19

Bierly PE, Chakrabarti AK (1996) Generic knowledge strategies in the US pharmaceutical industry. Strategic Manage J, 17: 123–135

Cohen M (1991) Individual learning and organizational routine: Emerging connections. Organ Sci, 2, 1: 135–139

Cohen MD, March J, Olsen JP (1972) A garbage can model of organizational choice. Admin Sci Quart, 17

Cohendet P, Llerena P (2001) A dual theory of the firm between transactions and competences. ETIC Conference, Strasbourg, Oct. 2001

Cohendet P, Llerena P (2005) A dual theory of the firm between transactions and competences: Conceptual and empirical considerations. Revue Econ Ind, 2005, 2

Collis DJ, Montgomery CA (1995) Competing on resources, strategy in the 1990's. Harvard Bus Rev, 1995, July-August: 118–128

Cook SDN, Brown JS (1999) Bridging epistemologies: the generative dance between organizational knowledge and organizational knowing. Res Pol, 10, 4: 381–400

Coriat B, Dosi G (1998) Learning how to govern and learning how to solve problems: On the co-evolution of competences, conflicts and organizational routines. In: Chandler AD, Hagström P, Sölvell O (eds.) The Dynamic Firm – The Role of Technology, Strategy, Organization, and Regions, Oxford University Press, Oxford

Cyert RM, March JG (1963/1992) A Behavioral Theory of the Firm. Blackwell, Oxford, 2nd edn.

D'Aveni RA (1994) Hypercompetition. The Free Press

Dosi G, Marengo L (1994) Some elements of an evolutionary theory of organizational competences. In: England R (ed.) Evolutionary Concepts in Contemporary Economics, 157–178, University of Michigan Press, Ann Arbor

Fransmann M (1998) Information, knowledge, vision and theories of the firm. In: Dosi G, Teece D, Chytry J (eds.) Technology, Organization, and Competitiveness – Perspectives on Industrial and Corporate Change, 147–192, Oxford University Press, Oxford

Grant RM (1996) Toward a knowledge-based theory of the firm. Strategic Manage J, 17, Winter Special: 109–122

Hamel G, Prahalad CK (1994) Competing for the Future. Harvard Business School Press, Boston, MA.

Hodgson G (1993) Economics and Evolution. Polity Press, Cambridge

Hyman A (1982) Charles Babbage: Pioneer of the Computer. Oxford University Press, Oxford.

Langlois R, Foss N (1996) Capabilities and governance: the rebirth of production in the theory of economic organisation. Kyklos, 52, 2: 201–218

Loasby BJ (1976) Choice, Complexity and Ignorance. Cambridge University Press, Cambridge

Loasby BJ (1983) Knowledge, learning and the enterprise. In: Wiseman J (ed.) Beyond Positive Economics?, 104–121, London, MacMillan

Loasby BJ (1999) Knowledge, Institutions and Evolution in Economics. Routledge, London

Loasby BJ (2000) Organisations as Interpretative Systems. Paper for workshop in Amiens

Marengo L (1994) Knowledge distribution and coordination in organizations: On some social aspects of the exploration vs. exploitation trade-off. Revue Internationale de Systémique, 7: 553–571

Marengo L (1996) Structure, competence and learning in an adaptive model of the firm. In: Dosi G, Malerba F (eds.) Organization and Strategy in the Evolution of the Enterprise, 124–154, Macmillan, Houndmills

Martens B (1999) The evolution of the concept of division of labour in economics. Max Planck Institut für die Erforschung von Wirtschaftssystemen, Discussion paper 18-99

Nelson R, Winter S (1982) An Evolutionary Theory of Economic Change. Belknap Press of Harvard University Press, Cambridge, MA

Nonaka I, Takeuchi H (1995) The Knowledge-Creating Company. Oxford University Press, New York, NY

Pagano U (1985) Work and Welfare in Economic Theory. Basil Blackwell, Oxford

Penrose E (1959/1995) The Theory of the Growth of the Firm. Oxford University Press, Oxford

Peteraf MA (1993) The cornerstone of competitive advantage: A resource-based view. Strategic Manage J, 14: 179–88

Porter ME (1980) Competitive Strategy. The Free Press, New York, NY

Porter ME (1985) Competitive Advantage: Creation and Sustaining Superior Performance. Havard Business School Press

Porter ME (1994) The role of location in competition. J Econ Bus, 1: 35–9

Prahalad CK, Hamel G (1990) The core competence of the corporation. Harvard Bus Rev, 1990, May-June: 79–91

Simon HA (1982) The economics of information processing. In: Simon HA (ed.) Models of Bounded Rationality Vol. 2, Behavioral Economics and Business Organization, 71–78, MIT Press, Cambridge, MA

Smith A (1776) The Wealth of Nations

Spanos YE, Lioukas S (2001) An examination into the causal logic of rent generation: Contrasting Porter's competitive strategy framework and the resource-based perspective. Strategic Manage J, 22, 10: 907–934

Spender JC (1996) Making knowledge the basis of a dynamic theory of the firm. Strategic Manage J, 17, Winter Special: 45–62

Stigler G (1991) Charles Baggage (1791 + 200 = 1991). J Econ Lit, 29, 3: 1149–1152
Teece DJ (1998) Capturing value from knowledge assets: The new economy, markets for know-how, and intangible assets. Calif Manage Rev, 40, 3: 55–79
Teece DJ, Pisano G (1994) The dynamic capabilities of firms: An introduction. Ind Corp Change, 3, 3: 537–556
Teece DJ, Pisano G, Shuen A (1997) Dynamic capabilities and strategic management. Strategic Manage J, 18, 7: 509–533
Wernerfelt B (1984) A resource-based view of the firm. Strategic Manage J, 5: 171–180
Williamson OE (1999) Strategy research: Governance and competence perspectives. Strategic Manage J, 20: 1087–1108
Winter S (1987) Knowledge and competence as strategic assets. In: Teece D (ed.) The competitive challenge - strategies for industrial innovation and renewal, 159–184, Ballinger, Cambridge/MA
Winter SG (1988) On Coase, competence and the corporation. J Law Econ Organ, 4, 1
Winter SG (1995) Four Rs of profitability: Rents, resources, routines, and replication. In: Montgomery C (ed.) Resource-based and evolutionary theories of the firm - Towards a synthesis, 147–178, Kluwer, Dordrecht

Part 2:

Entry of new firms and small firms

Massimo G. Colombo · Luca Grilli

Young firm growth in high-tech sectors: The role of founders' human capital

Abstract The aim of this paper is to analyze empirically the relation between the human capital of the founding team and the growth of new technology-based firms through the estimates of econometric models. For this purpose, we take advantage of a new data set relating to a sample composed of 391 young Italian firms that operate in high-tech industries in both manufacturing and services. In particular, due to a detailed description of the personal characteristics of firms' founders, we are able to disentangle the "wealth" and "entrepreneurial ability" effects of human capital. We also investigate the nature of the synergies that eventually arise from the presence within the founding team of complementary industry-specific professional skills.

Keywords Entrepreneurship · New technology-based firms · Firm growth · Founders' human capital
JEL Classification M13 · L25 · O33

1 Introduction

New technology-based firms (NTBFs) are of crucial importance for the efficiency and the rejuvenation of an economic system (Audretsch 1995). The study of the

The support of MIUR 2002 funds is gratefully acknowledged. We are indebted to Mario Calderini, Andrea Fumagalli, Luigi Orsenigo, Enrico Santarelli, Marco Vivarelli and participants in the 2003 EARIE Conference, the 2003 SMS Conference, the 2003 AiIG conference and the 2004 Schumpeterian Society Conference for helpful comments to this and related works. Responsibility for any errors lies solely with the authors. The authors are jointly responsible for the work. However, sections 1 and 2 have been written by Massimo G. Colombo, and the remaining sections by Luca Grilli.

M. G. Colombo
Politecnico di Milano, Department of Economics, Management, and Industrial Engineering, P.za Leonardo da Vinci, 32, 20133 Milan (ITALY), E-mail: Massimo.Colombo@polimi.it

L. Grilli
Politecnico di Milano

factors that determine their competitive success has attracted the interest of economists and managerial scientists, with special consideration being given to the human capital of the founding team (see for instance Storey 1994). The aim of this paper is to analyze empirically the relation between such a factor and the growth of NTBFs through the estimates of econometric models. From one side, in the early years after its foundation, the distinctive capabilities[1] of a NTBF are closely related to the human capital of the founders. Actually, "for a new, high-technology firm, the primary assets are the knowledge and skills of the founders. Any competitive advantage the new firm achieves is likely to be based upon what the founders can do better than others" (Cooper and Bruno 1977, 21). Therefore, in accordance with knowledge-based and evolutionary theories of the firm (see for instance Winter 1987; Conner and Prahalad 1996; Grant 1996), NTBFs established by highly qualified individuals should outperform other NTBFs because of their unique capabilities. From the other side, it is argued that capital market imperfections often make it difficult for NTBFs to access external financing sources, and this may seriously jeopardize their growth (see Carpenter and Petersen 2002a,b). Firms founded by wealthier individuals are less affected by financial constraints. Previous studies have shown that there is a positive relation between the human capital and the assets of individuals (see for instance Xu 1998; Åstebro and Bernhardt 1999). Hence, the positive relation between the growth rates of NTBFs and the human capital of their founders may simply highlight the presence of binding financial constraints.

As will be illustrated in the next section, several studies have gone beyond the classical Gibrat's law framework and have focused on the effect of founders' human capital on firm post-entry performance, finding a positive effect (see for instance Bates 1985, 1990, Brüderl et al. 1992, Brüderl and Preisendörfer 2000; see again Storey 1994 for a survey). Nonetheless, they generally have failed to distinguish empirically between the "entrepreneurial ability" and the "wealth" effects of human capital because of a lack of suitable data. In addition, while it is often claimed that heterogeneity of competencies within the founding team is a source of synergistic gains (Cooper and Bruno 1977; Eisenhardt and Schoonhoven 1990), the precise nature of such synergies has remained almost uninvestigated. In order to address such research questions, we take advantage of a new dataset relating to a sample composed of 391 young Italian firms that operate in high-tech industries in both manufacturing and services. The RITA (Research on Entrepreneurship in Advanced Technologies) database from which sample firms are extracted provides information on the human capital characteristics of each individual founder to a level of detail unparalleled by previous studies based on samples of similar size. In accordance with previous studies, the econometric results reveal a positive impact of founders' human capital on firm growth. More interestingly, they allow us to assert that this positive relation is not only caused by a "wealth" effect, but that there is also an "entrepreneurial ability" effect at work; in other words, the positive relation between human capital and growth cannot simply be explained by the greater asset endowment of highly qualified individuals. In addition, the results of the estimates highlight the synergies arising from the presence of complementary industry-specific professional skills within the founding team. In particular, while

[1] Distinctive capabilities (Winter 1987; Prahalad and Hamel 1990) can be defined as a firm's ability to select, mobilize and use tangible and intangible assets to perform tasks in a unique way. They express what a firm is able to do better than other firms.

NTBFs established by individuals with technical professional experience in the same sector as the new firm generally exhibit superior growth, founders' commercial industry-specific experience is found to have a positive effect on growth only if it is associated with technical experience.

The paper proceeds as follows. In next section, after a brief introduction to the early literature on firm growth related to Gibrat's law, we briefly synthesize the theoretical and empirical literature concerning the impact of founders' human capital on NTBF post-entry performance, and formulate the related research hypotheses. Then we describe the dataset, illustrate the variables that are introduced in the econometric models, and present the results of the econometric estimates. In the subsequent section a discussion of the main findings and of directions for future research concludes the paper.

2 Founders' human capital and firm growth: theory and hypotheses

2.1 The early literature on firm growth: Gibrat's law

Studies on the determinants of firm growth trace back to the well-known Gibrat's law of proportionate effect (LPE) which postulates, among other things, that firm growth rates are independent of initial firm size (see Gibrat 1931). Since then, several tests of the LPE have been run and, by now, they have constituted a conspicuous body of empirical research in industrial organization (see the two exhaustive reviews offered by Sutton 1997 and Caves 1998). Most of the early studies (see for instance Hart and Prais 1956; Simon and Bonini 1958; Hymer and Pashigian 1962), often dealing with samples of limited size and censored nature, generally found econometric evidence consistent with the LPE. Conversely, more recent investigations, based on more complete datasets, typically rejected the LPE, showing that small and young firms, conditional on survival, grow more rapidly than their larger and older counterparts (see for example Evans 1987a,b; Dunne et al. 1988, 1989; Hart and Oulton 1996).[2] Although an extensive examination of the issues concerning the test of Gibrat's law is beyond the scope of the paper,[3] it is worth stressing in our context that, even if a definitive answer on Gibrat's law has not yet been reached, the growing evidence that firm growth dynamics may not be governed by a random process has stimulated scholars to investigate factors that may explain systematic departures from the LPE. This has led to a large empirical literature exploring the determinants of firm growth. Recently, Barringer et al. (2005), starting from the mid-seventies, classified 106 articles, book chapters and books from the entrepreneurship, management and economics literatures, which deal with this topic. Besides firm-specific attributes, business strategies, human resources management practices, socio-economic characteristics of the environment in which firms are embedded, past literature individuate founder-specific characteristics as one of the main attributes that differentiate rapid-growth firms from

[2] Note, however, that there also exist contributions that reject the LPE by the reverse tendency of higher growth rates for larger rather than smaller firms (see for example Singh and Whittington 1975 and other studies mentioned in Sutton 1997).

[3] Colombo and Ermini (2005) find that LPE fails to hold among a sample of Italian NTBFs, similar to the one used in the present work. In particular, they highlight a negative relationship between growth and size.

slow-growth ones. In what follows, we focus on the link between founders' human capital and young firm growth in high-tech sectors, deriving the research hypotheses on the impact of founders' characteristics on NTBF growth and revising the main previous empirical contributions on the same topic.

2.2 The role of founders' human capital: research hypotheses

In this paper, we will use the term "founders' human capital" to refer to the technical, organizational and managerial skills of each individual founder, and their joint use within the new company so as to obtain synergistic effects. Founders' personal networks of relationships with potential customers and suppliers, business partners, and providers of external financing, a resource usually referred to in the literature as social or relational capital (see for instance Pennings et al. 1998; Lee et al. 2001), is also considered herein as part of human capital.

We expect NTBFs established by highly qualified individuals to exhibit superior growth. From one side, the primary role of a successful entrepreneurial firm is the integration of specialized knowledge so as to exploit neglected business opportunities in a way that is difficult for other firms to imitate (Knight 1921; Schumpeter 1934; Conner and Prahalad 1996; Hodgson 1998). Highly skilled individuals are in an ideal position to recognize the unexploited value of specialized knowledge (see Alvarez and Barney 2002). In addition, because of adverse selection and moral hazard problems, attraction, coordination and protection of specialized knowledge is more efficient if specialists are members of the founding team. From the other side, highly skilled individuals can often resort to personal savings to finance their entrepreneurial ventures. In fact, previous studies have shown that the human capital and net worth of individuals are positively correlated (see Xu 1998; Åstebro and Bernhardt 1999). Due to the presence of imperfections in capital markets that may inhibit external financing of NTBFs (Carpenter and Petersen 2002a), firms founded by wealthy individuals will exhibit higher growth rates, as they are less likely to be financially constrained.[4] Note also that much of high-tech investment is intangible or firm specific, and thus provides little inside collateral value. Since firms founded by high net worth individuals are in a better position to resort to outside collateral, they will also have easier access to the loan market (Bester 1985, 1987). This may again have a positive effect on growth.

In order to identify the relative explanatory power of the above arguments, one should consider in greater detail the nature of the human capital of entrepreneurs. In particular, following Becker (1975), a distinction may be made between *generic* and *specific* human capital. Generic human capital relates to the general knowledge acquired by entrepreneurs through both formal education and professional experience. Specific human capital consists of individual's capabilities that can directly be applied to the entrepreneurial role in the newly created firm. On the one

[4] Several studies have provided evidence in support of the view that new firms suffer from binding financial constraints; in particular, they have documented that the net worth of individuals has a positive effect on both the likelihood of establishing a new enterprise (see Evans and Jovanovic 1989; Evans and Leighton 1989; Meyer 1990; Holtz-Eakin et al. 1994a; Lindh and Ohlsson 1996; Blanchflower and Oswald 1998) and the post-entry performance of the newly created firms (see Holtz-Eakin et al. 1994b). For a diverging view, see Cressy (1996, 2000).

hand, previous works have shown that education and general professional experience are quite accurate predictors of individuals' wealth (Åstebro and Bernhardt 1999). Therefore internal financing and access to debt should be relatively easier for individuals with better educational attainments and greater work experience. On the other hand, it is often argued that such factors are unrelated to business success; what really matters is a deep understanding of firm organization and of the specificities of the business environment in which the new firm operates (see Storey 1994, 129). Accordingly, while the generic component of human capital reflects the "wealth" effect, it does not capture the "entrepreneurial ability" effect.[5] Consequently, it should have a less pronounced positive impact on the growth of NTBFs than the specific component which instead reflects both effects. The above reasoning leads to the following hypotheses.

Hypothesis 1 *NTBFs started by individuals with better education will enjoy greater post-entry growth.*

Hypothesis 2 *NTBFs started by individuals with greater professional experience will enjoy greater post-entry growth.*

Hypothesis 3 *NTBFs started by individuals with prior professional experience in the same industry in which the new firm operates will enjoy higher growth than those started by individuals with prior professional experience in other industries.*

As to the specific component of human capital, previous studies (Preisendörfer and Voss 1990; Brüderl et al. 1992; Brüderl and Preisendörfer 2000) differentiate between *industry-specific* and *entrepreneur-specific* human capital. From one side, founders' knowledge and capabilities are very much connected to what entrepreneurs learned in the organization in which they were formerly employed (Cooper and Bruno 1977; Cooper 1985). In particular, industry-specific experience yields valuable knowledge about technologies, customer needs, and the strengths and weaknesses of competitors which can profitably be used by the newly established firm. Moreover, founders' endowments of social capital are more valuable if the industry of the new firm and the one of the incubating organization coincide, as the new firm can more immediately exploit the network of personal relationships developed by the founders in their previous occupations. In this respect, it may be of interest to consider the nature of the specific professional experience gained by founders and, more specifically, to distinguish between the technical and the commercial one. Which one of the two has a greater impact on the growth of NTBFs is a matter of empirical testing, as stated in the following hypothesis.

Hypothesis 4 *NTBFs started by individuals with prior technical experience in the same industry in which the new firm operates will enjoy higher (lower) growth than those founded by individuals with commercial experience if founders' context-specific technical skills play a relatively more (less) important role for firm success than context-specific commercial skills.*

---

[5] Note, however, that the wage an individual commands in the labor market is positively associated with education and years of professional experience. As the opportunity cost of the entrepreneurial choice is greater for individuals endowed with greater generic human capital, it may be argued that those among such individuals who make the self-employment choice also have better entrepreneurial ability. In other words, because of a self-selection bias, generic human capital may indirectly signal entrepreneurial talent.

From the other side, one of the key functions performed by a successful entrepreneur is organizing and managing the work of employees. This includes giving directions to subordinates, delegating authority, designing incentives, and controlling results. Such activities are likely to benefit from learning-by-doing. Therefore, "leadership experience" (Brüderl et al. 1992) gained either through a managerial position in another firm or in prior self-employment episodes, should play a crucial role for the success of new firms.[6] Thus, the following hypothesis is formulated.

Hypothesis 5 *NTBFs started by individuals with prior managerial and entrepreneurial experiences will enjoy higher growth.*

Lastly, management of a business organization requires a range of complementary skills in different functional areas (e.g. skills in both marketing and technology). Due to their tacit, firm-specific nature, they cannot be outsourced by NTBFs. It also is very difficult for such firms that operate in a risky environment and lack a track record to attract highly skilled salaried individuals. Accordingly, it has been argued in previous studies (Cooper and Bruno 1977; Eisenhardt and Schoonhoven 1990) that, with all else equal, heterogeneity of competencies within the team of founders is likely to lead to better post-entry performance. This leads to the following hypothesis.

Hypothesis 6 *NTBFs started by individuals with complementary functional skills relating to industry-specific professional experiences, will enjoy higher growth than those started by individuals with similar functional skills.*

Hypotheses 1 and 2 mainly reflect the "wealth" effect of founders' human capital and are only indirectly associated with the "entrepreneurial ability" effect (see footnote 5). On the contrary, if the remaining hypotheses are accepted, they provide direct evidence supporting the argument inspired by evolutionary and knowledge-based theories of the firm, that the specific capabilities of founders are a key determinant of the growth of NTBFs.

2.3 The role of founders' human capital: the empirical literature

Numerous previous studies have analyzed the relation between founders' human capital and firm post-entry performance (i.e. likelihood of survival and growth of surviving firms), even though a smaller number have focused on NTBFs. In this section, we synthesize previous findings and explain why, in our view, there is a need for further research.

In empirical works (see Bates 1990; Stuart and Abetti 1990; Brüderl et al. 1992; Storey 1994; Westhead and Cowling 1995; Brüderl and Preisendörfer 2000; Almus 2002), generic human capital was proxied by education and general professional experience. The former is measured by education attainment, such as graduation and the achievement of a Ph.D. degree, or years of schooling; the latter is generally captured by the years of work experience before establishing the new firm or

[6] Note that we consider "leadership experience" irrespective of the sector in which it is gained as part of the specific component of human capital. In this respect, the first to acknowledge the non-sector specific nature of the entrepreneurial and managerial experience were Marshall and Marshall (1879, 51).

simply by the entrepreneur's age. As to specific human capital, previous works considered whether founders of a new firm have business experience in the same sector as a proxy of industry-specific human capital, while prior self-employment or managerial experiences was used as a proxy of entrepreneur-specific human capital.

In general, while the level of education of the founders is found to be negatively related to the hazard rate of newly established firms (see Bates 1990; Brüderl et al. 1992; Gimeno et al. 1997), results concerning growth are much less definitive. In a survey of empirical studies on the determinants of small firm growth, Storey (1994) highlighted the fact that only eight studies out of the seventeen surveyed found a strong positive relationship with entrepreneurs' education. More recently, Westhead and Cowling (1995) have shown that UK NTBFs whose founders have a bachelor degree grow more rapidly than other firms. Brüderl and Preisendörfer (2000) found that, in a univariate analysis, growth of newly established Bavarian firms is positively related to years of schooling of the firm's main founder; however the coefficient of this variable, though positive, is not significant in more complete multivariate regressions.[7]

As to the impact on firm post-entry performance of general professional experience, the empirical evidence is rather weak. Brüderl et al. (1992) found a U-shaped relationship between the founder's years of work experience at start-up time and the mortality rates of Bavarian firms. Brüderl and Preisendörfer (2000) showed that such a variable has no statistically significant impact on growth of such firms. According to findings by Almus et al. (1999), the fact that entrepreneurs have no professional experience at start-up time has a negative, though statistically insignificant, effect on the growth of innovative firms located in East Germany, while it negatively influences the growth of non-innovative firms.

On the contrary, findings relating to founders' industry-specific human capital are more robust. Brüderl et al. (1992), analyzing a sample of new Bavarian firms, showed that the failure rate is significantly lower if founders have business experience in the same sector as the new firm. Moreover, in a univariate analysis, the variable capturing founders' business experience is shown to have a significantly greater value for rapid-growth firms than for other firms; nevertheless, this relation disappears in a multivariate logit analysis (see Brüderl and Preisendörfer 2000). Cooper and Bruno (1977) studied NTBFs that in the '70s were located in the San Francisco peninsula. The comparison between rapid-growth and discontinued firms highlights the fact that the former were more likely than the latter to have been set up by individuals who came from incubating organizations that operated in the same sector as the start-up. Later studies in the '80s and '90s generally found similar results (see for instance Cooper 1985, Dunkelberg et al. 1987, Feeser and Willard 1990, Chandler and Jansen 1992, Siegel et al. 1993. For exceptions see Storey 1994).

However, the above mentioned studies usually have not considered the nature of the professional experience gained by founders. In this respect, one may wish

---

[7] Quite surprisingly, the nature of the education received by founders has almost been neglected in the empirical literature. An exception is work by Almus and Nerlinger (1999) on newly established West German firms. They provided evidence from estimates of multivariate econometric models that firms started by individuals who have a technical degree enjoy greater average yearly growth rates; this holds true in high-, medium- and low-tech industries. On the other hand, business education has a significantly positive impact only for low-tech firms.

to distinguish between technical and commercial experience. The former captures the context-specific skills of individuals in R&D, product and process design, and engineering, while the latter mirrors context-specific skills pertaining to marketing, sales and customer care activities; it is also related to social capital associated with relations with customers and suppliers. Stuart and Abetti (1990) made such a distinction in analyzing 52 new technical ventures located in the New York/New England area. They found that years of both technical and commercial experience have no statistically significant impact on a composite "growth and performance" index. Similar results are shown by Westhead and Cowling (1995). However, both studies considered general professional experience rather than industry-specific experience.

Turning to the entrepreneur-specific component of human capital, several studies have investigated whether prior founders' self-employment or managerial experience has a positive impact on firm post-entry performance, but they have provided mixed results. Bates (1990) and Brüderl et al. (1992), respectively, did not find evidence that individuals' prior managerial and self-employment experiences have any influence on new firm failure rates. Brüderl and Preisendörfer (2000) showed that such variables exhibit significantly higher values in the "rapid-growth firms" group than in the group that includes the remaining firms; in multivariate logit models the former variable has a positive, statistically significant coefficient, while the latter is positive, albeit insignificant. Stuart and Abetti (1990) found a strong positive correlation between NTBF performance and the entrepreneurial and managerial experience of founders. However, such results have not been confirmed by other studies (see for instance Chandler and Jansen 1992, Westhead and Cowling 1995; for a survey of earlier findings see again Storey 1994).

Lastly, in the previous section we pointed out that synergistic effects may be obtained by combining founders' different skills and experiences. In empirical work, the heterogeneity of founders' capabilities is usually proxied by the number of founders, with firms established by more than one individual generally exhibiting higher growth than single-person owned firms. Yet in this case, too, the evidence is not unanimous. For instance, Brüderl and Preisendörfer (2000) did not detect any significant relation between team foundation and the growth of new firms. Similar results were obtained by Almus and Nerlinger (1999) for high-tech new firms, even though they did find a positive effect of team foundation for low-and medium-tech firms. Actually, the number of founders is a quite unsatisfactory proxy: it is positively related to the amount of resources available to the new firm at start-up,[8] but it does not differentiate according to the nature of such resources. Whether particular combinations of founders' capabilities as regards especially their context-specific professional experience favor post-entry growth has not been directly investigated by previous empirical studies.[9]

[8] These include financial resources. So the positive effect on growth of team foundation may arise from the relaxation of financial constraints.

[9] Roure and Maidique (1986) considered the degree of team completeness, defined as the number of essential functions in a new company that are filled by founders at start-up time. They showed that successful companies have a more complete founding team. Eisenhardt and Schoonhoven (1990) explicitly considered the variety of industry experience among a firm's founders, measured by the standard deviation of the years of industry-specific experience of the entrepreneurs that started the new firm; they find that such a variable is linked with higher growth. Nevertheless, they did not examine the nature of the functional skills of founders.

3 The empirical analysis

3.1 Data

In this paper, we consider a sample composed of 391 Italian NTBFs (see also Colombo et al. 2004; Colombo and Grilli 2005). Sample firms were established in 1980 or later, were independent at start-up time (i.e. they were not controlled by another business organization, even though other organizations may have held minority shareholdings in the new firm) and operated in high-tech sectors, in both manufacturing and services.

The sample of NTBFs was extracted from the RITA database, developed at Politecnico di Milano. The RITA database was created in 1999, and was updated and extended in 2001. At the time of this study, the database contained detailed information on more than 400 Italian NTBFs and more than 1,000 of their founders. The development of the database went through a series of steps. First, Italian target firms that complied with the above mentioned criteria relating to age and sector of operations were identified. For the construction of the target "universe" a number of sources were used. These included lists provided by national industry associations, on-line and off-line commercial firm directories, and lists of participants in industry trades and expositions. Information provided by the national financial press, specialized magazines, other sectoral studies, and regional Chambers of Commerce was also considered. Altogether, around 2,000 firms were selected for potential inclusion in the database.

Second, a questionnaire was mailed to the target firms. The aim of the questionnaire was to collect both quantitative information relating to the activity, structure and performance of firms and the characteristics of their founders, and opinions of the founders on specific issues. The first section of the questionnaire provides detailed information on characteristics of the entrepreneurs, such as age, education, and prior work experience. The second section comprises further questions concerning the characteristics of the firms at start-up time and their post-entry performance, including the number of employees.

Lastly, answers to the questionnaire were checked and the questionnaires were completed if necessary by educated personnel through phone or face-to-face interviews with firm owner-managers. This final step was crucial in order to obtain missing data and ensure that answers were accurate.

The sample used in the present work consists of 391 NTBFs for which we were able to create a complete data set. The only exception concerns data on the previous entrepreneurial experiences of founders that were available only for a sub-sample of 260 firms. Note that there is no presumption here of a random sample. First, data provided by official national statistics do not allow us to obtain a reliable description of the universe of Italian NTBFs.[10] Second, the identification process of the target universe that was described above is likely to have led to the oversampling of growth-oriented firms, while micro-firms probably are underrepresented. Third, the sample was drawn in 1999, so only firms having survived up to the survey

[10] The main problem is that, in Italy, most individuals who are defined as "self-employed" by official statistics actually are salaried workers with atypical employment contracts. Unfortunately, on the basis of official data, such individuals cannot be distinguished from entrepreneurs who create a new firm.

date are included. Since it is widely documented that failure rates decrease with the human capital of founders, the impact of this factor on post-entry performance may be greater than the one highlighted by our empirical analysis due to sample selection bias. Nevertheless, the sample is sufficiently large and heterogeneous to provide adequate coverage of Italian NTBFs. In addition, the information relating to the human capital of the founding team is much more accurate than in previous data sets of similar size. The sample consists of 23 firms in the multimedia content sector (5.9%), 112 software houses (28.6%), 156 Internet and telecommunication service firms (39.9%), 19 firms in the biotechnology and pharmaceutical industries (4.8% of the sample), while the remaining 81 firms (20.7%) operate in the following manufacturing sectors: telecommunication equipment, electronic components, computers, optical, medical and electronic instruments, and aerospace.

3.2 Variables

In this study, we measure a firm's post-entry performance by the logarithm of the number of employees in the last year in which the firm is present in the RITA database. Since we add firm's age (*Age*) to the set of explanatory variables, the dependent variable is an indicator of the average employment growth in the period in which a firm is observed (for a similar approach see Westhead and Cowling 1995). The explanatory variables can be subdivided in three groups (for definitions see Table 1).

The first group encompasses variables aimed at analyzing the role played by founders' generic human capital as captured by education and professional experience. *Education* measures the mean number of years of education of founders and *Genworkexp* represents the average number of years of professional experience gained by founders in other sectors than the one of the NTBF. We also consider the nature of the education received by founders. In particular, as concerns graduate and post-graduate education, we distinguish between economic and managerial studies (*Ecoeducation*) and technical and scientific studies (*Techeducation*).[11] In addition, in order to facilitate comparison with previous works, we consider *Workexp* given by the mean number of years of work experience of a firm's founders independently of the sector of activity. Following Hypotheses 1 and 2, we predict a positive impact of all these variables on a firm's growth.

The second group of variables reflects the specific human capital of founders. The industry-specific component is captured by the mean number of years of professional experience gained by founders in the same sector of activity of the NTBF (*Specworkexp*). According to Hypothesis 3, this latter variable should have a greater positive effect on firm performance than *Genworkexp*. In order to gain further insights into the effects on firm growth of different functional skills associated with

[11] *Ecoeducation* measures years spent in the attainment of degrees in economics, management and political science, while *Techeducation* reflects years spent in obtaining degrees in engineering, physics, biology, chemistry, medicine, veterinary science, pharmaceutics, mathematics, statistics, geology and computer science. In order to properly judge the effective level of competencies of founders, we consider the minimum length of time necessary to attain a certain degree. In order to attain an Italian graduate degree in economics, law, management, political science and most scientific degrees, four years of studies are requested, while five years is the minimum time for a degree in engineering. Master and Ph.D. programs require one and three additional years, respectively, independently of the specific field.

Table 1 The explanatory variables of firm growth

Variable	Description
Education	Average number of years of education of founders
Ecoeducation	Average number of years of economic and managerial education of founders at graduate and post-graduate level
Techeducation	Average number of years of scientific and/or technical education of founders at graduate and post-graduate level
Workexp	Average number of years of work experience of founders before firm's foundation
Specworkexp	Average number of years of work experience gained by founders in the same sector of the start-up before firm's foundation
Techworkexp	Average number of years of technical work experience gained by founders in the same sector of the start-up before firm's foundation
Comworkexp	Average number of years of commercial work experience gained by founders in the same sector of the start-up before firm's foundation
Genworkexp	Average number of years of work experience gained by founders in other sectors than the one of the start-up before firm's foundation
DManager	One for firms with one ore more founders with a prior management position
DEntrepreneur	One for firms with one or more founders with a previous self-employment experience
Age	Number of years since firm's foundation
Founders	Number of founders
Infrastructure	Value of the index measuring regional infrastructures in 1989 (mean value among Italian regions=100; source: Centro Studi Confindustria 1991)[a]

[a] Index provided by Centro Studi Confindustria (1991) and calculated as the average in 1989 of the following indexes: per-capita value added, share of manufacturing out of total value added, employment index, per capita bank deposits, automobile-population ratio, and consumption of electric power per head.

founders' industry-specific professional experience, we also decompose *Specworkexp* in *Techworkexp* and *Comworkexp*. The former variable measures the average number of years of founders' work experience gained in R&D, process design, production and engineering departments of firms that operate in the same sector of the NTBF under consideration; the latter one is the corresponding measure for what concerns marketing, sale and customer care activities. As is stated in Hypothesis 4, whether technical or commercial skills are most critical for firm growth is a matter of empirical testing. Entrepreneur-specific human capital is captured by two variables, *DManager* and *DEntrepreneur*. They equal 1 if, prior to the establishment of the new firm, one or more founders had a managerial position and self-employment experience, respectively. Hypothesis 5 predicts a positive coefficient of both these variables. In addition, Hypothesis 6 posits that, due to the presence of synergistic effects, firms established by teams composed of individuals with complementary context-specific professional skills will exhibit greater growth. In order to capture such an effect, we introduce into the set of explanatory variables the interactive term *Techworkexp* × *Comworkexp* between the years of technical and commercial industry-specific work experience. We expect the coefficient of such term to be positive, suggesting that the marginal impact on firm growth of the technical

Table 2 Descriptive statistics of the explanatory variables of the econometric models

Variable	Min	Max	Mean	S.D.
Education	8	21.5	14.8661	2.5604
Ecoeducation	0	5	0.3695	0.9585
Techeducation	0	8	1.4936	1.9333
Workexp	0	54	12.4761	8.4737
Specworkexp	0	54	4.577	7.2346
Techworkexp	0	54	2.8321	5.8659
Comworkexp	0	26	1.289	3.8478
Genworkexp	0	45	7.8991	8.5289
DManager	0	1	0.2762	0.4476
DEntrepreneur	0	1	0.4038	0.4916
Age	0	22	7.7365	5.1009
Founders	1	12	2.8081	1.6506
Infrastructure	43.7	174.7	116.259	27.697

(commercial) skills of the entrepreneurial team is more positive the greater its commercial (technical) experience.

The last group includes control variables. *Founders* is the number of founders of NTBFs. Given the average level of human capital of the founding team, the larger the number of founders, the greater the tangible and intangible resources at their disposal. In particular, the financial constraints a NTBF may face should be less binding the greater the number of founders, due to greater internally available financial resources. So we predict a positive impact of such a variable on firm performance.[12] *Infrastructure* reflects the infrastructure endowment in the region in which the new firm is located. Location in an area with efficient infrastructure is likely to influence positively growth, since it enables a NTBF to benefit from the provision of complementary external assets of public good nature (e.g. transport system, public utilities). Finally, four industry dummies were introduced into the econometric models in order to control for industry-specific characteristics.

In Tables 2 and 3, we present descriptive statistics and the correlation matrix of explanatory variables, respectively. For the sake of synthesis, we omit control variables. Correlation across variables is generally low, suggesting the absence of any relevant problem of multicollinearity.

3.3 Results

Table 4 presents the results from ordinary least squares (OLS) regressions, corrected for heteroskedasticity. Model 1 includes years of education (*Education*) and of work experience (*Workexp*) of founders, allowing us to test Hypotheses 1 and 2. In Model 2, we split work experience into the specific and generic components, captured by *Specworkexp* and *Genworkexp*, respectively. Model 3 analyzes the role

[12] In previous works, the number of founders has often been considered as a proxy (actually, a quite unsatisfactory one) of the heterogeneity of founders' competencies. In this paper, we have a more direct measure of such effect, so the number of founders is a control variable.

Table 3 Correlation matrix of the explanatory variables of the econometric models[a]

	Variable	[1]	[2]	[3]	[4]	[5]	[6]	[7]	[8]	[9]	[10]
[1]	*Education*	1	0.2885	0.7398	-0.1616	-0.1101	-0.1284	-0.0414	-0.0671	0.0647	-0.0768
[2]	*Ecoeducation*		1	-0.1552	0.0176	-0.0567	-0.1271	-0.0474	0.0656	0.0418	0.003
[3]	*Techeducation*			1	-0.0851	0.018	0.0352	0.0019	-0.0999	0.0769	-0.0952
[4]	*Workexp*				1	0.4192	0.3032	0.1892	0.6379	0.3246	0.1472
[5]	*Specworkexp*					1	0.754	0.4705	-0.4317	0.2815	0.0545
[6]	*Techworkexp*						1	-0.0641	-0.3382	0.1036	0.0131
[7]	*Comworkexp*							1	-0.2111	0.2285	0.0695
[8]	*Genworkexp*								1	0.0837	0.0993
[9]	*DManager*									1	0.1062
[10]	*DEntrepreneur*[b]										1

[a] For the sake of synthesis we omitted correlation coefficients of control variables.
[b] Data available for a subset of 260 firms.

played by the nature of both education and industry-specific professional experience; for this purpose, we replace *Education* with *Techeducation* and *Ecoeducation*, and *Specworkexp* with *Techworkexp* and *Comworkexp*, respectively. In Model 4, we add the dummy variable *DManager* indicating whether founders had managerial competencies, while in Model 5 we also consider the variable *DEntrepreneur*, capturing previous entrepreneurial experiences. This latter information is available only for a subsample of 260 firms. Finally, Model 6 acknowledges the role possibly played by the heterogeneity of founders' industry-specific skills in explaining firm growth through the introduction of the interactive term *Techworkexp* × *Comworkexp*. Overall, estimation results appear to be highly stable throughout the models, lending support to the robustness of our findings.

Hypotheses 1 and 2 state that both education and professional experience of founders will positively influence firm growth. In accordance with such hypotheses, in Model 1 the coefficients of *Education* and *Workexp* turn out to be positive and statistically significant at 99%.

Hypothesis 3 argues that NTBFs started by individuals with prior work experience in the same sector in which the new firm operates will show better performances than those set up by entrepreneurs with prior professional experience in other industries. This argument is confirmed by the estimates of Model 2. While both *Specworkexp* and *Genworkexp* have positive coefficients, significant at conventional confidence levels, the magnitude of the former coefficient is found by the F-test to be significantly greater at 95% than that of the latter ($F_{(1)} = 3.97$). Moreover, note that in more complete specifications (see Models 3, 4, 5 and 6), the coefficient of *Genworkexp*, though positive, becomes insignificant, while the one of *Specworkexp* is always significant at conventional confidence levels.

Hypothesis 4 relates to the functional nature of the industry-specific professional experience gained by founders. The estimates of Model 3 indicate that only the technical component of industry-specific work experience, captured by *Techworkexp*, has a positive, statistically significant (at 99%) impact on firm growth, while the coefficient of the variable associated with commercial skills (*Comworkexp*) turns out to be insignificant. Accordingly, it is founders' technical rather than commercial capabilities gained in the same sector of a NTBF that provide the firm with a competitive advantage that allows it to grow faster than other firms. In Model 3, we also consider the nature of the education received by founders. Both *Techeducation* and *Ecoeducation* have positive coefficients, statistically significant at conventional confidence levels. However, the coefficient of *Ecoeducation* is greater, with the difference being statistically significant at 99% ($F_{(1)} = 9.45$).

Variables pertinent to Hypothesis 5 were introduced in two separate models (Models 4 and 5), as the information on founder's previous entrepreneurial experience was available only for 260 firms. In accordance with the hypothesis, both *DManager* and *DEntrepreneur* have positive, statistically significant coefficients. In other words, NTBF growth performances substantially improve if, within the founding team, there are individuals with prior managerial and entrepreneurial experience.

To test Hypothesis 6, in Model 6, we introduce the interactive term *Techworkexp* × *Comworkexp* which captures synergistic effects between the functional industry-specific capabilities of the founding team. The coefficient of the interactive term is positive and statistically significant at 99%, suggesting that, with everything

Table 4 The determinants of firm growth

	Model	1	2	3	4	5	6
a_0	*Constant*	-0.507 (0.421)	-0.543 (0.411)	0.316 (0.244)	0.343 (0.243)	0.460 (0.319)	0.337 (0.239)
a_1	*Education*	0.059 (0.022)**	0.060 (0.021)**	–	–	–	–
a_2	*Ecoeducation*	–	–	0.229 (0.063)**	0.222 (0.062)**	0.229 (0.076)**	0.225 (0.062)**
a_3	*Techeducation*	–	–	0.050 (0.028)†	0.043 (0.027)	0.058 (0.032)†	0.045 (0.027)
a_4	*Workexp*	0.018 (0.006)**	–	–	–	–	–
a_5	*Specworkexp*	–	0.028 (0.007)**	–	–	–	–
a_6	*Techworkexp*	–	–	0.026 (0.008)**	0.022 (0.008)**	0.018 (0.009)*	0.022 (0.008)**
a_7	*Comworkexp*	–	–	0.005 (0.012)	-0.003 (0.012)	-0.002 (0.018)	-0.006 (0.012)
a_8	*Genworkexp*	–	0.012 (0.007)†	0.007 (0.007)	0.004 (0.007)	0.001 (0.008)	0.005 (0.007)
a_9	*DManager*	–	–	–	0.247 (0.121)*	0.248 (0.139)†	0.250 (0.120)*
a_{10}	*DEntrepreneur*	–	–	–	–	0.231 (0.120)*	–
a_{11}	*Techworkexp* × *Comworkexp*	–	–	–	–	–	0.039 (0.009)**
a_{12}	*Age*	0.100 (0.011)**	0.100 (0.011)**	0.102 (0.011)**	0.103 (0.011)**	0.095 (0.013)**	0.104 (0.011)**
a_{13}	*Founders*	0.119 (0.029)**	0.121 (0.029)**	0.124 (0.030)**	0.117 (0.030)**	0.112 (0.040)**	0.114 (0.030)**
a_{14}	*Infrastructure*	0.005 (0.001)**	0.005 (0.001)**	0.004 (0.001)**	0.004 (0.001)**	0.004 (0.002)*	0.004 (0.001)**
	Adjusted R^2	0.226	0.232	0.241	0.248	0.226	0.251
	F-test	13.72 (9)**	12.84 (10)**	11.36 (12)**	10.90 (13)**	6.42 (14)**	10.35 (14)**

† $p < .10$; * $p < .05$; ** $p < .01$. All two-tailed tests. Huber-White standard errors and degrees of freedom for the F-tests in parentheses. Number of observations is 391, except for Model 5 where number of observations is 260. For the sake of synthesis we omit to report estimated coefficients of industry dummies.

else being equal, teams of founders with heterogeneous competencies run more successful NTBFs than teams characterized by similar functional skills. Note also that Model 6 replicates the results of previous models in that the coefficient of *Techworkexp* is positive and statistically significant at 99%, while the one of *Comworkexp* is insignificant. In other words, founders' marketing and sales skills have a positive impact on firm growth only if they are complemented by technical capabilities. By contrast, these latter have an autonomous positive effect, even though a more limited one than that arising from a founding team with both technical and commercial skills.

Lastly, let us briefly consider the effect of control variables. Both the number of founders and the infrastructure endowment of the region in which the NTBF is located always have a positive, statistically significant impact on firm growth. By contrast, with few exceptions, the industry dummies are insignificant, suggesting the absence of industry-specific fixed effects.

4 Discussion and conclusion

The aim of this paper was to examine the effects of the human capital of the founding team on the growth of NTBFs. For this purpose, we considered a sample composed of 391 young Italian firms operating in high-tech sectors, both in manufacturing and services. While controlling for firm age, we regressed the logarithm of the number of firm employees against a set of explanatory variables which provide a detailed description of the human capital of founders. We distinguished the generic and the specific components of human capital. The latter includes variables that reflect capabilities of founders that can directly be applied to the entrepreneurial job in the new firm; in particular, we considered the professional experience gained by founders in the sector of the new firm and variables indicating whether founders had prior managerial and entrepreneurial experiences. The former component relates to the general knowledge acquired by entrepreneurs through formal education and work experience; it was proxied by years of education and of work experience in different sectors than the one in which the new firm operates.

As was illustrated earlier in the paper, previous studies generally highlighted a positive effect of founders' human capital on firm growth. Nevertheless, because of lack of suitable data, they have failed to explain convincingly the reasons for such effects. Evolutionary and knowledge-based theories of the firm (Winter 1987; Conner and Prahalad 1996; Grant 1996) suggest that there is a close relation between the human capital of the founding team and the distinctive capabilities of the new firm. On the one hand, there is no market for entrepreneurial judgment, so the only way for an individual to exploit a smart business idea is to set up a new enterprise (see Knight 1921; Hodgson 1998; Loasby 1998). On the other hand, due to information asymmetries, it is quite difficult for NTBFs that operate in a risky environment and lack a track record to attract highly qualified salaried personnel. It also is difficult to prevent other firms from hiring them away. Accordingly, the unique skills entrepreneurs have learned in their professional careers are a key source of competitive advantage for a NTBF. Furthermore, it was shown by previous studies that the wealth of individuals is positively associated with their human capital Åstebro and Bernhardt (especially with education and work experience, see for instance 1999). In addition, NTBFs are likely to suffer from capital market imperfections that hinder external financing (Carpenter and Petersen 2002a). Firms established by highly qualified, allegedly wealthier individuals suffer less from financial constraints that may otherwise negatively affect growth.

Our findings suggest that the superior growth rates exhibited by NTBFs that are established by highly qualified individuals cannot be explained only by imperfections in capital markets. While all human capital variables considered in this work may be associated with a "wealth" effect, the "entrepreneurial ability" effect is more directly captured by the specific component of human capital. Our results show that both variables reflecting the generic and the specific components of founders' human capital are positively related to growth, but the latter component has a greater effect than the former one. In particular, the fact that, according to our estimates, the years of work experience of founders in the same sector of the new firm have a substantially greater positive impact on growth than those in different sectors is consistent with the view that the industry-specific capabilities of founders play a key role in providing newly created firms with a sustainable competitive advantage. A similar argument applies to founders' previous managerial and entrepreneurial experiences.

Furthermore, we investigated whether the functional nature of the industry-specific capabilities of founders matters for growth. The results again support arguments inspired by evolutionary and knowledge-based theories of the firm. More precisely, we considered the years of industry-specific experience of the founding team in R&D, process design, production and engineering departments, as opposed to that gained in marketing, distribution, and sale activities. Our findings indicate that it is only the former experience that is positively related to growth. This points to the technological nature of the competitive advantage enjoyed by successful high-tech start-ups. Lastly, we checked for the existence of synergistic effects arising from the presence within the founding team of complementary functional skills (i.e. technological and commercial skills). The econometric estimates highlight the fact that, when founders have both technological and commercial skills, the firms they establish perform better. In other words, absent technological skills, founders' commercial skills have a negligible impact on firm growth. Nonetheless, if these latter skills are present within the founding team, the contribution to growth provided by the simultaneous presence of sophisticated industry-specific technical skills sensibly increases.

In the authors' view, these results enlarge and deepen our understanding of the factors that drive the growth of NTBFs. The recent evidence on Gibrat's law indicates that growth is not a random process. In highlighting the role played by founders' human capital, this paper helps explain why. From this perspective, two directions for future research seem particularly promising.

First, we showed here that variables reflecting founders' generic human capital have a positive effect on firm growth. To the extent that such variables proxy founders' personal wealth, this evidence may be interpreted as indicative of the presence of binding financial constraints. Nevertheless, as to this aspect, an analysis of the different modes of financing used by NTBFs (i.e. internal financing, external private equity, bank loans, and public subsidies) and of their impact on growth would be much more informative; this issue is high on our research agenda.

Second, our findings indicate that the highest growth is obtained by NTBFs when both industry-specific technical and commercial skills are present within the founding team, witnessing the existence of synergistic effects. Among sample NTBFs, founders with industry-specific technical capabilities are much more common than those with commercial capabilities. This means that many new firms that were founded by individuals with sophisticated technical skills and thus may be considered as having a technology-driven competitive advantage, may nonetheless lack those complementary commercial skills needed to enjoy high growth rates. Therefore, such firms face the dilemma of whether and how to look for external support. Previous studies (see for instance McGee et al. 1995; Eisenhardt and Schoonhoven 1996; Lee et al. 2001) suggest that alliances with other firms, and especially larger corporations, may fulfill the lack of commercial capabilities suffered by NTBFs (in terms of sale force, brand name, customer care, distribution facilities). Nonetheless, to the extent that alliance partners are equipped with sufficient "absorptive capacity" (Cohen and Levinthal 1990), such partnerships also expose the new firm to the risk of unintended technological spillovers. In other words, partners may expropriate the technology-based competitive advantage of a NTBF and turn into powerful competitors (see Hamel 1991). Further research is needed to improve our understanding of the governance mechanisms of NTBF alliances and of the conditions under which these partnerships will benefit young and small high-tech firms.

References

Åstebro T, Bernhardt I (1999) The winner's curse of human capital. Working Paper CES 99-5
Almus M (2002) What characterizes a fast growing firm? Appl Econ, 34: 1497–1508
Almus M, Nerlinger EA (1999) Growth of new technology-based firms: Which factors matter? Small Bus Econ, 13: 141–154
Almus M, Nerlinger EA, Steil F (1999) Growth determinants of start-ups in Eastern Germany: A comparison between innovative and non-innovative firms. In: Oakey R, During W, Mukhtar SM (eds.) New Technology-Based Firms in the 1990s, 283–296, Elsevier, Amsterdam
Alvarez SA, Barney JB (2002) Resource-based theory and the entrepreneurial firm. In: Hitt MA, Ireland RD, Camp SM, Sexton DL (eds.) Strategic Entrepreneurship, 89–105, Blackwell Publishers, Oxford
Audretsch DB (1995) Innovation and Industry Evolution. MIT Press, Cambridge MA
Barringer BR, Jones FF, Neubaum DO (2005) A quantitative content analysis of the characteristics of rapid-growth firms and their founders. J Bus Venturing, 20: 663–687
Bates T (1985) Entrepreneur human capital endowments and minority business viability. J Human Res, 20: 540–554
Bates T (1990) Entrepreneur human capital inputs and small business longevity. Rev Econ Statist, 72: 551–559
Becker GS (1975) Human Capital. NBER, New York
Bester H (1985) Screening vs. rationing in credit markets with imperfect information. Amer Econ Rev, 75: 850–855
Bester H (1987) The role of collateral in credit markets with imperfect information. Europ Econ Rev, 31: 887–899
Blanchflower D, Oswald A (1998) What makes an entrepreneur. J Lab Econ, 16: 26–60
Brüderl J, Preisendörfer P (2000) Fast growing businesses: Empirical evidence from a German study. International Journal of Sociology, 30: 45–70
Brüderl J, Preisendörfer P, Ziegler R (1992) Survival chances of newly founded business organizations. Amer Sociol Rev, 72: 227–242
Carpenter RE, Petersen BC (2002a) Capital market imperfections, high-tech investment, and new equity financing. Econ J, 112: F54–F72

Carpenter RE, Petersen BC (2002b) Is the growth of small firms constrained by internal finance? Rev Econ Statist, 84: 298–309
Caves RE (1998) Industrial organization and new findings on the turnover and mobility of firms. J Econ Lit, 36: 1947–1982
Centro Studi Confindustria (1991) Indicatori Economici Provinciali. SIPI, Collana Industria e Territorio (in Italian), Rome
Chandler GN, Jansen E (1992) The founder's self-assessed competence and venture performance. J Bus Venturing, 7: 223–236
Cohen WM, Levinthal DA (1990) Absorptive capacity: A new perspective on learning and innovation. Admin Sci Quart, 35: 128–152
Colombo MG, Delmastro M, Grilli L (2004) Entrepreneurs' human capital and the start-up size of new technology-based firms. Int J Ind Organ, 22: 1183–1211
Colombo MG, Ermini B (2005) Crescita, dimensione ed età delle imprese high-tech italiane. L'Industria, forthcoming
Colombo MG, Grilli L (2005) Start-up size: The role of external financing. Econ Letters, 88: 243–250
Conner RK, Prahalad CK (1996) A resource-based theory of the firm: Knowledge versus opportunism. Organ Sci, 7: 477–501
Cooper AC (1985) The role of incubator organizations in the founding of growth-oriented firms. J Bus Venturing, 1: 75–86
Cooper AC, Bruno AV (1977) Success among high-technology firms. Business Horizons, 20: 16–22
Cressy R (1996) Are business start-ups debt-rationed? Econ J, 106: 1253–1270
Cressy R (2000) Credit rationing or entrepreneurial risk aversion? An alternative explanation for the Evans and Jovanovic finding. Econ Letters, 66: 235–240
Dunkelberg WC, Cooper AC, Woo CY, Dennis W (1987) New firm growth and performance. In: Churchill N, Hornaday J, Kirchhoff B, Krasner O, Vesper KH (eds.) Frontiers of Entrepreneurship Research, 307–32, Center for Entrepreneurial Studies, Babson College, Wellesly, MA.
Dunne T, Roberts MJ, Samuelson L (1988) Pattern of firm entry and exit in U.S. manufacturing industries. RAND J Econ, 19: 495–515
Dunne T, Roberts MJ, Samuelson L (1989) The growth and failure of U.S. manufacturing plants. Quart J Econ, 104: 671–698
Eisenhardt KM, Schoonhoven CB (1990) Organizational growth: Linking founding team, strategy, environment and growth among U.S. semiconductor ventures, 1978-1988. Admin Sci Quart, 35: 504–529
Eisenhardt KM, Schoonhoven CB (1996) Resource-based view of strategic alliance formation: Strategic and social effects in entrepreneurial firms. Organ Sci, 7: 136–150
Evans DS (1987a) The relationship between firm growth, size and firm age: Estimates for 100 manufacturing industries. J Ind Econ, 35: 678–582
Evans DS (1987b) Test of alternative theories of firm growth. J Polit Economy, 95: 657–674
Evans DS, Jovanovic B (1989) An estimated model of entrepreneurial choice under liquidity constraints. J Polit Economy, 97: 808–827
Evans DS, Leighton L (1989) Some empirical aspects of entrepreneurship. Amer Econ Rev, 79: 519–535
Feeser HR, Willard GE (1990) Founding strategy and performance: A comparison of high and low growth high tech firms. Strategic Manage J, 11: 87–98
Gibrat R (1931) Les Inegalites Economiques. Librairie du Recrueil, Paris
Gimeno J, Folta T, Cooper A, Woo C (1997) Survival of the fittest? Entrepreneurial human capital and the persistence of underperforming firms. Admin Sci Quart, 42, 4: 750–783
Grant RM (1996) Toward a knowledge-based theory of the firm. Strategic Manage J, 17, winter special issue: 109–122
Hamel G (1991) Competition for competence and inter-partner learning with international strategic alliances. Strategic Manage J, 12, summer special issue: 83–103
Hart PE, Oulton N (1996) Growth and size of firms. Econ J, 106: 1242–1252
Hart PE, Prais SJ (1956) The analysis of business concentration: A statistical approach. J Roy Statistical Society, 119: 150–191
Hodgson GM (1998) Competence and contract in the theory of the firm. J Econ Behav Organ, 35: 179–201

Holtz-Eakin D, Joulfaian D, Rosen HS (1994a) Entrepreneurial decisions and liquidity constraints. RAND J Econ, 25: 334–347
Holtz-Eakin D, Joulfaian D, Rosen HS (1994b) Sticking it out: Entrepreneurial survival and liquidity constraints. J Polit Economy, 102: 53–75
Hymer S, Pashigian P (1962) Firm size and the rate of growth. J Polit Economy, 70: 556–569
Knight F (1921) Risk, Uncertainty, and Profit. Houghton Mifflin, Boston
Lee C, Lee K, Pennings JM (2001) Internal capabilities, external networks and performance: A study on technology-based ventures. Strategic Manage J, 22: 615–640
Lindh T, Ohlsson H (1996) Self-employment and windfall gains: Evidence from the Swedish lottery. Econ J, 106: 1515–1526
Loasby B (1998) The organisation of capabilities. J Econ Behav Organ, 35: 139–160
Marshall A, Marshall MP (1879) The Economics of Industry. Macmillan and Co., London
McGee JE, Dowling MJ, Megginson WL (1995) Cooperative strategy and new venture performance: The role of business strategy and management experience. Strategic Manage J, 16: 565–580
Meyer BD (1990) Why are there so few black entrepreneurs? NBER Working paper n. 3537
Pennings JM, Lee K, Witteloostuijn AV (1998) Human capital, social capital, and firm dissolution. Acad Manage J, 41: 425–440
Prahalad CK, Hamel G (1990) The core competence of the corporation. Harvard Bus Rev, 1990, May-June: 79–91
Preisendörfer P, Voss T (1990) Organizational mortality of small firms: The effects of entrepreneurial age and human capital. Organ Stud, 11: 107–129
Roure JB, Maidique MA (1986) Linking prefunding factors and high-technology venture success: An exploratory study. J Bus Venturing, 1: 295–306
Schumpeter J (1934) The Theory of Economic Development. Harvard University Press, Cambridge, MA.
Siegel R, Siegel E, Macmillan IC (1993) Characteristics distinguishing high growth ventures. J Bus Venturing, 8: 169–180
Simon HA, Bonini CP (1958) The size distribution of business firm. Amer Econ Rev, 48: 607–617
Singh A, Whittington G (1975) The size and growth of firms. Rev Econ Stud, 42: 15–26
Storey DJ (1994) Understanding the Small Business Sector. Thomson Learning, London
Stuart RW, Abetti PA (1990) Impact of entrepreneurial and management experience on early performance. J Bus Venturing, 5: 151–162
Sutton J (1997) Gibrat's legacy. J Econ Lit, 35: 40–59
Westhead P, Cowling M (1995) Employment change in independent owner-managed high technology firms in Great Britain. Small Bus Econ, 7: 111–140
Winter SG (1987) Knowledge and competence as strategic assets. In: Teece D (ed.) The Competitive Challenge: Strategies for Industrial Innovation and Renewal, 159–184, Ballinger, Cambridge, MA
Xu B (1998) A reestimation of the Evans-Jovanovic entrepreneurial choice model. Econ Letters, 58: 91–95

Roberto Fontana · Lionel Nesta

Product entry in a fast growing industry: The LAN switch market

Abstract We provide empirical evidence on market positioning by firms, in terms of market niche, distance from technological frontier and dispersion. We focus on the switch industry, a sub-market of the Local Area Network industry, in the nineties. Market positioning is a function of the type of firms (incumbents versus entrants), market size and contestability as well as firm competencies. We find that incumbents specialize in high-end segments and disperse their product in a larger spectrum of the market. Instead, entrants focus on specific market niches. Market size, market contestability and firm competencies are also important determinants of product location.

Keywords Product introduction · Entry · LAN industry

JEL Classification L1 · O33

1 Introduction

This paper analyzes market positioning by firms, in terms of market niches, distance from the technological frontier and dispersion. It explores whether the strategy of market positioning differs between incumbents and entrants and whether it depends on the nature of the competition and firm capabilities.

The empirical work focuses on the switch industry, a submarket of the Local Area Network (LAN) industry in the 1993–1999 period. During this period, the switch market experienced a rapid and sustained growth accompanied by new

R. Fontana (✉)
Department of Economics, University of Pavia, and CESPRI - Bocconi University,
Via Sarfatti 25, 20139 Milan, Italy
E-mail: roberto.fontana@unibocconi.it

L. Nesta
Observatoire Français des Conjonctures Economiques,
Département de Recherche sur l'Innovation et la Concurrence, CNRS-OFCE-DRIC,
250, rue Albert Einstein, 06560 Valbonne, France
E-mail: lionel.nesta@ofce.sciences-po.fr

product entry due mainly to a very fast rate of technical change. We employ data from an original dataset of LAN equipment and firms, consisting of 704 new products marketed between 1993 and 1999. We develop indicators of distance among firms in the product space, product dispersion and distance from the quality frontier. These indicators are then regressed against a set of explanatory variables, which aim to capture the nature of the competition and firm capabilities.

The paper is structured as follows. Section 2 reviews some of the existing literature on product differentiation. Section 3 provides some background information on the technological evolution of the switch market and Section 4 describes both the indicators and the variables that are used in the empirical work. Section 5 presents the result and provides interpretation in terms of the nature of the competition and firm capabilities. Section 6 concludes and highlights the limitations of the analysis.

2 The determinants of product differentiation

Economic theory looks at product differentiation as a means for firms to compete on other grounds than mere price competition, allowing them to enjoy Schumpeterian rents from their temporary monopoly. In itself, product differentiation is usually analyzed in terms of location in a product and/or characteristics space. If prices are given, then the traditional Hotelling (1929) results hold according to which, in equilibrium, firms tend to locate products 'back to back' in the middle of a linear product spectrum. If, however, firms first locate and then compete in prices, there is the tendency to locate products away from each other (Shaked and Sutton 1982).

Two main factors influence product differentiation: the size of the product space and market contestability. When the size of the product space is large, firms attempt to appropriate parts of the demand, which is left unsatisfied, by entering into unexplored segments of the market. Conversely, when size is given, market contestability prevails. High contestability implies that the threat from potential entrants may persuade incumbents to introduce close substitutes for existing products in order to prevent firm entry. Incumbents with high market shares are likely to suffer from market contestability more than other firms because of the reduction in demand for existing product resulting from the introduction of their own substitutes. This phenomenon has been referred to as cannibalization (Spence 1976; Eaton and Lipsey 1979).

Evidence in favor of cannibalization is unclear. In the ready-to-eat cereal industry, incumbents have been found to employ a product proliferation strategy to fill existing niches before entrants (Schmalensee 1978). Looking at the PC industry over the 1981–1992 period, Buyus and Putsis (1999) found no evidence of such a strategy. The reasons for this mixed evidence are at least twofold. First, the extent of firm product diversification is likely to affect the choice of location. Product diversification in other markets may be a better strategy than proliferation when firms have not yet diversified (Bonanno 1987; Bhatt 1987) in order to spread their risky activities in several markets. Second, product introduction as analyzed by economic theory has put much emphasis on where firms are driven to by factors such as market size and contestability.

It has questioned less the capacity of firms to introduce innovative products in specific segments. Arguably, firm competencies as a whole may be important determinants of product location, other than market size and contestability. This has been repeatedly argued by the resource-based theory of the firm (Penrose 1959) in general, and the evolutionary theory of the firm in particular.

The evolutionary theory of the firm (Nelson and Winter 1982) is grounded on the very idea that firms make what they can rather than making what they want. This is due to the fact that the accumulation of productive knowledge is highly path dependent. Newly acquired knowledge at any point in time is a function of the firm knowledge base accrued in previous periods. As a result, knowledge accumulation inherently takes the form of successive small steps on related technological competencies rather than the form of stochastic moves across unrelated technological competencies (Teece et al. 1994; Breschi et al. 2003). More generally, the evolutionary theory of the firm puts much emphasis on the fact that firms are composed of heterogeneous knowledge and competencies, the exploitation and exploration of which is non-random and non-obvious, but above all non-immediate.

An additional argument for bounded opportunities rests on the idea that the product space is inherently rugged (Levinthal 1997). This is due to the fact that technologies form complex systems, that is, product components cluster around specific product architectures and a change in one component may imply changes in other components within the architecture or changes in the architecture itself (Henderson and Clark 1990). Thus what could appear as a small move for the firm may embody radical alterations of its technical competencies. In this context, previous experience with a specific market and/or technology may be a source of organizational inertia. It may also prevent established firms from understanding emerging customer needs and entering into new markets (Christensen and Rosenbloom 1995). Indeed, as technology race models predict, incumbents may enjoy either an advantage over entrants (Gilbert and Newberry 1982) or a disadvantage (Reinganum 1983; Lerner 1997), depending on whether the design of new products represents incremental improvements of existing ones or a radical departure, respectively.

The above implies that the mere identification of a new market opportunity represents only half the challenge. Firms must be able to modify their limited range of productive capabilities in order to introduce a supposedly more successful product. They do this in various ways: by renewing their human capital; by investing in new, more productive equipments; by undertaking systematic research activities; by identifying new suppliers or partners; by exploiting their complementary assets in a novel or more efficient way (Teece 1986; Mitchell 1989). For example, production and sales experience in existing markets provide firms with the possibility to compete efficiently in a related market niche by reducing entry costs (Klepper and Simons 2000; King and Tucci 2002). Within this context, incumbents may have advantages over entrants. In her study of product location in the PC market, Stavins (1995) finds that established firms disperse more of their products along the spectrum benefiting from learning and economies of scope. Looking at the automobile industry, Thomas and Weigelt (2000) find that incumbents and large firms use their experience to locate new products close to their existing ones and away from those of the rivals.

On the basis of the above discussion, we expect product location to reflect the nature of the competition, i.e. the size of the product space and market contestability, but also firm capabilities:

$$\Upsilon = f(\mathbf{M}, \mathbf{\Gamma}, \mathbf{Z}) \quad (1)$$

where Υ is a vector of variables describing market location by firms, **M** is a vector of variables describing the nature of the competition, $\mathbf{\Gamma}$ is a vector of variables on firm characteristics and **Z** is a set of control variables. In particular, we expect incumbents to have followed a segmentation strategy and to move incrementally in the product space. This is because incumbents can rely on past experience, leaving entrants to fill more distant and less crowded niches. However, we expect incumbents to produce a larger product line and to disperse their products in the product space more than entrants possibly to deter entry. Finally, we expect incumbents that produce close substitutes to locate in those market segments where customer preference for availability of complementary products, post marketing assistance and availability of continuous training tend to protect them from competition from new firms. Entrants would instead focus on new customers and locate their products in other segments.

3 The dynamics of the LAN Industry

We study the strategy of location in the product space by incumbents and entrants in the switch market. Switches are part of the infrastructure that constitutes Local Area Networks (LANs). Together with other types of equipment with which they have to interoperate, they are employed by users to send and to receive data packets. Switches were introduced in the context of data communication in 1990, first as a solution to congestion problems that existing equipment (i.e. hubs) could not deal with. Switches experienced a rapid diffusion in the first half of the last decade mainly as a consequence of the expansion of local networks and, subsequently, of the early diffusion of client server architectures within firms.

LANs are technical systems made up of different components (both hardware and software) that are combined to form an infrastructure to enable users located physically close to one another to send data and/or share common resources such as printers and other types of peripherals. Within the system, different types of equipment ensure that data transmission occurs: hubs that broadcast the data to many users located within a specific segment; switches that send data only to the user(s) who is supposed to receive it; routers which are able to choose the shortest path to destination and to embed additional management and security functions that switches do not have. In addition to the equipment, communications standards set the ‘rules’ and the speed at which data transmission occurs.

There are four different phases in the evolution of the LAN industry, each of them characterized by the introduction of new key equipment and the establishment of one or more specific communication standards.[1] The first phase, from the

[1] For the period breakdown we follow Christensen et al. (1995).

mid-1970s to the mid-1980s, is characterized by the absence of an official communication standard and the presence of many proprietary solutions marketed by different manufacturers. The second phase, from the mid-1980s until the end of the decade, saw the introduction of hubs and the official standardization of Ethernet and Token Ring as communication standards.[2] Hubs were introduced mainly to offer a unique point of concentration within firms in order to rationalize the cabling system. The third phase, from the end of the 1980s until mid-1990s, witnessed the growth of the hub market and the establishment of Ethernet as the dominant communication standards for LANs. As more and more nodes were added to the LAN, congestion increased, revealing the shortcomings of hubs in solving new bottlenecks. The introduction of the switch and the emergence of Fast Ethernet as the high-speed upgrade of Ethernet characterize the fourth phase. From a technological viewpoint, switches helped users reduce network congestion. Moreover, although more expensive than hubs, they provided users with additional features.[3]

Interestingly, the emergence and development of switches coincided with the entry of three types of manufacturers. First, we consider incumbents from within the LAN industry (both from the router and the hub market). Cisco, the future dominant firm in the switch market, came from the router market and was among the first firms to enter. More hesitant incumbents in the hub market (3Com, Bay Networks, DEC) were constrained by their previous investments as well as by the risk of cannibalizing their installed base. Second, we see incumbents from outside the industry but with previous experience either in the telecom industry or in the semiconductor industry. Third, we have start-ups searching for new opportunities especially during the fourth phase when high speed standards were developed.

The growth of the switch market was accompanied by developments in the equipment design. The first was the change in the design of the equipment following the transition to ASICs based architectures that occurred in the middle of the 1990s.[4] This development enabled high economies of scale to be obtained in the production of switches. The second was the improvement in the data management that made switches more sophisticated and similar to routers. As a result of these events, in the second half of the 1990s, two segments opened up in the switch market. The high-end switches, characterized by high performance, were targeted to customers with large networks. The low-end switches are low cost and generally less performing, support only one standard and are targeted to customers with small networks.

Importantly, the nature of the competition in the two types of market segments is different. In the low-end segment, manufacturers compete mainly on price. In the high-end segment, competition is usually influenced by the presence of substantial switching costs for customers. Switching costs derive mainly from the need to purchase equipment that is compatible with existing switches, and from the cost of

---

[2] Ethernet was originally developed at the Menlo Park Research Labs in the first half of the 1970s and had become the standard chosen by Xerox for its internal network. Token Ring was the standard of choice for IBM based networks. Other proprietary standards existed (i.e. ARCnet and StarLAN). See von Burg (2001).

[3] Switches possessed higher port density (i.e., they could connect more nodes) than hubs and were capable of forwarding data very fast and only to those user(s) who were supposed to receive them.

[4] ASICs (Application Specific Integrated Circuits) switches embody special purpose devices that speed up the forwarding of data packets and contribute to reduce network congestion.

learning to use equipment purchased from different manufacturers. Since LAN communication standards are open, in theory customers can mix-and-match equipment from different manufacturers in the same network. However, switches need software program to function and manufacturers often design proprietary software that makes the equipment incompatible with that of other manufacturers.[5] This practice has implications both for adoption choice and the strategy of product introduction by manufacturers.

On the demand side, previous investments in both hardware and software of specific manufacturers give a strong incentive for users to continue to buy compatible equipment from the same manufacturer, in order to avoid incurring the extra costs of switching. In a survey of the determinants of switch choice by users, Forman and Chen (2004) find evidence that the presence of an installed base of equipment from a particular manufacturer increases the likelihood of repurchasing from the same vendor if the customer decides to buy again. These issues become particularly important in the case of technologically sophisticated equipment such as high-end switches used in large networks of hundreds to thousand of users. Within these networks, costs for moving and adding new users are very high. Connecting and setting-up new configurations as well as troubleshooting are particularly complicated, and software that works for specific equipment may not work with others (Christensen et al. 1995).

On the supply side, manufacturers can exploit the presence of high switching costs to lock-in customers and to compete against rivals. Relying on an existing installed base, manufacturers may strategically set the timing of new product introduction in order to force customers to buy upgrades. Moreover, manufacturers tend to introduce families of products that work well together. These families typically include devices of different densities and configurations, and supporting different communication standards so as to be able to target the entire demand spectrum (i.e. going 'end-to-end') and/or to prevent rivals from entering specific niches. All major manufactures in the switch industry followed this strategy.[6]

The structure of the switch market has always been concentrated, although less than that of other markets within the LAN industry (i.e. routers). In 1994, the four biggest firms in the market (Cisco (Kalpana), 3Com (Synernetics), Alantec and Chipcom) accounted for 94% of revenue share. In the 2nd quarter of 1999, the last year of our sample period, the share of the four biggest companies had fallen to 81%, with Cisco maintaining the lead at 47%, followed by 3Com, Nortel Networks (Bay Networks) and Cabletron.[7]

---

[5] Pieces of equipment produced by the same manufacturer often work on a specific software 'platform' that is incompatible with a rival one. This is the case for instance of the Spectrum network management software developed by Cabletron, the Cisco Fusion architecture first and then the Internet Operating Systems (IOS) software developed by Cisco (Gawer and Cusumano 2002).

[6] In 1994, Cisco offered the first product of its Catalyst line of Switch equipment that was enriched in the following years by other high-end as well as low-end equipment. 3Com responded with the Superstack and Office Connect line of equipment, the former targeting big users, and the latter customers with smaller networks. Also, the other two big incumbents, Bay Networks and Cabletron, marketed an entire product line, the BayStack and Smartswitch, respectively.

[7] Sources: The Yankee Group (Network World: October 31, 1994). 1999 (2qt): Dell'Oro Group (Network World: October 18, 1999: p.30).

4 Data and measurements

The data used in this paper consist of information on product characteristics and prices of switches and hubs. They come from an original dataset of 1071 LAN products (535 hubs and 536 switches marketed between 1990–1999 and 1993–1999 respectively). The dataset was constructed using information from specialized trade journals (*Network World* and *Data Communications*) that periodically publish details on new product introductions. This information has also been double checked, whenever possible, with press communications and product announcements released by the manufacturers themselves.

In the original dataset, there are 199 different manufacturers. For the purpose of our analysis, we decided to consider only those manufacturers who marketed four or more products in the period 1990–1999. After consolidation, we were left with 82 firms and a subset of 704 products in total (464 switches and 240 hubs). As we can determine product location only when firms innovate, the number of observations in the sample, i.e., firms introducing at least one product for a given year in the switch market, is 176.

For each product, the dataset reports information on its technical characteristics, date of market introduction and list price. The chosen time span fully covers the year of development and growth of the switch market as well as the years of consolidation and decline of the hub market. By stopping in 1999, this analysis only partially accounts for the most recent period of the evolution of the switch market, characterized by the entry of telecom companies. Since most of these firms produce for the high-end part of the market, our data may lead to an underestimation of the impact of entrants on the location in this segment.

Figure 1 shows the pattern of new product introduction for both switches and hubs between 1993 and 1999.

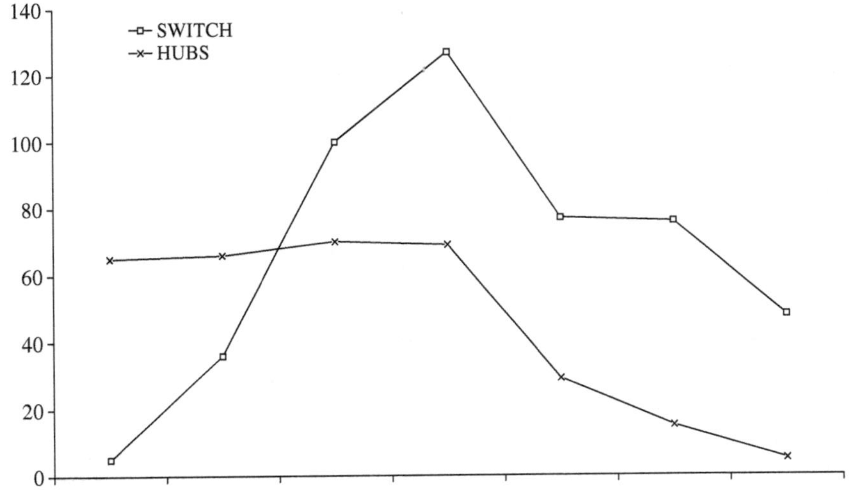

Fig. 1 Number of product innovations in the switch and hub markets

4.1 Dependent variables

We represent the location of product in a vertical or horizontal space using the generic technological characteristics of the products in the switch market and measuring distances across products. There is no authoritative way of measuring distances across products. The most challenging aspect of this analysis is deciding how products can be benchmarked. Following Stavins (1995), we do it in two steps.

In the first step, we reduce the multi-attribute structure (the technological characteristics) to a single dimensional measure of product quality. Assuming

Table 1 Ordinary least squares regression on hedonic prices

Dependent variable	Deflated product price
Backplane capacity	0.239
	[0.038]***
Number of ethernet ports	0.088
	[0.029]***
Number of fast ethernet ports	0.029
	[0.038]
Number of FDDI ports	0.030
	[0.060]
Number of token ring ports	0.129
	[0.045]***
Number of 100VG-AnyLAN ports	0.259
	[0.109]**
Number of ATM ports	0.120
	[0.044]***
Number of gigabit ethernet ports	0.361
	[0.058]***
VLANs capability	0.313
	[0.107]***
Modular configuration	0.906
	[0.136]***
Fixed configuration	−0.208
	[0.094]**
Intercept	8.044
	[0.438]***
Observations	469
R-squared	0.687
Log-*L*	−558.4

Robust standard errors in brackets
* significant at 10%; ** significant at 5%; *** significant at 1%
Year dummy variables omitted for clarity

independence across product technological attributes, we project them onto a linear scale as follows:

$$q_m = \sum_{l \neq k}^{D} \beta_j \cdot Z_{jm} \tag{2}$$

Eq. 2 suggests that the quality q of model m can be measured as the weighted sum of its characteristics. The weights β_j represent the marginal value of characteristics j that both consumers and producers place on the jth attribute. Such weights β_j are approximated by regressing observed prices, deflated into 1996 US dollars using the Implicit Price Deflator provided by the U.S. Department of Commerce, Bureau of Economic Analysis, on product characteristics:

$$\ln(P_{mit}) = \alpha + \alpha_t + \sum_j \beta_j \cdot Z_{jm} + \varepsilon_{mit} \tag{3}$$

where α is a constant and α_t is a time fixed effect. Table 1 provides the results from the hedonic regression. With 70% of the variance of prices explained, the overall fit is satisfactory enough, although a substantial part of the observed prices (30%) is due to factors other than those introduced in the regression. This may in turn be due to omitted product attributes and erroneous pricing reflecting changes in demand.

In Table 1, the estimated weights are constrained to be constant overtime, whereas the technology in the Switch market is likely to have evolved over time. This suggests that, depending on significant changes in product quality in the nineties, the pooled regression may produce inexact weights. We performed separate hedonic regressions in order to test for parameter equality over time by means of a Chow test. All years differ significantly and thus separate hedonic regressions were performed to produce β_{jt} instead of β_j.[8] These were used to calculate q, the predicted price by hedonic regressions. Whereas the observed prices embody error measurements reflecting various factors such as changes in demand, promotional discounts and other non-quality components (see Stavins 1995), the predicted price q reflects by construction the quality of the product. Ranking these prices is tantamount to ranking products according to their quality.

In the second step, we used the predicted price q to compute distances across product. We start by considering horizontal product differentiation by computing the mean Weitzman distance d^c_{mit} of a given model m from all models introduced the previous year:

$$d^c_{mit} = \frac{\sum_{n-1}^{N_{t-1}} \sqrt{(q_{mit} - q_{n,t-1})^2}}{N_{t-1}} \tag{4}$$

[8] As a result, quality of fit as measured by the r-square exhibits some positive trend, ranging from 0.55 in 1994 to 0.90 in 1999. The unexplained part of the variance, i.e. the error term, may combine two phenomena. First, the set of explanatory variables is more accurate for the most recent years, whereas some additional explanatory variables would have captured some of the variance of the observed price in the early years. Second, the market structure in early years allows for large firm mark-up, biasing the prices upwards as compared to what should be expected under perfect competition. Importantly, market power is supposedly larger in early years when incumbent firms enjoy a first mover advantage.

where N_{t-1} is the number of products at year $t-1$, q_{mit} is the quality of model m by firm i in year t and $q_{n,t-1}$ is the quality of model n in year t–1. Because of the squared differences $(q_{mit}-q_{n,t-1})^2$, Eq. 4 does not distinguish between low quality $(q_{mit}<q_{n,t-1})$ or high quality $(q_{mit}>q_{n,t-1})$ products. Thus, the product space as rendered by the Weitzman distance measure can be thought of as a *horizontal* scale in which products locate. The further from center c of the scale, the more peculiar the product is with respect to the representative product. Because firms can introduce several products in a given year, we calculated for every firm and every year the average distance $\overline{d}_{it}^{c}$.

Alternatively, we used the predicted price q to rank products on a *vertical* product space. To do so, we computed for every product its distance from the *quality* frontier as follows:

$$d^f_{mit} = \max(q_{it}) - q_{mit} \tag{5}$$

where q_{mit} is the quality of model m by firm i in year t. The higher d^f_{mit} the farther the product is from the quality frontier. Again, because firms can introduce several products in a given year, we computed for each firm the lowest distance from the technical frontier $d^f_{it} = \min\left\lfloor d^f_{mit}\right\rfloor_{it}$.

The last measure deals with product dispersion in order to understand whether incumbents spread their products over a wider range of the product space. We develop a measure of product dispersion in two steps (Stavins 1995). In the first step, we constructed a measure of product dispersion within the firm:

$$\sigma_{it} = \frac{\sum_{m=1}^{M_{it}} (q_{mit} - \overline{q}_{it})^2}{M_{it}} \quad \text{where } \overline{q}_{it} = \frac{\sum_{m=1}^{M_{it}} q_{mit}}{M_{it}} \tag{6}$$

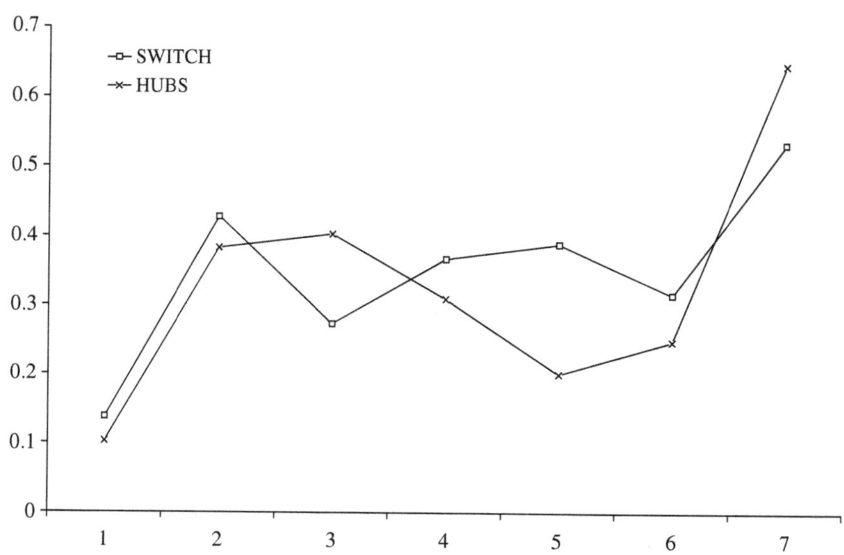

Fig. 2 Dispersion of model quality by firm age in the switch and hub markets

where M is the number of products introduced by firm i for a given year. In turn, the overall product dispersion is defined as:

$$\sigma_t = \frac{\sum_{n=1}^{N_t} (q_{nt} - \overline{q}_t)^2}{N_t} \quad \text{where } \overline{q}_t = \frac{\sum_{n=1}^{N_t} q_{nt}}{N_t} \tag{7}$$

where N is the number of products introduced by all firms for the same year. The relative dispersion index is defined as the ratio of the two dispersion measures σ_{it} and σ_t:

$$R_{it} = \frac{\sigma_{it}}{\sigma_t} \tag{8}$$

Figure 2 plots the mean relative dispersion index by firm age. It seems that firm age between 2 and 6 years on the switch or hubs market does not impact the dispersion of the product. However widening the age range from 1 to 7 years suggests a positive and non-linear relation between firm age and product dispersion.

4.2 Independent variables

Consistent with the empirical literature, we distinguish between entry and incumbency as a main determinant of innovative strategies for product location. Additionally, we argue that product location is a function of two main factors: the nature of the competition and firm capabilities. We decompose the notion of competition into two components: the size of the product space and market contestability. First, following Khanna (1995), we use Eq. 5 to develop a measure of the size of the product space as follows:

$$\text{Size of the Product Space} = SPS = \frac{\sum_N (\max(q_{it}) - q_{mit})^2}{N} = \frac{\sum_N \left(d^f_{mit}\right)^2}{N} \tag{9}$$

where N is the number of products for a given year and q_{mit} is the quality of model m by firm i in year t. Eq. 9 has the lowest value of zero, which means that the average distance is null and that all products introduced in a given year compete with the same characteristics. With a shrinking measure, the market is concentrated around a dominant quality and thus firms compete in a narrower product space. With a growing measure, the market expands with the introduction in the product space of novel product quality.[9]

[9] It is tempting to say that SPS is informative about the intensity of the competition between firms. However, as was rightly pointed out by one referee, this explanation should also consider vertical moves along the scale. Assume the following two examples: 1. All products are of the same quality, except one which is located one unit below the others; 2. All products are of the same quality, except one which is located two unit above the others. According to our interpretation, our measure SPS in example 1 should be lower than that of example 2. We would thus conclude that competition is lower in example 2 than in example 1. However, the second situation is likely to face fiercer competition since a one-unit move *above* other products is likely to be of much greater value than a one-unit move *below* other products. Thus, in our paper, SPS only exploits information along the *horizontal* scale and ignores movements along the *vertical* scale.

Figure 3 plots the computed measures of the product space for both the hub and switch markets. We observe two successive waves: the hub market, after having expanded until 1995, has been constantly concentrating around a narrower technological space; the switch market is still expanding, thus leaving room for potential entrants to enter. This is in line with our appreciative knowledge that the former is technologically older than the latter. Second, we develop a measure of market contestability as follows:

$$\text{Market Contestability} = MC = \frac{E}{F} \tag{10}$$

where E and F are for a given year the numbers of entrants and the total number of firms, respectively. This represents an ex post measure of market contestability based on the entrant counts, as opposed to ex ante measures such as, for example, barriers to entry. When the number of entrants is high, the ratio (E/F) is high and, therefore, market contestability is supposedly high. This variable varies between 0 and 1. A value close to unity indicates high contestability by entrants, whereas a low value indicates low contestability.[10] Figure 4 plots the accumulated number of firms F and the number of entrants E per year. The number of entrants increases until 1995 and drops to zero in 1998, stabilizing the number of firms in the switch market to around 80 from 1997 onwards.

Before turning to the issue of measuring firm capabilities, it is interesting to graph the number of product innovations by types of innovator. Figure 5 shows that incumbents play a major role in innovative activities in the switch market, since, from 1995 onward, most innovations were introduced by firms competing in the switch market for at least a year.

Firm capabilities range from technological competencies to capabilities in complementary activities such as distribution and commercialization. We measure these capabilities E_{it}^{switch} with the number of years for which the firm has been active in the switch market. Arguably, this should control for firm experience in both core competencies and complementary assets. In a similar fashion, we test the presence of economies of scope by controlling for firm experience E_{it}^{hubs} in the hub market. Firms may benefit from their experience in distinctive yet related markets, which in turn may potentially influence the capacity of firms to enter and to locate in particular segments of the switch market.

We can now rewrite Eq. 1 and model firm market location, in terms of distance from market center, distance from the technological frontier and dispersion, as a function of the type of innovator, the nature of the competition and firm capabilities:

$$\Upsilon = \alpha_0 + \alpha_1 \cdot I_i + \underbrace{\beta_1 \cdot SPS_{t-1} + \beta_2 \cdot MC_{t-1}}_{\text{nature of the competition}} + \underbrace{\delta_1 \cdot E_{it}^{\text{switch}} + \delta_2 \cdot E_{it}^{\text{hubs}}}_{\text{firm capabilities}}$$

$$+\alpha_2 \cdot T_t + \sum_i \alpha_{3i} \cdot D_i + \varepsilon_{it} \tag{11}$$

[10] We do not have information on product and/or firm exit, so that the number of firms does not account for manufacturers who decided to leave the industry. As a consequence, this variable may overestimate the effect of contestability on our indicators of product location.

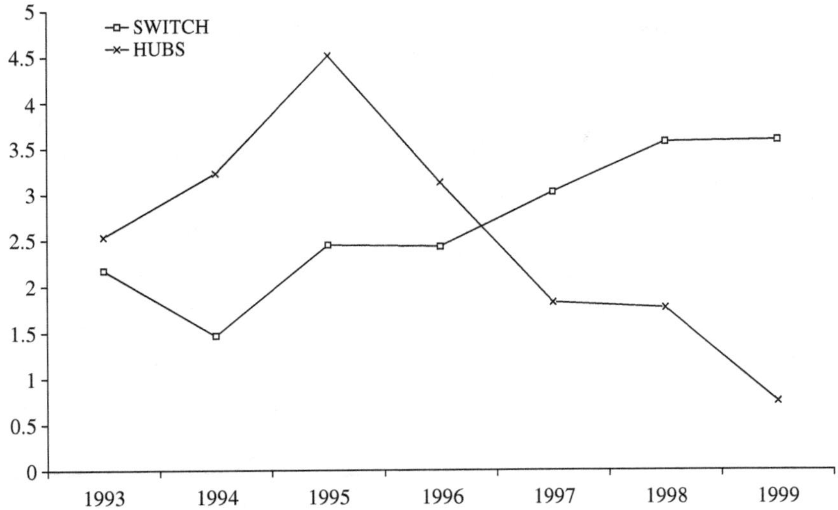

Fig. 3 Size of the product space in the switch and hub markets

where $\Upsilon = \{d^c_{it}, d^f_{it}, R_{it}\}$ is the vector of dependent variables describing market location by firms, I is a dummy variable set to unity if the innovator is an incumbent firm, 0 if it is an entrant; *SPS*, *MC* and *E* are defined as previously, *T* is a linear time trend and *D* is a vector of firm dummy variables catching unobserved heterogeneity across firms.

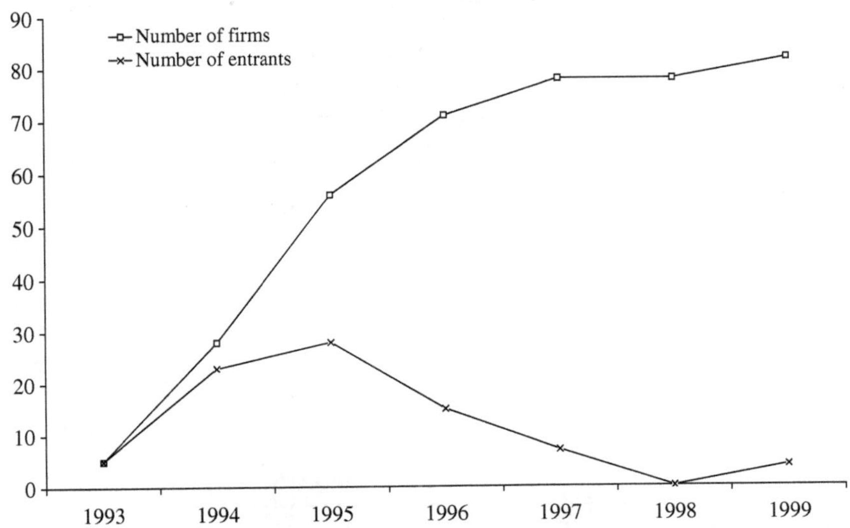

Fig. 4 Number of firms and entrants in the switch market

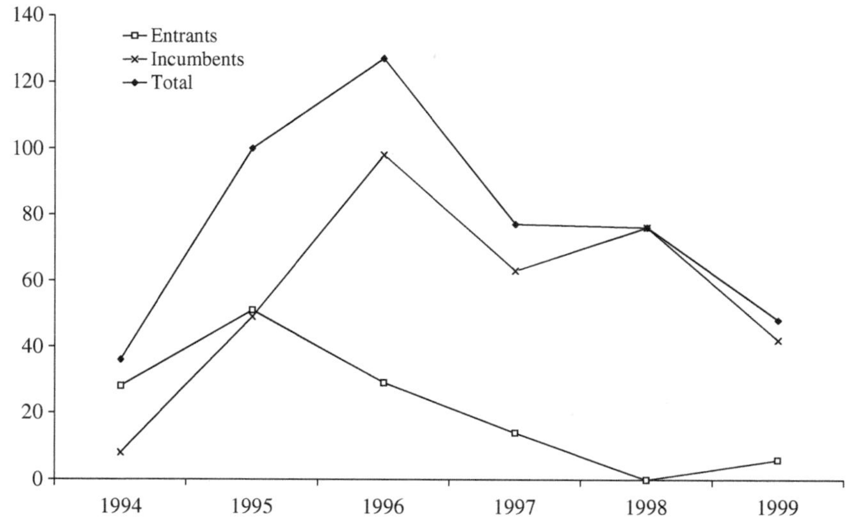

Fig. 5 Number of products introduced by entrants or incumbents in the switch market

Consistent with our analysis in Section 2 and the background knowledge of the switch industry provided in Section 3, we expect incumbency to impact negatively on product location, and positively on distance from the frontier as well as on product dispersion. The nature of competition should impact positively on product location and negatively on distance from the frontier. Finally, we generally expect firm capabilities to impact differently depending on whether we are considering

Table 2 Ordinary least squares on the location of innovation in the switch product space

Dependent variable	Mean Weitzman distance				
	(1)	(2)	(3)	(4)	(5)
Incumbents	−0.118	−0.196	−0.256	−0.193	−0.172
	[0.069]*	[0.085]**	[0.114]**	[0.109]*	[0.109]
Size of product space (t−1)		0.255	0.241	0.196	0.198
		[0.103]**	[0.097]**	[0.088]**	[0.089]**
Market contestability (t−1)		0.090	0.082	1.295	1.311
		[0.204]	[0.211]	[0.645]**	[0.653]**
Experience in switch			0.059	0.033	0.012
			[0.062]	[0.059]	[0.070]
Experience in hub			−0.046	−0.049	−0.040
			[0.023]**	[0.024]**	[0.028]
Trend				0.311	0.321
				[0.144]**	[0.153]**
Firm fixed effect	No	No	No	No	Yes
Observations	176	176	176	176	176
R-squared	0.017	0.099	0.126	0.163	0.189
Log-L	−105.5	−97.8	−95.1	−91.4	−88.5

Intercept not reported. Robust standard errors in brackets
* significant at 10%; ** significant at 5%; *** significant at 1%

experience in the same (i.e., the switch) or a related (i.e., the hub) market. Previous experience with switches should impact positively on product location, and negatively on distance from the frontier as well as dispersion. Experience with hubs, instead, should impact negatively on product location and positively on distance from the frontier as well as product dispersion.

5 Results

The results of OLS regressions are displayed in Table 2 Table 3 Table 4. In each model, we use several specifications to control for the robustness of the results. Because of the small number of observations (N=176 for Table 2 Table 3 Table 4.) and the unbalanced nature of the panel, the inclusion of firm dummy variables is limited to those firms innovating for at least 5 years.

Table 2 presents the results of the analysis of product location in the switch market. Five models have been estimated using the mean Weitzman distance as a dependent variable. In each of the proposed models, we sequentially introduce a new set of independent variables. Model (1) only takes into account whether the firm has previously introduced a switch, while models (2) and (3) look at the impact on firm location choice of technological competition, marker structure and previous experience in the switch and/or the hub market. We find a negative and significant coefficient of the incumbent variable, suggesting that established firms tend to locate their new products close to existing ones.

In model (2), we introduce the variables regarding the nature of the competition: the size of product space and market contestability. We lag these variables one period in order to account for their influence at the time when firm decisions occur. The positive and significant sign of the former suggests that, as the product space expands, firms introduce less close substitutes for existing products. In a context of expanding opportunities for product location, this strategy is consistent with entry deterrence behavior as predicted by Brander and Eaton (1984). Consistent with the result of Stavins (1995), market contestability enters positively, suggesting that the threat of new entrants leads firms to saturate the product space by locating new products far from existing ones. These results hold after controlling for additional factors such as firm capabilities (Model 3), time trend (Model 4) and unobserved heterogeneity between the most innovative firms (Model 5).

In model (3), we look at the influence of previous experience in the same and/or related markets and technologies. The coefficient of Experience in Switch is positive although not significant, whereas the coefficient of Experience in the Hub market is negative and significant. This suggests that the more experience firms have in this related market, the more they tend to locate their switches in crowded areas of the product space. Previous experience in a related market reinforces a segmentation strategy in the switch market using their existing distribution channels, assistance and training to serve the new customers. Moreover, for a hub maker, producing switches means producing a distant substitute for their products. The choice of locating close to other switch manufactures could be the consequence of a strategy aimed at filling empty spaces before competitors do in order to preserve customers.

Overall, these results show that established firms cluster their products in a relatively narrow and crowded portion of the product space (i.e., they segment). It implies that firms do not expect intense price competition either from existing rivals or from entrants given that, by producing more distant substitutes for existing products, incumbents might effectively deter entry by new firms. We question whether, in a context of market expansion, this type of behavior depends on the pattern of demand and the presence of high switching costs for users. If there are different categories of customers who, as argued in Section 3, have to sustain high costs of switching between manufacturers, then location in the product space becomes less a strategic decision for incumbents. Firms may produce close substitutes without worrying about price cuts from rivals or increasing competition from entrants.

Table 3 displays the results of the determinants of location in the vertical product space. Here the dependent variable is the lowest distance of each firm from the quality frontier (d_{it}^{f}), whereas the set of independent variables is identical to that of Table 2. Model (7) looks at the influence of the type of innovator and market structure on firm location. The sign of Incumbents is negative and significant, suggesting that established firms locate close to the frontier, i.e. incumbents target the high-end of the market, whereas entrants produce for the low-end. The sign of Incumbents does not change when controlling for other factors, although it becomes insignificant (Models 8 to 10).

The negative and significant coefficient of Size of Product Space indicates that the expansion of the product space increases technological opportunities to locate products close to the frontier (i.e. target the high-end market). This is surprising, given that a larger product space should be associated with opportunities to locate products further away from the frontier. A candidate explanation is that the product space moves upwards along the vertical scale, as a result of firms racing for first

Table 3 Ordinary least squares on the location of innovations in the switch product space

Dependent variable	Distance from quality frontier				
	(6)	(7)	(8)	(9)	(10)
Incumbents	0.035	−0.477	−0.328	−0.309	−0.352
	[0.196]	[0.219]**	[0.275]	[0.281]	[0.275]
Size of product space		−0.570	−0.544	−0.558	−0.539
(*t*−1)		[0.261]**	[0.248]**	[0.241]**	[0.233]**
Market contestability		−2.264	−2.190	−1.812	−1.744
(*t*−1)		[0.494]***	[0.499]***	[1.542]	[1.533]
Experience in switch			−0.187	−0.195	−0.086
			[0.114]	[0.115]*	[0.129]
Experience in hub			0.189	0.189	0.242
			[0.058]***	[0.058]***	[0.061]***
Trend				0.097	0.045
				[0.354]	[0.353]
Firm fixed effect	No	No	No	No	Yes
Observations	176	176	176	176	176
R-squared	0.000	0.129	0.182	0.183	0.298
Log-*L*	−291.7	−279.6	−274.0	−273.9	−260.6

Intercept not reported. Robust standard errors in brackets
* significant at 10%; ** significant at 5%; *** significant at 1%

mover advantage. In this case, the larger product space provides firms with incentives to race for vertical quality, thus introducing products nearer the frontier. Although this result agrees with our qualitative knowledge of the market, this explanation must be pursued further in our future work. The negative and significant coefficient of Market Contestability suggests instead that the threat from entrants leads firms to locate far from the frontier (i.e. target the low-end market). Overall, these results depict a situation in which opportunities to cater to high-end customers increase as the market grows but are seized upon mainly by incumbents. Since incumbents tend to segment, as shown in Table 2, we conclude that they tend to cluster their products mainly in the high-end segment.

Results from models (6) and (7) suggest that entrants target mainly low-end customers, implying that entrants do not leapfrog incumbents. The reasons for this are threefold. First, because incumbents saturate high-end segments of the market, there is little demand left for entrants. Second, evidence on adoption behavior in the switch industry shows that unattached users focus on the low-end segments (Forman and Chen 2004). Third, as argued in Section 3, the presence of barriers to entry in the form of economies of scale in production discourages new firms from entering at all in the market. The above remarks are reinforced when we focus on firm experience in the switch – and related – market.

Experience in Switch has a negative sign, suggesting that, as experience grows, incumbents continue to focus on the high-end segment, whereas Experience in Hub is positive and significant, indicating that the higher the previous experience in the hub market, the more firms position their products far from the frontier. These results are in broad agreement with the evolutionary theory of the firm, where the distinctive skills of firms in a high technology industry, such as the switch market, are related to their core technological competencies. Moreover, although hub makers could leverage on complementary assets, they do not possess the required technological competencies to manufacture high-end products.

Finally, we analyze product innovation in terms of product dispersion. Table 4 presents the results of the *Relative Dispersion Index* (*RDI*, Eq. 7) as a function of the type of innovator, the nature of the competition and firm experience. Most coefficients are not significant, the most robust result being the positive and significant sign of Incumbents across all the models (11)–(15). This suggests that incumbents disperse their products more than do entrants. It is consistent with the presence of demand side switching costs that give an advantage to firms that market full product lines. Market Contestability has a positive sign, suggesting that the threat from entrants provides impetus for incumbents to saturate the product space and engage in several segments of the market. This result is consistent with those of Table 2, where we argued that the presence of many firms in the market reduces the space for locating new products and lowers incentives to disperse new models.

Finally, we note that none of the variables capturing firm experience on product dispersion are significant.[11] This result suggests the absence of learning effects in explaining the behavior of incumbents, and reinforces the evidence that dispersion is mainly a consequence of the presence of demand side switching costs.

---

[11] One could argue that the reason for the non-significance of the variables related to firm capabilities is its co-linearity with the variable "Incumbents". These results hold even when regressing *RDI* on firm capabilities only.

Table 4 Ordinary least squares on the dispersion of innovations in the switch product space

Dependent variable	Relative dispersion index				
	(11)	(12)	(13)	(14)	(15)
Incumbents	0.221	0.250	0.246	0.292	0.327
	[0.065]***	[0.090]***	[0.119]**	[0.121]**	[0.123]***
Size of product space		−0.003	−0.006	−0.039	−0.044
(t − 1)		[0.094]	[0.097]	[0.098]	[0.095]
Market contestability		0.084	0.095	0.982	0.961
(t − 1)		[0.157]	[0.165]	[0.576]*	[0.587]
Experience in switch			−0.005	−0.024	−0.089
			[0.044]	[0.048]	[0.060]
Experience in hub			0.014	0.012	−0.009
			[0.022]	[0.022]	[0.025]
Trend				0.227	0.255
				[0.146]	[0.150]*
Firm fixed effect	No	No	No	No	Yes
Observations	176	176	176	176	176
R-squared	0.053	0.056	0.058	0.075	0.191
Log-*L*	−112.6	−112.3	−112.1	−110.5	−98.7

Intercept not reported. Robust standard errors in brackets
* significant at 10%; ** significant at 5%; *** significant at 1%

6 Conclusion

This paper has investigated the determinants of product location in the switch equipment industry. Two issues have been at stake: whether firms followed a specific strategy when introducing a new product; and whether previous experience with the same and/or related technology influenced the location decision. Concerning the first point, our analysis found that incumbents and entrants followed different product location strategies. We did not find evidence of a strategic pre–emptive behavior by incumbents, such as entering empty market niches. On the contrary, the overall results of the analysis of product location with respect to competing firms suggest that, in this market, incumbents have a tendency to segment. We also found that incumbents tend to segment in a specific portion of the switch market, namely the high-end. We interpret this evidence on the basis of the presence of demand side switching costs that shelter firms from (the reaction of) rivals and competition from entrants. The influence of demand side switching costs on product location choice is confirmed also by the choice of incumbents to disperse their product more in the product spectrum (i.e. they introduce wide product lines). Instead, entrants disperse less.

Concerning the second point, we found mixed evidence on the role of past experience on firm location choice. On the one hand, previous experience gained in related markets reinforces segmentation, especially in the case of firms targeting the low-end of the market. On the other hand, previous experience acquired in the same market does not affect the choice of product location. There seems to be a discontinuity in location in the high-end segment. Being an incumbent represents an advantage, but this advantage does not depend on past experience.

Overall, our results indicate that the nature of competition differs depending on whether the firm is an entrant or an incumbent. For entrants, competition consists of a succession of *moves*, aimed at climbing the ladder from the low-end to the high-end segments. In order to reach the top-end of the market, a firm must accumulate experience first, being active in another related market, for instance, and exploit it subsequently. For incumbents, competition consists of securing their *position*, by occupying the high-end of the market first, and then saturating the product space. These types of behavior coexist and plead in favor of the idea that firm type and, albeit in a less systematic way, firm capabilities also explain product location. Other strategic determinants of firm location such as market contestability and market size, although important, do not seem to be equally relevant to explain product differentiation in this industry.

The results of this paper capture stylized facts emerging from empirical analyses of the evolution of the switch industry. Although they are still preliminary, they are consistent with some of the predictions of the existing theoretical literature on product location (Brander and Eaton 1984) and entry deterrence in the presence of switching costs (Farrell and Shapiro 1988; Klemperer 1995). Finally, they also point to the importance of firm resources and capabilities in determining location choice and of market competition in boosting technical change.

Our results are subject to a number of limitations, mainly related to the lack of additional information on switch products. The most immediate is the absence of data on product sales. From the econometric viewpoint, the above estimations should be weighted for each product sale. From the theoretical viewpoint, product sales provide firms with market power, which in turn should prove influential in their strategy. The other issue concerns product exit. In markets where exit is frequent, market structure is highly affected by product exit. Our study should extend the analyses of product location to product exit, as done by Stavins (1995). The combined availability of product sales and exit would enable us to study the relationship between product location and product sales and exit. A further limitation concerns the difficulty to gather additional information on firms themselves in order to measure firm competencies in a more articulated way. Future work will explore ways of connecting unsystematic yet existing financial data on firms with information on firm patenting activity. Although challenging, the connection of these various datasets with the one explored in this paper should prove particularly fruitful in elucidating product location in fast growing industries.

References

Bhatt S (1987) Strategic product choice in differentiated markets. J Ind Econ 36(2):207–216

Bonanno G (1987) Location choice, product proliferation and entry deterrence. Rev Econ Stud 54:37–45

Brander J, Eaton J (1984) Product line rivalry. Am Econ Rev 74:323–334

Breschi S, Lissoni F, Malerba F (2003) Knowledge-relatedness in firm technological diversification. Res Policy 32:69–87

Buyus BL, Putsis WP Jr (1999) Product proliferation: an empirical analysis of product line determinants and market outcomes. Mark Sci 18:137–153

Christensen CM, Rosenbloom RR (1995) Explaining the attacker's advantage: technological paradigms, organisational dynamics and the value network. Res Policy 24:233–257

Christensen KJ, Haas LC, Noel FE, Strole NC (1995) Local area networks. Evolving from shared to switched access. IBM Syst J 34:347–374

Eaton BC, Lipsey RG (1979) The theory of market pre-emption: the persistence of excess capacity and monopoly in growing spatial markets. Economica 42:149–158
Farrell J, Shapiro C (1988) Dynamic competition with switching costs. Rand J Econ 19:123–137
Forman C, Chen P (2004) Switching costs and network effects in the market for routers and switches. Mimeo. Graduate School of Industrial Administration, Carnegie Mellon University
Gawer A, Cusumano M (2002) Platform leadership: how Intel, Microsoft and Cisco drive industry innovation. Harvard University Press, Boston
Gilbert R, Newberry H (1982) Pre-emptive patenting and the persistence of monopoly. Am Econ Rev 72:514–526
Henderson R, Clark K (1990) Architectural innovation: the reconfiguration of existing product technologies and the failure of established firms. Adm Sci Q 35:9–30
Hotelling H (1929) Stability in competition. Econ J 39:41–57
Khanna T (1995) Racing behaviour. Technological evolution of the high-end computer industry. Res Policy 24:933–958
King AA, Tucci CL (2002) Incumbent entry into new market niches: the role of experience and managerial choice in the creation of dynamic capabilities. Manage Sci 48(2):171–186
Klemperer P (1995) Competition when consumers have switching costs: an overview with applications to industrial organization, macroeconomics, and international trade. Rev Econ Stud 62(4):515–539
Klepper S, Simons KL (2000) Dominance by birthright: entry of prior radio producers and competitive ramifications in the U.S. television receiver industry. Strateg Manage J 21:997–1016
Lerner J (1997) An empirical exploration of a technology race. Rand J Econ 28(2):228–247
Levinthal DA (1997) Adaptation on rugged landscapes. Manage Sci 43(7):934–950
Mitchell W (1989) Whether and when? Probability and timing of incumbent's entry into emerging industrial subfields. Adm Sci Q 34:203–230
Nelson R, Winter S (1982) An evolutionary theory of economic change. Harvard University Press, Cambridge
Penrose E (1959) The theory of the growth of the firm. M.E. Sharpe, White Plains, NY
Reinganum JR (1983) Uncertain innovation and the persistence of monopoly. Am Econ Rev 73:741–748
Schmalensee R (1978) Entry deterrence in the Ready-to-Eat breakfast cereal industry. Bell J Econ 9:305–327
Shaked A, Sutton J (1982) Relaxing price competition through product differentiation. Rev Econ Stud 49:3–13
Spence AM (1976) Product selection, fixed costs and monopolistic competition. Rev Econ Stud 43:217–235
Stavins J (1995) Model entry and exit in a differentiated-product industry: the Personal Computer market. Rev Econ Stat LXXVII(4):571–584
Teece D (1986) Profiting from technological innovation: implication for integration, collaboration, licensing and public policy. Res Policy 15:285–305
Teece D, Rumelt JRP, Dosi G, Winter S (1994) Understanding corporate coherence: theory and evidence. J Econ Behav Organ 22:1–30
Thomas L, Weigelt K (2000) Product location choice and firm capabilities: evidence from the U. S. automobile industry. Strateg Manage J 21:897–909
von Burg U (2001) The Triumph of Ethernet. Stanford University Press, Stanford

Part 3:

Network externalities and networks of innovators

Daniel Birke · G. M. Peter Swann

Network effects and the choice of mobile phone operator

Abstract This paper explores the role of network effects in the consumer's choice of mobile phone operators in the UK. It contributes to the existing literature by taking a new approach to testing for direct network effects and by using individual-level data, which allows to analyse the impact that the immediate social network has on consumer choice in network markets. For our empirical analysis we use two sources of data: market-level data from the British telecommunications regulator OFCOM and micro-level data on consumers' usage of mobile telephones from the survey, *Home OnLine*. We estimate two classes of models which illustrate the role of network effects. The first is an aggregate model of the comparative volume of on-net and off-net calls. This finds that the proportion of off-net calls falls as mobile operators charge a premium for off-net calls, but even in the absence of any price differential between on-net and off-net, there is still a form of *pure* network effect, where a disproportionate number of calls are on-net. The second is a model of the individual consumer's choice of operator. This finds that individual choice shows considerable inertia, as expected, but is heavily influenced by the choices of others in the same household. There is some evidence that individual choice of operator is influenced by the total number of subscribers for each operator, but a much stronger effect is the operator choice of other household members.

We are grateful to Hilary Anderson from OFCOM, the Institute for Social and Economic Research, University of Essex, the ESRC Data Archive and Nicoletta Corrocher for help with data. We are also grateful to Chris Easingwood, Gautam Gowrisankaran and Francesco Lissoni, as well as participants at the Schumpeter Society conference in Milan, Italy and seminar participants at Manchester Business School, Chimera and the University of Nottingham for helpful comments. Daniel Birke would like to acknowledge financial support from the University of Nottingham Business School and the ESRC. The analysis is conducted using the econometrics package STATA.

D. Birke (✉) · G. M. P. Swann
Doctoral Programme, Jubilee Campus, Nottingham University Business School,
Wollaton Road, Nottingham, NG8 1BB, UK
E-mail: mail@daniel-birke.de

Keywords Network effects · Social networks · Mobile telecommunications · Discrete choice

JEL Classification D12 · L96 · M31

1 Introduction

The objective of this paper is to explore the role of network effects in the consumer's choice of mobile phone operators. After the seminal article of Rohlfs (1974), and the influential papers of Katz and Shapiro (1985) and Farrell and Saloner (1985), there has been a plethora of theoretical studies into the nature of network effects and by now network effects theory has reached a rather mature state. However, empirical work in this area has been slow to keep track with the advances in theory, and it is only comparatively recently that such studies have appeared in any numbers. Recent empirical studies include Goolsbee and Klenow (2002) on home computers, Berndt et al. (2003) on anti-ulcer drugs, Dranove and Gandal (2003) on DVD-players, Rysman (2004) on yellow pages and Gowrisankaran and Stavins (2004) on electronic payment.

The literature on network effects usually distinguishes between two types of network effects: direct network effects and indirect effects. Direct network effects refer to the case where users benefit directly from other users of the same network. In mobile communications, a direct network effect arises when the user can call a larger set of persons. Indirect network effects, on the other hand, arise, because bigger networks support a larger range of complementary products and services. In second generation mobile networks (e.g. GSM), indirect network effects are only of second-order significance, but they will play an increasing role after the introduction of third generation networks (e.g. UMTS), where usage will be heavily influenced by the availability of data services. During the period that we study, third generation networks have not been introduced yet and direct network effects are therefore the predominant form.

While it is widely acknowledged that network effects are a key feature of telecommunications industries, and indeed that telecommunications networks provide perhaps the leading example of network effects, relatively few studies have analysed the empirical importance and extent of network effects in the telecommunications market. Gruber and Verboven (2001) have studied the related topic of benefits to standardisation in telecommunications. The UK Competition Commission (2003) examined the role of network externalities in their enquiry on prices for mobile call termination in the UK: here, a stated preference approach has been taken to estimate the value of network externalities. Directly related to our research are the studies from Kim and Kwon (2003) on network effects in the Korean market, from Grajek (2003) on network effects in the Polish market and Doganoglu and Grzybowski (2004) on network effects in the German mobile telecommunications market.

The aim of our research is to empirically analyse the determinants of network effects and its importance for operator choice. To resolve this question, we follow the approach in Swann (2002). There, it is argued that many (or perhaps even, *all*) other users are a potential source of network effects, but some users matter more than others.

Most of the empirical literature on network effects so far has focused on markets with indirect network effects. Markets with direct network effects present the researcher with an additional challenge. A similar approach to the one typically taken for estimating indirect network effects would be to regress current network size on previous network size. However, economies of scale then results in an obvious endogeneity problem.

The first aim of this paper is therefore to test for network effects in the mobile telecommunications market and thereby demonstrating an alternative way of testing for direct network effects. We overcome the aforementioned problem by not analysing the adoption of mobile phones per se, but by analysing usage patterns. By doing so, we intend to demonstrate that there are network effects which are induced by mobile phone operators, i.e. we will test whether consumers react to the price differential between calls to other networks (off-net calls) and calls to the same network (on-net calls). Charging higher prices for off-net calls is a strategy pursued by network operators in many countries and these induced network effects are closely related to what Laffont et al. (1998) call "tariff-mediated network effects."

The second aim of this paper is to probe the determinants of network effects a bit deeper. This is done by analysing the question of which network is relevant to the consumer when choosing between different network operators. Is it the total number of subscribers of a particular network or is it rather the choice of people in the immediate social network? This question is of particular relevance, as an often assumed equivalence between direct and indirect network effects hinges on the assumption that only overall network size matters and not *who* is on the network.

The paper is organised as follows: Section 2 gives a brief introduction to the mobile telecommunications industry in the United Kingdom and discusses some of its characteristics that are important for our analysis. Section 3 introduces the data used in our analysis. It consists of market level data from OFCOM and micro-level data on consumers' usage of mobile telephones from the survey, *Home OnLine*. Sections 4 and 5 contain our statistical results. Section 4 describes a model of the comparative volume of on-net and off-net calls using aggregated market level data to test whether network effects are present in mobile telecommunications. Section 5 describes a discrete choice model of operator choice using survey data which examines whether network effects not only affect usage behaviour, but also play a role in individual choice of operator. Section 6 discusses the results and highlights some interesting possibilities for future research.

2 The mobile telecommunications industry in the UK

The analysis of this paper is conducted based on data from the UK telecommunications industry and more specific on data from the four main GSM-operators Vodafone, O_2, T-Mobile and Orange.[1,2] In the United Kingdom, O_2 (Cellnet) and Vodafone started operation in 1985 with analogue mobile networks.

[1] See Valletti and Cave (1998) for an analysis of the UK market from 1985 to 1998 and more background information.

[2] The other two important operators today are Virgin, which started at the end of 1999 and uses T-Mobile's network and "3" which started in 2003 and is building up its own 3G-network. At the end of our study period, Virgin had over half a million subscribers, but accounted for less than 2% of the market.

There had been relatively slow growth until the entry of T-Mobile (One-to-One) and Orange after 1993 introduced stronger competition to the market. However, the market really took off with the widespread use of prepaid cards, which made mobile telephony attractive for the mass market and especially for low-usage consumers. Although a first prepaid tariff was launched by Vodafone in September 1996, prepaid usage became popular only after mid 1998.

Especially interesting for our analysis is the development of market shares (see Fig. 1). At the end of 1998, the market was dominated by the incumbent operators O_2 and Vodafone, which together accounted for almost 70% of the market. However, by the beginning of 2001 this lead has dissipated and subscriber market shares have been levelled. Today, the market is about equally split between the four GSM operators.[3]

The ability of T-Mobile and Orange to catch up with the incumbent operators is somewhat unique to the UK market and is different from, for example, the German market in which the two biggest operators (T-Mobile and Vodafone) still control about 80% of the market and reported stable market shares for the last years. In most European markets, number of subscribers shows a convergence tendency. However, this has not resulted in a complete levelling of market shares like in the UK and in some cases (e.g. Blu in Italy), smaller companies have stopped operating altogether. With strong network effects present in the market, one would regard a development as in the UK as unlikely, as network effects result in a strong tendency towards higher market concentration. It could be argued that the development in the UK market is due to the high compatibility between networks. However, as our analysis will show, network effects do play an important role in the adoption of mobile telephones and in operator choice. An alternative explanation would be the existence of high switching costs in the market, which make it more profitable for incumbents to exploit their existing locked-in customer base.

3 Data

3.1 Data sources

This study uses market data from the UK telecommunications regulator OFCOM (formerly OFTEL) and a survey entitled *Home OnLine* with data on individual mobile phone usage. The OFCOM data consists of quarterly time-series market data on number of subscribers and prices for the four main operators. Number of subscribers are reported on a voluntary basis by the four main UK GSM-network operators and prices are based on market research commissioned by OFCOM. Furthermore, we have aggregate price and volume data (number of minutes) for on- and off-net calls, which we use for Section 4.

The *Home OnLine* survey was conducted in three waves (October to December 1998, January 2000, and February 2001) by the Institute for Social and Economic Research, University of Essex and was sponsored by BT. The survey data (Brynin 2002) can be accessed from the UK Data Archive (http://www.data-archive.ac.

[3] Note that this holds for subscriber market shares. Although there has been a similar trend in revenue market shares, Vodafone still boasts the highest revenue, as its customers generate a higher average revenue per user (ARPU).

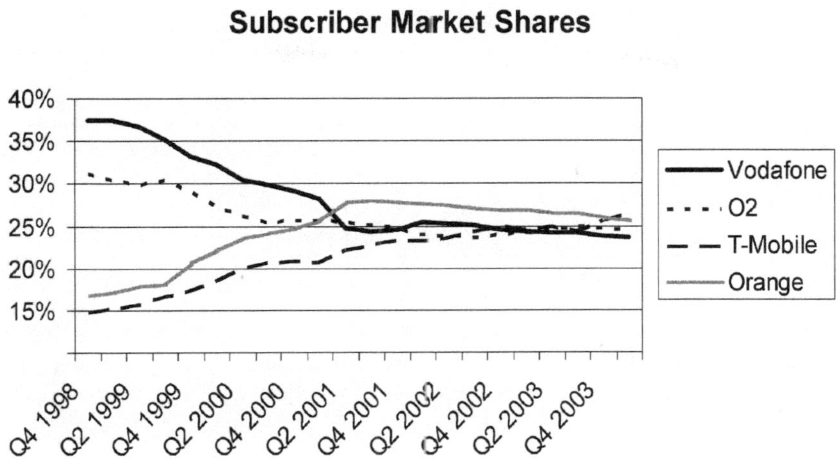

Fig. 1 Development of subscriber market shares

uk/). It consists of data on information and communications technology (ICT) access and usage behaviour from 1,000 households. For each wave, there are about 1,500 individuals from these households that were 16 or older and for whom there is data on mobile phone usage available. The most interesting variables for our purpose are information on mobile operator chosen, demographic information (socio-economic group, age, sex etc.) and household constitution.

The sample of the *Home OnLine* survey was selected using two-stage clustered sampling with stratification to ensure the inclusion of geographically clustered areas with representation of different social strata close to that of the population. Selection of households was random within these clusters. However, there is one important exception. BT required overrepresentation of homes with a PC. The

Table 1 Market shares and penetration rate

	Wave 1		Wave 2		Wave 3	
	Home OnLine[a] (%)	Actual (OFCOM) (%)[c]	Home OnLine (%)	Actual (OFCOM) (%)	Home OnLine (%)	Actual (OFCOM) (%)
Vodafone	36.0	37.5	29.2	32.3	25.4	28.3
O_2	31.4	31.1	31.6	27.2	29.9	25.7
T-Mobile	11.0	14.8	15.2	18.5	15.6	20.7
Orange	18.9	16.6	23.9	22.0	29.1	25.4
Penetration rate[b]	26.5	27.0	53.7	46.0	70.9	67.0

[a] Operator market shares do not sum up to 100% as 12 respondents reported that they have another operator in the first wave
[b] OFCOM penetration rates are based on market research. Aggregated figures can not be used, as some users may subscribe to more than one operator
[c] January 1999

Table 2 Demographic statistics for wave 1 (October to December 1998)

Wave1	Sample (N=1,720)	Vodafone (N=164)	O_2 (N=143)	T-Mobile (N=50)	Orange (N=86)
AGE	44.70 (17.94)	39.55 (12.08)	40.53 (13.43)	36.3 (11.99)	34.48 (11.38)
MALE	0.46 (0.50)	0.59 (0.49)	0.58 (0.50)	0.54 (0.50)	0.57 (0.50)
MRSCODE	2.72 (1.29)	2.03 (1.07)	2.17 (1.14)	2.41 (1.16)	2.19 (1.00)
HHINC	1,997 (1,570)	3,238 (2,088)	2,831 (1,607)	2,238 (1,678)	2,846 (1,357)

Figures in brackets are standard errors

selection process ensured that the sample contained 50% of homes with a PC. Interviews were conducted face to face in wave one and by telephone in waves two and three. A full description of the sampling process can be found on the UK Data Archive website.

3.2 Descriptive statistics

Market penetration has risen considerably in recent years and the strongest growth happened in the period from the end of 1998 to the beginning of 2001, which is also covered by the survey we use. During this period, the penetration rate of mobile telephony in our sample increased from 26.5 to 70.9% (see Table 1). Whereas, users of mobile telephones were still a minority at the end of 1998, mobile phones were almost ubiquitously used two and a half years later. Table 1 compares the market shares observed in our sample with the market shares reported by OFCOM. Market shares are roughly the same, although T-Mobile users are slightly underrepresented and Orange users are slightly overrepresented in the Home OnLine sample.

Table 2 presents summary statistics for the individual-specific variables that we use in our model in Section 5. Apart from monthly household income (HHINC) there are few missing values for these variables. Looking at Table 2, it can be stated that at the end of 1998, adopters were superproportionally young, male and from a higher socio-economic group.[4] Furthermore, customers of T-Mobile have a significantly lower household income (and come from a significantly lower socio-economic group) than Vodafone customers. On the other hand, O_2 and Vodafone customers are on average older than their counterparts at T-Mobile and Orange, which does not surprise given the longer time in the market for O_2 and Vodafone.

Similar tendencies persist also in the second and third wave (see Table 3 and Table 4 respectively). Most interesting is that more and more consumers come from lower socio-economic groups and that the predominance of male adopters is getting smaller.

[4] Socio-economic group (MRSCODE) can take on values from 1 (AB) to 5 (E) with 5 being the lowest group.

Table 3 Demographic statistics for wave 2 (January 2000)

Wave2	Sample (N=1,522)	Vodafone (N=236)	O_2 (N=255)	T-Mobile (N=123)	Orange (N=193)
AGE	44.90 (16.83)	39.17 (13.40)	41.29 (14.30)	36.83 (14.92)	41.03 (14.98)
MALE	0.45 (0.50)	0.50 (0.50)	0.52 (0.50)	0.43 (0.50)	0.47 (0.50)
MRSCODE	2.91 (1.32)	2.50 (1.16)	2.64 (1.26)	2.80 (1.20)	2.69 (1.16)
HHINC	2,008 (3,292)	1,934 (1,406)	3,448 (6,695)	1,675 (1,247)	2,180 (1,131)

Figures in brackets are standard errors

4 Aggregate model of network effects

4.1 Introduction

Mobile networks are highly compatible with each other from a technological point of view. Those using one operator can without difficulty call those using another operator. It would therefore be wrong to say that choice of a popular operator yields greater network effects simply because that operator has a larger network of users. Network effects in this market are typically *induced* by network operators when they charge higher prices for off-net than for on-net calls. As noted above, the network effects can be called 'tariff-mediated' network effects. The mobile phone user whose friends and contacts all use the same mobile operator will enjoy low bills for the calls (s)he makes to this group. By contrast, the mobile phone user whose friends and contacts all use different operators will suffer higher bills for these calls.

Most of the empirical literature on network effects so far focused on the analysis of markets with indirect network effects (like CD-Players and CDs, VHS-recorders and cassettes etc.). Mobile telecommunications is at the moment still characterised by a prevalence of direct network effects, which makes an estimation of network effects more complicated. As mentioned earlier, one approach to estimating network effects is to regress current network size on past network size. However, network size and prices are closely interrelated, as bigger network sizes drive down prices (economies of scale) and lower prices increase demand. This makes this approach to measuring network effects prone to estimation errors. Furthermore, there are also no good instruments readily available for prices. Input prices, for example, arguably play a smaller role than in other industries and these prices are also difficult to measure due to economies of scale and fast technological advances in the supplier market.

When network effects are mediated by tariffs, as described above, another approach to measuring network effects is to see whether the relative proportions of on- and off-net calls are associated with the relative prices of on-net and off-net calls.[5] If these induced network effects work as expected, we can expect a stronger ratio of on-net to off-net call volumes when the price of off-net calls is relatively high. This could arise for two related reasons. Firstly, when users are aware that off-net calls are more expensive than on-net calls, they may try to limit the frequency and length of off-net calls—for example, by switching to landlines for longer calls. Secondly, users may try to coordinate the choice of mobile operator with their

[5] Off-net calls are calls made to other mobile networks. Landlines are not included in the analysis.

Table 4 Demographic statistics for wave 3 (February 2001)

Wave3	Sample (N=1,503)	Vodafone (N=262)	O_2 (N=308)	T-Mobile (N=161)	Orange (N=300)
AGE	46.49 (17.29)	43.20 (15.00)	42.20 (15.51)	41.51 (15.67)	41.79 (15.70)
MALE	0.45 (0.50)	0.54 (0.50)	0.47 (0.50)	0.47 (0.50)	0.45 (0.50)
MRSCODE	2.94 (1.35)	2.62 (1.25)	2.66 (1.23)	2.92 (1.35)	2.70 (1.21)
HHINC	2,089 (2,195)	2,391(1,989)	2,781 (3,500)	1,884 (1,410)	2,154 (1,517)

Figures in brackets are standard errors

friends and associates in order to be on the same network as most of their peers. This effect might not have been very strong in the early years of mobile telephony when the majority of calls were made to landlines, but is likely to have grown in later years.

4.2 The model

The model developed below describes the ratio of off-net to on-net calls, using OFCOM data. In particular, the model compares the actual ratio of off-net to on-net calls, and the ratio we would expect if calls from one operator were spread evenly across the other operators' networks.

Suppose that there are four operators (i=1,...,4) and that the *actual* pattern of calls between networks is given by a 4*4 matrix with elements w_{ij} (where w_{ij} is the share of volume of calls from network i to network j). Now we do not have *actual* data on all the elements of this matrix, but we do have data on the actual volume of on-net calls (the sum of the elements on the main diagonal) and the actual volume of off-net calls (the sum of the off-diagonal elements). If we express these as percentages of the total volume of calls, then the *actual* proportion of on-net calls is given by:

$$Vol_{on}^{actual}(\%) = \sum_{i=1}^{4} w_{ii} \tag{1}$$

and the *actual* proportion of off-net calls is given by:

$$Vol_{off}^{actual}(\%) = \sum_{i=1}^{4} \sum_{\substack{j=1 \\ j \neq i}}^{4} w_{ij} \tag{2}$$

Furthermore, suppose that the market share of each operator is given by m_i. If there are no differences in prices for off-net and on-net calls and if the calls from one operator's network are spread evenly across the other networks (in proportion to the sizes of these networks) then we can calculate the *expected* pattern of calls as shown in Table 5.

Table 5 Expected shares (by volume of calls)

		To network:			
		1	2	3	4
Calls from network:	1	m_1m_1	m_1m_2	m_1m_3	m_1m_4
	2	m_2m_1	m_2m_2	m_2m_3	m_2m_4
	3	m_3m_1	m_3m_2	m_3m_3	m_3m_4
	4	m_4m_1	m_4m_2	m_4m_3	m_4m_4

From this matrix, we see that the *expected* proportion of on-net calls is given as:

$$Vol_{on}^{\exp ected}(\%) = \sum_{i=1}^{4} m_i m_i \tag{3}$$

and the *expected* proportion of off-net calls is:

$$Vol_{off}^{\exp ected}(\%) = \sum_{i=1}^{4} \sum_{\substack{j=1 \\ j \neq i}}^{4} m_i m_j \tag{4}$$

If there are no network effects present, we would expect the actual on-net call volume percentage to be close to the sum of the main diagonal of the matrix. During the period we study, this expected on-net call volume percentage is between 25% and 28%. Figure 2 shows the actual development of on- and off-net calls since the beginning of 1999. Whereas about an equal amount of on- and off-net calls were made in the beginning, the on-net call volume increases considerably faster afterwards. For the whole period the percentage of on-net calls is above 50%—far higher than the expected values calculated.

Figure 3 depicts the development of the ratio between prices for off-net calls and for on-net calls. In early 1999, off-net calls were about twice as expensive as

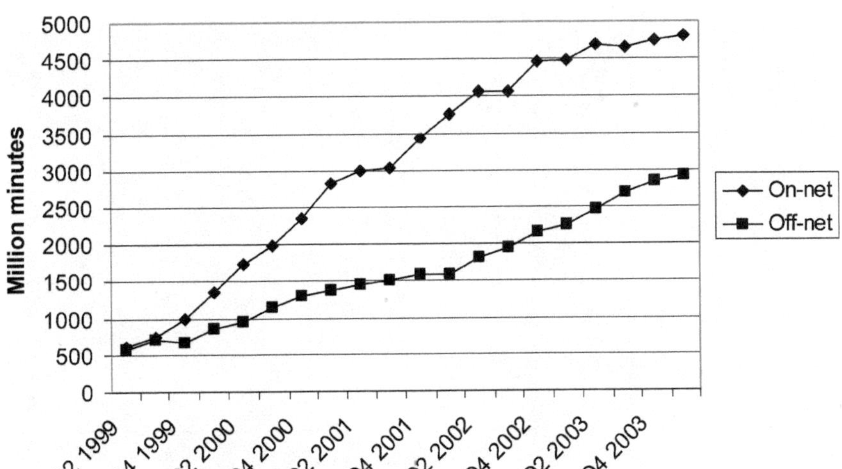

Fig. 2 Development of on- and off-net call volumes. Based on data from OFCOM

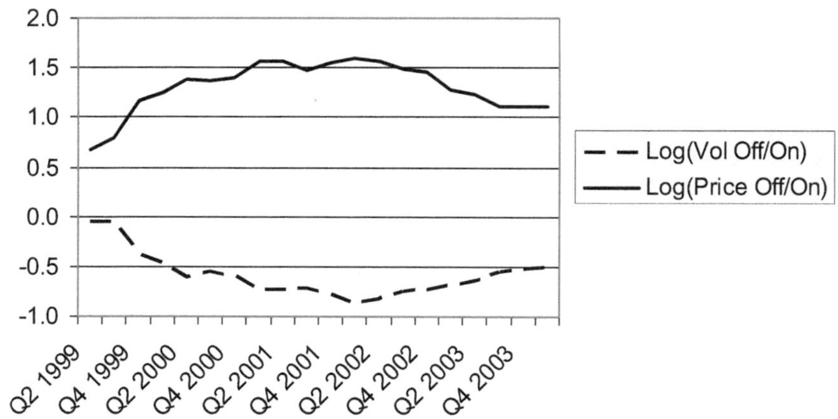

Fig. 3 Prices and the ratio between off- and on-net calls

on-net calls (19 pence per minute compared to 10 ppm). Two years later, off-net calls were about five times more expensive (26 ppm compared to 6 ppm). Afterwards, a decrease in the price ratio can be observed, but prices for off-net calls were still about three times higher in early 2004 (16 ppm compared to 5 ppm).

In a first step, we can plot the difference between the actual and the expected ratio of on-and off-net call volumes $\log\left\{\frac{Vol_{off}}{Vol_{on}}\right\}^{actual} - \log\left\{\frac{Vol_{off}}{Vol_{on}}\right\}^{expected}$ (y-axis) against the price ratio between on- and off-net calls $\log\left\{\frac{P_{off}}{P_{on}}\right\}$ (x-axis, see Fig. 4).

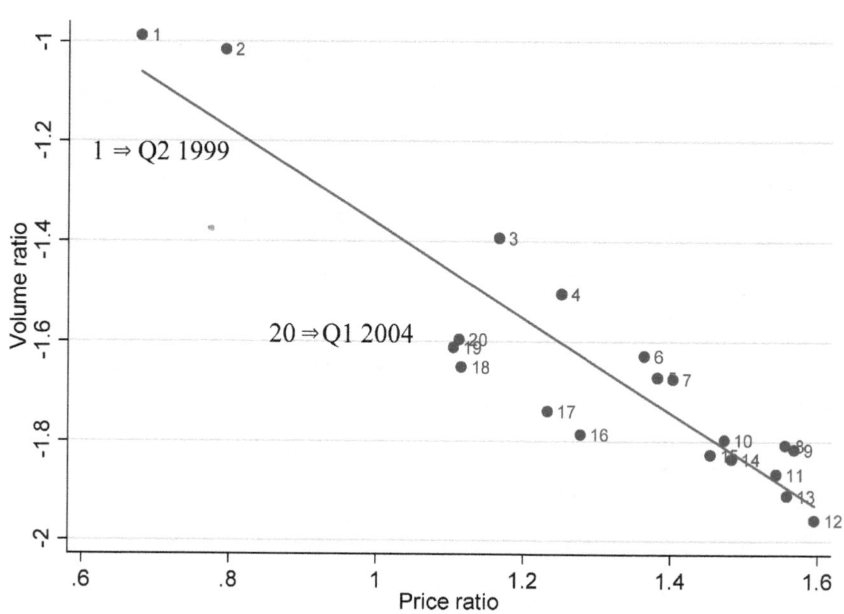

Fig. 4 Impact of relative price change in off-net calls on call volumes

There obviously is a very strong relationship and we then examine this relationship in the following regression:

$$\log\left\{\frac{Vol_{off}}{Vol_{on}}\right\}^{actual} - \log\left\{\frac{Vol_{off}}{Vol_{on}}\right\}^{\exp ected} = \alpha - \beta_1 \log\left\{\frac{P_{off}}{P_{on}}\right\} \tag{5}$$

This demand function is a very simple one. It takes no account of income or quality (e.g. network coverage and reliability). However, arguably neither of these vary much over the short period from which this sample is drawn. Moreover, even if call volumes in total increase with income, it is not clear why income should have much effect on the ratio of off-net to on-net calls.[6]

The parameter β_1 is a sort of price elasticity (relating to the premium for off-net calls). It describes how the off-net share is expected to fall as the premium for off-net calls rises. If there is no premium for off-net calls, then the ratio P_{off}/P_{on} is 1, and hence the right hand side of Eq. (5) reduces to α. If there is only a direct effect of prices on off-net calls, then we would expect α to be zero. If however, α were negative, that would imply that even in the absence of any price differential between off-net and on-net calls, a disproportionately large number of calls are on-net. This would be suggestive of a "pure" network effect, unrelated to the existence of price premia for off-net calls, which could be due to households already having coordinated their operator choice.

4.3 Estimation results

A regression based on Eq. (5) is less likely to suffer from a strong simultaneity bias than would be a regression using total network size as the dependent variable. Table 8 (Model 1) shows the results of this model using Ordinary Least Squares (OLS).

The DW statistic indicates that there may be an issue with autocorrelation in the model. To account for a possible dynamic misspecification of the model, we include the first lag of the dependent variable in the regression and test:

$$y_t = \log\left\{\frac{Vol_{off}}{Vol_{on}}\right\}^{actual} - \log\left\{\frac{Vol_{off}}{Vol_{on}}\right\}^{expected} = \alpha - \beta_1 \log\left\{\frac{P_{off}}{P_{on}}\right\} + \beta_2 y_{t-1} \tag{6}$$

The results are shown in Table 6 (Model 2). As the error term is heteroskedastic, we used robust standard errors. Using the alternative Durbin–Watson and the Breusch–Godfrey-test, we now cannot reject the null hypothesis of no serial correlation. The price elasticity β_1 is much lower than in Model 1, but is still highly significant. The value of 0.460 means that a 1% increase in the price premium for off-net calls leads to a 0.460% reduction in the dependent variable. The long-run

---

[6] We might however expect this ratio to depend on quality—for example, if the reliability of on-net calls is higher than the reliability of off-net calls. As we will argue later, reliability is comparable across networks.

Table 6 Regression results for off-/on-net call volumes

	Model 1	Model 2	Model 3
Constant α	−0.419 (0.12)**	−0.274 (0.095)*	−0.356 (0.042)**
Log (P_{off}/P_{on})	0.945 (0.096)**	0.460 (0.107)**	0.856 (0.032)**
y_{t-1}		0.482 (0.075)**	
t			−0.017 (0.001)**
Number of observations	20	19	20
R^2	0.843		0.985
F-test	96.66	130.38	538.56
Durbin–Watson (2,18)	0.175	Not applicable	1.562
Alternative DW F/Prob.>F		0.511/0.4859	
Breusch–Godfrey F/Prob.>F		0.625/0.4414	

Figure in brackets are standard errors
*Significant at 5% level
**Significant at 1% level

effect of the price ratio is given by $\frac{\beta_1}{1-\beta_2} = .888$, which is very close to the estimated coefficient β_1 of model 1. In short, the observed ratio of off-net to on-net calls is reasonably sensitive to the price premium for off-net calls. The significant lag of the dependent variable demonstrates the inertia present in the consumers' adoption process towards price changes. This could be due to imperfect information and the switching costs present in the market.

Even more interesting than the interpretation of β_1, however, is the interpretation of α. α is significant and negative, which means that even if there is no price premium for off-net calls, there would still be a disproportionately large share of on-net calls. This suggests that there is a pure network effect here, unrelated to price.

Visual inspection of the simple regression from model 1 (see Fig. 4) also suggests a third model. If we fit a linear line based on Model 1, we get a very good fit. However, most of the early values lie above this line and most of the later values below, which suggests that there might be a time trend. Reestimating Eq. (5) and including a time trend, leads to the results displayed in Table 6 (Model 3).

All coefficients are highly significant and the overall fit is best. The DW-statistic is just outside the 95%-confidence interval (1.10;1.54) and the error term is homoskedastic. The significant time coefficient suggests that over time, we would expect the ratio of off-net to on-net calls to grow. The time coefficient can be seen as a proxy for an underlying process of users aligning their operator choice with their peers. We cannot forecast how long this process will work, but this process might not be easily revertible due to the switching costs present in mobile telecommunications. In other words, even after the price differential has vanished, we would expect a far higher share of calls to be on-net than off-net, which would also be a possible explanation for the significant coefficient α.

One limitation of our analysis of aggregate data is obviously given by the limited number of observations available to us. Because we have data on individual adoption of mobile telephony available, we can further probe into the determinants of network effects in mobile telecommunications, which will be done in the following section.

5 Network effects and operator choice

5.1 Introduction

Section 4 has shown that network effects play a role in the extent to which consumers make on- and off-net calls. We have suggested that consumers try to avoid high costs of off-net calls by coordinating operator choice with their peers. To get an intuition of the importance of intra-household coordination in the *Home OnLine* sample, Table 7 compares the number of operators per household to the total number of adopters per household.

For wave 3, we altogether have 1,502 respondents that we could include in our analysis. Of these 1,502, 472 did not use any mobile network operator and are not displayed in Table 7. Furthermore, 373 respondents live in a household with only the respondent having a mobile (many of those are one-person households). For those individuals, we do not have any data regarding their social network. The majority of the rest of the respondents seems to coordinate operators within other household members: Out of the 472 respondents living in households with two mobile phone users, 320 (68%) use the same operator and only 152 (32%) use different operators. As Table 7 shows, results are similar even if there are more than two adopters within a household.

Table 8 displays the distribution of operators under the assumption that operators are chosen independently within households. For a household with two adopters, we would only expect 124 (26%) people to live in households with only one operator instead of the 320 people (68%) observed.

As Table 7 and Table 8 show, results are similar even if there are more than two adopters within a household, which is a strong indication that operator choice is coordinated within households. This is also confirmed by a χ^2-test, which yields a test statistic of 1,045. With six degrees of freedom, the critical value is 12.6 and the hypothesis of independent choice is strongly rejected. In the following, we will set out to test this hypothesis more formally.

Table 7 Number of operators per household compared to total number of adopters per household (wave 3)

Number of adopters living in a household	Number of operators chosen per household			
	1	2	3	Total
1	373	0	0	373
2	320	152	0	472
3	56	75	18	149
4	20	4	12	36
Total	769	231	30	1,030

Table 8 Number of operators per household compared to total number of adopters per household (wave 3) if household members choose independently

Number of adopters living in a household	Number of operators chosen per household				
	1	2	3	4	Total
1	373	0	0	0	373
2	124	348	0	0	472
3	11	53	86	0	149
4	1	3	20	13	36
Total	509	404	106	13	1,030

5.2 The model

Assume that consumers can choose between four different networks. Suppose there are N people (j=1, ..., N), and for any person i, the total length of calls (s)he makes to person j is n_{ij}. Let $r(j)$ describe the mobile operator chosen by person j (r ε {1,...,4}). Let $p_s(r(j))$ describe the cost per minute incurred by someone using operator s in making a call to the network chose by person j. Furthermore, let $C_{\text{is,other}}$ be the cost for individual i using operator s for making all other calls (to landlines etc.). Then, the cost to i of making the calls (s)he wishes using operators s=1, ..., 4 is given by:

$$C_{is} = \sum_{j=1}^{N} n_{ij} p_s(r(j)) + C_{is,other} \tag{7}$$

In practice, we could not expect to know an individual's pattern of calls in such detail. For simplicity, three basic types of calls can be distinguished: off-net calls, on-net calls and other calls. Consequently, the consumer faces a vector of prices, depending on what kinds of calls are made. The induced network effects of mobile telecommunications affect the price difference between on- and off-net calls.

Let $V_{i,mobile} = \sum_{j=1}^{N} n_{ij}$ describe the total call volume that individual i makes to mobile phones, let $V_{i,\text{other}}$ be the volume of other calls (to landlines etc.) and let $\rho_{s,\text{off}}$ be the percentage of off-net calls. Note that $V_{i,\text{mobile}}$ and $V_{i,\text{other}}$ do not depend on operator choice, but $\rho_{s,\text{off}}$ does. Assuming that off-net calls have the same price regardless of which (other) network they terminate in, the total cost to i of making the calls (s)he wishes can be written as:

$$C_{is} = \rho_{s,off} V_{i,mobile} p_{s,off} + \left(1 - \rho_{s,off}\right) V_{i,mobile} p_{s,on} + V_{i,other} p_{s,other} \tag{8}$$

We do not have detailed enough data to take into account that the price ratio between on- and off-net calls may differ between operators. Consequently, we assume that the relation between p_{other}, p_{off} and p_{on} is constant across operators and rewrite Eq. (8) as:[7]

$$C_{is} = p_s\left(V_{i,mobile}\left\{\rho_{s,off}\pi_{off} + \left(1-\rho_{s,off}\right)\pi_{on}\right\} + V_{i,other}\pi_{other}\right) \tag{9}$$

where p_s is a price index per operator and π_{other}, π_{off} and π_{on} are indices capturing the relative price of the three call types. The most interesting parameter here is $\rho_{s,\text{off}}$, the percentage of off-net calls out of the total mobile-to-mobile calls. $\rho_{s,off}$ varies across operators and person i can minimise C_{is} with the use of two basic strategies:

1. by choosing the operator with the lowest general price level p_s or
2. by choosing the operator, for which $\rho_{s,\text{off}}$ (the percentage of off-net calls) is lowest.

The optimal choice for a consumer could be made, if (s)he had perfect information on all the persons j that (s)he wishes to communicate with. That is unrealistic even in a static framework, as it is not conspicuous, which operator is chosen by a particular person. In a dynamic framework neither the network choice of the communication partners nor the communication partners themselves are constant. However, the consumer has two observable proxies influencing the choice: First, the expected network size for each operator. Second, every consumer may have information on network choice of a limited number of family members and close friends.[8]

5.3 Model estimates

We use the data we have available on the price level per operator p_s, the network size per operator and the data on operator choice of other members of the same household without further specifying the exact structure of the influence of these variables on call expenses. Our dependent variable is operator choice and has four different unordered outcomes. Therefore, we use a mixed conditional logit/ multinomial logit model. On the one hand, we have data on characteristics of the choice. For every operator, we have data on the number of subscribers (SUBSCRIBERS) measured in millions, price level (PRICE)[9] and a counter indicating which networks were chosen how often by other members of the same

---

[7] It can be expected that smaller operators gain least by having large price differences between on- and off-net calls, as this would deter consumers from choosing these (smaller) networks. Indeed, when entering the UK market, "3" heavily advertised that there is no difference between on- and off-net calls for some of their tariffs. For the time considered in this study, there is no new entrant into the GSM market and consequently our assumption seems tenable.

[8] A third possibility to reduce total call expenses would be to reduce the length and/or frequency of calls to persons using a different network. This is only a relevant choice factor if this adoption process is not the same for every communication partner. For simplicity, we assume that this process does not affect operator choice.

[9] PRICE gives an indication of the price level of different operators as calculated by OFCOM. However, in reality consumers face a large variety of different price plans targeted at different consumer types and different usage behaviour. The PRICE variable therefore might have a relatively low signal to noise ratio.

household (HOUSE_NETWORK). This choice-specific data can be used in a conditional logit model as first introduced by McFadden (1973).

Note that there is no quality parameter in our model. The main variables we expect to have an impact on quality are network coverage, international roaming, reliability and customer service. Until 1998, network coverage was lower for Orange and especially T-Mobile.[10] Likewise, international roaming was possible in far more countries for Vodafone and O_2 users. However, for the period of study and especially for the third wave in 2001, all networks cover over 97% of the population and international roaming is possible in most countries. Reliability measures the percentage of successful call completion. Especially in the early years of mobile telephony, reliability was lower, because of a shortage of capacity. OFCOM reports that reliability now typically is higher than 95%; the only pronounced difference being a lower reliability of T-Mobile in Scotland and Wales. We therefore conclude that network quality is roughly equivalent between operators and do not include a quality measure in our analysis. This might bias the estimates of the price parameter, but prices are only of second order importance for our model.

Consumers are highly heterogeneous and there are large differences between highly mobile business users and users who have a mobile phone just for emergency purposes. We try to account for this variety through the inclusion of individual-specific characteristics in the multinomial logit part of our model. The parameters we included are socio-economic group (MRSCODE), age (AGE) and sex (MALE). Combining the MNL and the conditional logit part results in the following estimation equation (see also Maddala 1983):

$$\Pr\left(y_i = \gamma | x_i, z_i\right) = \frac{\exp\left(\gamma z_{ia} + \eta_a x_i\right)}{\sum_{j=1}^{4} \exp\left(\gamma z_{ij} + \eta_j x_i\right)} \tag{10}$$

where z_{ia} is the vector of choice-specific characteristics for the four alternatives a and x_i being the vector of individual-specific characteristics. Note that γ is constant across individuals, whereas η is not. This means that we get a separate estimated coefficient of the individual-specific characteristics for all four choice alternatives.

The model is estimated separately for each wave. Results for the three waves are similar. Table 9 (Model 1) shows the regression results for the third wave (beginning of 2001).

The results from the multinomial part of Table 9 (Model 1) reinforce the results obtained from the descriptive statistics in Section 3: there are some significant differences between operators with regard to age, sex and socio-economic group of their customers. T-Mobile users, for instance, come from significantly lower socio-economic groups than Vodafone users (the comparison group for the multinomial logit part) and Orange has significantly more female users than Vodafone.

Most interesting are the results from the conditional logit part. Although SUBSCRIBERS and PRICE have the expected signs, neither is statistically significant. However, the counter HOUSE_NETWORK capturing the impact of operator choice from other household members on the own operator choice is

[10] Data on network coverage and international roaming are taken from the trade journals *What Cellphone* and *What Mobile*.

Table 9 Determinants of operator choice

	Model 1	Model 2
HOUSE_NETWORK	1.440 (0.077)***	1.275 (0.082)***
SUBSCRIBERS	0.086 (0.162)	0.138 (0.051)***
PRICE	−0.016 (0.065)	−0.035 (0.022)
LAST_CHOICE		2.172 (0.109)***
MRSCODE_O2	0.058 (0.071)	
MRSCODE_ORANGE	0.114 (0.072)	
MRSCODE_T_MOBILE	0.192 (0.088)**	
AGE_O2	0.003 (0.005)	
AGE_ORANGE	−0.001 (0.005)	
AGE_T_MOBILE	−0.014 (0.007)*	
MALE_O2	−0.332 (0.181)*	
MALE_ORANGE	−0.384 (0.183)**	
MALE_T_MOBILE	−0.348 (0.217)	
No. of observations	1,030	1,030
Pseudo R^2	0.179	0.339
Log L	−1,172.6	−944.0
LR χ^2 (*df*)	510.5 (12)***	968.2 (4)***

Figures in brackets are standard errors.
*Significant at 10%-level
**Significant at 5%-level
***Significant at 1%-level

highly significant. If not only network choice but also call volumes are strongly influenced by household members, then this would support our hypothesis from Section 4 that network effects play an important role in determining the ratio between off- and on-net calls and that these network effects operate through peers choosing the same network operator in order to minimise the costs incurred by off-net calls.

As an alternative to the above regression model, we use a model where we drop the individual-specific characteristics and instead include a variable (LAST_CHOICE) indicating the operator choice in the previous wave (13 months earlier). This variable should capture most of the inter-individual heterogeneity, as it can be expected that choice determinants, like for example age, sex, socio-economic group, do not change dramatically from one wave to the next. This approach also captures some of the inherent inertia in operator choice: consumers are not completely free to choose their network operator at every wave, because they might be bound contractually and because of other switching costs present in mobile communications (number portability etc.).

Table 9 (Model 2) shows the results from this model estimation. The parameters already used in model 1 have consistent signs and stable coefficients. The estimate for LAST_CHOICE is highly significant and also the HOUSE_NETWORK parameter keeps its high explanatory power. The estimates of the coefficients for PRICE and SUBSCRIBERS are slightly higher in the second model, whereas the standard errors are lower. Although PRICE is still not significant (which might be due to the problem discussed in footnote 9), the SUBSCRIBERS variable is significant at the 1%-level. However, according to this estimate, roughly 9.2

million subscribers to a network have the same impact as one additional member from the same household being on the same network. This would mean that consumers not only take into account the choice of other household members, but also prefer networks with a higher number of subscribers. The estimates for Model 2 resulted in a higher log-likelihood ratio (LR) and a higher R^2 and Model 2 is therefore preferable to Model 1.

There is a wide variety of possible combination of operators used in the same household, but to get a more intuitive idea of how we can interpret the HOUSE_NETWORK results, Table 10 shows the predicted probabilities for a household member choosing O_2 depending on the number of other household member using O_2.

If there is not other O_2 user in the same household, we would predict that a respondent chooses O_2 with a probability of 23.4%. If there are three more O_2 users in the same household, this predicted probability increases to 93.3%. The variation within the first row (due to price and number of subscribers) is far lower than the variation within the first column (due to the choice of other household members). This indicates that operator choice of other household members influences operator choice much stronger than overall network size.

6 Discussion

The results obtained from the previous sections give a strong indication that network effects play an important role in mobile telecommunications. This role does not only have an impact on the adoption of the technology per se, but also on usage patterns of mobile phones and on operator choice. The results from Section 4 and the similar quality levels of the four operators further suggest that the induced network effect rather than an information contagion process leads to the coordination of operator choice. Furthermore, whereas learning effects played an important role in the computer adoption study of Goolsbee and Klenow (2002) and might also be important for the *adoption* of mobile phones, it is not clear why they should significantly affect *operator choice.*

This paper has hopefully shown the richness of impacts caused by network effects. Network effects in the mobile telecommunications market seem to work both at the aggregate level (overall network size matters) and at the micro-level (choice of peers matters). These different impacts are currently only partially explained by the existing economic literature. Especially the standard assumption

Table 10 Predicted probabilities of operator choice

	O_2(%)	Orange (%)	Vodafone (%)	T-Mobile (%)
No other O_2 users in household	23.4	27.6	27.7	21.3
One other O_2 users in household	52.2	17.2	17.3	13.3
Two other O_2 users in household	79.6	7.3	7.4	5.7
Three other O_2 users in household	93.3	2.4	2.4	1.9

The example shows the probabilities, if no other operator apart from O2 is used in the household and if the household member newly adopted a mobile. The results for the other operators are very similar that network effects grow linearly with aggregate network size rests on very strong assumptions (Swann 2002) and is not supported by our empirical evidence. The utility that an individual consumer derives from using mobile telecommunications rather depends on who of her potential calling partners has already subscribed and to which network. As mentioned earlier, roughly 9.2 million subscribers to a network have the same impact as one additional member from the same household being on the same network. This casts doubt on an equivalence between indirect network effects and direct network effects. Whereas in markets with indirect network effects consumers do not care who is on the same network, we have shown that consumers do care who is on the same network in a market with direct network effects.

The results are also especially interesting for network operators currently introducing third generation mobile networks. Although network effects will be even more complex for these networks, and will include indirect network effects arising from services offered over the networks, the strong reaction from consumers to changes in the price ratio of off- and on-net calls suggest that inducing network effects by operators has been a successful strategy. It can in particular be used by the incumbent operators to fend off challenges by new entrants, such as "3" in the UK and also by any operator gaining a lead over the other operators. Furthermore, we have shown how strongly operator choice is coordinated within households. This suggests that operators gaining an early lead in the 3G market have an advantage over later entrants and that it is important for operators to support this choice behaviour by their customers through appropriate pricing strategies.

From a regulatory perspective, network effects are seen as a reason for higher termination charges, as users of mobile networks benefit from additional users in every other network (Competition Commission 2003). High termination charges and high costs for off-net calls have been regarded in a recent ruling by the UK regulator OFCOM as being the result of significant market power that operators have on its individual networks. As our results suggest, the high price of off-net calls cannot only be a *result* of market power, but can be a significant *source* of market power, which can especially be used to pre-empt entry by new competitors. If high switching costs are present in mobile telecommunications, this market power would be highly stable once consumers have aligned their operator choice even after the price differential between on- and off-net calls has been lowered.

We see two main shortcomings of our current research. Firstly, we would expect operator choice not only to be influenced by members of the same household, but a consumer's social network in general. Getting a full picture of these effects would require data which includes information on relationships to people outside the household. Secondly, using a narrow definition of network effects, one could argue that household members share common characteristics—which are unobserved by the econometrician. We do not think that this is the reason behind the coordination of operator choice within household because of the results of Section 4 and because mobile phone networks are relatively homogeneous products (little quality differences, relatively similar brand images etc.) Nevertheless it would be interesting to compare the coordination of operator choice to the coordination of other consumer products. In a follow-up study (Birke and Swann (2005)), we aim at further exploring these two issues. We are indeed able to demonstrate that the social network matters for consumer choice and that consumers coordinate operator choice, but don't seem to coordinate choice of mobile handset brands.

References

Berndt ER, Pindyck RS, Azoulay P (2003) Network effects and diffusion in pharmaceutical markets: antiulcer drugs. J Ind Econ 51(2):243–270

Birke D, Swann GMP (2005) Social networks and choice of mobile phone operator. Nottingham University Business School, Industrial Economics Division Occasional Paper Series, No. 2005–14

Brynin M (2002) Home OnLine, 1998–2001 [computer file]. Colchester, Essex, UK Data Archive [distributor], SN: 4607

Competition Commission (2003): Vodafone, O_2, Orange and T-Mobile: reports on references under section 13 of the Telecommunications Act 1984 on the charges made by Vodafone, O_2, Orange and T-Mobile for terminating calls from fixed and mobile networks. Available at: http://www.competition-commission.org.uk/rep_pub/reports/2003/475mobilephones.htm (Accessed 14-06-05)

Doganoglu T, Grzybowski L (2004) Diffusion of mobile telecommunication services in Germany: a network effects approach. Munich: Ludwig Maximilian University, Available at: http://www.lrz-muenchen.de/~grzybowski/de-mobiles.pdf (Accessed 14-06-05)

Dranove D, Gandal N (2003) The DVD vs. DIVX standard war: empirical evidence of vaporware. J Econ Manag Strat 12 (3):363–386

Farrell J, Saloner G (1985) Standardization, compatibility, and innovation. RAND J Econ 16 (1):70–83

Goolsbee A, Klenow PJ (2002) Evidence on learning and network externalities in the diffusion of home computers. J Law Econ 45(2):317–343

Gowrisankaran G, Stavins J (2004) Network externalities and technology adoption: lessons from electronic payments. RAND J Econ 35(2):260–276

Grajek M (2003) Estimating network effects and compatibility in mobile telecommunications, WZB Discussion Paper SP II 2003-26, Available at: http://www.wz-berlin.de/mp/dps/abs/2003/ii2003-26.en.htm (Accessed 14-06-05)

Gruber H, Verboven F (2001) The evolution of markets under entry and standards regulation—the case of global mobile telecommunications. Int J Ind Organ 19(7):1189–1212

Katz ML, Shapiro C (1985) Network externalities, competition, and compatibility. Am Econ Rev 75(3):424–440

Kim HS, Kwon N (2003) The advantage of network size in acquiring new subscribers: a conditional logit analysis of the Korean mobile telephony market. Inf Econ Policy 15(1):17–33

Laffont JJ, Rey P, Tirole J (1998) Network competition: II. Price discrimination. RAND J Econ 29(1):38–56

Maddala GS (1983) Limited-dependent and qualitative variables in econometrics. Cambridge University Press, Cambridge

McFadden D (1973) Conditional logit analysis of qualitative choice behaviour In: Zarembka P (ed) Frontiers of econometrics. Academic Press, New York, pp 105–142

Rohlfs J (1974) A theory of interdependent demand for a communications service. Bell J Econ Manage Sci 5(1):16–37

Rysman M (2004) Competition between networks: a study of the market for yellow pages. Rev Econ Stud 71(2):483–512

Swann GMP (2002) The functional form of network effects. Inf Econ Policy 14(3):417–429

Valletti TM, Cave M (1998) Competition in UK mobile communications. Telecommun Policy 22 (2):109–131

R. Cowan · N. Jonard · J.-B. Zimmermann

Evolving networks of inventors

Abstract In this paper we develop a model of R&D alliance formation. Pairs of firms combine their knowledge in an attempt to innovate. Whether this attempt is successful depends in part on whether the pair has been successful in the past: accumulated experience teaches a pair of firms how to innovate together, but at the same time increases the similarity of their knowledge stocks. A tension exists between the desire for a familiar partner, and desire for a partner with complementary knowledge. How this tension is resolved depends on the nature of the innovation process itself, and the elasticity of substitution of different types of knowledge inputs in knowledge production. From the alliance-innovation process, a variety of networks form. In different parts of the parameter space we observe isolated agents, a dense, connected network, and small worlds.

Keywords Innovation · R&D collaboration · Networks · Knowledge · Computational experiment

JEL Classification L14 · Z13 · O3

1 Introduction

One of the long-standing legacies of Schumpeter is the view that innovation plays a central part in shaping the evolution of a dynamic economy. Also pivotal is Schumpeter's idea that innovation consists largely of the recombination of existing

---

R. Cowan (✉)
MERIT, University of Maastricht, P.O. Box 616,
6200 MD Maastricht, The Netherlands
E-mail: r.cowan@merit.unimaas.nl

N. Jonard
CNRS, CREA Ecole Polytechnique, 1 Rue Descartes,
75005 Paris, France
E-mail: nicolas.jonard@shs.polytechnique.fr

J.-B. Zimmermann
CNRS, GREQAM, EHESS, 2 Rue de la Charitè,
13002 Marseille, France
E-mail: Jean-Benoit.Zimmermann@univmed.fr

knowledge. The practical force of this idea is that if agents have access to more and a wider variety of knowledge or information, innovation and thus growth will be fostered.[1] Recently, changes in the technological landscape have made this issue more important than ever before. It has been argued that technologies both being used and being produced involve technological expertise that covers a much broader range of disciplines than has hitherto been the case.[2] What this implies is that the types of knowledge necessary to innovate and compete successfully can lie outside a firm's main area of expertise. One strategy to address this challenge is to look outside the firm's boundaries. Collaboration with other firms or other institutions more generally can serve to transmit knowledge, either codified or tacit. It is no surprise, then, that we observe a rapid increase in the number of inter-firm strategic alliances in recent decades.

Nevertheless, collaboration is risky in the sense that it is marked by uncertainty relative to both the skills of the partner and his reliability (Powell 1990, p. 318, provides a discussion of those risks). Successful collaboration demands mutual knowledge and sharing of routines, representations and ways of thinking, that is, a form of proximity that the experience of collaboration permits to build (Garcia-Pont and Nohria 2002). Partners have to learn to collaborate. The repetition of interactions creates information that permits partners to reduce uncertainty and increase predictability regarding each others' behaviour.[3] Also, through the repetition of interactions a common language develops, an intermediary-level knowledge specific to the partners, whose availability increases the efficiency of collaboration (see Galison 1999, for a study of the field of experimental physics, and how the dialogue between theorists, experimenters and instrument makers was made possible by the emergence of a jargon specific to their purpose).

This discussion suggests that in a variety of ways, after a successful interaction the two parties will find each other more attractive than they did before. Successful interactions increase partners' understanding of each others' motives, increase the awareness of each others' skills, and generate more common tacit knowledge which improves communication. Empirically, firms with a history of partnering are more likely to have alliances than those without (Powell et al. 1996), and two firms that have worked together in the past are more likely to choose each other as partners than to choose new firms (Roijakkers 2003).

At the same time, repeated collaboration tends to increase partners' similarity, possibly reducing their mutual attractiveness. Nooteboom (2004) notes that "... ongoing interaction will yield a reduction of cognitive distance". Mowery et al. (1998, p. 517) find that "... technological overlap between joint venture partners after alliance formation is greater than their pre-alliance overlap". (See also Dyer and Nobeoka 2000.) But increased similarity means that partners can have less to contribute to each other. In the extreme, if they become identical in what they know there is no reason for them to collaborate, no matter how much they trust each other or how smoothly they can interact.

While the strategic alliance tends to be a bilateral relationship, firms can have, over time, more than one partner. Thus in an industry in which strategic alliances

[1] On an application of that idea to a growth model, see for instance Weitzman (1998).

[2] See the discussion of that trend in Powell et al. (1996), Grandstand and Sjolander (1990), Grandstand (1996), and Teece and Pisano (1989).

[3] On that issue see Coriat and Guennif (1998), Sako (1991), Dodgson (1996) and the model by Klos and Nooteboom (2001).

are common, networks arise: collections of firms, each of which has some non-market relationship with a small number of firms in the industry. There has been considerable empirical interest in the structure or architecture of the networks that arise in different industries in different periods (see for example Powell et al. 2005 for a recent contribution).

In addition to this empirical interest, there is now a growing literature in economic theory on network formation, but the majority of it treats the problem in game-theoretic terms, looking for stable structures that emerge from agents' one-time strategic decisions about which links to form. In general in that literature it is difficult to provide an exhaustive description of the stable structures (a notable exception being Goyal and Joshi 2003), the studies tend to focus on the stability of some specific architectures (the star, the wheel, the complete network for instance) and a tension between stability and efficiency is often identified (for a recent survey, see Dutta and Jackson 2003). The results rest upon strong assumptions about what agents know (that is, everything) and very poorly address the adaptive, path-dependent nature of network formation and operation. By contrast, the model developed in this paper continues work using another approach, centered on the idea that agents are continually forming and breaking links with each other, modifying their own characteristics, and that this ongoing activity is what underlies the networks that we observe. Agents choose with whom to form a link in order to achieve some immediate goal which modifies their properties. The repetition of these actions results in the formation of a network that evolves over time.[4]

In this paper we incorporate effects of collaboration in a simple model of network formation. The goal is to understand network formation as the consequence of individual firms creating bilateral alliances in which they innovate. We abstract from many features of alliance formation and focus only on the effects that follow from the goal of knowledge production.[5] Firms combine their knowledge to create new knowledge, and the amount they can create is determined by the complementarities in their knowledge stocks. In addition, repeated interaction between a pair of firms increases the probability of success, but at the same time can change partners' abilities to complement each other. By repeated alliance formation and dissolution a network emerges and continuously evolves. How this structure and the nature of the economy's knowledge stock change in response to parameters governing innovation are the issues that we explore.

2 The model

Consider an industry in which each period a finite number of firms form alliances, with the goal of creating new knowledge. An alliance having formed, the partners pool their knowledge to create a joint stock and, by combining elements of that joint stock, seek to innovate. If the project succeeds, new knowledge is created and added to each partner's knowledge stock. Then the alliances are dissolved and the process is repeated.[6] Formally, the population of firms is denoted $S = \{1,\ldots,n\}$ (n is

[4] For related approaches to adaptive network formation, see for instance Cowan et al. (2003), Kirman and Vriend (2001), Plouraboué et al. (1998), and Weisbuch et al. (2000).

[5] See Oliver (1990) or Narula (1999) for discussions of the motives underlying alliance formation.

[6] The processes of knowledge pooling and production, described in detail below, are similar to those we use in Cowan et al. (2004b).

even). Each $i \in S$ is characterized by a knowledge endowment of $\ell \geq 2$ types of knowledge represented as a vector $\alpha(i)=(\alpha_l(i);l = 1,...,\ell)$, where $\alpha_l(i)$ represents the amount of knowledge of type held by i. Firms are thus treated as located in an ℓ-dimensional knowledge space.

2.1 Knowledge production

There are many ways to characterize knowledge, none of them without its pitfalls. The approach proposed here is simple and flexible. It obeys the following intuitive requirements. First, when innovation is jointly conducted, it is natural to expect that the post-innovation knowledge stocks held by the partners are larger than their pre-innovation stocks. Second, after innovation, the similarity of the knowledge profiles of the partners (that is the relative distance between them in the underlying knowledge space) has fallen.

Formally, when a partnership, ij, forms, i and j first combine their knowledge into a joint vector $\alpha(ij)$. This is done in each category $l = 1,...,\ell$ through

$$\alpha_l(ij) = (1-\theta)\min\{\alpha_l(i), \alpha_l(j)\} + \theta\max\{\alpha_l(i), \alpha_l(j)\}. \quad (1)$$

The joint knowledge vector becomes the input to the knowledge production process.

In Eq. 1, θ expresses the nature of the knowledge pooling which the knowledge creation task demands. If it is possible to separate the sub-tasks in the innovation process agents will specialize, each agent doing some sub-tasks, and bringing the results together at the end to create the complete innovation. The better econometrician will do the econometrics, the better game theorist will do the game theory, and the joint knowledge vector will consist of the maximum level of knowledge of each type. This is captured by setting θ equal to one. By contrast, if the tasks are not separable, and both partners must be involved in every sub-task, then the weaker partner will be a bottleneck: joint knowledge is the minimum of each type, and θ will approach zero.

The motivation for θ has been the separability of the innovation task, which maps nicely into the range $0 \leq \theta \leq 1$. However, there is a second interpretation of θ, which connects explicitly with the characteristics of partnerships. Do I look for partners who are different from me, or similar to me? We can address this by asking whether an agent prefers joint or autarchic innovation. If θ is small, joint knowledge is driven by minimum values. If j is worse than i in any knowledge type, then the joint profile looks worse than i's own profile. Similarly if i is worse than j anywhere, the joint profile is worse than j's profile. A weaker partner pulls me down. In the extreme, if θ is zero, I will only be interested in partners who are in no way worse than me. Since this is reflective, only identical firms can be partners. By contrast, if θ is large, no partner can make me worse off (since if you are worse than me anywhere, we use my knowledge there). Again this is reflective, thus the most natural partner is one who is strong where I am weak, and vice versa. By this reasoning, θ indicates a taste for dissimilar partners. In the literature there is some disagreement about whether innovative success increases or decreases with the distance between partners, i.e. whether a firm prefers similar or dissimilar partners. Mowery et al. (1998) find a U-shaped relationship, a conclusion also supported by

Nooteboom (2000). In the models by van Alstyne and Brynjolfsson (1997) and Peretto and Smulders (2002) on the other hand it is assumed that the relationship is monotonic, either increasing or decreasing. With our formalism, this effect is parametrized, and explicitly linked to the nature of the innovation task. Thus there is the possibility of a relatively complex relationship due to the multi-dimensional nature of firms' knowledge.

The joint knowledge vector serves as the input to the innovation process. To formalize this, we use a standard constant elasticity of substitution production function

$$\phi(\alpha(ij)) = \left[\sum\nolimits_l \alpha_l^{\beta}(ij)\right]^{1/\beta}. \tag{2}$$

The parameter $0<\beta\leq 1$ is an inverse measure of the elasticity of substitution across knowledge types, which is defined as $1/(1-\beta)$. When β is very small it is difficult to substitute between different types of knowledge, and the CES function approaches a fixed-coefficients production function: output is determined by the minimal input. Here, if an agent has a weakness in one knowledge type, he has strong incentives to find a partner who is strong there, in order to create a flatter input profile. By contrast, if β is large, it is relatively easy to substitute between different types of knowledge. If an agent has a weakness in one type, this can be offset if he is strong somewhere else. The incentive to find a partner is thus weaker. This suggests that incentives to find partners decrease with β, and consequently so should the amount of networking we observe.

2.2 Innovative success and experience

Innovation is an inherently uncertain activity. A project may fail, and the anticipated new knowledge may not be created. It also seems reasonable to suppose that firms will not undertake projects with extremely low probabilities of success. Thus, as an initial step, we bound the success probability of any project between $\underline{\pi}$ and $\overline{\pi}(0 < \underline{\pi} < \overline{\pi} < 1)$. However, the success probability of a partnership can change over time and here history is important, because project success is driven in part by familiarity of the partners, as discussed above.

Consider period t has just elapsed, and firms are to make decisions for the new period $t+1$. Formally, for all $1\leq s\leq t$ and all $i\neq j$, define the binary variable $\eta_s(ij)$ as taking the value 1 if there has been a successful collaboration between i and j, at time s and 0 otherwise. Then define

$$\gamma_t(ij) = \sum_{1\leq s\leq t} \eta_s(ij)\rho^{t-s}, \tag{3}$$

where $0<\rho<1$ is a discount factor, to be the measure of i and j's historical success. $\gamma_t(ij)$ is a discounted sum of past experiences. A history of joint success will improve a pair of firms' ability to work together, and so increase the probability of future successes. But experiences further in the past weigh less heavily in future success, and so are discounted. Now define $\pi_t(ij)$ as the probability that a collaborative attempt by i and j in period $t+1$ is a success, conditional on their history up to t. Success, as measured by $\pi_t(ij)$, should increase in $\gamma_t(ij)$ while remaining bounded by the minimum and maximum success probabilities, $\underline{\pi}$ and $\overline{\pi}$. There is no obvious choice for the functional relationship between $\pi_t(ij)$ and $\gamma_t(ij)$, so to maintain a gradual impact of $\gamma_t(ij)$ on $\pi_t(ij)$ we use a simple linear form.[7] Noting that

$$\gamma_t(ij) \leq \sum_{1 \leq s \leq t} \rho^{t-s} = \frac{1-\rho^t}{1-\rho}, \tag{4}$$

an upper bound to $\gamma_t(ij)$ is $1/(1-\rho)$. The functional form we assume for $\pi_t(ij)$ is then

$$\pi_t(ij) = \underline{\pi} + \gamma_t(ij)(\overline{\pi} - \underline{\pi})(1-\rho). \tag{5}$$

On this formulation, the first time two firms attempt to innovate together, because they have no history, their success probability is $\underline{\pi}$. By contrast, when a firm attempts to innovate alone, since there is no issue of learning to cooperate with oneself, we assume that its success probability is the maximum, $\overline{\pi}$.

Finally the expected amount of knowledge produced by a collaboration between i and j at time t can be expressed as

$$F(ij) = \pi_t(ij) \cdot \phi(\alpha(ij)). \tag{6}$$

This represents the quantity produced by a successful venture, multiplied by the probability of success.[8]

If the partnership succeeds in innovating, the new knowledge is added to the partners' respective knowledge stocks. The general intuition is that as an agent uses knowledge or is exposed to it, he will assimilate at least part of it, and thereby change the precise area of his expertise. We here assume that there is some path dependence in knowledge production, so output will resemble input. To simplify we assume that one type of knowledge is produced, but it resembles the input probabilistically:[9] the probability of the new knowledge being of type m is

$$\frac{\alpha_m(ij)}{\sum_l \alpha_l(ij)}. \tag{7}$$

2.3 Pair formation and equilibrium

We draw on the literature on matching problems for our basic model of pair formation. We have a single population of agents, and wish to allow for the

[7] An exponential specification was also tried, but the effect of changes in $\gamma_t(ij)$ was too marked, though similar behaviour could be seen to emerge. The linear form, on the other hand, produces smoother changes in the aggregate statistics.

[8] Rating potential partners according to their expected values is formally similar to the score approach in Klos and Nooteboom (2001). In their model the score of i is increasing with my trust in i's intention to honour agreements to the best of his ability (p. 514). This subjective belief however does not correspond to an objective risk, as in their model transactions never fail. We abstract from this sort of difficulty by simply viewing $\pi_t(ij)$ as a measure of i and j's ability to work together.

[9] This assumption regarding the distribution of new knowledge is not critical. For example, one might suppose that each partner creates slightly specific knowledge – partly associated with the joint profile, and partly associated with his idiosyncratic profile. Provided the weight on the joint profile is not trivial, the results are qualitatively identical.

possibility of autarchic innovation, or in this context, self-matching. Thus we have a generalization of the Roommate matching problem (Gale and Shapley 1962). Every agent ranks every agent in the economy (including himself) as a potential partner. Based on these rankings partnerships form, and the population is partitioned into q single agents and $(n-q)/2$ pairs. We use μ to denote such a matching, where $\mu(i)$ is the partner of i.[10] A matching is stable if there is no pair of agents who would block it. Formally, writing $j \succ_i k$ to express that j is preferred by i to k, a matching μ is stable if there is no pair $ij \notin \mu$ such that $j \succ_i \mu(i)$ and $i \succ_j \mu(j)$. That is, there is no pair ij who are not matched under μ but who prefer each other to their respective partners in μ.

As our agents are interested in knowledge creation, a natural way to rank prospective partners is by the expected amount of knowledge a partnership would produce: i prefers j to k if the partnership ij produces more knowledge (in expectation) than ik. Formally $j \succ_i k$ if and only if $F(ij) > F(ik)$, a calculation any agent is able to make to create a complete transitive ordering.[11]

In general, in a roommate matching problem, stable matchings do not always exist, and when they do they are not guaranteed to be unique. Our model, however, is a special case (though still relatively general) and a unique stable matching always exists due to the fact that both parties to a partnership assign it the same value (see Proposition 1 in Cowan et al 2004a). The general idea is that of all potential partnerships, there is one that yields the highest payoff (and both partners agree what the payoff is). This partnership must be part of any stable matching. A simple recursion of this observation generates the unique stable matching.

3 Computational experiment

We study a population of n=100 firms each endowed with a five-category knowledge vector. At the outset, individual knowledge endowments are independently and identically distributed over the unit interval. Each period, firms form pairs (or decide to stay alone) to conduct innovative activities. This list of pairs and singletons constitutes a stable matching in which the value of a pair (or singleton) is equal to the expected amount of knowledge produced by that pair (or singleton). Any new knowledge created within the partnership is added to both firms' existing knowledge. In addition, firms record their experience with their partners. Partnerships dissolve, and the next period begins. We iterate this process for 1,000 periods.

We are interested in understanding the effects of knowledge pooling (θ), substitutability in the production of new knowledge (β), and whether or not history, or learning about a partner, matters in success probabilities. We construct a 100×100 grid of equally spaced points in the (β, θ)-space (the unit square) and record data for the entire history of the industry, under the two learning regimes (history does or does not affect success probabilities) for which we set $\underline{\pi} = 0.9$ to $\overline{\pi} = 0.99$.

We examine two aspects of the emergent structure: the properties of knowledge accumulation and distribution; and the properties of the network. Our characterization of knowledge is as a multi-dimensional resource, thus to generate

---

[10] A matching is thus a bijection $\mu : S \rightarrow S$ such that $\mu(\mu(i))=i$ for all $i \in S$.

[11] In case of a tie we apply an arbitrary rule to guarantee that the score function is still strict: if $F(ij)=F(ik)$ then $j \succ_i k$ if and only if $j<k$.

statistics, it is useful to reduce this vector to a scalar. Since knowledge production is the assumed goal of our firms, the scalar we use is a firm's innovative performance $\phi(\alpha(i))$. By simply treating a firm's knowledge stock as the quantity the firm could produce in isolation, an accessible presentation of the results is obtained.

The results on network structure are slightly more involved. In any period the static network consists of q isolated agents and $(n-q)/2$ disconnected pairs, as given by the stable matching μ_t. To study the properties of the dynamic network, we record the list of connections active over time. This generates a weighted graph, in which the weight of an edge indicates how frequently the two partners have interacted. Formally, in each period t denote A_t the adjacency matrix associated with the stable matching μ_t, that is $A_t(i,j)=1$ iff $j=\mu_t(i)$. The weighted network recording past interactions is then denoted B_t, where $B_t(i,j)=\sum_{1 \leq s \leq t} A_s(i,j)/t$. To move from B_t to a proper adjacency matrix again, distances must be computed first. Define the distance $d(i,j)$ between two firms i and j as the number of edges in the highest frequency path linking them. Any path $i_0, i_1, \ldots, i_\xi$ with $i_0 = i$ and $i_\xi = j$ has an associated frequency $\prod_{l=1,\ldots\xi} B_t(i_{l-1}, i_l)$ and a length $\xi \geq 1$. There is a finite number of such paths, and thus a path with maximum frequency exists: its length ξ is defined as the distance $d(i,j)$. Two agents are neighbours if the distance between them is one, and the adjacency matrix unfolds. Using this reduction to an unweighted graph permits us to use standard descriptive statistics to characterize network structures.

The results are presented in two-panel figures, where each panel is a shaded contour plot. Such a plot reads as a map in an atlas: darker grey scales imply higher values on the z axis. Contours are added to make patterns clearer. This provides a compact display of the relationship between the substitutability of knowledge types in production (β), the pooling parameter (θ), and the performance measures we are concerned with. Left panels correspond to situations where past experience has no effect on future success (the initial or a priori success probability is $\overline{\pi} = 0.99$). Right panels display effects of the same parameters when past success increases the probability of future success, with a success probability increasing from $\underline{\pi} = 0.9$ to $\overline{\pi} = 0.99$ as past successes accumulate.

3.1 Knowledge

Firms are primarily interested in knowledge creation, and form partnerships precisely for that purpose. At the aggregate level, this leads us to ask how aggregate knowledge grows, whether knowledge is evenly distributed over firms, whether firms become specialists, and finally whether firms all become similar. Regarding the first question, any pattern in aggregate knowledge growth is driven entirely by the parameter β. Indeed, as β increases from 0 to 1 the marginal product of the knowledge inputs falls quite dramatically, particularly near zero. Thus aggregate knowledge as measured by total innovative performance simply increases as β falls.

3.1.1 Knowledge distribution

Over the history of the economy, each agent creates a certain amount of new knowledge. As agents begin this history with similar knowledge levels, final levels

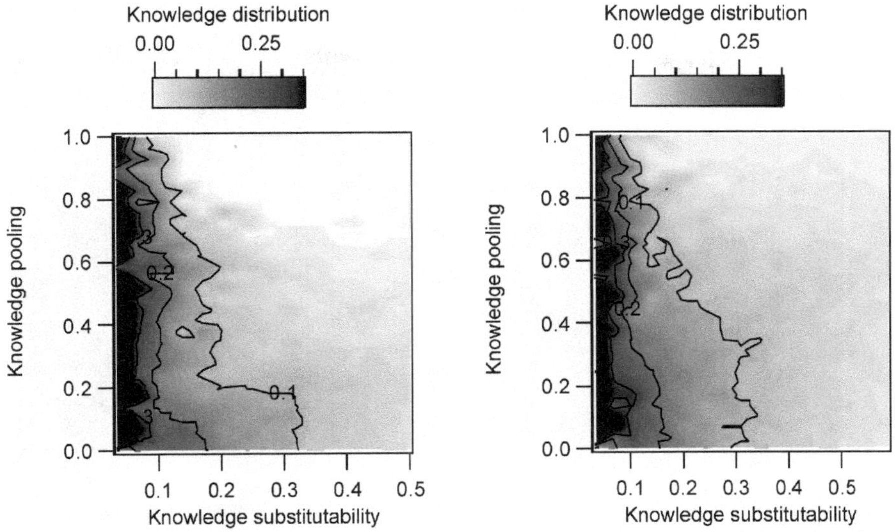

Fig. 1 Coefficient of variation of innovative performance in the (β,θ)-space (*left*: no learning; *right*: learning)

can be used to examine the distribution of knowledge generated by innovation. Figure 1 shows the relationship between the coefficient of variation of final knowledge levels and the pair (β, θ).

The general patterns are the same when experience matters and when it does not. Inequality in knowledge distribution is decreasing along the β axis with a sharp peak when β is close to 0, with only a marginal effect of θ. The effect of β is explained through the nature of the CES production function. As noted above, a small β implies a very high marginal product. Thus when β is small, well-endowed agents will make large innovations. They will therefore grow faster than less well-endowed agents, which will preserve, and often magnify initial differences.

3.1.2 Specialization

A generalist firm can be thought of as one having roughly equal amounts of all types of knowledge. A specialist, by contrast, will have an identifiable area of expertise – having noticeably more knowledge of one (or a few) types than the others. To measure specialization at the firm level, we compute the coefficient of variation of the firm's endowment. For a single firm this ranges from 0 (for the perfect generalist) to 2 (for the perfect expert, with only one type of knowledge). Figure 2 reports the population average of individual coefficients of variation at each (β, θ) pair.

The effect of θ is clear, with the degree of specialization falling as pooling moves towards the maximum of the partners' endowments. Agents become highly specialized when they innovate largely as individuals, which happens increasingly as θ decreases. What drives this results is that the type of knowledge produced is probabilistically the same as the knowledge input. Thus an agent is likely to innovate where he has most knowledge. In expected value, this process will lead to an agent innovating always in the same knowledge type, and so drive extreme

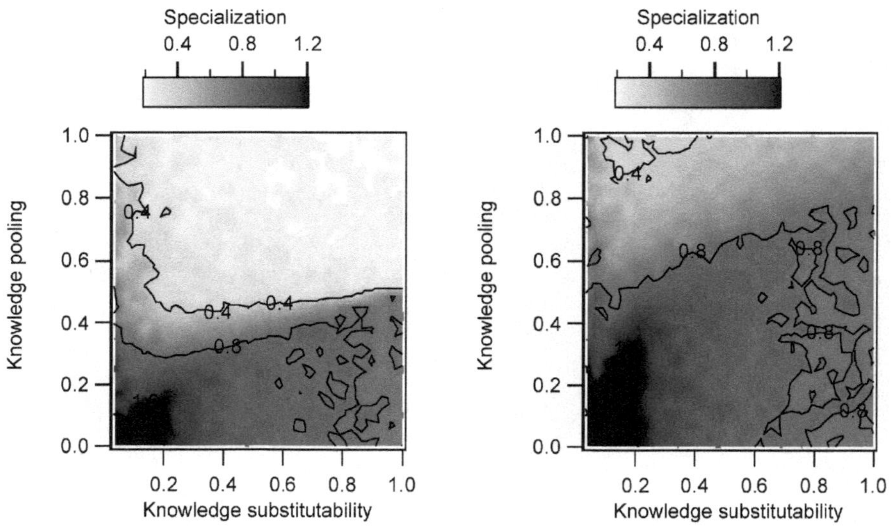

Fig. 2 Average agent specialization in the (β,θ)-space (*left*: no learning; *right*: learning)

specialization. When alliances form in a more systematic way (larger θ) generalist profiles quickly emerge. Sometimes an agent will innovate in his specialty, sometimes in his partner's, and for large θ, an agent will have many different partners. This will both smooth the agent's profile, and possibly even shift his area of expertise. This sort of mixing produces much flatter profiles, and the more partners an agent has, the more this mixing will take place. The effect of β is comparatively much weaker: as we move from right to left, initial differences in knowledge profiles are increasingly amplified, yielding the strongest specialization when β is close to 0.

3.1.3 The homogenization of firms

Quantifying convergence in knowledge profiles of firms can be done by computing the average angle between a firm and its partners in knowledge space. (This average is weighted by the frequency of interaction.) The larger this angle is, the more different (or complementary) is the expertise in the pairs that have formed. Figure 3 shows the population average of those weighted means in the (β, θ)-space.

The general intuition is that dissimilarity will be large when firms look for partners different from themselves (complementary, i.e. forming large angles) which happens for high θ. For low θ by contrast only pairs of similar people will form, contributing to a low angle over history. Thus dissimilarity is expected to fall when θ falls. This is the effect we observe in Fig. 3. In addition there is another, more mechanical effect. Given firms have constant opportunities for partnerships, those with a smaller number of partners will have lower average angle to them than those with a large number of partners. Number of partners is also expected to fall when θ falls. So in this case the strategic and mechanical effects act in the same direction to produce the patterns seen in Fig. 3.

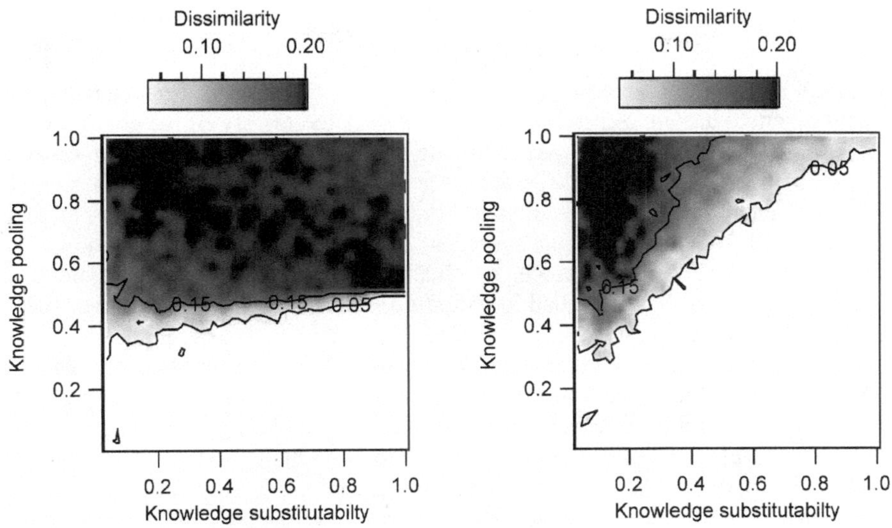

Fig. 3 Weighted average dissimilarity between partners in the (β,θ)-space (*left*: no learning; *right*: learning)

3.2 Network

Three statistics are used to describe the networks that emerge from the process of collaborative R&D. Defining Γ_i as the set of neighbours to whom i is directly connected and letting n_i be the number of those neighbours ($n_i = \#\Gamma_i$), the average degree of the graph is

$$\frac{1}{n}\sum_i n_i,$$

and measures the density of the interaction structure. Characteristic path length is defined as

$$\frac{1}{n(n-1)}\sum_i \sum_{j\neq i} d(i,j)$$

and simply measures how distant vertices are on average. Finally, neighbourhood clustering is the share of active links between the neighbours of any given vertex, that is, the proportion of existing over possible triangles in a firm's neighbourhood. Average clustering is then written as

$$\frac{1}{n}\sum_i \sum_{j,l\in\Gamma_i} \frac{\omega(j,l)}{n_i(n_i-1)},$$

where $\omega(j,l) = 1$ if $j \in \Gamma_l$ and 0 otherwise.

3.2.1 Degree distribution and economic performance

Figure 4 shows the number of connections held by the average firm, that is, how many distinct partners a firm has on average over the history of the economy.

The first observation is that including the effect of experience on success changes the order of magnitude by a large factor. When experience matters, success in a partnership increases the relative attractiveness of those two firms to each other, and this can be a strong source of inertia in partnership formation. The process is described, for instance, by Gulati and Gargiulo (1999): firms tend to stick with partners they have had in the past to reduce search costs and the risk of opportunism in inter-firm ties.

Both panels show a frontier between autarchic and joint innovation: for low values of θ (implying that knowledge is pooled on the minimum) firms prefer to innovate in isolation. When θ approaches 0, and knowledge pooling uses the minimum level of the two partners, a firm cannot be made better off by having a partner. If my partner is worse than me in any category, my knowledge production falls, since the minimum of our knowledge serves as the input. Thus for low values of θ, firms prefer isolation. The strength of this effect depends, though, on a firm's ability to compensate for its own weakness with its own strength, that is, on the elasticity of substitution in production $1/(1-\beta)$. If it is difficult to substitute internally (small β), the need for a partner to fill a hole in my own knowledge is strong, and it may be worth accepting a partner who reduces the input in one category because he increases it in another. Thus the threshold value of θ, below which no partnerships form, increases in β.

The effect of θ on the density of the network operates through the nature of optimal partnerships and the dynamics of knowledge. If dissimilar partners are the norm (as is the case when θ is large), repeated partnering with the same partner drives firms too close together and they soon switch to another. Firms thus have many partners. By contrast, when similar partners are desirable (θ is small), sticking with the same partner is a good thing, thus firms tend to have few partners.

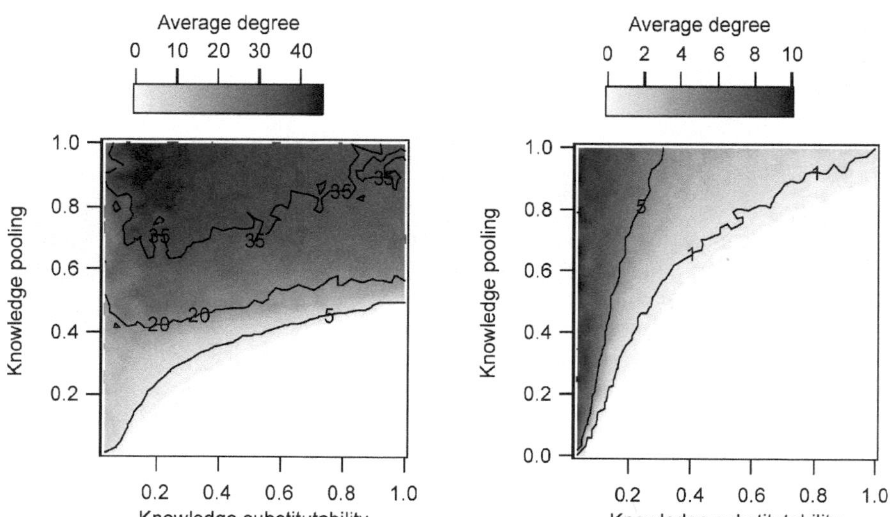

Fig. 4 Average degree of the network in the (β,θ)-space (*left*: no learning; *right*: learning)

Table 1 Correlation between degree and performance at the firm level

		β						
		0	0.17	0.33	0.50	0.67	0.83	1
No learning								
	1	**0.875**	−0.054	−0.191	−0.030	−0.019	−0.372	−0.028
	0.83	**0.935**	**0.073**	**0.093**	−0.181	−0.121	**0.161**	−0.407
	0.67	**0.836**	**0.745**	**0.442**	**0.496**	**0.107**	−0.112	0.104
θ	0.50	**0.738**	**0.684**	**0.401**	**0.478**	**0.416**	**0.076**õ	
	0.33	**0.704**	**0.353**	−0.008	−0.413			
	0.17	**0.449**	−0.139					
	0							
Learning								
	1	**0.731**	**0.007**	**0.103**	**0.195**	**0.319**	**0.014**	−0.265
	0.83	**0.531**	**0.448**	**0.278**	**0.443**	**0.156**	−0.254	
	0.67	−0.207	**0.299**	**0.391**	**0.245**			
θ	0.50	**0.327**	**0.322**	**0.176**				
	0.33	**0.571**	**0.221**					
	0.17	**0.291**						
	0							

This effect explains the increased density of the network as θ increases. The corollary of the increase in density is that as θ increases, networks are more likely to be connected.

Finally, comparing the two panels, we see that when experience matters autarchy is more attractive: it takes place in a larger area of the parameter space. The explanation is in terms of the likelihood of innovative success. We have assumed that when experience does not matter, success probabilities are high, so there is no penalty (from that source) in forming a partnership. By contrast, in a partnership, success probabilities are by definition lower than for autarchic innovation. Thus this penalty has to be overcome before a firm will consider joint innovation.[12]

Switching focus from the macroscopic properties of networks to economic performance, we can ask whether the firms who engage in many partnerships are better off than the ones who do not.[13] We thus calculate the correlation coefficient between degree and performance as measured by innovative potential. The results are summarized in Table 1, in which θ values decrease from top to bottom, to mimic the figures.

Over most of the parameter space where networking takes place there is a (strong) positive association between degree and performance. The firms with largest innovative value are also those with larger degree. Partnerships are thus essentially a valuable asset. There are two regions where negative correlations

12 This result depends entirely on the assumption that success probabilities are high when experience does not matter: it would be reversed had we assumed that success probabilities are low when experience does not matter.

13 Note that being involved in more alliances rather than fewer does not imply higher costs, the cost of forming an alliance being assumed negligible. Adding a cost to creating and maintaining partnerships would render autarchy preferable over a larger parameter region.

occur, though. The first is along the autarchy frontier. In this region firms that have relatively "good" endowments tend to innovate alone, those with weaknesses in their own stock are forced to look for partners. Here, partnership signals a weakness that needs be overcome. The second region is in the top right triangle in the table, an area where both θ and β are high. From Fig. 2 we can see that this corresponds to the region in which firms tend to be generalists. Here, the negative correlation is driven by firms that have fewer than average partners over history. Because of the convergence caused by innovation, a long-lived pair must consist of two specialized firms (otherwise they converge quickly and part company). When specialists come together, they gain a lot from the partnership and so will make relatively large innovations, particularly compared to a pair of generalist firms. Long-lived pairs of specialists will create more knowledge than pairs of generalists (which are common in this region) and this creates a negative correlation between density and innovation performance.

3.2.2 The emergence of local and global structure

Figure 5 displays the average distance between agents in the network. Because a disconnected graph has some agents who are infinitely distant from each other, the averages are computed only for networks that are connected. Thus this figure also indicates when a single connected component emerges, namely when θ is large enough for given β. Comparing the two panels, this critical θ is much larger when experience with partners matters. As discussed above in Section 3.2.1.

In both panels the contour line indicating an average degree of 5 (from Fig. 4) has been added. Though the parameter region over which a unique component forms is much larger when experience does not affect success probabilities, both panels contain a region where the degree is approximately 5, but the network is

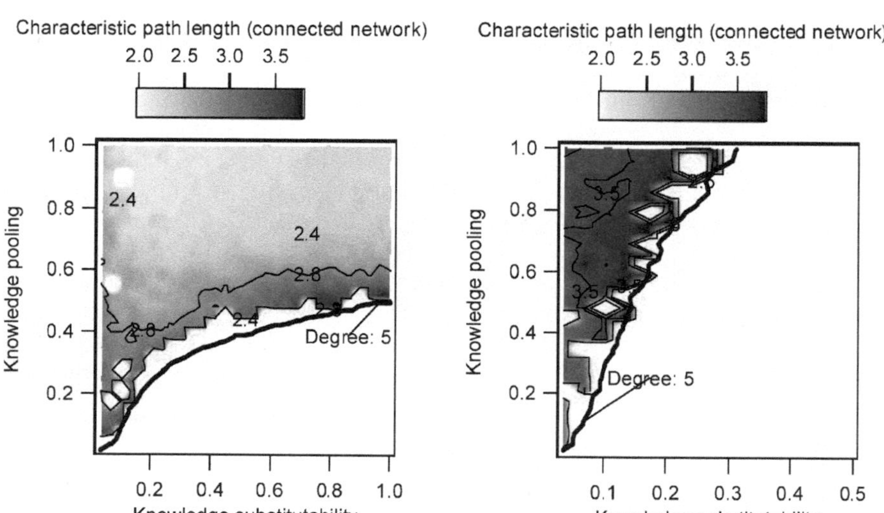

Fig. 5 Characteristic path length in the (β,θ)-space (*left*: no learning; *right*: learning)

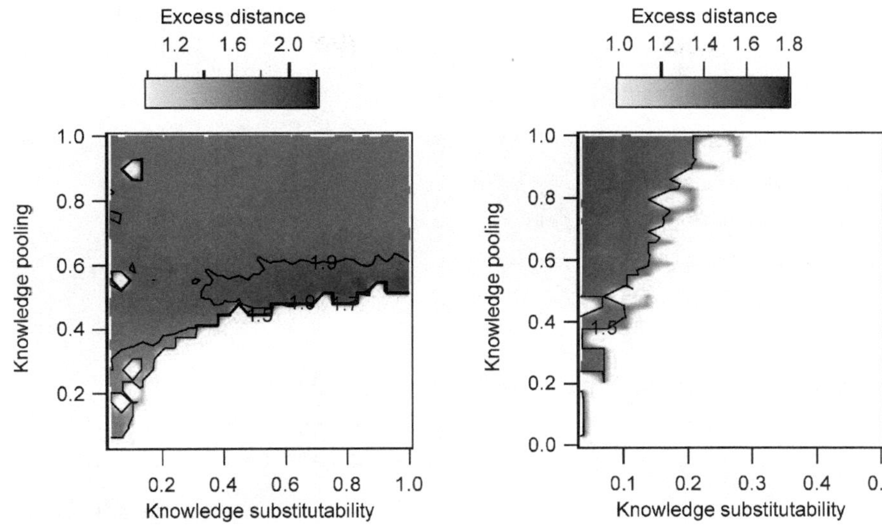

Fig. 6 Excess characteristic path length in the (β,θ)-space (*left*: no learning; *right*: learning)

disconnected. This region lies near the autarchic innovation border, and here the network consists of connected sub-networks. The pattern in the (β,θ)-space that Fig. 5 shows is essentially symmetric to the one observed in Fig. 4: as the average degree increases the average distance falls mechanically.

To get a finer idea of the extent to which structure is emerging, a useful benchmark is a random graph of the Erdös and Rényi (1960) type, in which a constant (average) degree is assumed for each vertex. In such a graph, when average degree is Δ, the expected path length is well-approximated by $\ln n / \ln \Delta$.

In Fig. 6 the ratio of the observed characteristic path length to that of the Erdös and Rényi benchmark is shown. In both learning regimes path lengths are slightly longer than those in equivalent random graphs. The networks that emerge from the processes we model are not random, but have a greater prevalence of localized connections.

Consider now clustering as a measure of local order in the network. Figure 7 shows the population average clustering. The relationship between clustering and the two parameters θ and β is very similar to that for degree. Since path length does not change dramatically as links are added to the network (Fig. 5), this implies that as firms find new partners they are not creating shortcuts to distant agents (which would reduce path length), but are rather reinforcing local coherence. (The white regions at the bottom of the graphs correspond to the region in parameter space in which firms tend to innovate as individuals, yielding zero clustering by definition.)

Again clustering is driven to a very large extent by the degree of the graph. As agents acquire more links, even if they are acquired at random, the network becomes denser locally. Thus it is necessary to compare clustering also with the Erdös and Rényi benchmark, where clustering is approximately Δ/n. Figure 8 shows the ratio of observed clustering over the benchmark. Values significantly larger than 1 would indicate a structure richer than a random graph.

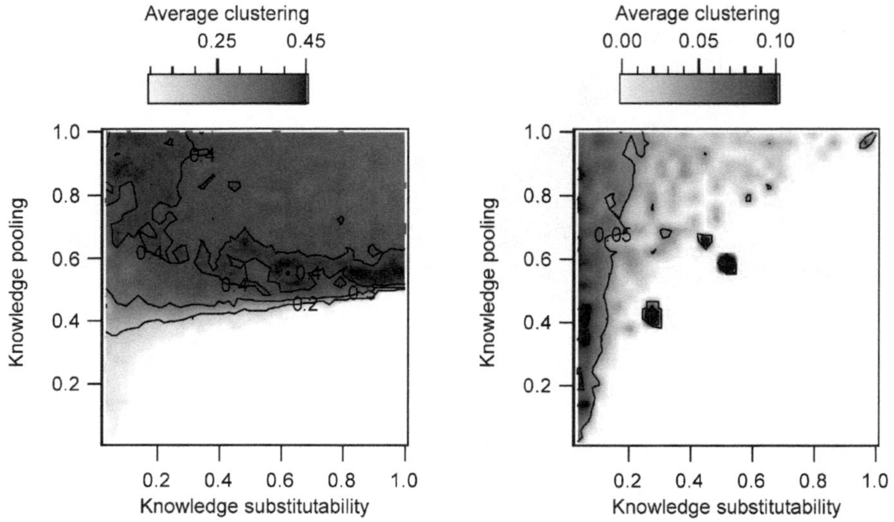

Fig. 7 Clustering in the (β,θ)-space (*left*: no learning; *right*: learning)

In the left panel, over almost the entire parameter space there is more clustering in our networks than there is in random graphs. In the right hand panel, clustering is lower than or similar to that of random graphs in the upper left corner of the parameter space, and larger elsewhere. The interesting feature is that in both panels, excess clustering is present, and strong along the autarchy frontier. In this region, θ is just above the threshold at which networking begins, stable matchings

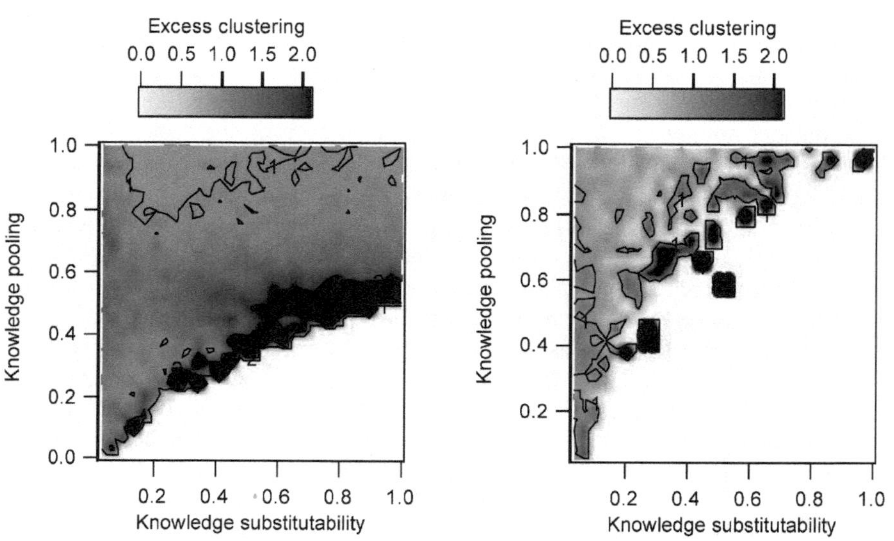

Fig. 8 Excess clustering in the (β,θ)-space (*left*: no learning; *right*: learning)

unite agents who are similar to each other. Here a great degree of similarity between partners is tolerated. Thus a pair of agents tends to stay together for a long time, and converge toward each other in knowledge space. But partners can be too close to each other. When this occurs they search for new, but still relatively similar partners. It is likely that this area of the parameter space will engender partner-swapping. There is, then, a suggestion that cliques tend to form here, which together with the existence of longer path lengths (as compared to a random situation) is evidence partly supporting the small world conjecture.[14] By extension, since the desire for similarity falls as θ increases, partnerships dissolve more rapidly, and when they do, firms search for very different partners. This takes them away from firms with whom they have already partnered, and also away from partners of partners.[15]

4 Conclusion

In this paper we have modelled the processes of knowledge growth and network formation. Both result from firms' decisions to form bilateral alliances in which they pool their existing knowledge and use it to create new knowledge. For a single firm, repeated interaction with the same partner has a positive effect by increasing the probability of successful innovation, but also has a negative effect through generating similarity in knowledge stocks. Familiar partners, by having similar knowledge profiles, will have small innovations with high probability; unfamiliar partners, with dissimilar profiles, will have larger innovations with lower probability. The resolution of this trade-off determines the nature of the networks that emerge, and this depends both on substitutability of inputs in knowledge production, and on whether the innovation task is decomposable into independent sub-tasks.

Specifically, two central parameters drive the model: the ease with which knowledge from different fields can be substituted in the knowledge production function (β), and whether the pooled knowledge vector lies closer to the maximum or the minimum of the partner's vectors, (θ), which is determined by the extent to which the innovative task is decomposable.

The most striking result regarding knowledge accumulation is the decline of specialization as separability increases. Separability implies that a division of labour is possible in innovation, and one might expect that this will make it possible for firms to specialize. The contrary result that we find arises from the assumption that, probabilistically, the type of knowledge created mirrors the joint vector of knowledge inputs to innovation. When innovation is thus separable, firms look for complementary partners, and their joint knowledge vector will be relatively flat. On average, this flattens firms' own stocks as the alliance-innovation

[14] There is no explicit value from transitivity in the model, so we should not expect too much clustering in general.

[15] Firms move close to their partners in knowledge space, and this "close to" relationship is transitive. So as a firm is close to its partners, it will also be (slightly less) close to its partners' partners. In that case the search for partners who are different from oneself will prevent the creation of closed triangles, and cliquishness will be small.

process proceeds.[16] When experience with partners influences the probability of success, the decline of specialization with θ is less marked. Partnerships are more persistent because the experience of collaboration creates additional value by lowering the risk of failure, and thus an ongoing partnership can develop its own area of expertise. A similar effect exists at the population level. When innovation separability is strong, firms have many partners. Convergence following a successful innovation drives a firm from one to another partner, since complementary rather than similar partners are desirable. This partner churning, and period by period convergence between existing partners, has the effect of generating a population of homogeneous firms. Care must be taken in this interpretation, however, since even if there is homogenization at the aggregate level, it remains that at the level of the individual firm, similarity between a firm and its partners falls as task separability increases.

Our second interest lies in the emergent networks themselves. Given the recent literature on the subject, a natural question to ask is whether we observe small worlds: a collection of densely connected sub-groups joined by clique-spanning ties? (On that issue, see for instance Baum et al. 2003 and the references therein.) First, we find that there is a critical frontier which partitions the parameter space into two regions, one of in which agents always innovate in isolation, and one in which networking takes place. This critical frontier $\theta^*(\beta)$ is increasing, and exists both in the presence and absence of learning about partners. Above the alliance-autarchy frontier, the number of distinct partners a firm has over time increases with θ, and except when the number of partners gets very high, firms with more partners tend to accumulate more knowledge over the history of the economy.

As the innovation task is easier to separate (as θ increases) firms seek partners different from themselves, and as a consequence have more of them. In a fairly natural way then characteristic path length falls and average clustering rises. However we find more clustering and a higher characteristic path length than exists in a comparable random graph. This suggests that our networks have more structure. We can be more precise. As we move away from the frontier of autarchic innovation, while local structure remains in the form of excess clustering, the networks come more and more to resemble random graphs. Below the frontier, agents spend their histories in isolation, so the network cannot really be said to exist.

However, in the region around the autarchy-alliance frontier, more structure is present. There, what we observe are densely connected subgroups, with sparse connections between them (sometimes too sparse to connect the entire graph, but dense enough to create connected components). Thus small worlds do emerge from the processes we have modelled. In the narrow region where they are present, the innovation process is separable enough to allow partners to retain some differentiation in their knowledge profiles, but still joint enough to permit mutual learning and transfer.

Two properties of innovation are central in this model: the ability of a firm to compensate for its weakness with is own strengths, which is driven by the nature of substitution in knowledge production; and the way in which firms are able to combine their knowledge stocks, which is driven by separability in the innovation

[16] In Cowan et al. (2003) a situation in which firms where innovating most often in their own domain of expertise was analysed, and there specialization increased as task division got easier.

process. What we observe is a strong similarity between the effects of these properties on knowledge variables and on network variables. This suggests very strongly that the two processes, of knowledge generation and network evolution, are strongly linked. It would, in general, be a mistake to try to understand one without examining the other. In addition, we observed a strong effect of θ, the parameter capturing issues of how firms pool their knowledge, and how the innovative process works. Thus to understand network formation it is necessary to know the details of the processes by which firms jointly innovate. For reasons of simplicity and parsimony, the model presented has been very stylized in its treatment of knowledge and innovation. Nevertheless, it makes the point that the interplay between the decomposition of tasks in knowledge creation, collaborative learning through repeated interaction, and the general properties of innovation as knowledge recombination are central in understanding how innovation networks, and industry structures more generally, emerge and evolve.

Acknowledgement Jonard and Zimmermann acknowledge support from the French Ministry of Research (ACI Systèmes complexes en Sciences Humaines et Sociales, SCo0061).

References

Baum JAC, Shipilov AV, Rowley TJ (2003) Where do small worlds come from? Ind Corp Change 12:697–725

Coriat B, Guennif S (1998) Self-interest, trust and institutions. In: Lazaric N, Lorenz E (eds) Trust and economic learning. Edward Elgar, London

Cowan R, Jonard N, Zimmermann J-B (2003) La dynamique conjointe de la connaissance et des réseaux. Rev Econ Ind 103:253–274

Cowan R, Jonard N, Zimmermann J-B (2004a) On the creation of networks and knowledge. In: Gallegati M, Kirman A, Marsili M (eds) The complex dynamics of economic interaction. Lecture notes in economics and mathematical systems. Springer, Berlin Heidelberg New York, p 531

Cowan R, Jonard N, Zimmermann J-B (2004b) Networks as emergent structures from bilateral collaboration. MERIT Research Memorandum, RM2004-017

Dodgson M (1996) Learning, trust and inter-firm technological linkages: some theoretical associations. In: Coombs R, Richards A, Saviotti PP, Walsh V (eds) Technological collaboration, the dynamics of cooperation in industrial innovation. Edward Elgar, London

Dutta B, Jackson M (2003) On the formation of networks and groups. In: Dutta B, Jackson M (eds) Networks and groups: models of strategic formation. Springer, Berlin Heidelberg New York

Dyer JH, Nobeoka K (2000) Creating and managing a high performance knowledge sharing network: the Toyota case. Strateg Manage J 21:345–367

Erdös P, Rényi A (1960) On the evolution of random graphs. Publ Math Inst Hungarian Acad Sci 5:17–61

Gale D, Shapley L (1962) College admissions and the stability of marriage. Am Math Mon 69:9–15

Galison P (1999) Trading zone: coordinating action and belief. In: Biagioli M (ed) The science studies reader, Routledge, New York, pp 137–160

Garcia-Pont C, Nohria N (2002) Local versus global mimetism: the dynamics of alliance formation in the automobile industry. Strateg Manage J 23:307–321

Goyal S, Joshi S (2003) Networks of collaboration in oligopoly. Games Econom Behav 43:57–85

Grandstand O (1996) Towards a theory of the technology-based firm. Res Policy 27:465–489

Grandstand O, Sjolander S (1990) Managing innovation in multi-technology corporations. Res Policy 19:35–60

Gulati R, Gargiulo M (1999) Where do inter-organizational networks come from? Am J Sociol 40:619–652

Kirman A, Vriend N (2001) Evolving market structure: an ACE model of price dispersion and loyalty. J Econ Dyn Control 25:459–502
Klos T, Nooteboom B (2001) Agent-based computational transaction cost economics. J Econ Dyn Control 25:503–526
Mowery DC, Oxley JE, Silverman BS (1998) Technological overlap and interfirm cooperation: implications for the resource-based view of the firm. Res Policy 27:507–523
Narula R (1999) Explaining strategic R&D alliances by European firms. J Common Mark Stud 37:711–723
Nooteboom B (2000) Learning and innovation in organizations and economics. Oxford University Press, Oxford
Nooteboom B (2004) Learning and governance in inter-firm relations. Rev Econ Polit Janvier-Février: 114
Oliver C (1990) Determinants of inter-organizational relationships: integration and future direction. Acad Manage Rev 15:241–265
Peretto P, Smulders S (2002) Technological distance, growth and scale effects. Econ J 112:603–624
Plouraboué F, Steyer A, Zimmermann JB (1998) Learning induced criticality in consumer? Adoption pattern: a neural network approach. Econ Innov New Technol 6:73–90
Powell WW (1990) Neither markets nor hierarchies: network forms of organization. Res Organ Behav 12:395–336
Powell WW, Koput KW, Smith-Doerr L (1996) Inter-organizational collaboration and the locus of innovation: networks of learning in biotechnology. Adm Sci Q 41:116–145
Powell WW, White DR, Koput KW, Owen-Smith J (2005) Network dynamics and field evolution: the growth of inter-organizational collaboration in the life sciences. Am J Sociol 110:1132–1205
Roijakkers N (2003) Inter-firm cooperation in high-tech industries: a study of R&D partnerships in pharmaceutical biotechnology. PhD thesis, Maastricht University, Maastricht, Datawyse/ Universitaire Pers Maastricht
Sako M (1991) The role of trust in Japanese buyer-supplier relationships. Ric Econ 45:449–474
Teece D, Pisano G (1989) Collaborative arrangements and global technology strategy: some evidence from the telecommunications equipment industry. In: Burgelman RA, Rosenbloom RS (eds) Research on technological innovation, management and policy, vol 4. JAI, Greenwich, CT, pp 227–256
van Alstyne W, Brynjolfsson E (1997) Electronic communities: global village or cyberbalkans?. Working Paper, MIT Sloan School
Weisbuch G, Kirman A, Herreiner D (2000) Market organization and trading relationships. Econ J 110:411–436
Weitzman M (1998) Recombinant growth. Q J Econ 113:331–360

Paul Muller · Julien Pénin

Why do firms disclose knowledge and how does it matter?

Abstract In this paper we propose a simulation model aimed at describing the emergence and the dynamics of innovation networks, with particular emphasis on the role of open knowledge disclosure. It is argued that firms that widely disclose knowledge to other firms are more likely to enter innovation networks and to acquire a central position within these networks. By disclosing knowledge, firms increase their reputation, which indicates to other organizations that they are competent, and that it is worth starting a partnership with them. The higher a firm reputation, the higher its probability of entering new R&D partnerships with other firms. Our model provides, therefore, a rationale for behaviors of open knowledge disclosure by showing that such strategies, although risky in the short run, may pay in the long run by enabling firms to access external sources of knowledge more easily.

Keywords Open knowledge disclosure · Inter-firm cooperation · Reputation · Network simulation

JEL Classification D85 · L14 · L16 · L23

1 Introduction

The aim of this paper is to provide a theoretical framework describing the formation and evolution of innovation networks. Specific attention is paid to the emergence of central agent(s) within innovation networks and to the role of knowledge disclosure in giving rise to the emergence of these central players. In

P. Muller (✉) · J. Pénin
BETA, University Louis Pasteur, 61 avenue de la Foret-Noire, 67085 Strasbourg, France
E-mail: paul.muller@cournot.u-strasbg.fr

J. Pénin
Département des Sciences Économiques, Université du Québec à Montréal, case postale 8888, Succursale Centre Ville, Montréal, Québec, Canada, H3C 3P8
E-mail: penin@cournot.u-strasbg.fr

short, we argue that disclosing information not only provides competitors with useful information, but also allows disclosing firms to improve their external reputation. This, in turn helps them to develop R&D collaborations with other firms or with public institutions, enabling disclosing firms to increase the efficiency of their research.

We start from the observation that many firms often widely disclose parts of their knowledge, which then become available to other firms, including competitors. Firms allow their researchers to publish in scientific journals and to present their research at conferences, apply for patents while not planning to use the exclusive property right associated with a patent, release information on their web sites, etc. (Allen 1983; Hicks et al. 1994; Hicks 1995).

Since such behaviors of open knowledge disclosure are costly and do not give rise to direct rewards, they have often been misunderstood by mainstream economics. But, nowadays, reasons underlying open knowledge disclosure are well accounted for by standard economic theory (Allen 1983; von Hippel 1987; De Fraja 1993; Harhoff 1996; Boivin 2000; Eaton and Eswaran 2001; Pénin 2003; Harhoff et al. 2003). In particular, behaviors of open knowledge disclosure may be motivated by reputation concerns. Acquiring a reputation as an innovator may facilitate firms' access to financing, may make it easier to get contracts, grants and subsidies, or may enable firms to hire the best researchers. Furthermore, a reputation as an innovator may allow a given firm to find competent partners with whom to collaborate more easily in R&D. These effects stem from the fact that open knowledge disclosure (the fact that firms publish in scientific journals, for instance) enables firms to signal their competences to academic and industrial communities, i.e. by openly disclosing knowledge, firms indicate to potential partners that it may be worth collaborating with them.

The present paper studies the effects of strategies of open knowledge disclosure on the formation of R&D collaborations by making use of numerical simulation. Our model of network morphogenesis is as follows. At the beginning of the simulation, a population of firms is located on an empty graph. Each firm has a scalar knowledge endowment. At each timestep, firms engage in R&D activity in order to produce new knowledge. The probability of producing a new piece of knowledge depends on the amount of knowledge (external or internal) to which the firm has access. Second, they consider whether or not to disclose part of the knowledge they hold. Disclosing knowledge decreases current profit, but also allows the firm to build up its reputation. Third, they consider whether to link up with other firms. Such a connection process is driven by reputation effects. Collaborations open access to partners' stock of knowledge and thus increase the stock of knowledge of connected firms, which in turn enhances their R&D performance.

Firms therefore face the following trade-off: Either they choose frequently to disclose knowledge to all other firms, thus decreasing their current profit, but also increasing their reputation (which may enable them to enjoy in the future more inter-firm R&D cooperative relationships and to have access to extra sources of external knowledge), or they decide to keep their knowledge secret, thus favoring their current profitability, but eventually impeding their ability to create new connections (since they will suffer from a lack of reputation).

A major result of our simulation is the setting of a particular network shape in which some firms acquire a central position in the innovation network while others adopt more "peripheral" positions. Moreover, we find that status within innovation networks and profitability are highly sensitive to the strategy of open knowledge disclosure (low or high frequency) they adopt. It may not only be the firms endowed initially with higher amounts of specific knowledge that acquire a central status but also the most active ones in terms of knowledge disclosure. Adopting a reasonable strategy of open knowledge disclosure enables them to catch up with firms initially endowed with higher amounts of specific knowledge.

The paper is structured as follows: in the next section, we explain the link between open knowledge disclosure, the initiation of formal and informal R&D collaborations and the problems that arise from incomplete information. Section 3 depicts the theoretical model. In Section 4 we test this model by using numerical simulations and we discuss the results. We conclude with remarks and indications for future research.

2 Open knowledge disclosure, R&D collaborations and incomplete information

2.1 Innovation as a collective process

Innovation is usually perceived as a group of activities involving interaction and knowledge exchanges between people and organizations. It is the outcome of interpersonal and inter-organization interactions that take place continuously and everywhere. An innovative firm interacts with many different actors: other firms (competitors, suppliers, customers), universities, research institutes, venture capitalists, banks, patent offices, government agencies, etc. In other words, knowledge production is a collective process. In order to be innovative, economic agents often have to cooperate to set up formal research joint ventures (RJV) or more informal innovation networks, in which they can exchange knowledge and share specific competences. An innovation network can be defined as a reasonably stable set of partners that collaborate in order to improve their research. More precisely, it is: "A set of reciprocal, reputational or customary trust and cooperation based linkages among actors that coalesces to enable its members to pursue common interests, in this case on respect of innovation" (Cooke 2001, p. 953; see also Maskell and Lorenzen 2003). Formal or informal R&D collaborations with other firms or with universities are important for two main reasons:

(1) They increase the efficiency of the research, since they allow firms to share the tasks, the costs and the risks of doing research. R&D cooperation allows a more efficient division of research by enabling firms to develop synergies with complementary partners;

(2) They open access to technical opportunities and to external sources of knowledge, first by helping to recruit young researchers and, second, by providing access to new technologies and to sticky knowledge. Companies who are members of innovation networks keep intellectual alert and are able to move quickly into new areas.

Being a member of innovation networks is specifically important for accessing valuable knowledge held by other firms (Kogut 2000). Knowledge is usually not a public good that, once produced, benefits other agents in the economy instantly. Rather, knowledge is a collective good, a club good in the sense that it flows only within the communities or networks that produced it and is transferred through complex interactions among the members of those networks. Hence, if firms that are members of an innovation network have a privileged position for acquiring knowledge, outside firms often cannot access the knowledge produced within the network. Networks *exclude* as well as they *include*. As argued by Gibbons (1994), in a collective form of knowledge production, knowledge production and knowledge appropriation converge: in order to access the produced knowledge, outsiders must first be granted access to the network.

2.2 How can firms identify potential fruitful R&D partnerships?

An important question that firms involved in this collective process of innovation have to address is related to the identification of the most adequate partners with whom to collaborate. It is very important for them to select good partners because collaboration is costly. It usually means opening access to the most precious knowledge of the firm to collaborators, which may often be rivals, too. Collaboration also requires sometimes investment in the construction of common, specific assets (Williamson 1975), exposing firms to risks of hold-up. Furthermore, even if these considerations about the cost of the collaboration could be neglected, it remains true that firms are clearly more interested in collaborating with firms that are at the forefront of the technological frontier. All these elements support the view that firms must wisely select their partners.

However, the innovation process takes place in an environment of incomplete information. Adverse selection problems impede collaborations among actors of innovation. Most of the time, firms aiming at developing collaborations do not know the exact characteristics and, more specifically, the competences of other organizations. Hence, how can they identify potential fruitful partnerships? How (on which criteria) can they discriminate between profitable collaborations and less profitable ones? Those problems are at the core of the collaboration problem and, as such, at the core of the innovation process itself. But curiously, little economic literature has focused on them. A notable exception is the work of Grosseti and Bès (2002), who explored the ways through which 110 R&D collaborations were initiated in France between firms and public labs.

Grossetti and Bès considered three main ways through which R&D collaborations may start. The first is called the "logic of market", when collaboration results from a link established through scientific publications or public conferences. The second is called "the logic of personal network", when collaboration results from former existing personal relationships between the two parties (see for instance, the work of Granovetter 1973 and 1974, dealing with the importance of weak ties in connecting people).[1] The third way to initiate

[1] In this case, the personal network is already structured and nothing is said about how it was formed. It is however likely that the logic of market played an important role in developing a firm's personal network.

collaboration is the "logic of institutions"; in this case, collaboration is organized and structured by public institutions that connect the different actors.[2]

Our interest in this paper lies precisely in "the logic of market". We consider that the choice of a partner with whom to collaborate is mainly based on reputation (Muller 2003). Since reputation mitigates the uncertainty associated with a first interaction, firms aiming at connecting with an other business tend to give priority to firms enjoying higher levels of reputation. This central role played by reputation may provide a convincing explanation to open knowledge disclosure behaviors, which has often been misunderstood by mainstream economics.

2.3 On open knowledge disclosure

Open knowledge disclosure is defined here as when a firm chooses voluntarily to disclose knowledge, without being directly rewarded for this disclosure and without being able to prevent some other firms from accessing the disclosed knowledge (Pénin 2004). Channels through which firms may openly disclose knowledge are scientific publications, patents, conferences, and the Internet.

Since open knowledge disclosure cannot pretend to any direct remuneration and may involve important costs[3], conventional wisdom suggests that such behaviors remain rare and that firms try most of the time to keep their research confidential. Still, there is convincing evidence that many firms often decide to openly disclose part of their knowledge (Allen 1983; Koenig 1983; Hicks et al. 1994, 1996; Hicks 1995; Cockburn and Henderson 1998; Lim 2000). For instance, Hicks (1995) displayed figures indicating that in many sectors that are firms publishing massively. She concludes that "Firms such as Philips, Hitachi, ICI, Ciba, Siemens, Sandoz, Roche, Hoechst and Toshiba contribute as much to the public literature as medium sized universities" (Hicks 1995, p. 403). Furthermore, these publications by firms contain, at least to some extent, valuable knowledge, since they are as much quoted as publications written by researchers in universities. Our explanation why firms publish is very close to the view developed by Hicks. She suggests that such publications may be targeted to signal unpublishable resources to the academic and industrial communities.

Mainstream economics usually refer to the notions of altruism, irrationality or to intrinsic motivations to explain open knowledge disclosure. Yet, there exist many rational explanations. A wide range of indirect mechanisms make open knowledge disclosure rewarding for firms. For instance, firms may want openly to disclose part of their knowledge in order to induce other firms to reciprocate (von Hippel 1987; Eaton and Eswaran 2001), to trigger network effects (Boivin 2000) or pecuniary spillovers (Harhoff et al. 2003), to keep bright researchers working for

[2] It is possible to add to this list a fourth way to initiate R&D collaborations: the logic of serendipity. In every searching process there is an important part of chance playing for matching partners.

[3] The most important cost is the one that stems from the communication of information to potential competitors, which can thus absorb and use this knowledge in order to compete more fiercely with the disclosing firm. Other costs involve the codification of the disclosed knowledge (which must be articulated, expressed in a comprehensible language and written down on a support) and the fact that the means through which diffusion is operated may not be free (for instance, patent application is very expensive).

them, to enhance their reputation, to expand production in downstream sectors (Harhoff 1996), to shorten patent races (De Fraja 1993), etc.

In this paper, we focus on reputation effects triggered by open knowledge disclosure. We argue that there is a relationship between open knowledge disclosure, the formation of R&D collaborations and problems of incomplete information. Firms may decide openly to disclose knowledge in order to signal to potential partners that it is worth collaborating with them, i.e. in order to break adverse selection problems impeding collective innovation. Indeed, open knowledge disclosure is a way to show to potential partners what the firm is doing and what it is looking for. Hence, open knowledge disclosure may help firms to trigger R&D partnerships. It may be a powerful device to improve firms' position on the 'market to find R&D collaborations'. This was emphasized, for instance, by Gambardella (1992), Callon (1998), Henderson and Cockburn (1994), Hicks (1995) and Grosseti and Bès (2002).[4]

3 A model of innovation networks morphogenesis

3.1 Basic assumptions of the model

At time t, n firms are located on a graph $G_t = (V, \Gamma_t)$, where $V = \{1, \ldots, n\}$ is the set of firms (vertices) and $\Gamma_t = \{\Gamma_t^i, \forall i \in V\}$ is the list of connections in the graph. $\Gamma_t^i = \{j \in V / \{ij\} \in G_t\}$ constitutes the set of R&D agreements enjoyed by firm i at time t, and $\{ij\}$ represents the tie binding two firms i and j at time t. For $t=0$, $\{\Gamma_0\} = \{\varnothing\}$, the graph forming the innovation network is empty. At the beginning of each simulation, firms may differ through two characteristics: the quantity of specific knowledge they hold and their strategy of open knowledge disclosure.

The model relies on the coexistence of two types of knowledge. Knowledge of the first type is qualified as "general" or public because it is accessible to all firms. It is made up of knowledge produced by public institutions but also, and this is a main focus of the model, by knowledge produced within firms and openly disclosed to all the other firms (e.g. through scientific publications). The second type of knowledge is assumed to be proper to each firm and may be qualified as firm specific.[5] It consists in particular know-how, pieces of knowledge subject to secrecy policies, etc.

Each firm, while developing R&D partnerships, is assumed to make a share β of its specific knowledge available to other members of its collaboration network. We qualify this type of knowledge as collective, since only firms that are members of R&D partnerships share it. However, for sake of simplicity, we consider this knowledge as being specific in the sense that we assume that firms' stock of specific knowledge does not decrease when they make some of their specific knowledge available to R&D partners.

[4] Grossetti and Bès found that 42 collaborations out of 110 resulted from a "logic of market". The "logic of market" is especially relevant when firms and labs are not located in the same region (in this case it works in 60% of collaborations).

[5] In this model, "specific knowledge" does not mean knowledge only usable in the context of the firm, but pieces of knowledge held secret by firms (or communicated only to some partners). Specific knowledge can be used by other firms if the owner chooses to make it available to them.

The total knowledge each firms can access while carrying out its R&D activities comes from three sources: specific knowledge that has been internally produced, public knowledge that has been absorbed and specific knowledge held by partners and made available through joint-ventures or any other type of cooperation agreement. It is assumed that pieces of knowledge produced by firms are perfect complements. There are no redundancies between knowledge held by each firm and their partners. It follows that the total knowledge a firm may mobilize to perform R&D is given by:

$$K_{i,t}^{\text{Tot}} = K_t^{\text{Gen}} + K_{i,t}^{\text{Spe}} + \beta \sum\nolimits_{j \in \Gamma_t^i} K_{j,t}^{\text{Spe}}, \quad 0 \leq \beta \leq 1$$

where K_i^{Gen} represents the general knowledge of the economy at time t (this stock of public knowledge is common to all firms), $K_{i,t}^{\text{Spe}}$ the specific knowledge held by firm i at time t, β the share of specific knowledge made accessible by firm i partners to its R&D collaborators and Γ_t^i firm's i set of R&D partners at time t. It is assumed that β is set as fixed and equal for every firm. Furthermore, for the sake of simplicity we do not introduce the absorption capacity of external knowledge (Cohen and Levinthal 1989), which would imply that firms cannot mobilize the totality of the stock of external knowledge but only a fraction, depending on their absorption capacity. In our model, firms can absorb all the external knowledge that is made available to them.

Each firm $i \in V$ is characterized by an open knowledge disclosure strategy, which is assumed to be given and fixed over time. This corresponds to the frequency φ_i at which firms choose openly to disclose knowledge. Openly to disclose knowledge means, for a given firm, to make this knowledge available to all the other firms. Hence, specific knowledge that is openly disclosed automatically becomes public.

Lastly, the objective of each firm is to generate profits. Revenues mainly stem from the monopolist rent provided by the use of the specific knowledge. Such rent may come from licensing agreements, patent rights, etc. We assume that profits are a linear increasing function of the amount of specific knowledge held by the firm. Moreover, we assume that profits are not a function of the general knowledge firms have access to. Since public knowledge is, by definition, available to each firm, it is assumed to give no competitive advantage and hence it does not increase profit. But, as we will see below, it increases the probability to innovate and therefore to generate further specific knowledge in the next periods.

Firms have to bear costs related to the production of knowledge. Costs of the first type gather internal R&D expenditure, administrative costs and all other type of costs related to the activity of the firm. Costs of the second type stem from inter-firm partnerships. This issue has been widely discussed, notably in the frame of Transaction Costs Economics (Williamson 1975). Any inter-firm relationship involves costs for developing compatibility, building common assets, etc. We assume that firms must bear a constant unitary cost per collaboration α. The profit function of firm i at time t is given by:

$$\Pi_{i,t} = \gamma K_{i,t}^{\text{Spe}} - \alpha k_{i,t} - c$$

where $k_{i,t}$ represents the number of partnerships of firm i at time t. The parameters γ, α and c represent, respectively, the income stemming from one unit of specific knowledge, the unitary cost of an inter-firm relationship and fixed costs (R&D

expenditure, administrative costs, etc.). Those parameters are set as fixed and equal for every firm at any time.[6]

3.2 Dynamics of the model

Our model incorporates the fact that, for knowledge intensive firms, signalling competences through open knowledge disclosure constitutes a way to improve the performance of their R&D by easing the formation of R&D partnerships and, hence, by allowing enlarged access to knowledge held by partners. As summarized in Fig. 1 below, the dynamics of the system decomposes into three main steps: knowledge production, knowledge disclosure and partnership binding.

First step: Production of specific knowledge. At each time step, each firm undertakes R&D activity. Firms mobilize their total stock of knowledge $K_{i,t}^{\mathrm{Tot}}$ in order to increase their stock of specific knowledge. Firms having a higher stock of total knowledge are more likely to produce specific knowledge. At each period, we assume that a new unit of specific knowledge is created with a probability $p_{i,t}$, which is increasing with the firm's total stock of knowledge at this period, but with decreasing returns to scale. This probability is given by the distribution function of an exponential law:

$$p\left[K_{i,t}^{\mathrm{Spe}} = K_{i,t-1}^{\mathrm{Spe}} + 1\right] = F\left(K_{i,t-1}^{\mathrm{Tot}}\right) = 1 - \exp\left(-\lambda K_{i,t-1}^{\mathrm{Tot}}\right), \text{ with } \lambda > 0,$$
$$\frac{\partial F}{\partial K} > 0 \text{ and } \frac{\partial^2 F}{\partial K^2} < 0$$

Second step: disclosure of specific knowledge. During the second stage of the process, each firm faces the decision whether openly to disclose or not a fraction δ of the specific knowledge it holds. Open knowledge disclosure triggers two opposite effects for disclosing firms. On the one hand, it decreases their instant profit since, once disclosed, the fraction δ of specific knowledge turns instantly general. On the other hand, knowledge disclosure induces an increase in the reputation R_t^i of the emitting firm, which may enable it to enjoy in the future more inter-firm R&D relationships and hence to access extra sources of external knowledge. A central assumption of the model, which will be discussed in more depth later, is that, whatever the amount of specific knowledge that is disclosed by firm i at a moment t, the reputation of this firm increases one unit from period t to $t+1$.

During that stage of knowledge disclosure, one important constraint is put on firms, namely their funding after having disclosed specific knowledge must remain positive. For a given firm i, its funding at time t_0 is given by the sum of its past profits and its initial funding $\Psi_{i,0}$ such as $\Psi_{i,t_0} = \Psi_{i,0} + \sum_{t=1}^{t_0} \Pi_{i,t}$, where $\Pi_{i,t}$ stands for firm i's profit at time t and $\Psi_{i,0}$ for firm i's initial funding. Firms not

[6] An important feature of the model is that the action of making knowledge available to partners has no negative effect on profit. Since only specific knowledge serves to determine profit and since knowledge made available to partners does not decrease the stock of specific knowledge, disclosing knowledge to partners does not affect current profit. Yet, this is not true for knowledge openly disclosed. As soon as it is openly disclosed, specific knowledge becomes public, i.e. the action of openly disclosing knowledge has a direct negative effect on disclosing current profit.

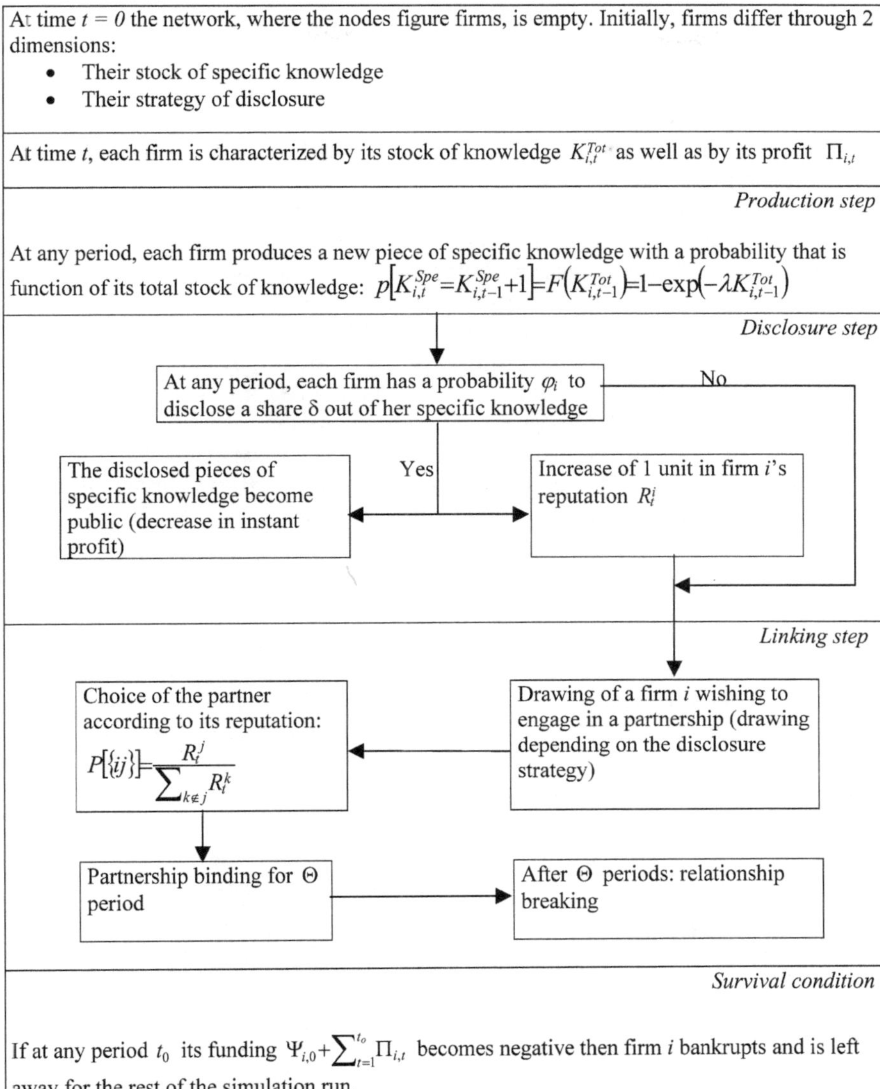

Fig. 1 Dynamics of the model

complying with this survival condition are considered bankrupt and removed from the simulation.

The parameter φ_i plays here a crucial role by influencing the decision to engage the knowledge disclosure process. φ_i corresponds to the probability for firm i openly to disclose knowledge at any period, which means that firms having adopted higher φ_i values are likely to disclose knowledge more frequently than firms that have adopted lower φ_i values. The fraction δ of specific knowledge that is openly disclosed is assumed to be the same for each firm. Thus, we assume that at any period, each firm has a probability φ_i openly to disclose a share δ of its specific knowledge, which hence becomes public.[7]

Third step: Formation of R&D partnerships. Periodically, the third stage of the process, corresponding to partnership binding, takes place. During that stage, one firm has to link up a partnership, which will last over Θ periods, with another firm. The formation of R&D partnerships occurs as follows:

First, we randomly draw from the global population of n firms a firm i that will initiate the partnership. The probability of picking firm i depends on φ_i. We assume that firms endowed with an active disclosing strategy are more likely to be drawn than less active firms. Indeed, the main goal of an active knowledge disclosure strategy is to increase the firm's reputation, which in turn increases the likelihood of binding numerous inter-firm partnerships. But the same type of reasoning may be turned the other way round. Whereas an active firm may receive more partnership proposals than a less active one, the former may also be more motivated to initiate new partnerships than the latter. Thus, it is expected that firms that actively disclose knowledge are more willing to initiate R&D partnerships. The probability $P_{\text{ini}}(i)$ for a firm i to initiate a new relationship is given by:

$$P_{ini}(i) = \frac{\varphi_i}{\sum_{j \in V} \varphi_j}, \quad \text{where } i \in V$$

Second, the firm that has been drawn to initiate the R&D partnership has to select among all other firms (excluding current partners) a potential partner with whom to collaborate in R&D. The choice of the partner is made according to firm reputation, i.e. in function of their disclosing strategy. For a firm i initiating a new inter-firm relationship the probability to choose another firm j that does not already belong to firm i's set of current partners is given by:

$$P[\{ij\} \in G_t / \{ij\} \notin G_{t-1}] = \frac{R_t^j}{\sum_{k \in V - \{T_{i,t-1}\}} R_t^k}$$

With R_t^j the reputation of firm j at time t. Finally, the partnership is assumed to last Θ periods.

3.3 Numerical analysis: settings

Since network models are particularly hard to deal with in an analytical way, we apply the methodology provided by numerical simulation. We focus on the evolution of an innovation network and the condition under which, starting from a situation where no firms are connected to each other, such a network develops (through the accumulation of inter-firm partnership agreements). We focus specifically on the effect of open knowledge disclosure on firms' connections and profitability.

[7] φ_i can be interpreted either as the frequency at which firm i discloses an amount of knowledge identical for all firms or as the amount of knowledge disclosed by firm i at a frequency identical for all the other firms. In the following, we will use both terminologies and refer to φ_i as reflecting alternatively the frequency of the disclosure and the amount of disclosed knowledge.

We consider an economy with 100 firms. Each simulation runs for 60,000 periods during which knowledge production and open knowledge disclosure may occur at any period. Periodical inter-firm relationships bindings are performed every 40 periods, on average. The reason behind such a periodicity lies in the fact that, for the simulation to provide significant results, the graph has to feature the sparseness condition.[8] The parameters we vary are three: the proportion of high disclosing firms, the frequency at which firms openly disclose knowledge, and firms' initial endowments in specific knowledge. The value of all parameters and the initial value of the variables are summarized in Table 1 below.

In a first series of simulation runs, we set the initial endowments in specific knowledge as equal for all firms. As such, firms only differ in their disclosing strategies. We consider only two types of firms: firms of the first type (HD firms as High disclosing firms) adopt an active strategy of knowledge disclosure, giving thus rise to high values for φ_i, whereas firms of the second type (LD firms as low disclosing firms) tend to adopt a more passive strategy associated with low φ_i values. As our results will confirm, firms adopting the latter strategy are giving priority to the maximization of current profit, whereas firms of the first type favor a more long-term strategy. In this first series of simulations, we also vary the proportion of high and low disclosing firms. One can therefore study the evolution of the industry and the individual performances of firms according to different values of the gap between HD and LD firms and according to different proportions of these two types of firms. Specifically, we aim to study what happens when HD firms intensify their activity of knowledge disclosure and to observe whether there is a threshold beyond which knowledge disclosure becomes unbearable, i.e. HD firms all disappear in the first periods of the simulation.

In a second series of simulation runs, we compare the effects of open knowledge disclosure on firm performance with those stemming from initial endowments in specific knowledge. To this end, we set the proportion of HD firms equal to 30% and we allow firms to differ in their initial endowments in specific knowledge. Again, we consider two types of firms with respect to their initial endowments in specific knowledge: firms highly endowed with specific knowledge and firms slightly endowed. During this second series of simulation, we are therefore confronted with four types of firms: Small (in the sense of having low initial endowments in specific knowledge) high disclosing firms (SH firms in the following), small low disclosing firms (SL firms), big (in the sense of having high initial endowments in specific knowledge) low disclosing firms (BL firms) and big high disclosing firms (BH firms). The proportion of firms with respect to the previous typology is given in Table 2.

4 Simulation results and discussion

4.1 When firms only differ in their disclosure strategy

Figure 2 depicts the average number of R&D partnerships contracted by HD firms versus LD firms by displaying the evolution of the ADD (Average Degree

[8] The sparseness condition states that $D_{\text{full}} >> D_{\text{sparse}}$, where D_{full} corresponds to the density of a fully connected graph and D_{sparse} to the density of a sparsely connected graph (Watts 1999).

Table 1 Simulation parameters

Parameters	Definition of the parameters	Values
Model parameters		
N	Number of firms	100
	Number of time steps for each simulation run	60,000
	Average number of time steps between two relationship bindings	40
Θ	Duration of R&D partnerships (in time steps)	7,500
Knowledge parameters		
$K_{i,0}^{\text{gen}}$	Firm's i ($\forall i$=1,...,100) endowment in general knowledge at time $t=0$	500
β	Share of specific knowledge made accessible to partners through R&D agreement	0.5
δ	Fraction of specific knowledge openly disclosed to the industry	0.1%
λ	Coefficient for the exponential law providing the probability of generating 1 new unit of specific knowledge through R&D	10^{-5}
	Frequency of open knowledge disclosure for low disclosing firms	0.1%
Financial parameters		
γ	Marginal income stemming from the use of specific knowledge	0.75
α	Unitary cost for maintaining an inter-firm relationship	0.3
C	Fixed costs that each firm must support at any period	5
$Fund_{i,0}$	Firm's i ($\forall i$=1,...,100) monetary endowments at time $t=0$	2,000
Parameter values specific to the first runs of simulation (Section 4.1)		
	Initial endowment in specific knowledge	5
	Frequency of open knowledge disclosure for high disclosing firms	0.5–1–2–3–5–10%
	Proportion of high disclosing firms in the economy	10–30–50%
Parameter values specific to the second runs of simulation (Section 4.2)		
	Initial endowment in specific knowledge for slightly endowed firms	5
	Initial endowments in specific knowledge for highly endowed firms	10–100–500
	Proportion of highly endowed firms	30%
	Frequency of open knowledge disclosure for high disclosing firms	1–3–10%
	Proportion of high disclosing firms in the economy	30%

Difference) statistics.[9] Whatever the proportion of HD firms and the frequency of the disclosure of HD firms, the ADD statistics remains always positive and displays an increasing trend, indicating that HD firms tend to develop more R&D partnerships than LD firms and that this feature is reinforced through time. As the simulation goes on, the difference between the average number of R&D agreements set up by HD firms increases compared to the average number of partnerships set up by LD firms. The ADD statistics increases sharply at the beginning of the simulation and then, after approximately 10,000 periods, increases only slightly. This break is due to the connection breaking condition, which starts to be effective after 7,500 periods. At the beginning of the simulation, HD firms

[9] The ADD statistics computes at any period t the average degree for high disclosing firms minus the average degree for low disclosing firms: $\text{ADD}_t = \frac{1}{n-m}\sum_{i\in I} k_{i,t} - \frac{1}{m}\sum_{j\in J} k_{j,t}$ where k_i represents the number of R&D partnerships contracted by firm i, I is the set of firms that disclose a high amount of knowledge (n–m firms are in this set) and J the set of firms that reveal a low amount of knowledge (m firms).

Table 2 Proportion of firms by type

		Strategy of open knowledge disclosure	
		High	Low
Initial endowments in specific knowledge	High	9%	21%
	Low	21%	49%

contract R&D partnerships and do not break any partnerships. Then, after 7,500 periods, partnerships contracted earlier in the simulation come to an end, implying a slow down in the increase of the ADD statistics, although HD firms still contract many partnerships.

Figure 2 also stresses the role played by the frequency of open knowledge disclosure on the ADD between HD and LD firms. Overall, the ADD statistics increase when the frequency of the disclosure for HD firms passes from 0.5% to 10%. The more a firm discloses knowledge, the higher the number of R&D agreements in which it is involved. The effect of the frequency of knowledge disclosure on the ADD statistics is also very sensitive to the proportion of HD firms. The ADD increases only very slowly with the frequency of knowledge disclosure when the proportion of HD firms is high (30 and 50%) but increases sharply when the proportion of HD firms is low (10%).

Furthermore, the value of the ADD statistics decreases with the proportion of HD firms. Whatever the frequency of the disclosure, the less numerous the HD firms, the more they develop R&D partnerships compared to LD firms. This point can easily be explained in the frame of our model: we consider open knowledge disclosure as a signal of competence that aims at easing the identification of potential partners with which to cooperate in R&D. Hence, when too many firms disclose knowledge, it is harder for disclosing firms to be distinguished from other firms, i.e. disclosing firms will attract fewer partnerships. Conversely, if only a few firms reveal, these firms will be more easily identified by other firms and, on average, will develop more R&D partnerships.

To summarize, HD firms tend to develop more research agreements than LD firms, i.e. openly to disclose knowledge which has a positive effect on the formation of R&D partnerships. Does this tendency of HD firms to develop more R&D agreements materialize into profit? Are disclosing firms also more profitable than LD firms?

Figure 3 describes the average profit of HD firms as compared to LD firms, by displaying the evolution of the APR (Average Profit Ratio) statistics.[10] Whatever the frequency of the disclosure and the proportion of HD firms, open knowledge disclosure is not profitable in the short run. At the very beginning of each simulation the average profit ratio is always less than one, indicating that HD firms

[10] The APR statistics displays at any period t, the average profit for high disclosing firms divided by the average profit for low disclosing firms (Average Profit Ratio—*APR*) $\mathrm{APR}_t = \frac{\frac{1}{n-m}\sum_{i\in I}\Pi_{i,t}}{\frac{1}{m}\sum_{j\in J}\Pi_{j,t}}$ where i is the set of firms that disclose a high amount of knowledge and j the set of firms that reveal a low amount of knowledge.

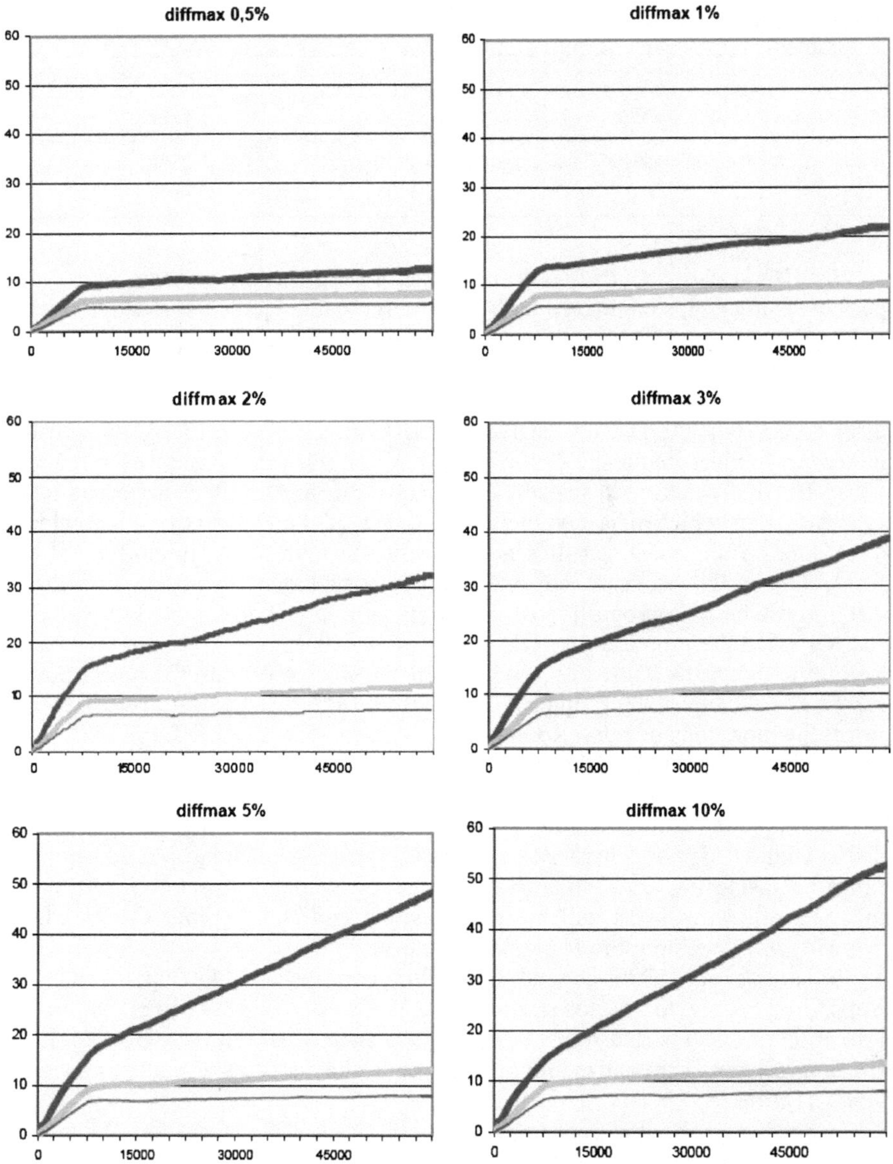

Fig. 2 ADD of high disclosing firms versus low disclosing firms. Legend: proportion of HD firms: *thick and dark*: 10%: *thick and clear*: 30%: *thin and dark*: 50%. Note: Each simulation was repeated 50 times and results displayed here represent the average outcome of the 50 simulations

are less profitable than LD firms. This profit ratio is sometimes negative, meaning that HD firms may even experience losses in the short run.

Although open knowledge disclosure is a risky strategy in the short run, after several periods it can become profitable. In all our simulations, after approximately 12,000 time steps, HD firms become more profitable than LD firms, provided that HD firms adopt a frequency of disclosure lower than 2%. This threshold of 12,000 periods is affected neither by the proportion of HD firms nor by the frequency of

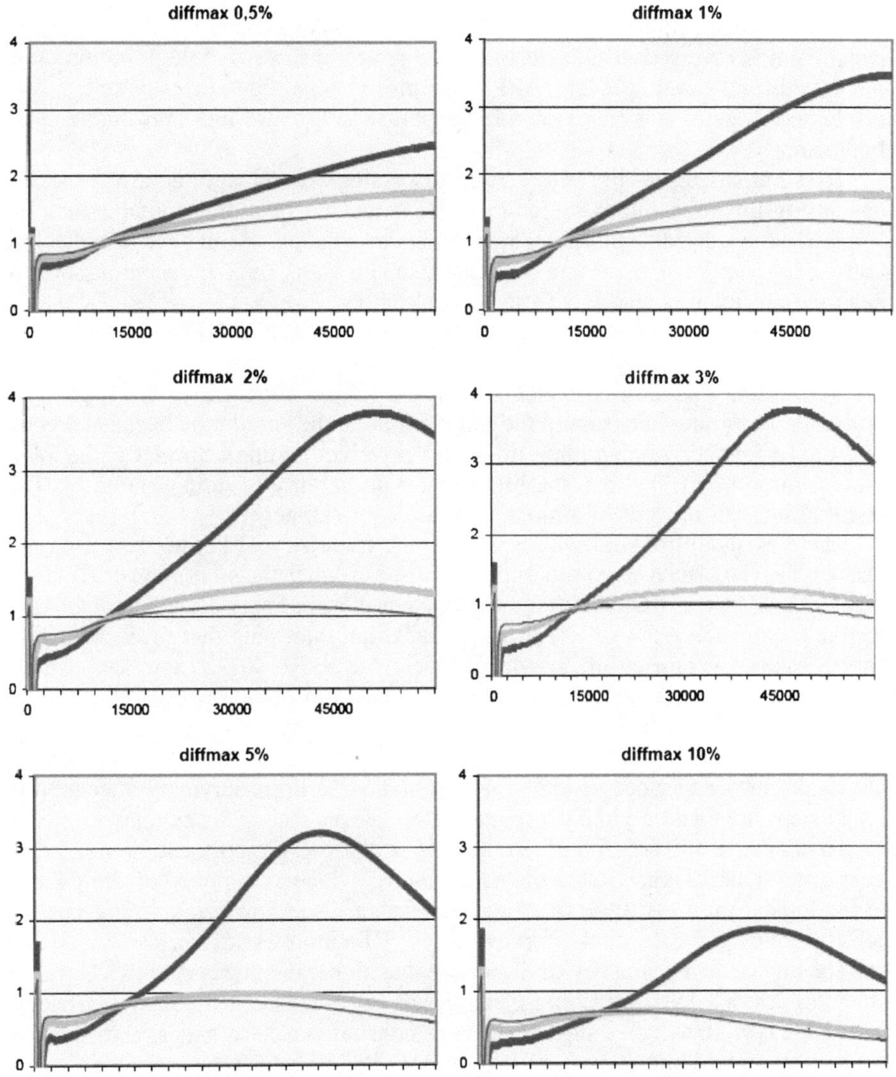

Fig. 3 APR of high disclosing firms versus low disclosing firms. Legends: proportion of HD firms: *thick and dark*: 10%: *thick and clear*: 30%: *thin and dark*: 50%. Note: Each simulation was repeated 50 times and results displayed here represent the average outcome of the 50 simulations

the disclosure. In other words, whatever the frequency (yet lower then 2%) and the proportion of HD firms, after approximately 12,000 periods, disclosing knowledge tends to be profitable for firms. But as soon as HD firms disclose too frequently their knowledge, their profitability decreases compared to LD firms. When the frequency of disclosure becomes higher than 5%, the APR is always lower than 1 (unless the proportion of HD firms remains very low), indicating that disclosing knowledge is no longer profitable.

The profitability of adopting a strategy of high knowledge disclosure depends also on the proportion of firms that disclose knowledge. The less numerous the HD firms, the more profitable they are in the long run, compared to LD firms. Yet, this conclusion is reversed in the short run, where we observe that the less numerous the HD firms, the lower the APR. Therefore, when few firms adopt a high disclosing strategy, it is more risky in the short run but also more profitable in the long run.

To summarize, in the short run, disclosing knowledge is always profit decreasing for firms, but in the long run, HD firms can be more profitable than LD firms if they do not disclose too much knowledge. Moreover, to disclose knowledge can be a risky strategy since, in the short run, the accumulation of negative profits may lead HD firms to bankruptcy.

Table 3 displays statistics related to the mortality rate of HD and LD firms. It indicates that, on average, HD firms are more likely to disappear than LD firms in the short run. The more frequently firms disclose knowledge, the higher the probability of going bankrupt in the first periods of the simulation because they do not satisfy the survival condition (to have a positive funding). Similarly, the lower the proportion of HD firms, the higher their probability of going bankrupt. This result points out the risk of a too active disclosure strategy.

There exists a threshold with respect to the frequency of knowledge disclosure above which HD firms all go bankrupt. For instance, when the proportion of HD firms is set up to 10% and when those firms adopt a disclosure frequency higher than 40%, then approximately 95% of HD firms go bankrupt, indicating that firms can hardly survive with a frequency of knowledge disclosure above 40%. There also exists a threshold in time above which firms do not go bankrupt. Bankruptcy occurs early in the simulation. For instance, when the frequency of disclosure is lower than 10% then, on average, 90% of all the bankruptcy occurs in the first 2,000–2,500 periods, and the last bankruptcy never occurs after 5,500 periods. HD firms surviving after the first 5,000 steps are therefore likely to remain alive for the rest of the simulation.

To conclude this first run of simulations, we also explored the evolution of the aggregate stock of knowledge of the economy, which is a proxy of the pace of technological progress. Overall, open knowledge disclosure has a strong positive effect on the aggregate stock of knowledge.[11] The more firms disclose knowledge and the higher their frequency of disclosure, the higher the increase of the aggregate stock of knowledge. Interpreting this result as evidence that open knowledge disclosure is welfare increasing and, as such, constitutes a socially desirable behavior may be premature but, at least, this result does not contradict this possibility.

[11] We do not reproduce the figure describing the evolution of the aggregate stock of knowledge of the industry. Yet, two phenomena are to be observed. First, the evolution of the total amount of knowledge in the industry follows an exponential trend, which is in line with the theoretical view that knowledge production follows increasing returns (Scotchmer 1991). Second, open knowledge disclosure accelerates this process of knowledge accumulation. Higher disclosure intensity and higher proportion of disclosing firms enhance the production of knowledge in the industry. However, this is true only in the limit of the survival condition. There is a threshold in the frequency of knowledge disclosure above which disclosing knowledge may have a negative effect on the aggregate stock of knowledge since, although firms that remain alive produce more knowledge, fewer firms survive.

Table 3 Percentage of bankruptcy for high disclosing firms

Frequency of the disclosure (%)	Proportion of high disclosing firms								
	10%			30%			50%		
	Percentage of bankruptcy (%)	Average time to the last firm's death	Average time to 90% of realised bankruptcy	Percentage of bankruptcy (%)	Average time to the last firm's death	Average time to 90% of realised bankruptcy	Percentage of bankruptcy (%)	Average time to the last firm's death	Average time to 90% of realised bankruptcy
0.5	17.44	3,500	2,000	7.13	4,250	1,750	4.67	3,000	1,750
1	27.99	4,250	2,250	10.49	4,000	2,000	5.81	4,250	2,000
2	39.35	5,000	2,500	12.78	3,000	1,750	6.42	3,000	1,750
3	48.07	4,500	2,500	13.99	3,500	2,000	6.99	3,000	2,000
5	50.71	6,250	2,000	14.32	4,000	2,000	7.36	3,000	2,000
10	56.19	5,500	2,750	18.70	3,250	2,000	10.45	4,000	2,000
20	67.95	14,250	4,000	–	–	–	–	–	–
30	83.37	17,750	5,750	–	–	–	–	–	–
40	95.54	18,250	4,750	–	–	–	–	–	–

Note: Each simulation was repeated 50 times and results displayed here represent the average outcome of the 50 simulations

4.2 When firms differ both in their disclosure strategy and in their initial endowments of specific knowledge

In the second run of simulations we explore the consequences of open knowledge disclosure on firm performance and, more specifically, we explore whether this strategy can allow firms to compensate lower initial level of specific knowledge. Initial level of specific knowledge should matter for long run profitability. Firms endowed with higher initial level of specific knowledge have a higher probability of producing further specific knowledge, which provide them with a self-reinforcing advantage through time. We expect firms endowed with higher initial level of specific knowledge to perform better both in terms of profitability and of R&D partnerships than other firms.

Our interest here lies in the comparison of the performances of firms endowed with low initial level of specific knowledge, but that adopt an active policy of open knowledge disclosure (SH firms) as compared to firms endowed with higher initial level of specific knowledge (BL and BH firms). This comparison may allow us to draw some conclusions regarding the efficiency of a strategy of open knowledge disclosure in order to overcome the initial disadvantage coming from lower endowments in specific knowledge. Hopefully we will show that, whereas enjoying low initial endowments in specific knowledge is indeed a handicap for long run profitability, adopting an active strategy of open knowledge disclosure can balance it.

Figure 4 displays average profitability according to firm type. Three cases can be distinguished: when high disclosing firms (BH and SH firms) adopt low frequency of knowledge disclosure (lower than 2%), when they adopt a frequency of approximately 2 or 3% and when they adopt a frequency higher than 3%.

First, when high disclosing firms adopt a frequency of knowledge disclosure lower than 2%, long run profitability is mainly determined by their disclosure strategy (either low or high). Initial endowments in specific knowledge play a role in the short run but not in the long run. Overall, both BH and BL firms exhibit higher profits than SH and SL firms in the first periods of the simulation, but firms that disclose a low amount of knowledge are outperformed by those that disclose a high amount. In the long run, initial endowments in specific knowledge do not affect profitability. BH and SH firms, on the one hand, and BL and SL firms, on the other hand, have exactly the same average profits. Moreover, in the long run, the differential of profits between low and high disclosing firms increases slowly but steadily (this may not appear as obvious on the graph since the scale is logarithmic).

Initial endowments in specific knowledge affect the time needed for SH firms to offset BL firms and to catch up with BH firms. The smaller the initial difference in specific knowledge, the faster the catching up. For instance, when initial endowments in specific knowledge are set up to ten for highly endowed firms, high disclosing firms quickly become more profitable. After approximately 5,000 periods, SH firms outperform BL firms, and after 10,000 time steps, SH firms converge toward BH firms and BL firms converge toward SL firms. In comparison, when initial endowments in specific knowledge are set up to 500 for highly endowed firms SH firms need 30,000 periods to catch up and to outperform BL firms and 40,000 periods to catch up with BH firms. To sum up, when the frequency of disclosure is lower than 2%, disclosing firms catch up with highly

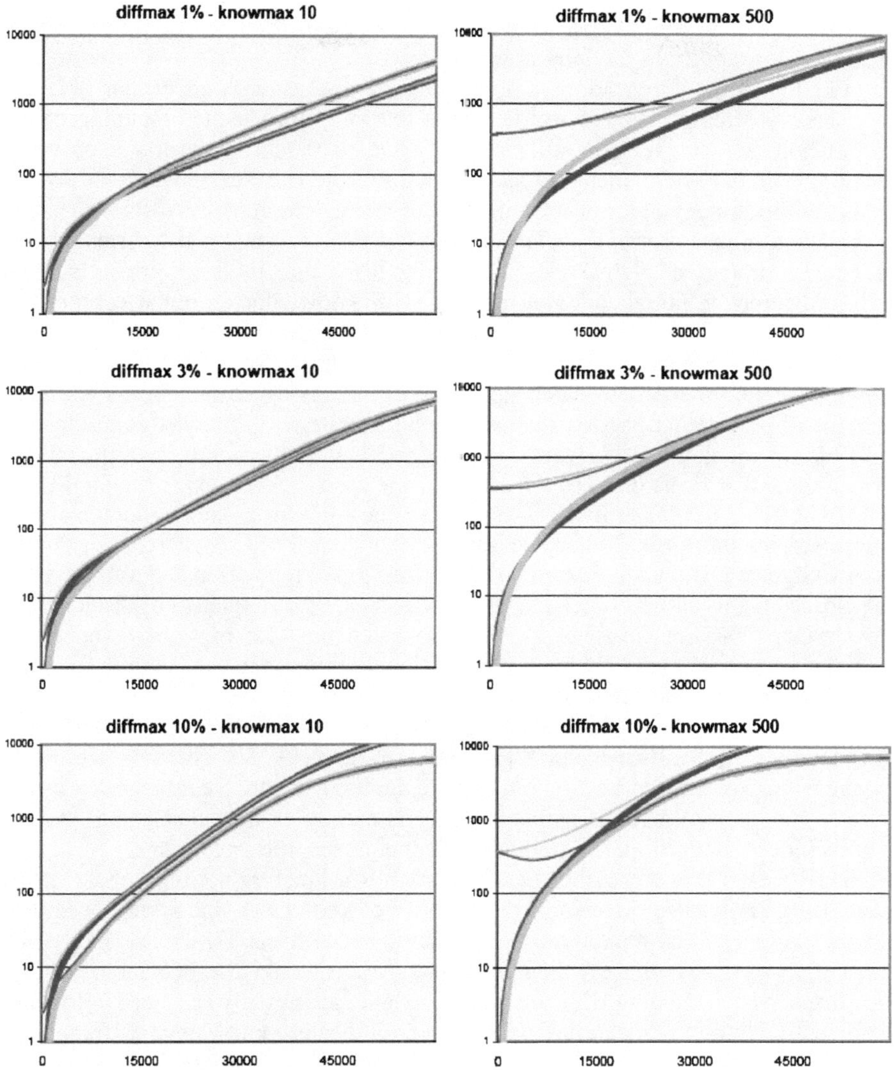

Fig. 4 Firms' profits according to the frequency of knowledge disclosure and to initial endowment in specific knowledge. Legend: Type of firms: *thin and dark* = *BH*; *thick and dark* = *SL*; *thick and clear* = *SH*; *thin and clear* = *BL*. Note: The scale is logarithmic. Each simulation was repeated 50 times and results displayed here represent the average outcome of the 50 simulations

endowed firms. The initial difference in specific knowledge does not affect long run profits, but only the time needed to achieve long run positions.

This conclusion changes when firms adopt a frequency of knowledge disclosure of approximately 2 or 3%. In the short run, firms highly endowed with specific knowledge still remain more profitable than firms slightly endowed, the latter still catch up with the former in a relatively short time span, which depends on the difference in initial endowments in specific knowledge, but then all the four types of firms follow the same trend. In this case, neither initial endowments in specific knowledge nor the strategy of disclosure influence significantly profitability in the long run.

Finally, when firms adopt a frequency of disclosure higher than 3%, the conclusions stressed above are reversed in the long run. At the beginning of the simulation, BH and BL firms still perform better, but the initial advantage provided by higher initial endowments in specific knowledge is quickly overcome by the effect of the strategy of knowledge disclosure. In the long run, low disclosing firms experience higher profits than high disclosing firms, meaning that firms having adopted a strategy of high level of knowledge disclosure are less competitive than others. Moreover, initial endowments in specific knowledge do not affect profit in the long run, but only the time needed for SH and SL firms to catch up with BH and BL firms. The higher the difference in endowments in specific knowledge, the longer the time needed to catch up. Differences among firms' profits are only explained by whether or not firms adopt a high strategy of knowledge disclosure.

Prior to concluding, it is worth stressing an important assumption of the model. We assume that firms disclose a fraction of the specific knowledge they hold, but that whatever the "quantity" of knowledge they disclose, their reputation only increases by one unit. Hence we assume that it is only the act of disclosing knowledge that counts in order to build the firm's own reputation and not "what" is disclosed.[12] This assumption can explain why, when the frequency of disclosure is lower than 2%, SH firms always tend to catch up with big firms. Indeed, if reputation is only based on the act of disclosing, it implies that reputation of SH and BH firms, on the one hand, and of BL and SL firms, on the other hand, follows exactly the same trend, which means that BL firms are likely to develop as many partnerships as SL firms and BH firms as many as SH firms (since R&D partnerships are mainly function of firms' reputation). Figure 5 confirms that firms' degree (i.e. their number of partners) depends mainly on their disclosing strategy and is independent of their initial endowments in specific knowledge.

However, if it was assumed that reputation is a function not only of the act of disclosing knowledge but also of the quantity of knowledge that is disclosed, BH firms would see their reputation grow faster than SH firms. Hence it is likely that the latter would have more difficulties to catch up. It would also become harder for SH firms to catch up with BL firms, since the latter are initially endowed with more knowledge, implying that, although they do not disclose knowledge as frequently as SH firms, the quantity of knowledge they disclose may be more important than the quantity of knowledge disclosed by SH firms. Hence BL's reputation may also grow faster than SH firms' reputation.

5 Conclusion

This paper aimed to provide a theoretical model describing the formation of innovation networks by allocating a central role to behaviors of open knowledge

---

[12] In order to support this assumption, it can be argued that firms, when they assess other firms' reputation, consider only the act of disclosing knowledge because to consider other indicators would be more time consuming and more expensive.

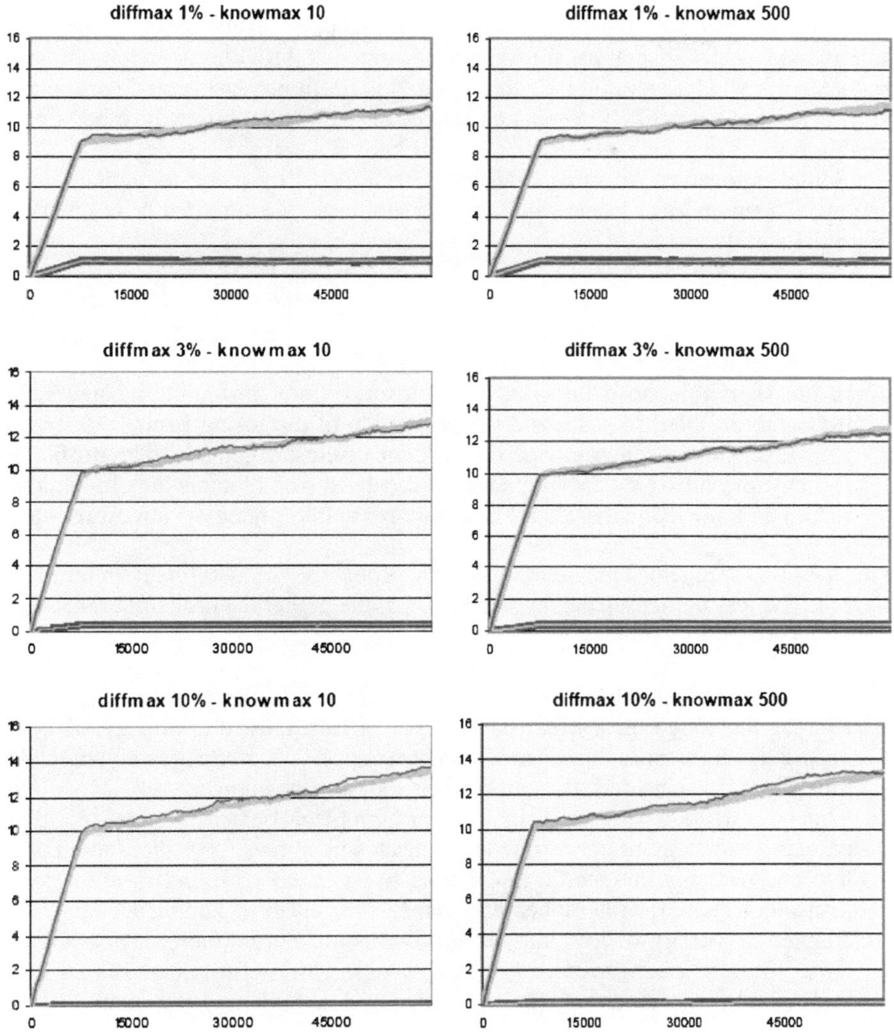

Fig. 5 Firms' degree according to knowledge disclosure intensity and to initial endowments in specific knowledge. Legend: *thin and dark=BH*; *thick and dark=SL*; *thick and clear=SH*; *thin and clear=BL*. Note: *BL* and *SL* curves on the one hand and *BH* and *SH* curves on the other hand follow exactly the same path, which explains why only two curves appear clearly on the graph. Each simulation was repeated 50 times and results displayed here represent the average outcome of the 50 simulations

disclosure. More generally, the model developed in this paper was concerned with the fact that a strategy of open knowledge disclosure constitutes not only a gift made by disclosing firms to other firms, but also contributes to easing the formation of R&D collaborations. Firms that widely disclose knowledge to other firms are more likely to enter innovation networks and to acquire a central position within those networks because the act of disclosing knowledge increases their reputation, which constitutes the main criterion for a firm's decision to start a cooperating with other firms.

Firms are therefore confronted with the following trade-off: actively disclosing knowledge is penalising on the one hand, since it provides competitors with valuable knowledge, but, on the other hand, it is profit increasing since disclosing firms are also more likely to engage into R&D collaborations with other firms, enabling them to access external sources of knowledge.

Some implications of our model have been tested by using numerical simulations. To summarize, let us emphasize the following results that have emerged from the simulations:

(1) Open knowledge disclosure increases the number of R&D partnerships contracted by high disclosing firms. The less numerous the disclosing firms, the higher the number of R&D partnerships they are involved in;
(2) In the short run, open knowledge disclosure is not profitable, whatever the frequency of the disclosure and the proportion of disclosing firms;
(3) Conversely, in the long run, open knowledge disclosure can become a profitable strategy, provided that the frequency of the disclosure remains low. In the long run, open knowledge disclosure is a more profitable strategy if few firms adopt it;
(4) Open knowledge disclosure increases the probability of bankruptcy in the short run. The less numerous the disclosing firms, the higher the probability that they go bankrupt;
(5) In the short run, firms highly endowed with specific knowledge always perform better than firms slightly endowed, whatever the frequency of the disclosure. But, in the long run profitability is only affected by the strategy of open knowledge disclosure. Initial endowments in specific knowledge only affect the time firms need in order to achieve their long run position;
(6) When the frequency of the disclosure for high disclosing firms is not too high, adopting a strategy of open knowledge disclosure allows firms that started with low endowments in specific knowledge to catch up with and to outperform firms that started with higher endowments in specific knowledge and that adopted a strategy of low knowledge disclosure. Furthermore, it also allows those firms to catch up with (but not to outperform) firms that started with higher endowments in specific knowledge and that adopted a strategy of high level of knowledge disclosure;
(7) Conversely, for disclosure frequencies higher than 3%, firms that adopted a strategy of low level of knowledge disclosure perform better, thus indicating that, if disclosing knowledge can be profitable in the long run, disclosing knowledge too frequently is harmful.

Our model provides therefore a rationale to behaviors of open knowledge disclosure by showing that such strategies, although risky in the short run, may pay in the long run because they enable firms to access external sources of knowledge more easily. Yet, many points will deserve a more in depth treatment. Here are three suggestions that may be worth including in future developments:

First, we may consider a different reputation function that would take into account the quantity of knowledge that is disclosed and not only the act of disclosing. This, as discussed shortly above, may greatly affect our conclusions regarding the convergence among firms initially endowed with high and low levels of specific knowledge.

Second, it may be necessary to include firms' absorption capacity in the model. Firms' ability to absorb external knowledge depends on the knowledge they already hold (Cohen and Levinthal 1989). By introducing an absorption capacity, we may therefore favor firms initially endowed with a higher level of knowledge because those firms may be more able to absorb external knowledge. In turn, this may slow down the process of catching up among SH firms and BH and BS firms.

But above all, it may be central to lay more emphasis on behaviors of knowledge trading, which were considered as exogenous in our model. The interaction between open knowledge disclosure and knowledge trading stands at the heart of the innovation process. It is therefore essential to explore how these two types of knowledge disclosure co-evolve in the development of innovation networks.

Acknowledgements This paper was presented at the Druid 2004 summer conference in Copenhagen and at the International Schumpeter Society 2004 conference in Milan, where the authors were awarded the ISS best poster prize, sponsored by *Industrial and Corporate Change*. We are particularly grateful to Uwe Cantner for helpful comments.

References

Allen RC (1983) Collective inventions, J Econ Behav Organ 4:1–24
Boivin C (2000) Protection et partage des connaissances dans l'economie du savoir. Paper presented at *L'Economie du Savoir: Enjeux Economiques et Enjeux de Gestion*, École des HEC, Montréal, 5-6 mai
Callon M (1998) Les pouvoirs publics doivent-ils soutenir la recherche Académique? Ann Mines :59–64
Cockburn I, Henderson R (1998) Absorptive capacity, coauthoring behavior, and the organisation of research in drug discovery. J Ind Econ 46:157–182
Cohen WM, Levinthal DA (1989) Innovation and learning: the two faces of R&D. Econ J 99:569–596
Cooke P (2001) Regional innovation systems, clusters and the knowledge economy. Ind Corp Change 10:945–975
De Fraja G (1993) Strategic spillovers in patent races. Int J Ind Organ 11:139–146
Eaton BC, Eswaran M (2001) Know-how sharing with stochastic innovations, Can J Econ 34:525–548
Gambardella A (1992) Competitive advantage from in-house scientific research: the US pharmaceutical industry in the 1980s. Res Policy 21:391–407
Gibbons M (1994) The new production of knowledge: the dynamics of science & research in contemporary societies. London
Granovetter M (1973) The strength of weak ties. Am J Sociol 78:1360–1380
Granovetter M (1974) Getting a job: a study of contacts and careers. Harvard University Press, Cambridge, MA
Grosseti M, Bès M-P (2002) Proximité spatiale et relations science-industrie: savoirs tacites ou encastrement ('Polanyi ou Polanyi')? Rev Econ Rural Urbaine 5:777–788
Harhoff D (1996) Strategic spillovers and incentives for R&D. Manage Sci 42:907–925
Harhoff D, Henkel J, von Hippel E (2003) Profiting from voluntary information spillovers: how users benefit by freely revealing their innovation. Res Policy 32:1753–1769
Henderson R, Cockburn I (1994) Measuring competence? Exploring firm effect in pharmaceutical research. Strateg Manage J 15:63–84
Hicks D (1995) Published papers, tacit competencies and corporate management of the public/private character of knowledge. Ind Corp Change 4:401–424
Hicks D, Ishizuka T, Keen P, Sweet S (1994) Japanese corporations, scientific research and globalization. Res Policy 23:375–384

Hicks D, Isard PA, Martin BR (1996) A morphology of Japanese and European corporate research networks. Res Policy 25:359–378
Koenig M (1983) A bibliometric analysis of pharmaceutical research. Res Policy 12:15–36
Kogut B (2000) The network as knowledge: generative rules and the emergence of structure. Strateg Manage J 21:405–425
Lim K (2000) The many faces of absorptive capacity: spillovers of copper interconnect technology for semiconductor chips. working paper 4110, MIT Sloan School of Management
Maskell P, Lorenzen M (2003) The cluster as market organization. Druid working paper 2003-14
Muller P (2003) The role of authority in the governance of knowledge communities. Paper presented at DRUID Winter 2003 PhD Conference. Hotel Comwell Rebild Bakker, Aalborg, Denmark, January 16–18
Pénin J (2003) Endogénéisation des externalités de recherche: le rôle de la capacité d'emission des connaissances. Rev Econ Ind 102:7–28
Pénin J (2004) Open knowledge disclosure: an overview of the empirical evidences and economic motivations. Presented at the third EPIP meeting in Pisa, 2–3 April
Scotchmer S (1991) Standing on the shoulders of giants: cumulative research and the patent law. J Econ Perspect 5:29–41
von Hippel E (1987) Cooperation between rivals: informal know how trading. Res Policy 16:291–302
Watts DJ (1999) Small worlds: the dynamics of networks between order and randomness. Princeton University Press, Princeton
Williamson O (1975) Markets and hierarchies: analysis and antitrust implications. Free, New York

Alicia Giacchero · Nora Donnini · Fanny Martin

Innovation and competitiveness in local SMEs: Characteristics of firms, entrepreneurs, environment and their interrelationships

Abstract In this paper, we evaluate the capacity of a local SME's productive system to incorporate scientific and technological innovation. We identify the principle external and internal conditions of enterprises required to form a critical mass in order to fulfil their task in this process. To find empirical support, we look at the features of firms and local entrepreneurs through surveys of Bahía Blanca enterprises that produce goods and services in different productive sectors: food, wood and furniture, plastics, chemicals and metal mechanics. Additionally, we consider factors external to the firm: R&D, the educational system, and research centers and their interconnections.

Keywords Local innovation system · Public policy · SME · Entrepreneur
JEL Classification O3 · O54 · H7

1 Introduction

Technological change is considered a Schumpeterian process with three moments, not always strictly distinguishable, these being invention, R&D and innovation. This traditional model of innovation suggests that knowledge takes only one direction, from research centers to productive sectors. However, there exist other forms of knowledge diffusion and innovation of higher importance to the relationships between agents than their ability to gather information and knowledge. These relationships between agents, and between agents and institutions, become an important element in the creation of dynamic competitive advantages, based on the formation, transmission and evolution of knowledge. This implies that the essential support of a productive system is not to further this system directly, but rather lies in promoting the institutions, making available knowledge to firms, the entrepreneurial environment and the productive framework within which they act.

A. Giacchero · N. Donnini · F. Martin
Departamento de Economía, Universidad Nacional del Sur, Bahía Blanca, Rep. Argentina, E-mail: giacche@criba.edu.ar

Thus, the specific location itself becomes a knowledge generating factor, where local institutions seem to play a more active role related to regional economic activities.

With this emphasis on local conditions in mind, a new strategy becomes relevant: that of decentralized politics, with a central role played by the local authority and agents, without neglecting the interaction with the central government. This setting, which constitutes a policy target, depends on historical aspects such as preceding linkages as well as on the types of agents and institutions acting within the region. In this sense, initial conditions as well as the local evolution impose constraints on the opportunities at improving the productive environment. Policy actions aimed at forming a local system of entrepreneurs must consider aspects such as the innovative enterprise, the institution of networks and the improvement of entrepreneurial links with universities and technological centers. In short, they have to focus on a system driven by the capacity to access and absorb knowledge, establishing and improving individual or collective activities. Innovation in this context is understood as a result of the evolution in learning, where the consolidation of previous knowledge as well as the use of new concepts allows for the building up new competences.

An important issue within this context is that every design of technological support to SMEs must take into account that their network is weak or non-existent. Therefore, political support should act not only upon the nodes of the network, but upon those accessory nodes which have sporadic links with agents, counsellors, the whole network and the system. Thus, it is evident that any player in the net must adopt the role of a generator of or contributor to a favorable network environment for ongoing innovative activities.

This paper attempts to evaluate the capacity of a local SME productive system to incorporate scientific and technological innovation, both seen as dynamic factors for progress. We identify the principle external and internal conditions of enterprises required to form a critical mass in order to fulfil their task in this process. In this respect, we rather briefly review a few theoretical considerations about the relationship between the size of firms with innovative activities and their application to SMEs. To find empirical support, we look at the features of firms and local entrepreneurs through surveys of Bahía Blanca enterprises that produce goods and services in different branches corresponding to different productive sectors: food, wood and furniture, plastics, chemicals and metal mechanics. Additionally, we consider factors external to the firm: R&D, the educational system, universities and research centers and their interconnections. Finally, we identify the dynamic interaction between actors who generate knowledge, and investigate in which sense these processes may lead to or contribute to regional agglomerations or corridors. We attempt to identify whether an innovative and competitive system exists in a local environment through the educational, productive, scientific and technological structures, as well as through their interactions with the social organization.

2 Innovation and the size of the firm

Research on the differences in economic development has always been a center of interest. Different theories have placed emphasis on the endowment of existing resources, on the predominant cultural values, on the nature of the economic organization, on the incorporation of new techniques, and in any combination of these elements. Some recent approaches have emphasized the growing importance of the vitality of the creative and network oriented businessman.

In these approaches, the competitiveness of a firm is based less and less on scale and capital but more on the motivation and abilities of businessmen. Accepting this argument of the importance of business management and the creative use of modern technology, SMEs in the third world suddenly are provided with opportunities to regain competitiveness. In this framework, vital importance is placed on the attitudes of businessmen. Using the assumption of perfect information, conventional economic thought left no role to the businessman who appeared as a mere arbitrator of capital, work and technology.

It was Schumpeter who introduced the concept of the entrepreneur in order to shift the focus on innovative activities and innovations. The businessman is the one who carries out with new combinations. Here Schumpeter ([1935] 1912) defines new combinations or innovations as:

1. The introduction of a new good or a new type of good in the market,
2. The introduction in an industry of a new method of production
3. The opening of a new market in a country
4. The attainment of new sources of supply of raw material or not completely manufactured
5. The implantations of a new organizational structure

In view of this, different types of innovations can be identified. First, *technical innovations* relate to products and services, technical productive process and service operations. Second, *organizational innovations* relate to the structure and administrative direction of the organization, processes and human resources. Third, it is possible to distinguish *process innovations*, restricted to the specific company in which it takes place, and the *product innovations* that allow the satisfaction of demand and the expansion of market share of the company. In addition, we can define an innovation as *radical* if fundamental changes can be realized in the activities of the firm; an innovation is called *gradual* when those changes are marginal with respect to habitual practices.

For Schumpeter, the execution of new combinations is a special function and constitutes the privilege of a type of agent. Schumpeter utilized this concept to explain structural change, economic growth and economic cycles, combining psychological and economic ideas. This approach of Schumpeter can be summarized in three pairs of oppositions: a) the tendency of the economic system to equilibrium as opposed to disruptions from inside; b) the static theoretical apparatus as opposed to the dynamic one; c) the "manager" versus the "businessman". A fourth opposition can be added: d) that of the "inventor" as opposed to "innovative businessman".

As to the last, we should distinguish among "economic leadership" and "invention". Inventions lack economic importance if they are not put in practice. The application of any improvement is a completely different task from the invention, and requires different aptitudes. Though businessmen may be inventors, as they can be capitalists, they do so coincidentally and not by nature.

Schumpeter himself induced a discussion about the influence of firm size on innovative success. When in 1912 he focused on the entrepreneur, in Capitalism,

Socialism and Democracy (1942) he emphasized the large firm running a research laboratory and seeking systematically for new combinations. The discussion of the so-called Neo-Schumpeter-hypotheses takes up this issue and seeks the relation between firm size and innovation. However, no clear result about the intensity and direction is found in the literature. Many authors (Sullivan and Kang 1999; Damanpour 1992; Dewar and Dutton 1986; Ettlie et al. 1984) find a positive relationship. Consequently, the high stock of technical knowledge in large-sized firms, their diversity and abundance of resources, allow them to be relatively more innovative, as well as to surpass the risks and failures of such activities more easily (Camisón Zornoza et al. 2002)

For those firms, innovation management attempts to control the relation between innovation-input and innovation-output: the innovation input is often measured by the expense in R&D related to sales, the R&D intensity. The result of innovation is measured by the number of products developed and commercialized or by the frequency or intensity of innovation. The management of innovation takes into account the level of integration in organizational mechanisms: functional coordination, tools coordination and external connections. Whereas traditional innovation theory states that R&D activities constitute the central aspect of innovations efforts, later developments imply that the process must be realized in an integrated organization (Parthasarty 2002).

This indicates an emphasis on innovation as a result of a systemic scheme, by acknowledging the importance of functionally integrated organizations strongly connected to the intensity of innovation activities. However, a high external orientation and integration of those activities may be an important aspect when a firm shows a low R&D intensity. In this sense, firms could be protected from internal organization failures by relying more on external sources of information and know-how. We have to be aware of this issue when we analyze the innovative behavior of SMEs, as this type of firms has not been shown (Dichiara 2002) to pursue innovation in a system that relies much on R&D activities. Thus, external links become important, as the firm can replace internal activities by strong external interlaces.

Notwithstanding a positive relation between firm size and innovation capacity, we must not forget the effect of other variables, such as entrepreneurial attributes, that fortify the development of several innovations but will debilitate other ones. Besides this, we have to consider the influence of factors such as promoters and inhibitors to innovation adoption.

3 Firm Characteristics

For Argentina, several papers (Yoguel 2003; Angelelli et al. 1999) have pointed out that SMEs face serious difficulties due to scarcities in innovative capacities. Those firms are framed in highly disarticulated nets, isolated supporting instruments, without an adequate systematization. Several analyses (Milesi 2002; Yoguel and Rabetino 2002; Yoguel and Boscherini 2001; Bisang and Gutman 2001) show that these difficulties of SMEs, principally working in technology oriented sectors are related to their internal and external environment as well as their links to other firms and institutions. As a consequence, we can detect as a feature of SME-based industrial systems a rather low number of firms with high innovative capacity, a low integration of production into their respective local systems, and a very low degree of connections due to supplying directly political instruments. Those elements together provide for an important asymmetry, with the consequence that innovative capacities, work training and connections to large firms are weak and the possibility of learning by doing within the network disappears.

Hence, the weak or non-existing connections between the public and private sphere, the weak linkages between firms and scientific or technical institutions or programs, cause learning processes to break down. Consequently, a loose coupling between supply and demand due to ignorance or non-utilization of existing programs is to be observed. In addition, an informal interaction between private agents aiming at (product) innovation becomes less likely to work out.

Modernization in SMEs becomes difficult, as most show a lag in the use of modern manufacturing processes and new technologies. Such a lag is evident in hardware as well as in software and management technologies. Moreover, SMEs rarely provide training or qualification to their workers, and do not participate in public efforts or adopt external technologies. Another remarkable disadvantage present in the case of Argentina is that innovative activities in many fields require large investment. This can only be provided by large-sized companies, since SMEs have neither their own funds nor - due to their low (or null) possibilities for guaranteeing repayment - have easy access to credits. Innovations of SMEs are generally a consequence of the "spillover" of knowledge from the large companies.

Based on these points, the process of incorporating new technologies into SMEs is slow in Argentina. Different studies (Burachik 2002; Sanchez and Dichiara 2002; Salloum 2002) affirm that the process of investment in these firms is weak and proportionally smaller than the large companies.

4 The SME manager

The success and growth of a small and medium sized company depends on many factors, among which several are decisive in each business. It is the task of the firm to identify and to evaluate these in order to develop them further to reach a better competitive position in its sector. There are basic activities an entrepreneur has to bear in mind. His success depends on the correct application of the respective techniques. Among the most important activities are planning, organization and director training. However, the small firm owner usually is too occupied with daily tasks to take time in planning the future of his company. Nevertheless, he would be conscious that planning would allow him to face future problems.

It is very important to equip a business with an efficient organization and structures that allow for division of labor and a corresponding allocation of responsibilities. In the small and medium company, human resources are relatively scarce and directive tasks are assigned to a single person or a small group. In effect, in many small firms, the same person is owner, chief of directorate, administrative director, sales and operation director. This functional diversification causes problems of two types: the difficulty to acquire a deep knowledge in all the fields of enterprise management, and the lack of time to fulfil all these obligations simultaneously.

Personal characteristics relevant for a successful fulfilment of these tasks are found in the literature. Screening the material available on SMEs (Giacchero and Valentini 2003), dominant characteristics of successful manager are:

- Self confidence;
- push and energy;
- long term commitment;
- precise goal setting;
- optimal resource usage;
- fighting against self-imposed standards ;
- moderation of risk;
- dealing with failure:
- use of retro-information;
- leadership and responsibility;
- personal and internal firm control;
- tolerance with ambiguity and uncertainty.

As to strategy, there are different kinds applicable to small enterprises. We observe that a great deal of small firm owners operate without strategies. Their organization is reactive and their decisions are only responses to actions received from the environment. Other firms work as if their decisions follow entirely the intuition of their owners. Finally, there are organizations which operate within a well-known and transmitted strategic frame which provides a shared vision for the firm and its employees.

Certain principles that SMEs must be aware of aim at the basic conditions of competitiveness that must be regarded as essential to survival and to the achievement of success. Some of the problems that must be solved by a firm are related to management and qualification areas and their entailment with the goals set by the management.. One of the keys to solve these problems is a modern and flexible structure embedded in a global conception of management. It is for that reason that beyond the inherent skills of the SME owners, their natural ability to face the daily challenges, their agility and their sense of smell for the business are today indispensable in their transformation into a modern manager.

According to Schumpeter, when innovations appear, they do so in a discontinuous way and in swarms. Innovating industrialists appear in the same way. New industrialists facilitate others, making the innovation more general and propitiating its acceptance.

Taking these points together, an innovating industrialist can be characterized by a combination of innate talent and a propitious atmosphere that allows taking advantage of that talent, which has been generated and/or promoted within a respective institutional environment.

5 The new cooperative-competitive forces

The new strategies for the competitive development of SMEs focus on the creation of networks of innovators, where institutions and companies are related by knowledge flows influencing each other. These knowledge flows are based on cooperative behavior. This would constitute a new frame of analysis in the style of the competitive forces by Porter that allows a firm in an extended scheme of competition to achieve a favorable or disfavorable performance. In this new frame of analysis, the central forces are of a cooperative nature in order to obtain a higher competi-

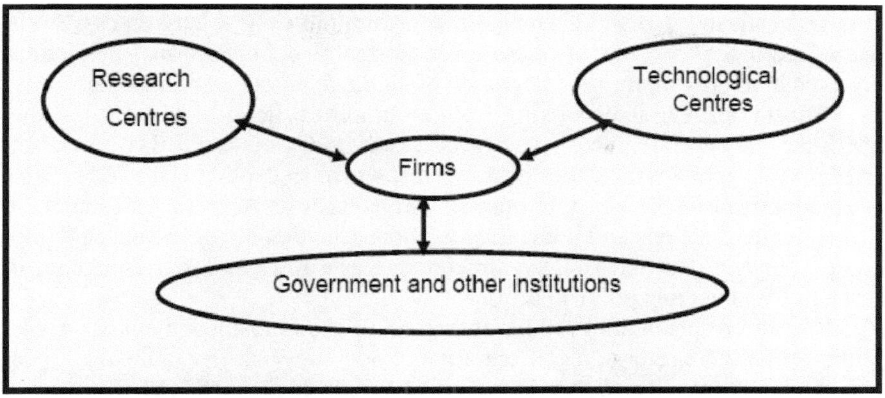

Fig. 1 Cooptitive forces

tiveness, which can be denominated cooperative-competitive or cooptitive. Those cooptitive forces appear in Figure 1.

In this scheme of analysis, three characteristics have to be emphasized. In the first place, there is a normative question as to how those relations have to be designed. This means that technological research centers would have to develop structures of technical attendance and formation, for which it is necessary that there exists a disposition to generate new forms of interaction with other firms and institutions. In addition, it is necessary that the state generates a frame suitable to the development of these relations. This requires not only contributing the legal frame but also offering stimuli to incorporate and to develop innovations, on the one hand, and to reinforce or generate relationships and cooperation, on the other hand. Thus, a state is needed that stimulates the cooptitive relations among members of the system, in addition to generating own relationships.

The second characteristic of the relations in Figure 1 points to these dynamic features as relations are marked by a culture shared among the participants, resulting from history and previous structures generated by the actors. Therefore, the cooptitive forces obey a scheme of repeated games in which the participants' decisions are influenced by the experience from the past. With this history dependence they have specific ties to the past, which provide for a particular scheme of cooperation in each locality or region. Consequently, if one has to develop a cooptitive strategy, this cannot be of the general type: One must take into consideration the concurrent scheme combined with the conditions that allow the system to evolve, to innovate and to learn.

The third key element in this scheme that arises from the previous ones is the institutional frame within which these cooptitive forces are to be developed. This frame has to allow for a suitable management of knowledge that is to be integrated. Hence, when we think here about the role of government, one may ask whether it is the national, provincial or municipal government which should act. In fact, the lines of policy arise from the highest level. Therefore, if we consider financing on a large scale, the necessary resources in general will not be available at the level of local government. Contrariwise, when we think about management and the control of resources, the local capabilities to intervene directly and the local experience become invaluable assets. This means that institutions exist at least on two levels, one at the macro and one at micro level. Formal and / or informal institutions involved here provide for the integration of all the relations in the scheme.

With this background, looking from an empirical point of view at the way SMEs are embedded regionally, several cases can be distinguished. SMEs in Brazil or Mexico fit into dense institutional weaves, where an explicit vision of policies of competitiveness shows up. In other countries, such as Argentina, Colombia or Chile, frequent actions and a low degree of coordination are observed. In a third group countries, we find isolated structures (Costa Rica, Ecuador, Uruguay) or structures without any political actions.

To be more specific in the case of Argentina, the fundamental characteristics of the SMEs there refer to a concentration of activities in a few branches, a geographical concentration as they in general supply local demand. These operate in branches that are not SME branches, which implies that they compete with large firms. This last characteristic means that they are not isolated productively, but they are isolated at the time of making decisions.

Looking at the system of political support aiming at SME's general environment, several aspects appear negative. There often is a mismatch between policy organs and instruments applied, combined with a lack in the diffusion of these policies. Moreover, there is little adjustment of these instruments to the necessities of the firms, and the induced effect is much reduced. Moreover, a bias is observed towards getting companies of larger size. Thus, for Argentina, one can state that there is a scarce use of support policies towards SMEs. Also reflected in several studies (Yoguel 2003; Moori-Koenig 2000; Chudnovsky and Lopez 1997; Galante et al. 2001) is a fact that subsidies are devoted more to the demand side, whereas they are of very little or minor importance for the use of scientific technological services or of those offered by specific organs (such as Instituto Nacional de Tecnología Industrial, CONICET, etc.)

The aspects absent in these support systems constitute a key nucleus in their structure: The most important ones are SMEs. In the economic system, they would serve as a subsidiary device, acting in an intense network of contacts and providing for pivotal relations and contacts therein. As to any policy invention, the instruments applied are expected to design these relations. Doing so, the heterogeneity of the segment focused on has to be recognized and to be taken into account. In this respect, the decentralization of support measures in local organizations seems to be appropriate to cope with any specific problems companies face in their business environment. The policy measure would be in general driven by a vision about any improvement in the interaction of firms, the resulting network structures and the bonds with other institutions. Taken all these points together provides the third element of the scheme in Figure 1, that is an adequate institutional frame within which a **lo**cal **n**etwork of **d**evelopment and **i**nnovation (LONDI) (red local de desarrollo e innovación, RELDI, in Spanish) can work effectively.

6 The local LONDI

The City of Bahía Blanca is located in the south west of the province of Buenos Aires and constitutes a nodal point of great importance in the country. The economic structure of the city, as part of the urban conglomerates in the pampas, emerges in a context of intense connections between rural sectors and urban economic activities. Due to a less complicated geography, this area was always open for effective occupation and allowed gradual adjustments and growth from immigration. For that reason, Bahía Blanca already had a prominent role. The surrounding area provides farming products, which over time lead to the emergence of a center of commercialization of cereals and bovine cattle. Due to an adequate sea port structure, the city has experienced a strong exporting dynamic, confirming the spatial organization model suggesting a strong capital accumulation in coastal cities.

Within that frame, commercial, industrial and services structures have allowed the city to supply demands from the south of the country, which in turn caused it to develop as a node of redistribution of goods and services, as well as a site of numerous administrative jurisdictions. Not the least, this is also due to its function as a geographic node where several national and provincial routes meet.

The farming sector provides for the pre-eminent economic profile of the region. The three main categories of basic raw materials available in Bahía Blanca and its hinterland are those from farming, gas and petroleum. The availability of primary resources has been the source of a tremendous productivity development within the agro-alimentary and petrochemical local industry.

The establishment of industrial firms was bound to the specialization of the pampas region in farming. Flour mills, oil products, alimentary suppliers and firms dedicated to the provision and repair of rural machinery and of tools occupy a dominant share in the industrial sector. Since the '70s, the tendency towards the specialization in foods and drinks has been accentuated, as has the development of the metal -working industry. In addition, a raising share in the manufacturing of chemical products derived from petroleum has been observed. During the ′90s, in a new social context, an economic and political crisis led to stagnation, causing the destruction of numerous jobs. According to data of the last Economic Census (INDEC 1994), Bahía Blanca was made up of 1135 industrial establishments, 985 of which were grouped in manufacturing branches, with the remaining 150 belonging to construction and to the provision of energy. The more prevailing economic activities are food and beverages, manufacture of metallic devices, construction and printing. The value of production is strongly concentrated in petrochemical large-sized enterprises gathered in a chemical complex.

In different studies (Gorenstein and Burachik 2002; Dichiara et al. 2002) on the regional productive structure based on economic censuses and surveys, several characteristic features were identified showing a concentration of the number of firms combined with an increase of unemployment.

As a dominant characteristic we find the almost total orientation of production towards internal and especially regional markets. Export operations have been limited to a reduced group of companies with high amounts of orders at hand. Firms have been showing a low need for qualifications allowing them to improve their tasks and an attitude towards cooperation which weakens and expose them to unstable situations, preventing them from taking advantage of their geographic proximity. These firms have a higher than six year annuity of capital equipment and their main source of financing is through reinvesting profits.

The Strategic Plan (Municipalidad de Bahía Blanca, Dirección de Planificación Estratégica, "Bahía Blanca: Diagnóstico y Escenarios Territoriales", 1998.) issued from the municipal government concerning local industrial SMEs is based on several perceptions about the economic environment, such as the loss of market share by local SMEs combined with a lack of competitiveness. A low entrepreneurial profile is perceived in the development of productive activities, and there is a low potential for cooperation in different productive sectors.

In the present paper, the surveyed SMEs display characteristics that can be summarized in three aspects: personal characteristics, industrial attitudes and relationships with the environment in searching for information or knowledge.

Regarding personal characteristics, the survey shows that SMEs principally are managed by men with a higher than average age, who have formed or belong to familiar firms, and who have an average educative level not necessarily related to the activity they carry out. They are to a large extent industrialists not professionally trained (university or tertiary), and they scarcely use professional personnel in their company (0.65 professionals by firm). Industrialists of the professional type participate in activities of the more successful firms. Before founding their company, they knew about production and commercialization, whereas they lacked knowledge mainly in accounting and legal matters. Thus they frequently consult suitable professionals on these subjects. Limited knowledge about the details of foreign trade, suggested by surveys, is a result of the rather local or regional orientation.

As far as attitudes are concerned, the industrialists share risks with partners and are very conservative. This tendency is accentuated with increasing age. A deficiency of entrepreneurial attitudes seems to exist, since new business are not generated as a consequence of the striving for independence or for new challenges, but by changes in trade names impelled by the concurrent situation. New small businesses are initiated by young industrialists, based on a natural search for better opportunities, greater confidence and persistence as compared to older firms. Usually with SMEs authority is strongly concentrated, their focus is on local markets and any own activities in developing the qualifications of the personnel of the company are absent. It has to be emphasized that the fact of being a professional industrialist does not necessarily imply higher qualifications for personnel. At the present time, they consider their own qualification as much more essential than that of their personnel. They even show contradictory views in their concept of qualification as well as in those of quality, of strategy and of effective action. They are moderately persistent in goal seeking, a characteristic that diminishes with the increasing age of the industrialists. Those goals and objectives are clear, attainable and realistic, but not too challenging, due to their high degree of conservatism. These objectives arise from planning, and they do not worry too much about a systematic pursuit of projects. They have a dose of self-confidence, which can be described as average as far as their beliefs in themselves and their abilities to achieve the goals are concerned. It is to be emphasized that they frequently do not use more modern administrative tools to accomplish their tasks.

With respect to the search of information and knowledge, as a basic characteristic we detect a scarce preoccupation in the search for information for decision making. Besides, an investigation about technical information or commercial opportunities is not made through formal or institutional contacts, although a growth of contacts with some type of entrepreneurial association is observed. Although the city provides research centers and specialized educative institutions, they have no contact with them. We could not find use of meetings with other industrialists as a source of information, although in general the relation with their partners is good. Neither could we find contacts with those associations in order to develop projects in common or to create support networks as a kind of survival strategy.

Fundamental characteristics of industrial SMEs of Bahía Blanca are the concentration of activities, especially in supplying local demands, and isolated decision making. This situation is not a detail of minor or marginal importance. The theoretical scheme described before shows the cooperative attitudes between companies as an important element in the fortification of abilities of a company. Nevertheless, the survey did not detect any important cooperative behavior, either between the local companies or with the block of MERCOSUR companies; they were only detected as some isolated attempts of agreements that, however, did not last for long.

Another element perceived in this LONDI is the distance between the SMEs and the universities or institutes of education, as well as of other institutions that could provide a network of beneficial contacts. However, about 60 % of SMEs receive technical information in a systematic way. These companies mainly belong to the plastic and chemistry sector as well as the sector of final goods, which acknowledge the greater importance of this type of information. That technical information is not obtained through the network of contacts, but is received through specialized publications. Besides informal elements, the system offers information on commercial opportunities.

Finally, information about technical assistance is obtained through different channels:

1. public or private research centers;
2. public or private consulting;
3. suppliers of equipment;
4. other companies;
5. customers.

Suppliers have a higher importance in providing information to the local industrial SMEs from which 17 % use research centers and consultants. It is important to emphasize that 24 % of the SMEs do not make use of any contact.

If we compare firms with respect to the evolution of their competitive performance, that is, their reactions to unfavorable conditions and to opportunities in their environment, two types of companies can be distinguished. The first type shows a better development, in the sense of a higher capacity to react to changing conditions and an explicit aptitude towards innovation. The rest of the companies show a less favorable competitive structure, with low capacity or willingness to cope with change or even to induce it. Concerning the search for information and learning, as well as the advantage of the use of a LONDI shown in the scheme of Figure 1, this is a usual practice by the first group of companies. However, we cannot conclude that this attitude is generally observable in all firms without observing a deep difference between the two types mentioned. In those sectors, a reflection about information and learning is required, through an abilities improving network of firm linkages, considering two different goals: dynamic firms generate strengths that allow them to take advantage of opportunities, while, in the less entrepreneurial ones, it would be still more urgent to establish the links mentioned before confronting the internal weaknesses or the external threats.

As far as the support system for SMEs is concerned, negative elements are prominent, such as the mismatch between political organs and instruments, the lack of policy diffusion, low responsiveness in adjusting instruments to firm necessities and a reduced accessibility. Moreover, more emphasis is laid on larger firms. In the specific case of municipal policies in the city of Bahía Blanca, research on "Municipalities and SMEs" (Calzada Percivale et al. 2000) provides a long list of incentives based on the industrial promotion that allows implementation of active policies. Among these are the elimination or reduction of taxes, subsidies for commercial trips, delays in the repayment of credits used for investments that improve the environmental quality, and prizes awarded for improvements in the quality of processes. Nevertheless, data on the use of these benefits are not available, nor is it known of the impact they had in the respective sector.

7 Recommendations

The discussion above emphasized the absence of a suitable LONDI in Bahía Blanca that promotes the formation of industrialists and the development of entrepreneurial initiatives. A policy approach has to be designed in order to generate a flexible and innovation-inducing institutional atmosphere causing investments to be carried out by entrepreneurial subjects accepting the risks that every entrepreneurial activity entails.

In developed nations, institutional environments are effective in the sense that a collaborative environment is initiated. By contrast, in Latin American countries, macroeconomic instability, low levels of entrance, legal insecurity, limitations on capital and financial markets, as well as the lack of a strongly cooperative culture, conspire against the development of innate entrepreneurial activities. There entrepreneurs often lack the professional training needed to plan businesses, and, in many cases, to finish developing their activities in informal sectors of the economy. In some cases, it is indeed a less favorable environment that locks up the opportunities for people who show a suitable proclivity in initiating entrepreneurial activities.

The generation and fortification of local systems has to constitute the target of policies of different governmental spheres in which SMEs operate, by means of:

- development of networks among firms and institutions;
- improvement of the interrelation between universities and technological centers with firms;
- promotion of the development of technological competitions of firms and human resources;
- revelation of lacking competences;
- creation and improvement of the financial and regulatory frame in support of innovation;
- promoting active participation of firms in support policies or in the action of the local government (for example participatory local budgets).

It becomes necessary to rely on public actions and, of course, on public budgets to implement these high-priority tasks. Otherwise, a sustained systematic technological policy focused on the construction and enrichment of production networks may not work. In some sense, when authorities design and apply such public policies, they will have to become themselves innovating industrialists committed to an efficient public management.

References

Angelelli P, Gatto F, Yoguel G (1999) Restricciones asociadas al desarrollo de competencias. Consultoría y Capacitación. In: Yogel, Moori-Koening (eds.) Los problemas del entorno de negocios. El desarrollo competitivo de las PyME argentinas, FUNDES

Bisang R, Gutman G (2001) Netwoks agroalimentarios y acumulación en los países del Mercosur. Jornada Anual de Investigación, 3

Burachik G (2002) Cambio tecnológico y dinámica industrial en América Latina. In: Dichiara R (ed.) Competitividad de Pequeñas y Medianas Empresas Industriales y Desarrollo Regiona, EDIUNS, Bahía Blanca

Calzada Percivale AM, et al (2000) Municipios y PYMe. VI Congreso de la Pequeña y Mediana Empresa ,CPCECF, Buenos Aires

Camisón Zornoza C, Alcamí RL, Ciprés MS, Navarro MB (2002) Marco conceptual de la relación entre innovación y tamaño organizativo. www.madridmasd.org

Chudnovsky D, Lopez A (1997) Política tecnológica en la Argentina: ¿hay algo más que laissez-faire? CENIT, Documento de Trabajo N°20

Damanpour F (1992) Organizational Size and Innovation. Organ Stud, 13, 3: 375–402

Dewar RD, Dutton JE (1986) The Adoption of Radical and Incremental Innovation: An Empirical Analysis. Manage Sci, 32, 11: 1422–1433

Dichiara R (2002) Competitividad de Pequeñas y Medianas Empresas Industriales y Desarrollo Regiona. EDIUNS, Bahía Blanca

Dichiara R, Sanchez O, D´Amore M (2002) Pequeñas y medianas empresas industriales. Empleo y desarrollo regiona. In: Dichiara R (ed.) Competitividad de Pequeñas y Medianas Empresas Industriales y Desarrollo Regiona, EDIUNS, Bahía Blanca

Ettlie JE, Bridges WP, O'Keefe RD (1984) Organization Strategy and Structural Difference for Radical versus Incremental Innovation

Galante O, Muñoz I, Vivori A (2001) El Programa de consejerías tecnológicas, un instrumento argentino de promoción a la innovación orientado. PYME, SEPCYT

Giacchero A, Valentini G (2003) El perfil del empresario Pyme. El caso del sector industrial de agroalimentos en Bahía Blanca. UNS

Gorenstein S, Burachik G (2002) Empleo, pequeñas empresas locales y estrategias de desarrollo endógeno. Experiencias en la Argentina. In: Dichiara R (ed.) Competitividad de Pequeñas y Medianas Empresas Industriales y Desarrollo Regiona, EDIUNS, Bahía Blanca

INDEC (1994) Instituto Nacional de Estadísticas y Censos

Milesi D (2002) Del ajuste macro a la competitividad micro: el caso de las pequeñas y medianas empresas industriales. In: Lugones G, Bisang R, Yoguel G (eds.) Apertura e Innovación en la Argentina. Para desconcertar a Vernon, Schumpeter y Freeman, Edición Grupo Redes

Moori-Koenig V (2000) El caso de Argentina. In: Zevallos E, Zamora A, González ET (eds.) Experiencias internacionales sobre políticas para la empresa media, FUNDES- McGraw-Hill

Parthasarty R (2002) Product innovation input and outcome: moderating effects of the innovation process. Journal of Engineering and Technology Management, 19: 75–91

Salloum C (2002) El financiamiento a los proyectos innovadores. In: Dichiara R (ed.) Competitividad de Pequeñas y Medianas Empresas Industriales y Desarrollo Regiona, EDIUNS, Bahía Blanca

Sanchez O, Dichiara R (2002) Estrategias de inversión e Innovación tecnológica: Estudios de las PyMEs industriales. In: Dichiara R (ed.) Competitividad de Pequeñas y Medianas Empresas Industriales y Desarrollo Regiona, EDIUNS, Bahía Blanca

Schumpeter JA ([1935] 1912) The Theory of Economic Development / Theorie der wirtschaftlichen Entwicklung. Duncker & Humbolt, Berlin

Schumpeter JA (1942) Capitalism, Socialism and Democracy. Unwin, London

Sullivan P, Kang J (1999) Quick response adoption in the apparel manufacturing industry: Competitive advantage of innovation. Journal of Small Business Management, 37: 1–13

Yoguel G (2003) Innovación y aprendizaje: Las redes y los sistemas locale. In: Aportes para una estrategia Pyme en Argentina, CEPAL. UN

Yoguel G, Boscherini F (2001) El desarrollo de las capacidades innovativas de las firmas y el rol del sistema territorial. Revista Desarrollo Económico, 161

Yoguel G, Rabetino R (2002) La incorporación de tecnología en la industria manufacturera argentina en los 90: los factores determinantes. In: Lugones G, Bisang R, Yoguel G (eds.) Apertura e Innovación en la Argentina. Para desconcertar a Vernon, Schumpeter y Freeman, Edición Grupo Redes

Part 4:

Innovation and industrial transformation in various sectors

Stefano Brusoni · Giorgia Sgalari

New combinations in old industries: The introduction of radical innovations in tire manufacturing

Abstract The literature on long waves gives great emphasis to the process of the diffusion of new pervasive technologies that raise productivity and growth, first in the sectors that generate them, then in those sectors that progressively adopt them. This paper explores this process of diffusion, relying on an ongoing analysis of the diffusion of robotized, modular production process in the international tire industry. The analysis captures three key features of the diffusion process of the long wave. First, we look at the emergence of radical innovations in mature sectors through the adoption of pervasive technologies, i.e. robotics. Second, we see that the adoption of the new process goes hand in hand with the adoption of new organizational routines at the level of both design and manufacturing processes, modularity. Third, this study highlights the role played by key individuals—entrepreneurs—who develop new ways of doing business.

Keywords Radical innovation · Modularity · Tire manufacturing · Long waves

JEL Classification 030 · 033

1 Introduction

The chief aim of this paper is to illustrate the micro-level processes of diffusion that underpin the upswing of a 'long wave' of growth and development (Freeman 1977; Freeman and Louça 2001; Perez 2003). Our key objective is methodological in nature. Rather than demonstrating any specific point, we want to emphasize what

S. Brusoni (✉)
(CESPRI & CRORA) and Silvio Tronchetti-Provera Foundation, Bocconi University, Milan, Italy
E-mail: stefano.brusoni@unibocconi.it

G. Sgalari
Competitive Analyst, Pirelli Tyres, Milan, Italy
E-mail: giorgia.sgalari@pirelli.com

kind of data and information are needed to operationalize some of the fundamental concepts put forward by long wave theory. The literature on long waves gives great emphasis to the process of the diffusion of new pervasive technologies that raise productivity and growth, first in the sectors that generate them, then in those sectors that progressively adopt them. Moreover, this literature stresses the fact that this process of technological diffusion ought to be accompanied by a parallel process of adaptation of the economic institutions that use the new technologies. The presence of a 'techno-economic mismatch' (Freeman and Perez 1988) is the fundamental friction that slows down the upswing of the wave.

This paper explores this process of diffusion, relying on an ongoing analysis of the diffusion of robotized, modular production process in the international tire industry. Through this procedure, we can capture three key features of the diffusion process of the long wave. Note first that it focuses on the implementation of a revolutionary breakthrough. This is not a case of a process of incremental improvements, however rapidly these may be introduced (Levinthal 1998). Second, the adoption of the new process goes hand in hand with the adoption of new and different 'design rules' (Baldwin and Clark 2000) that brings about different organizational routines at the level of both design and manufacturing processes. Third, this study highlights the role played by key individuals—entrepreneurs—who develop new ways of doing business (Witt 2000).

Tire manufacturing, though a mature industry, is still characterized by a high rate of innovation, and constant improvements in all phases of the production process as well as periodic breakthroughs. We will focus on the latter, examining the defining features of the R&D activities of the major tire manufacturers, how these can be related to traditional core businesses and company cultures, and to competitors and market interactions. We will focus in particular on the most recent breakthrough in the production process, with the introduction of fully automated manufacturing systems. The analysis of the way in which the various tire manufacturers developed the new production systems represents an exemplary case of how even a mature sector can be characterized by revolutionary innovations with enormous impacts on costs, labor organization and product properties. The flexibility required by an increasingly diversified market will be a key element in interpreting these phenomena.

The paper is structured as follows. Section 2 will discuss theoretical issues related to the rejuvenation process of mature industries, building upon life cycle and long wave theories. Section 3 will analyze the main technological differences between convention and innovative tire manufacturing. On this basis, Section 4 focuses on the driving forces underpinning radical innovation in this industry, and on the heterogeneous strategies followed by the main players in the industry. Section 5 looks at the role of intrapreneurs and Section 6 on the mismatch between technology and organization that followed the introduction of the radical innovation. Finally, Section 7 concludes.

2 Setting the scene

2.1 Life cycle approaches

Life cycle theories have given considerable attention to the early evolution of the international tire industry. These approaches look at the relationship between innovation and the dynamics of entry, exit and concentration. Klepper and Simons (1997) identified three different approaches that identify innovation as the key engine of industrial evolution (and of shake outs leading to concentration). Specifically, these approaches are the innovative gamble model (Jovanovic and MacDonald 1994), dominant design theory (Utterback and Suarez 1993), and increasing returns to continuous innovation (Klepper 1996). All theories identify innovation as the key engine that explains patterns of entry and exit in this and other industries, but differ in terms of what type of innovation drives the process, when innovation occurs with respect to shake out, what kind of firm is more likely to survive with respect to time of entry, and what happens after the shake out. In the literature, two specific periods of the evolution of the tire industry are given particular attention: the 1920s and the 1970s. The former period corresponds to the completion of the original wave of entry and the shake out that led to concentration of production in the hands of a few global producers (e.g. Klepper and Simons 1997). The latter captures the demise of the large US producers located in the Akron district and their acquisition by foreign producers (e.g. Sull 1999). This paper focuses instead on recent developments in tire manufacturing, and on the underlying firm-level dynamics. It looks at what happens after the shake out has occurred, focusing on the dynamics of both product and process innovation.

As pointed out by Klepper and Simons (1997), the three approaches propose different predictions about what type of innovative strategies survivors will follow after the shake out. The innovative gamble model simply posits that all firms that survive the shake out converge over time to the same level of output. The dominant design approach, as with the increasing returns story, argues that, after shake out (induced by the emergence of a dominant design), firms should shift from product to process innovation. The increasing returns story also predicts that the largest firms will play a leading role in introducing process innovations. Both dominant design and increasing returns to R&D accepts the idea that, over time, product variety will decrease. Overall, all approaches share the idea that after the shake out occurs, the evolution of the competitive landscape of mature industries tends to slow down: firms focus on incremental, process-oriented changes; scale economies and price competition prevail. As a result, product variety stabilizes or decreases.

The approach in this paper, unlike those summarized above, focus on the analysis of what happens to industries where the shake out had already occurred. The starting point of our analysis is the empirical observation that, in recent times, the tire industry, as well as other mature sectors, has exhibited remarkably pro-active innovative strategies which often fall under the radar screen of R&D and patent indicators. Just considering the past decade, the industry has witnessed a remarkable acceleration of the pace of *both* product and process innovation, with the latter preceding in most cases the former. If we look at the 1990s (the second half of the decade in particular), we can indeed say that this period was characterized by a consistent effort to overcome the bottlenecks of the traditional manufacturing process, not through further incremental innovation, but rather through the introduction of radically different production principles such as: flexibility, customization, automation, and modularization. Conversely, product ranges have begun to increase again, with some of the key manufacturers beginning to offer 'custom-designed' tires to the key carmakers.

There are at least three empirical trends that need to be reconciled with existing explanations of industrial evolution. First, despite what the theory predicts, firms in the tire industry are engaged in radical process innovation. What could be achieved through improvements at the margin in the traditional process has already been achieved. Second, product segmentation does play a key role in explaining the competitive dynamics of this industry. Further, this role has increased significantly in recent years (particularly at the top end of the market). Third, firms consistently differ in terms of the product and process innovation strategies they pursue.

2.2 From life cycles to long waves

While developments in downstream industries and environmental pressures have impacted the entire industry, enormous heterogeneity can be observed at firm level in terms of the adoption of new, robotized equipment (see Section 4 below). Throughout the 1990s, tire makers exhibited a wide range of reactions to the opportunities and threats posed by changes in their main customer base, from the incremental adoption of specific pieces of equipment within the context given by the traditional production process, to the leap into radically different ways of producing tires. In order to understand the persistent vitality of this seemingly rather mature business, we argue that one has to look beyond life cycle theory (which provides nevertheless an extremely fruitful starting point) to frame the recent evolution of the tire industry as unfolding of 'long waves' of economic and technological development (Freeman 1977; Freeman and Louça 2001; Perez 2003; Tylecote 1993).

As stressed in Perez (2003):

> 'Maturity does not inevitably end in the marginalization of a system, nor is it necessary that a radical innovation in the core product itself should come to the rescue and replace the previous mature product. Both can occur and sometimes do occur. What is more likely to take place, especially at those times—such as the 1970's—when many inter-related systems tend to come to maturity more or less simultaneously, is that a general solution appears in the form of a technological revolution. What happens then is the diffusion of a new set of generic technologies, capable of rejuvenating and transforming practically all existing industries, together with the creation of a group of new dynamic industries, at the core of radically new technology systems. These are the technological revolutions described by Schumpeter (1939) as "creative gales of destruction." They have occurred about every fifty or sixty years and it is this phenomenon that lies at the root of the so-called long waves in economic growth.'

We suggest that the tire industry is undergoing one of these waves of dramatic change. Such change has been enabled by the diffusion of new technologies and design strategies for both products and processes. Such diffusion follows the emergence of new pervasive technologies (e.g. robotics and software engineering) that are bringing about a radical shift in the way this industry works. Following again Perez (2003), we might be witnessing what she calls a 'change [in] the "common sense" criteria for engineering and business behaviour across the board'. According to the long wave approach, the key bottleneck that slows down the diffusion of the new, generic technology (or technologies) is the need to re-organize the socio-institutional framework within which economic activities unfold.

At the micro-level, this means that adopting firms and industries need to change their own organizational and managerial processes. Adoption, in these situations, requires more than acquiring 'equipment'. Indeed, the long wave approach posits that the early stage of the new wave is characterized by a systematic techno-economic mismatch. Changing organizational practices may require even more effort than introducing new technologies. Entrepreneurs, then, play a key role also during the diffusion process. Adoption rates are likely to rise slowly at the beginning, and then pick up speed once one pioneering organization has proved the viability of the new 'common sense'. Thus, during the early stages of diffusion, one should expect high firm-level heterogeneity in terms of adoption rates, as well as the modality of adoption. Moreover, one should expect adoption of path breaking technologies to be accompanied by in house R&D efforts aimed at understanding and adapting the new technologies.

Second, the diffusion of the new 'common sense' can also be traced to the interplay of product and process innovation. We argue here that both product and process innovation can be the key elements of the radical change that leads to a new wave. In fact, introducing process innovations, particularly radical ones, entails a major shift in the mindset of the people who design products and run production facilities. In many ways, production processes (particularly the capital intensive ones) are likely to be managed in a rather conservative manner. Hence, one should expect mature industries to react first by introducing incremental improvements in their existing products. However, when a new 'wave' is surging, radical changes in the production processes will need to be implemented, and only afterwards will product innovations allowed by the new production technologies be developed.

Another factor that has to be taken into account when discussing the differing rates of process vs. product developments lies in the 'black box' nature of the process. It is in fact open to inputs from the outside world of suppliers (of materials, energy and manpower) and clients and to some control coming from laws regarding working conditions and environmental emissions, but it is not open to public inspection. Unlike the product, which ends up in the hand of the public and can then be analyzed in all its facets, the process of production is a "Company Secret," only partially divulged in press releases and patents. In this industry, the number of process patents is in fact far less than the actual innovations developed in a company. The protection offered by patent issuing is considered in general not worth the revelation of the secret to the public, unlike the case of patents of product innovations. Sometimes even when patenting a process innovation, companies try to protect their processes by using screen societies (in the tire manufacturing world we discuss in this paper, we see for example that Michelin used the Societè Anonyme SEDEPRO to deposit most of the key patents for its innovative process).

To conclude, we think that the development of new automated production processes for tire manufacturing show some of the salient features of the diffusion process typical of the upswing of a new wave. In the next two sections, we shall try to argue that, first, we are witnessing a radical shift in the way in which tires are produced. Such shift derives from the adoption of new, pervasive technologies (i.e. robotics) and design practices (i.e. modularity) in the industry. Second, diffusion is led by pioneering organizations (and by key people within these organizations) that at some point decide to 'leap forward', as the new production processes cannot be reached by incrementally improving the pre-existing technologies. Third, different firms, facing similar environmental constraints and opportunities, will react differently, as they rely on technological skills, mindsets and market positions.

3 A revolution: automated tire manufacturing processes

Over the last ten years, the tire industry has witnessed the introduction of a radical transformation in the manufacturing process. We will briefly summarize in this section the principles of conventional tire manufacturing and the technological content at the heart of the new processes. We will then try to analyze the way in which the various manufacturers developed their innovative processes and how the general success of these processes and the various implementations can be related to common driving forces or to specific company characteristics/histories.

3.1 Conventional tire manufacturing

Unlike most processes in the chemical industry, tire manufacturing has traditionally been a discontinuous process during which the various raw materials used (polymers, chemicals and fillers for the rubber compounds) and fabrics and steel

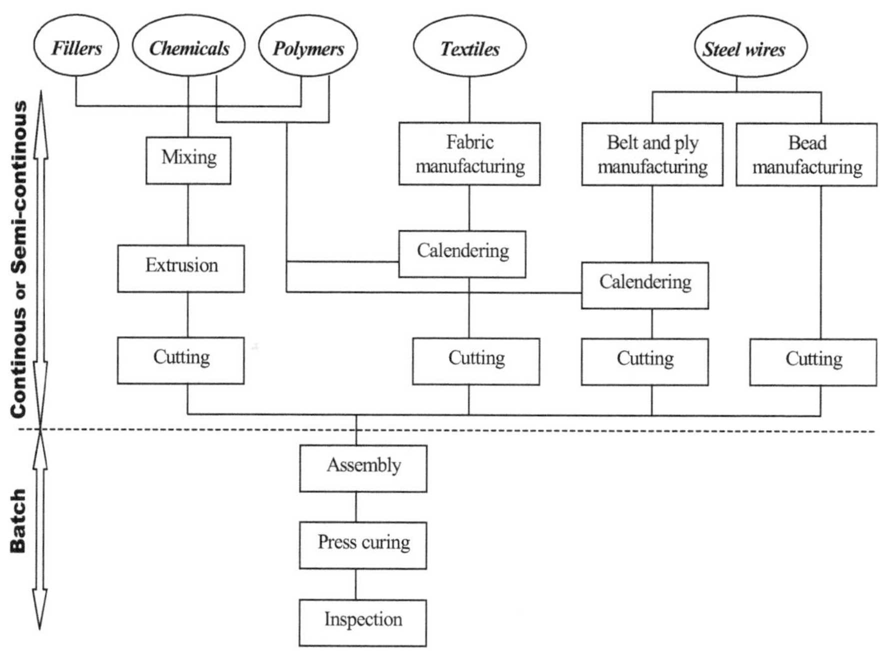

Fig. 1 Conventional tire manufacturing workflow

reinforcements are preprocessed in a usually continuous fashion (with eventual batch storage), then cut or shaped in discrete components which are then assembled to create each individual green tire and then cured.

A convention workflow diagram of a tire manufacturing process is shown below (Fig. 1).

We see that various rubbery polymers are mixed with fillers—e.g., carbon black or silica—and other chemical additives to make the various rubber mixes for the different parts that constitute a tire—sidewalls, tread, bead areas, liner, inserts, etc. (see Fig. 2)—and that are characterized by widely different physical properties.

The mixing is usually carried out by Banbury mixers,[1] after which the various rubbers are extruded to specific contours and cut to length. Textile fibers are, in the meantime, twisted and woven into fabrics that are impregnated with adhesives and then calendered and coated with rubber. The calendered fabric is cut to specific widths and angles in order to be used as reinforcement plies (carcass or breaker). High tensile steel wire is similarly treated and coated with rubber and cut to predetermined widths and angles to be used for reinforcement plies (carcass or belts) and for the bead. All the individual components are deposited sequentially onto assembly drums and pressed into shape by automatic rollers. This assembly is done either in a two-phase sequence or, more recently, on single machines. After this step, the green tire is transferred onto a curing bladder that supports the tire during the vulcanization phase, where rubber curing and molding of tread designs are simultaneously performed.

3.2 Innovative tire manufacturing

In the innovative process, we see that production is characterized by a flow-line system that makes all components in their final shape so that conventional batch processes are eliminated. Each building process is focused around a rigid torus, fixed in size and design. All rubber parts are directly extruded onto the rigid drum in thin rubber strips. Similarly, textile plies are knitted in situ around the tire, while bead wires, belts and all reinforcement plies are deposited onto the drum as pre-extruded tapes of rubber-coated cords. The sequence of deposition of the various parts (Fig. 3) is accomplished in a single multifunctional station (C3M, i.e. Michelin's process), or as the drum travels on a trolley on rail through various stations (BIRD, i.e. Bridgestone's process), or as is taken around by robots (MIRS, i.e. Pirelli's process). When the green tire is completed with the deposition of the rubber tapes constituting the tread, the torus is brought to the curing phase. In some of the innovative process, e.g. Michelin C3M (and maybe Bridgestone BIRD), the curing is done by electrical heating, while in others, e.g. MIRS, curing is done with internal injection of steam or nitrogen so that the molding of tread designs is accompanied by inflation as in conventional manufacturing. When the curing cycle

[1] The Banbury mixer (introduced in 1916) was the innovation that, according to Jovanovic and MacDonald, sparked the 1920s shake out. Other key innovations proposed as 'candidate' triggers of the shake out are the drum tire machine and the core-tire building machine (Klepper and Simons 1997). All innovations were introduced by the mid 1910s and quickly diffused throughout the industry. With modifications, they are still in use in most of the traditional manufacturing processes.

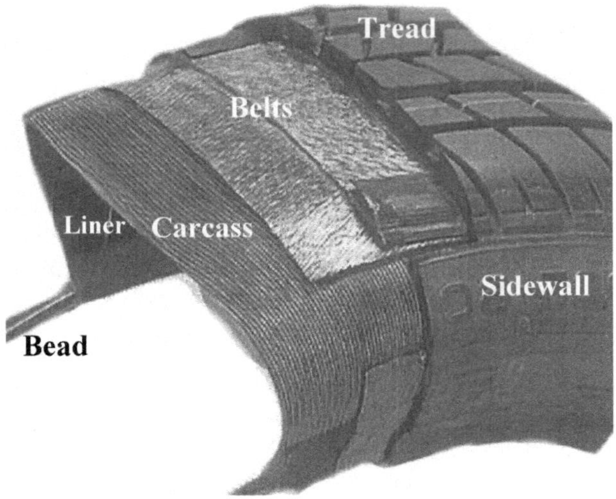

Fig. 2 Tire section with its main constituents

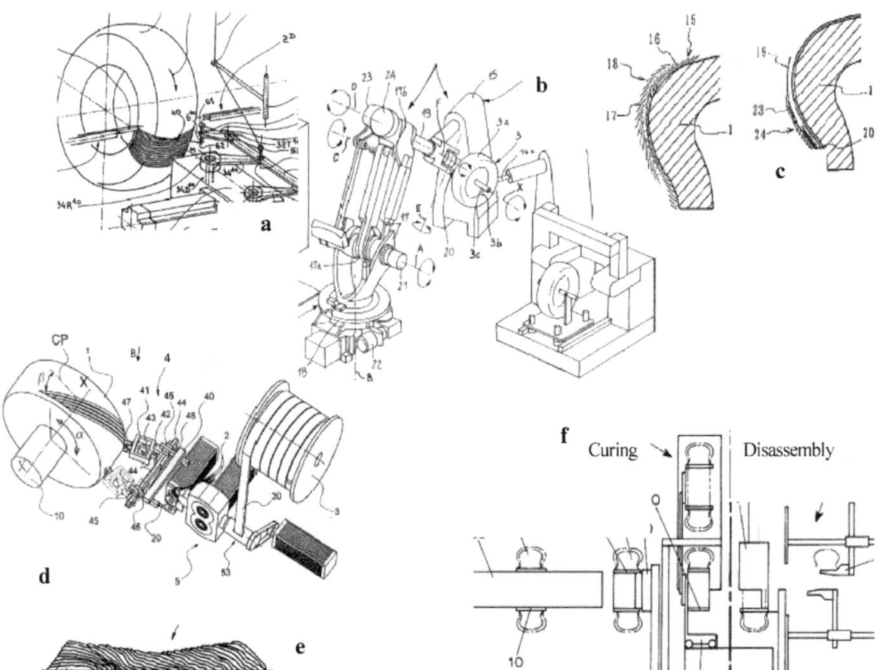

Fig. 3 **a–f** Innovative Process Machinery from Patent Publications. **a** Deposition of carcass cords (Michelin 2001). **b** Robotized deposition of strips (Pirelli 2001). **c** Deposition of rubber strips to make sidewall and bead area (Bridgestone 2001). **d** Deposition of strips (Michelin 2003). **e** Tread part made of specified rubber strips (Sumitomo 2000). **f** Final steps of the flow line: curing and drum disassembly (Michelin 1998). When a different tire size and model has to be made, the central software database reprograms the various machines with new balances of rubber ingredients to the mixers, numbers and characteristics of rubber tapes to extruders, characteristics of textiles plies and wires to the winding machines, and heating times for the vulcanization apparatus

ends, the toroidal core, which is formed of a number of separate sectors, is disassembled, thus allowing its removal from the finished tire.

4 The driving forces underpinning the introduction of robotized processes

What has pushed tire makers to look for alternative ways of producing tires? First, the tire industry is reacting to developments in its main user industries, namely automotive. For the best part of the past twenty years, carmakers have busied themselves restructuring their production activities, design processes, supply chains, and marketing strategies (Womack et al. 1990). Finally, tire makers have been asked to join the struggle to offer increasingly customer-based solutions, in a more timely and efficient manner. In particular, increasing market segmentation downstream, the emergence of new product niches altogether (such as Sport Utility Vehicles, SUVs) and the need to add flexibility to manufacturing processes has provided tremendous pressures on tire makers to abandon the traditional business model focused on the 'tire as commodity' principle. Second, environmental considerations are pushing the entire industry to rethink the way tires are produced, adopting approaches that take into consideration the entire life cycle of the product, which, of course, leads them to look more closely at the way tires are produced: what materials are used, how they are processed, what and how much waste is generated, and so on. Increasingly, production processes have become a key concern for environmental regulations. Third, developments in supplier industries have enabled (some) firms in the tire industry to adopt equipment developed upstream that embodies state-of-the-art developments in robotics and software engineering to automate production stages traditionally handled through labor-intensive processes. The latter provides us with our entry point into the analysis of the tire industry.

Processes in the tire industry are intrinsically resistant to changes, since it is the repeatability and the standardization of the activities that allow for the achievement of maximum efficiency. So, it is not surprising that industrial manufacturing has always been an area with very little permeability to the forces acting in the outside world and a strong inertia toward innovation. Only when exogenous factors reach a critical strength can they produce radical breakthroughs. One of such factor is represented by recent advancements in automation and robotics. Especially in Europe (Germany and Italy in particular), and Japan, manufacturing production has traditionally represented an important economic sector. They control, respectively, 52 and 20% of the world market (Fabbi 2003). This has made available to the European and Japanese tire manufacturers a set of automatic production techniques and devices. In such an environment, radically innovative processes for tire production have become feasible and economically advantageous. Interestingly, some of the ideas at the heart of these processes are not completely new. For instance, the idea of direct rubber deposition as small tapes had already been used after the war by Firestone-Canada but was then put aside. Similarly, Pirelli declared that a large part of MIRS technological content dates back more than twenty years and that what was then impractical became possible thanks to today's robots (Gorcelin 2003).

While the availability of equipment played a key enabling role, as in many other sectors two major economic factors provided the trigger to invest in the development of the new processes: cost reduction, on the one hand, and flexibility and quality improvement on the other. The more innovative processes were able to satisfy these requirements. The new plants are characterized, in fact, by strong savings in terms of operating costs, manpower, energy, as well as productivity rate, and they allow a range of options in tire design and construction that were undreamed of in conventional manufacturing, opening roads for large improvements in tire performance.

Regarding cost reductions, we report in the table below a summary of the various tire manufacturers declarations for the innovative processes:

Concerning the issue of flexibility, we have to highlight the fact that tires, even though traditionally considered a commodity product, have become, in the last decade, more and more influenced by market forces. Car manufacturers' requests for tires specifically designed for single car models, even more than final user's desire for a wide range of tire models and sizes, have had a strong influence on the emergence and success of production processes that allowed extreme flexibility. This is, of course, a major issue, particularly for those tire producers (such as Pirelli) historically focused on the top end of the market (i.e. on vehicles such as Mercedes, BMW, Ferrari, etc.).

C3M, MIRS and BIRD are the prime examples of processes characterized by this flexibility in production with minimal switching time between different tire sizes or models. Related to this issue, Michelin claims C3M machines can be transported around the world with minimum notice and costs. The choice of electrical curing in fact frees C3M from the traditional vulcanizing equipment with steam lines that is bulky and complex in installation. This ease of transportation gives Michelin the possibility of creating local manufacturing capacities in quick response to market requests.

In terms of product quality, we see that the revolutionary idea of building on a hard core implies that no conformation is needed as in conventional manufacturing, where the initially flat drum is expanded to shape. This, in turn, means that no stretching is applied to the tire and thus possible non-uniformities induced in the tire by the deformation are avoided. Volume and geometry, in other words, are far more stable, and the chances of failure are reduced. This is even more true for the processes that use electrical curing that eliminates, as previously mentioned, the inflation during moulding.

The improvements in quality just described, as well as the repeatability allowed by fully automated plants and other technical possibilities offered by the tape-wise depositing solution, which we will not discuss here, are indeed enormous and it is not surprising that the natural application of the new plants is in high performance tire production. It is in fact in that market niche that the new automated lines can offer the most important competitive advantage: production of the intelligent automated machines is plainly not profitable for small mass-produced tires. Michelin's initial usage of C3M lines for low range tires was only for precautionary reasons since they wanted to wait for a perfectly reliable process before attacking the real target of top range tires. Not surprisingly, the tire manufacturers that actually implemented the new lines are those for which the presence in the UHP (ultra-high performance) market is the strongest. Michelin and Bridgestone are tire market leaders with 20% of world sales share, and Pirelli, only 6th in the general tire market, is also considered one of the key players in the top end market, with a tradition of providing fitments for prestigious car makers such as BMW, Audi,

Jaguar etc. This market orientation is of key importance in interpreting tire manufacturing strategies, since the performance tire segment has increasingly become one of the largest and most thriving ones: 12-15% of world tire market with a growth rate of 37% during the last 3 ars (in a general economic slump, McCarron 2003; Infomotori 2004). In Europe alone, 30% of new cars are equipped with UHP tires and the forecast is for 37% in 2009 (Infomotori 2004).

4.1 Persistently different firm-level strategies

Having described in the previous sections the major changes that are affecting the international tire industry, we now focus on the specific paths that different companies have followed in exploring the new trajectory. Michelin C3M (Confection Mandrin Monofilament Michelin) is the first example of such a fully automated tire manufacturing process. The first plant of this type was built in 1993 and started producing commercial tires as early as 1995, even though Michelin delayed using it for full-scale production until 1998 (AutoBusiness Reports 2003). The decade of activity was marked by a continuous refinement of the process, accompanied by a change of application of the innovative lines, from the production of standard tires to high performance and top range tires.

Meanwhile, Pirelli developed MIRS (Modular Integrated Robotized System), from the beginning a high technology production process characterized by full automation in which, characteristically, the rigid drum is carried along through the various assembly steps by robots. The MIRS process is unique also in that it presents an integration of the production line from the Compounding step, with its CCM (Continuous Compound Mixing) technology, down to vulcanization. From the beginning, MIRS was dedicated to the higher end of the tire market, with mainly UHP tire production.

With a four to five year delay, all the major tire manufacturers of the Far East started developing similar innovative processes. The second largest tire manufacturer worldwide, Bridgestone, recently announced its BIRD (Bridgestone Innovative and Rational Development) system, which is the first system to integrate logistics, manufacturing and inspection systems thanks to its unique fully automated quality control system (or AIMS—Automated Inspection Modular System). BIRD is undergoing its final trials and is going to be implemented in many of Bridgestone's plants worldwide in 2004. For the manufacturing step, the realized version of BIRD appears to be a blend of the innovative process and conventional manufacturing, with the first phase of tire assembly (i.e. carcass deposition) still done in flat (Gorcelin 2003). Bridgestone has also worked, like Michelin and Pirelli, on full rigid core manufacturing, with its first patents in the area dating back to 1997, but it is unclear whether fully innovative processes are going to be implemented by this company.

In the same period, the other large Japanese tire manufacturer, Sumitomo, was also depositing a large number of key patents. Last year, Sumitomo announced its innovative system, Taiyo (Japanese for "Sun"), which, as with Bridgestone's BIRD, appears to be a blend of innovative concepts (in particular, concerning rubber deposition) and the latest versions of conventional processes. The system started producing prototype tires in 2000, and, since 2002, the company employed the process for regular car production in Japan. Sumitomo is still developing newer generation processes and plans to extend Taiyo to other plants and to more comprehensive production ranges, from run-flat to motorcycle and truck tires.

In the last few years, another Japanese company has proposed another innovative production process, Fuji Seiko. The rights to this ultra-compact tire building system, which offers advantages similar to Michelin's C3M and Pirelli's MIRS, were bought by Toyota and then licensed to Yokohama Rubber. Lastly from the Far East, the Korean tire manufacturer Kumho announced at the end of 2003 that it also would produce tires according to a fully automated production process. Details have still to be unveiled.

This burst of activity in the Far East was not mirrored by the two major Western competitors of Michelin and Pirelli, viz., Goodyear and Continental. These two companies both introduced innovative production systems, IMPACT (Integrated Manufacturing Precision-Assembled Cellular Technology) and MMP (Modular Manufacturing Process) respectively, which represent an optimization of conventional technologies for cost-reduction, an optimization of production efficiency and/or flexibility, but are not as revolutionary as the manufacturing processes that characterize the above mentioned systems.

We show in the table below (Table 1) a summary of the various new processes in order of increasing innovative content. The table also shows information about two key features of the new processes. First, column 3 highlights how the two most innovative processes (C3M and MIRS) target explicitly the high end of the market (e.g. HUP). This is not accidental. Indeed, the new robotized processes offer incredible advantages in terms of lead time and flexibility, features that are key competitive factors in the high end of the market, where the core automakers expect tire suppliers to be able to provide a wide range of tires tailored to the specific needs of each new model. Second, column 4 stresses that the processes developed by Michelin and Pirelli represent major breakthroughs with respect to the traditional manufacturing process, as the new process technologies cannot be integrated within the existing plants. This is not the case for the processes developed by the other main competitors (Table 2).

5 The role of intrapreneurs

While normally considered a low tech sector, tires manufacturers have consistently invested in Research and Development activities. Figure 4 below reports some aggregate data for the main players in the industry. Sharp differences emerge not so much in terms of the levels of R&D undertaken, but more clearly in terms of area

Table 1 Major new process developments in the international tire industry

Process [1]	Year [2]	Product segment [3]	Integrability with existing plants [4]
MMP (Continental)	1997	All	Yes
IMPACT (Goodyear)	1998	All	Yes
TAIYO (Sumitomo)	2002	All	Yes
BIRD (Bridgestone)	2002	Special/All	Yes
C3M (Michelin)	1992	Special	No
MIRS (Pirelli)	2000	Special	No

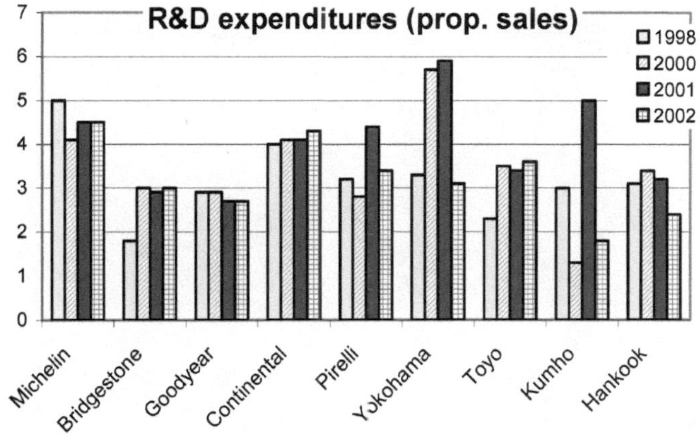

Fig. 4 R&D expenditures of the major tire manufacturers. Data from Tire Business 2003

and organization. In respect of the former, firms differ to the extent to which they invest in rubber chemistry projects. Michelin and Goodyear, for example, are traditionally more active than Pirelli in this area. However, as long as the development of the robotized process is concerned, key differences emerge with respect to differences in how R&D activities are organized.

Generally speaking, we have to consider that R&D expenditures vary by company and destination. Michelin has always been characterized by the largest investments in R&D, and Pirelli has also had very large ratios of R&D spending relative to sales. Bridgestone, by contrast, has typically had lower R&D ratios—which nonetheless corresponds to large absolute values of R&D spending—but we should not forget that BIRD, at least in the form that has been unveiled to the public, has not been implemented in its more radical form but rather integrated with aspects from conventional tire manufacturing. Other companies, such as Continental or Yokohama, also saw in recent years a strong increase in R&D funding but these were not devoted so strongly to process changes. Continental, for example, has been investing heavily in developing the synergy with Continental Teves to create the first examples of integration modules of tires, suspensions, brakes and automotive electronics.

Concerning R&D organizational structures, Michelin, Pirelli and Bridgestone have managed, with slightly different interpretations, to create a suitable organizational environment for the development of creative process research. Michelin took advantage of the traditional attention to research that has always

Table 2 Key performance indicators of the new production processes

	C3M	MIRS	BIRD
Production (ty/day)	100–120	315	500
Savings (% vs.Conventional):			
Productivity	90	80	96
Space	80	80	65–75
Energy	50	25	40
Plant cost (million\$)	35–40	45–50	20
Floor space (mq/unit)	30	350	450 (+150 Inspect.)

a

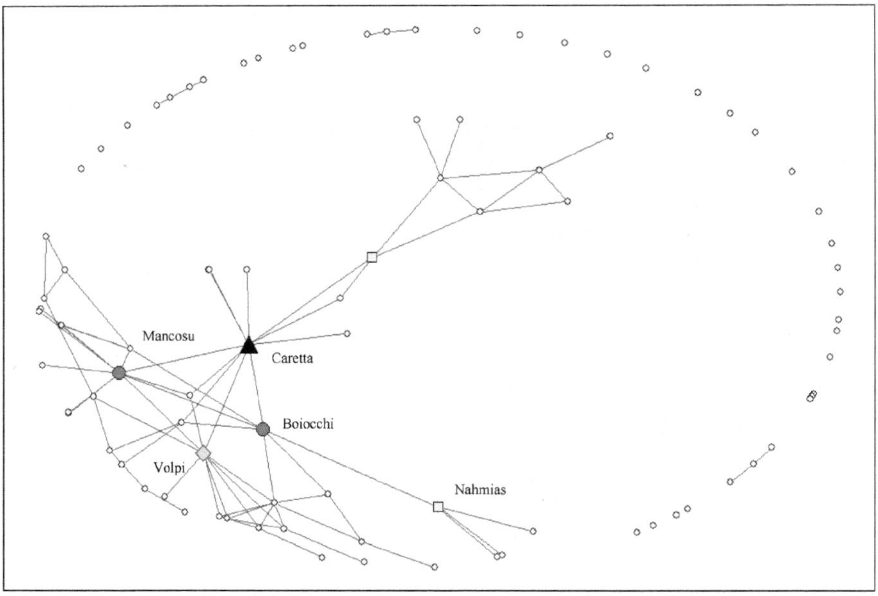

b

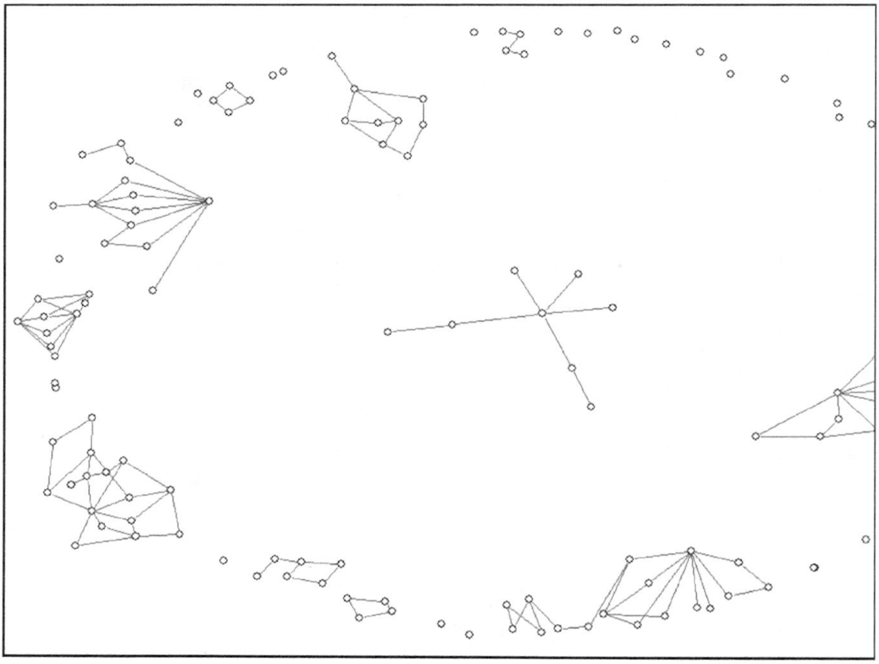

Fig. 5 **a** Pirelli's network of inventors. Source Lissoni 2004. **b** Michelin's network of inventors. Source Lissoni 2004

permeated the company, and in its R&D headquarters at Clermont-Ferrand, a team of engineers worked on the project, responding directly only to François Michelin (AutoBusiness Reports 2003).

Pirelli and Bridgestone, by contrast, centered work around key R&D personalities — Mr. Caretta in Pirelli-Bicocca and Mr. Ogawa in Birdgestone-Kodaira — who proved capable in unifying and integrating ideas and resources. This organizational solution is evident in a network analysis on the companies' process patents (Lissoni 2004). We report first the graph for Pirelli (Fig. 5a). The graph was constructed by selecting first all the firms within the Pirelli group involved in the tire business. Then all the EPO patents applied for by these companies were used to identify their inventors. Finally, links among inventors were established, looking for co-inventors (i.e. two inventors are assumed to be linked if their names appear on the same patent). What clearly emerges from the graph is the central role played by Mr. Caretta, whose role in the network is not only characterized by the number of patents on which he appears as inventor, but for his connecting role among otherwise disconnected areas.

It is striking to compare Pirelli's network of inventors with Michelin's (Fig. 5b). Unlike the Italian manufacturer, the French company's network seems to lack any agent of such a high visibility. Different technological areas seem to be developed in a much more decoupled way, without any one technical figure to integrate them. Besides, in Pirelli, the working groups on innovative process were given by their respective directors a certain freedom from the daily need of efficient process, or short-term, developments. Pirelli's CEO Tronchetti Provera words were, for example, "Do whatever you want, but come back with something!" (Caretta 2004; personal communication).

We can speculate a bit more about the role played by such central actors. A large body of literature in the innovation management and organization science field provides insights in our understanding as to why the absence of central individuals was important in Michelin's failure to lead the introduction of the robotized process. Focal individuals play central roles in developing connections among different domains. The importance of focal individuals is consistent with studies that found successful product and process innovations require *champions*. Research has illustrated the importance of champions, such as executives and senior engineers, acting as boundary spanners in redefining linkages across domains within firms and between firms and the environment to introduce radical innovations (Allen 1977; Roberts 1987; Rothwell 1992). This line of enquiry needs to be enriched by a more precise theoretical understanding of what such champions actually do. For example, technology brokers recognize, store, blend, and transform technologies (Hargadon and Sutton 1997; Hargadon 2003). We contend that, besides boundary-spanning, focal individuals also play an *integrating* role, acting as the channels of transmission of innovation across domains and the development of new bodies of knowledge and expertise. In so doing, they facilitate the introduction of new combinations of technological elements which lead to major changes not only in production processes, but also in business strategies. In fact, it was the presence of such focal individuals that allowed companies, such as Pirelli, to solve a number of problems which emerged in the process of introducing the new manufacturing technology.

6 The mismatch between technology and organization

The new process enabled—and forced—tire designers to change radically the way they work. First of all, the new production process enabled designers to gain unparalleled control over the manufacturing process. In the process of designing tires, they now also set the process parameters that will activate the robotized production process. This streamlines enormously the interactions between the design center and the manufacturing facilities. This is, of course, linked to the shorter lead-time of the new process with respect to the traditional process. In the traditional process, the interaction between designers and plant operators and managers had to be managed through appropriate organizational and IT interfaces. The former had control over the design process; the latter had control over the manufacturing process. With the new process, designers have gained control of both design and manufacturing. They now have the advantage of being able to fine-tune the manufacturing equipment to the specific needs of the specific tire they are developing. Figure 6 summarizes the main organizational differences between the traditional and the innovative process.

Second, the design variables that the tire designers can manipulate to achieve a certain performance (in terms of, e.g., handling, rolling resistance etc.) can now be more precisely fine-tuned to the specific characteristics they aim for. This is made possible by the process through which the raw materials are disposed onto the drum that moves along the line. Unlike the traditional process, here the different layers which make up the final product (e.g. liner, carcass, sidewall, etc.) are applied by the automated extruders one by one. Moreover, each layer is in a sense modularized into a finite set of thin bands of homogeneous material. In a way, the whole tire has been decomposed into a finite set of modules, each of which can be manipulated by the designer. To push the point, modularity is not only a property of the tire production process, but also of the tire. This extends to the process of tire design the potential gains deriving from the adoption of modular design strategies widely discussed in the modularity literature.

Indeed, one of the key problems that the developers of the robotized process Pirelli and Michelin had to face concerns the need to alter the traditional division of labor between tire designers, process engineers and plant operators. Whereas the

<table>
<tr><th colspan="3">TRADITIONAL PROCESS</th><th colspan="3">INNOVATIVE PROCESS</th></tr>
<tr><th>Activities</th><th>Responsible</th><th>Duration</th><th>Activities</th><th>Responsible</th><th>Duration</th></tr>
<tr><td>Initial tire design</td><td>Product engineer</td><td rowspan="4">One day</td><td rowspan="2">Initial tire design</td><td rowspan="6">Tire designer (single point of responsibility) + specialists (e.g. mould designers)</td><td rowspan="6">One day</td></tr>
<tr><td>Initial design of components (e.g. sidewall)</td><td>Product engineer</td></tr>
<tr><td>Choice of materials</td><td>Product engineer</td><td>Choice of materials</td></tr>
<tr><td>Tire design: sizing, tread, etc.</td><td>Product engineer</td><td>Tire design: size, tread, etc.</td></tr>
<tr><td>Definition of building process</td><td>Process engineer</td><td rowspan="2">One week</td><td rowspan="2">Definition of building process. Final design of components</td></tr>
<tr><td>Final design of components</td><td>Process engineer</td></tr>
<tr><td>Building, phase 1</td><td>Plant operator</td><td rowspan="2">Two/three days</td><td rowspan="2">Building</td><td rowspan="3">Plant operator</td><td rowspan="3">Half a day</td></tr>
<tr><td>Building, phase 2</td><td>Plant operator</td></tr>
<tr><td>Vulcanization</td><td>Vulcanizer operator</td><td>One day</td><td>Vulcanization</td></tr>
</table>

Fig. 6 Phases of design and production

traditional process was based upon a clear allocation of tasks and responsibilities to specialist engineers and workers, the innovative process has required engineers to develop a much more integrated understanding of both process and product issues. Indeed, the integration of product and process know-how was the key challenge that developers of the robotized process had to face. In a related paper (Brusoni and Prencipe 2005), whose conclusions are summarized in this section, this very process is studied in deeper detail. The main point is that the validation of the new process, from a technological point of view, did not present unexpected challenges to developers. It was instead the development of the new design skills required by the new process, and the related introduction of a new way of organizing the design and manufacturing activities (as sketched in Fig. 6) that proved to be the most challenging step in the production process (Brusoni and Prencipe 2005).

Obviously, given the ongoing nature of the phenomena under analysis (and secrecy issues), this section can provide only a rather sketchy summary of the changes accompanying the diffusion of the robotized production process. The point we want to stress here is that, consistent with the argument put forward by Perez, the diffusion of the 'technology' ought to go hand in hand with other organizational variables. In this case, besides changes in the manufacturing process, we also need to consider changes in the design process and how it is organized.

7 Discussion and conclusions

The literature on innovation management gives privileged attention to the innovative activities carried out in industries and organizations based upon new, fast moving bodies of knowledge, such as microelectronics (since the mid 1970s) and biotechnologies (since the 1980s). This paper has focused on the innovative dynamics of a 'traditional' sector, characterized by unpredictable, rapid and original processes of technological innovation. This paper has also argued that managerial choices and strategies play an extremely critical role in such industries because all, or most, of the 'low hanging fruits' have already been picked. In a way, innovation in traditional sectors can only be 'radical' if it has to deliver sustainable results (Grant 1996); and management play a key role in changing firms' established 'traditions' (Achilladelis et al. 1990) in innovative ways.

We have developed this line of reasoning building upon the case of the international tire industry. While the results are still exploratory, partial, and largely descriptive, we think that this type of analysis has a great potential to shed light on the micro-level processes of diffusion of both technologies and organizational practices that underpin the upswing of the a long wave of economic development. This research needs to be further extended in several ways. First of all, the analysis of the development and implementation of new 'design rules' (Baldwin and Clark 2000) and related organizational practices have only been sketched here. Second, more and better data are needed to evaluate the competitive positions of firms in the tire industry to explain why Pirelli led the way toward the new production process, while Michelin (seemingly with the advantage in the early 1990s) lost its leadership position. Third, we need to extend this type of analysis to other mature sectors undergoing pervasive and radical technological and organizational changes.

To conclude, we wish once again to stress the explorative nature of this study. We have analyzed the case of revolutionary changes in a rather mature line of business. We have argued that, by looking at cases such as tires, we can shed light on the joint dynamics of technological and organizational practices that underpin the emergence of a new wave of economic development. We have focused on a subset of the recent dynamics of the tire industry. Besides major changes at the process level, tires are also changing rapidly. New functionalities are being added to the old 'rubber ring' through the introduction of sensors and other electronic devices capable of making the tire 'intelligent', i.e. capable of detecting the possible emergence problems (i.e. pressure drops, excessive usage) and communicating them to the driver before problems result in catastrophic events. Again, such processes of rejuvenation require firms to approach in innovative ways the physical and social technologies upon which they have relied for many years.

References

Achilladelis BG, Schwarzkopf A et al. (1990) The dynamics of technological innovation: the case of the chemical industry. Res Policy 19(1):1–34

Allen TJ (1977) Managing the flow of technology: technology transfer and dissemination of technological information within the R&D Organization. MIT, Cambridge, MA

AutoBusiness Reports (2003) Trends in the global tire industry

Baldwin C, Clark K (2000) Design rules, Vol I

Brusoni S, Prencipe A (2005) Making design rules: a multi-domain perspective. Prepared for the workshop 'Organizing the search for technological innovation', Copenhagen Business School

Fabbi F (2003) Smart manufacturing: EU research sparks revolution in tools, robots and automation. European Commission Innovation Reports, 22-10-2003

Freeman C (1977) The Kondratiev long waves, technical change and unemployment. In: Structural determinants of employment, ii. OECD, Paris, pp 181–196

Freeman C, Perez C (1988) Structural crisis of adjustment: business cycles and investment behaviour. In: Dosi G, et al. (eds) Technical change and economic theory. Pinter, London

Freeman C, Louça F (2001) As time goes by—from the industrial revolutions to the information revolution. Oxford University Press, Oxford

Gorcelin (2003) Bridgestone BIRD takes flight. Le Pneumatique 74:24–27

Grant R (1996) The Richardson Sheffield story: revolution in the cutlery industry. In: Baden-Fuller C, Pitt M (eds) Strategic innovation: an international management casebook. Wiley, London

Hargadon A (2003) How breakthroughs happen: the surprising truth about how companies innovate. Harvard Business School Press, Cambridge, MA

Hargadon A, Sutton R (1997) Technology brokering and innovation in a product development firm. Adm Sci Q 42:716–749

Infomotori (2004) BFGoodrich Profiler, 15–03

Jovanovic B, MacDonald GM (1994) The life cycle of a competitive industry. J Polit Econ 102:322–347

Klepper S (1996) Entry, exit, growth and innovation over the product life cycle. Am Econ Rev 86:562–583

Klepper S, Simons K (1997) Technological extinctions of industrial firms: an inquiry into their nature and causes. Ind Corp Change 6(2):379–460

Levinthal DA (1998) The slow pace of rapid technological change: gradualism and punctuation in technological change. Ind Corp Change 7:217–247

Lissoni (2004) Pirelli's network of inventors—a preliminary analysis. CESPRI internal report. Università Bocconi, Milan (I)

McCarron K (2003) Special report: not just a niche now. Many tire makers competing for a piece of HP tire market. TireBusines 21 July

Perez C (2003) Technological revolutions, paradigm shifts and socio-institutional change. Forthcoming in: Reinert E (ed) Evolutionary economics and income inequality. Edward Elgar, London. Seen at http://www.carlotaperez.org/papers/basic-technologicalrevolutionparadigm.htm#comienzo, last accessed 13 May 2004 at 14.00

Roberts E (1987) Generating technological innovation. Oxford University Press, New York

Rothwell R (1992) Successful industrial innovation: critical factors for the 1990s. R&D Management 22(3):221–239

Schumpeter JA (1939) Business cycles: a theoretical, historical, and statistical analysis of the capitalist process. McGraw-Hill, New York London

Sull DN (1999) Why good companies go bad. Harvard Bus Rev (July–August):42–52

Tylecote A (1993) The long wave in the world economy—the present crisis in historical perspective. Routledge & Kegan Paul, London

Utterback JM, Suarez FF (1993) Innovation, competition, and industry structure. Res Policy 22:1–21

Witt U (2000) Changing cognitive frames–changing organizational forms: an entrepreneurial theory of organizational development. Ind Corp Change 9(4):733–755

Womack JP, Jones DT, Roos D, Carpenter DS (1990) The machines that changed the world. MIT, Cambridge, MA

Simone Ferriani · Gino Cattani · Charles Baden-Fuller

Fitness determinants in creative industries: A longitudinal study on the Hollywood film-making industry, 1992-2003

Abstract It is often overlooked that fitness is a multidimensional concept, and that its components are context-specific. The multifaceted nature of fitness is most evident in cultural/creative industries, because firms are confronted with the challenge of balancing seemingly conflicting needs: artistic performance and commercial imperatives have to be satisfied for long term survival. In this study we examine two important component-traits that make up the fitness function for the Hollywood motion picture industry, which we argue are human capital and network capital. Although many studies have recognized the critical role of 'creative' human capital – which is typically embedded in individuals and groups – and network capital – that is, inter-organizational networks – we do not have many studies that empirically analyze their complex relationships using large scale data sets. We situate the analysis within the period 1992-2003, one in which we have good data and

All authors contributed equally. We gratefully acknowledge financial assistance from the EU - Marie Curie Intra European Fellowship, the Mack Center for Technological Innovation at Wharton, the Stern School (NYU), and ESRC-Evolution of Business Knowledge program for their support of this work. We also acknowledge helpful comments from our colleagues at Cass, particularly Vincent Managematin and Jing Zhang, and from participants of the 2004 Schumpeter Society, 2004 EGOS, and 2005 Academy of Management Conferences. We are especially grateful to Franco Malerba, Uwe Cantner and anonymous reviewers. The authors contributed equally to this work. All errors remain our responsibility only.

S. Ferriani
Marie Curie Research Fellow, Cass Business School, City University, 106 Bunhill Row, London EC1Y 8TZ, UK
and Management Department, University of Bologna, Via Capo di Lucca, 24 – 4126 Bologna, Italy, E-mail: rashomon@economia.unibo.it

G. Cattani
Department of Management & Organizations, The Stern School – NYU, Tisch Hall – Suite 7-14, New York, NY, 10012
Tel: +1 212-998-0264, E-mail: gcattani@stern.nyu.edu

C. Baden-Fuller
Cass Business School, City University, 106 Bunhill Row, London EC1Y 8TZ UK
Tel: +44 (0) 20 7040 8652, E-mail: C.Baden-Fuller@city.ac.uk

the industry appears relatively stable and very productive. While still exploratory, our paper shows how such human capital and network resources interact with the structure of the industry and influence different dimensions of the fitness function. We show how the traits work differently on the different dimensions of fitness to create a dynamic tension between creativity and performance.

Keywords Fitness · Human capital · Network resources · Commercial and artistic performance
JEL Classification L82 · D83 · D85

1 Introduction

Studying the determinants of organizational fitness has a long-standing presence in the management and organization literatures. Structural contingency theory, for instance, maintains that fitness differences originate from the level of internal consistency or fit between organizational structure and strategy (e.g. Chandler 1962; Learned et al. 1965), environmental conditions (e.g. Burns and Stalker 1961; Lawrence and Lorsch 1967; Pennings 1987) or technology (Scott 1975; Barley 1990). Similarly, the strategy-structure-performance paradigm, a sub-stream of structural contingency, argues that firm fitness – as captured by firm performance – essentially depends on the alignment between strategy, structure and the environment (e.g. Child 1972; Miles and Snow 1978; Porter 1980; Doty et al. 1993).

Despite its widespread acceptance, however, this stream of research suffers from several shortcomings such as the reliance on single contingencies (for a comprehensive review, see Galunic and Eisenhardt 1994). Although, recent research has attempted to tackle some of these problems (e.g. Levinthal 1997; Rivkin 2000; Siggelkow 2001, 2002; Zajac et al. 2000), a few issues deserve further investigation.

First, the choice of the traits or attributes that are presumed to affect organizational fitness is not always adequately related to the selection environment. It is unclear why certain attributes, and not others, are to be treated as fitness determinants without a finer-grained qualification of the context in which firms operate. In this paper, we take it as axiomatic that fitness is a complex multifaceted concept that is 'context-dependent' (e.g. Hawley 1950; Hannan and Freeman 1989; Hodgson 1996; Levinthal 1997; Knudsen 2004). As in biology, the nature of the selection environment determines which traits or attributes affect organizational fitness (e.g. De Jong 1994).

Second, while the identification of the relevant organizational attributes is context-specific, their relative importance is likely to be contingent upon the particular dimension of the fitness measure being chosen. Reliance on a single fitness dimension can be misleading (Meyer and Gupta. 1994), and different fitness dimensions may interact. Success along one dimension may interfere with another.

The objective of this paper is to look at the relationships of each dimension with organizational traits that are typically viewed as critical in a particular environment. We chose a research site, the film production companies of the Hollywood movie industry, which we traced over the period 1992-2003. In this industry, artistic performance is said to be both necessary and antithetical to commercial success: creativity helps underpin future commercial success and the survival of the genre, but attributes that significantly affect box office receipts are less relevant for artistic performance. Our twelve-year time slot allows us to study the tension between these two dimensions of fitness. To this end, we distinguish between fitness as a measure, i.e., box office and artistic performance, and fitness as a function, i.e., the specific functional form that relates those attributes to the fitness value (Hannan and Freeman 1989; Venkatraman 1989; Knudsen 2004).

While film production companies have to control many dimensions, we concern ourselves with two critical traits: *human capital* and *network resources*. We focus on these two traits for both theoretical and empirical reasons. Theoretically, a large body of research has shown the way in which human capital and network resources critically affect firm performance (e.g. Nahapiet and Goshal 1998). Moreover, as we shall see, movie production companies are essentially project-based organizations, and human capital and network resources are among the most salient factors. On the human capital side, fitness stems from the ability to attract and retain 'creative' resources, typically embedded in individuals and groups (Caves 2000); on the network resource side, the creation of social networks allows consistent access to creative and financial resources (e.g. Jones and DeFillippi 1996). We therefore build on past work to create a more integrated perspective. Our purpose is to study the impact of human capital and network resources on the previous fitness dimensions, i.e., commercial and artistic performance.

The paper is organized as follows. In the next section, we elaborate on the notion of fitness. We then examine our research question in the context of creative/cultural industries. In particular, we describe the context-specific organizational traits that affect organizational fitness, and advance some conjectures that we explore in the empirical section. We then describe the data, the model, the methods used in the analysis, and the results of the statistical analysis. We conclude with the main implications of the findings and the conclusions that follow from them, and we identify important topics for future research.

2 Theory

As Stearns (1976, 4) puts it, fitness is "something everyone understands but no one can define precisely." Despite a wealth of diversity in fitness definitions, full agreement is hardly achieved. In biology, fitness has been alternatively used to indicate a short-term measure of reproductive success or a measure of general *adaptedness*. A common shortcoming of these fitness definitions is that they amount to "a description of natural selection, not an explanation, thus ... rejecting any use of fitness that refers to an innate quality or a good design" (De Jong 1994, 4).

The determinants of fitness are, of course, difficult to assess. A first strategy to study them is to "describe the association between a trait considered to be of interest and fitness, or, more usually, a fitness component" (De Jong 1994, 6). This association corresponds to a fitness function, which relates the value of a trait to the value of fitness, however measured. A potential problem of this approach is that "the relation between trait and fitness might be developmental and incidental, in that trait differences accompany fitness differences, but do not give rise to fitness differences" (De Jong 1994, 6). To get around this issue, an alternative approach is to study the association between traits of interest and fitness components with respect to a specific context (De Jong 1994; Metcalfe 1998; Knudsen 2004). From this perspective, fitness is defined as a mapping between amounts of a trait at two different points in time, and is used to measure whether the 'genetic' makeup of particular types is superior to the makeup of other types, but with respect to a specific context or selection environment (e.g. Sober 2000).

In this paper, we use the second approach, that is, we map traits onto performance dimensions. We ground the analysis in a well defined empirical setting on the premise that the nature of the selection environment determines which traits are likely to affect the degree of organizational fitness at any given point in time. The next section provides a more detailed discussion of the determinants of fitness in the selected context, the Hollywood film-making industry. The objective is to pin down which traits seem to be related to organizational fitness – a preliminary but necessary step before estimating their relative weight or contribution.

3 The selection environment of the film industry

The film industry is an ideal setting for studying the multifaceted nature of organizational fitness. As is often the case with the "art world" (Becker 1982) and "creative industries" (Caves 2000), filmmaking exhibits an intrinsic tension between artistic and economic or commercial success (Baker and Faulkner 1991; Lampel et al. 2000). Commercial considerations are often negatively correlated with artistic performance because "in an effort to achieve greater market success, the producers of culture tend to aim their offerings at the lowest common denominator of mass acceptance ..." (Holbrook 1999, 144). On the other hand, though firms need enough financial resources to produce the next movie, no enduring success can be attained without producing movies of good artistic quality. We first explain more clearly the firms in question and then the selection environment.

The Hollywood film-making industry consists of several categories of players that operate across a complete value chain. Along with audiences and distributors (or majors), production companies are among the key industry players. Since audiences pay for the producer's goods (i.e., the movies) and the distributors create a market for them, producers "make the most sound and profitable decisions for their pictures and their companies if they hold themselves accountable to these two categories. Audiences and distributors are the producers most beneficial and crucial sources of business checks and balances" (,see also Lee 2000; Honthaner 2001, 7,).

Production companies can be classified into two groups. First come the production company-subsidiaries of the 8 major firms, seven of which are established players – i.e., Metro-Goldwyn-Mayer, Paramount Pictures, Sony Pictures Entertainment (Columbia-Tristar), Twentieth Century Fox, Universal Studios, Walt Disney Company, and Warner Brothers – and one of which is a newcomer Dreamworks (Scott 2002, 961). The other group consists of a large number of independent production companies. These firms "interact with one another in complicated ways as any given motion-picture production project moves through it three main stages of development, namely: (a) pre-production [...]; (b) production [...]; and (c) post-production [...]" (Scott 2002, 961). This independent sector "consists of hundreds of very small production companies of a few key employees each" (Litman 1998, 37). These companies are typically established around few key figures (producers, budgetary personnel and maybe some creative people) whose work essentially consists of assembling packages of scripts, talents (director, actors, cinematographer, etc.) and other assets to present to the studios in the hope of securing a distribution agreement and, quite possibly, some financing.

There are three approaches to feature-film production. The majors produce projects autonomously, fully relying on their financial and physical assets (with varying degrees of vertical integration). Since the break-up of the vertical structure of the industry some years ago, this is not a common practice. Most commonly, the large firms work closely with independent production companies, with the latter assuming primary responsibility for organizing overall production tasks. Finally, independent companies may work autonomously, approaching the majors as distributors rather than partners.

In understanding the selection environment it is important to realize that the friction between artistic aspirations and commercial needs has shaped the dynamics of the movie industry virtually from the outset (Vogel 1998; Svejenova and Alvarez 2002). Caught in the middle of a ceaseless tension between the studios' urgency to capitalize on large and highly uncertain investments and the creative ambitions of imaginative filmmakers, production companies have to make movies that are the expression of relentless and strenuous attempts at reconciling opposing imperatives. A distinctive feature of this search is the classic struggle between the producer and the artists, where "the producer's interests (making a film that will attract a large audience that is also in time and on budget) are in conflict with the director's (creating a work of art) and the director can misrepresent actual expenditures of time and money" (Baker and Faulkner 1991, 286).

Artists usually desire the freedom to generate a creative outcome without being constrained by financial or market-related considerations. In contrast, the production company's financial backers seek to maximize returns on their investments (Glynn 2000). As the popular film producer and director Roger Corman noted (Corman and Jerome 1990, 237), filmmaking "is a compromised art form. It's a 50-50 split, art and commerce. Maybe that's why Americans are good at it. In a time when American industry is falling behind other countries, the American film industry is by far the most successful in the world. That's what we're good at–art and commerce, compromised." The reward for those who know how to reconcile such contrasting tensions is great, leading to worldwide recognition and success.

In his attempt to explain the way that selection operates, Lee (2000) argues that companies resolve these tensions by adopting one of the following two styles: creative protectionist and balanced. The first style reflects the stance of most of the independent producers and is characterized by its creative protectionist attitude toward distributors. More specifically, producers who take this stance in a film are so "intensely focused on the creative aspects of their pictures that distributors appear to represent a threat to the artistic integrity of their pictures and are commonly blamed if their pictures financially under-perform. This attitude often creates a love-hate relationship with their distributors" (Lee 2000, 8). But the industry also comprises a relatively small group of producers – e.g., Imagine Entertainment, Intermedia and Phoenix Pictures – whose objective is to strike a balance between their focus on creative movies and the need to please an audience and generate profits. Unlike creative protectionists, balanced producers tend to produce lower-risk movies by avoiding overemphasizing one aspect (creativity) at the expense of the others (income and audiences).

We do not use Lee's distinction to classify companies because we observe that companies change their stance over time and because the classification is overly crude in terms of understanding the fitness-dimensions and how traits operate. However, the point made regarding the tensions between creativity and financial performance is clearly correct. Firms are in fact expected to balance between these opposing imperatives (Hirsch 1972; Lampel et al. 2000). As a result, in their pursuit of the mass entertainment goal, firms "should not lose sight of artistic values. If artistic values dominate, commercial survival dictates that market realities cannot be ignored indefinitely" (Lampel et al. 2000, 265). These tensions map out in complex ways as this industry is beset by high levels of uncertainty and high mortality.

If commercial success and artistic performance are critical dimensions of fitness, what are the organizational attributes that influence them? While several traits could be taken into consideration, our analysis focuses on the roles of *human capital* and *network resources* both for theoretical and empirical reasons. The next two sections further elaborate on this point.

3.1 Traits: human capital and organizational fitness

Human capital refers to the talent, skills, knowledge and creativity of individuals working within an organization (Becker 1993). A large body of research in the resource-based tradition, for instance, emphasizes that human capital is among the firm's most critical resources (see Penrose 1959; Reed and DeFilippi 1990; Castanias and Helfat 1991; Coff 1997). The focus on "people" as a source of fitness is hardly new (e.g. Pfeffer 1994; Gimeno et al. 1997). However, human capital has become even more salient on the grounds that knowledge that lies at the core of an organization's competitive advantage (Grant 1996) is largely embedded in individuals (e.g. Nonaka and Takeuchi 1995).

Several studies have shown that the firm's stock of human capital bears a strong relationship with fitness. In particular, a firm's stock of human capital accounts for stable performance (Hitt et al. 2001; Bailey and Helfat 2003) and survival differences among firms (Bröcheler et al. 2004; Pennings et al. 1998). Not surprisingly, in knowledge intensive industries, organizations strive to retain their members (Coff 1997; Argote and Ophir 2002) or to expand their knowledge base by 'poaching' skilled individuals from rival organizations (Baty et al. 1971; Flides 1990; Rao and Drazin 2002).

The quality of human capital is undoubtedly important for both the artistic and economic performance of any creative industry, whereby the long-term survival of firms heavily depends on the ability to replenish 'creative' resources (White 1970). Talent is essential for the production process because creative activities critically bear on the imaginative and unique inspiration of gifted people who concern themselves with the originality, technical prowess, and harmony achieved in a creative act (Caves 2000). The quality of human capital also signals the value of goods that can hardly be assessed *ex-ante* because of the experiential nature of consumption in cultural/creative industries.

However, as creative resources are typically embedded in individuals and groups, firms usually have limited control over them (Stearns et al. 1987; Robins 1993; Saundry 1998). Consequently, they are faced with the need to "recruit and motivate individuals who seem to possess the insight and intuitive understanding of how creative resources can be discovered and nourished. Their competitive advantage depends on finding these individuals and also on developing structures which leverage creative resources without at the same time stifling them" (Lampel et al. 2000, 265).

This is a vital challenge in any industry such as the movie industry, where individual talents, skills, values and creative experiences are the keystones around which critical decisions are made (Morley and Silver 1977). As the world acclaimed director Peter Weir said: "... behind movies shown worldwide there are just men and their talent." This point is also shared by Michael Medavoy, Phoenix Pictures chairman: "... it is the object of every producer to work with more talented filmmakers... because in the final analysis, you're in their hands. It is their vision that makes the film whatever it finally becomes" (in Squire 1992). By the same token, Sidney Lumet noted: "The contributions by the director, actors, music, sound, camera, set, editing, are so crucial that the film keeps changing continuously ... all the efforts from the various professionals add up to a final result that is far better than any single part."

3.2 Traits: network resources and organizational fitness

While human capital is arguably a major fitness component, firms are also embedded in a dense system of inter-organizational ties and linkages. Firms can leverage these networks and their structural position in them as strategic resources. Drawing from the structural embeddedness perspective, Gulati (1999) defines *network resources* as resources that accrue to firms as a result of their position within the networks to which they belong. McEvily and Zaheer (1999) likewise assume network resources to coincide with the informational advantages associated with a firm's structure of ties. In essence, a firm's network resources or interfirm ties expose it to new ideas, information, and opportunities enhancing its capacity to compete. Along this line, a number of organizational theorists have demonstrated the impact of such positional traits as network centrality and/or density on firm innovativeness (Ahuja 2000) and performance (Powell et al. 1999, missing: Powell et al., 1999). Taken together, these studies suggest a link between an organization's position in a network structure and its fitness.

The importance of network resources in this industry is supported by several related arguments. First, the film industry is organized around projects, so the structuration of the field heavily rests on inter-organizational and personal networks. That is why organizational scholars often refer to this industry as a project-network organization (Jones 1996; Alvarez and Svejenova 2002). And indeed: "It all works like a network – noted an experienced film production manager – everyone knows everyone. If you do not know them, you normally know about them. If you don't know you can find it out" (Jones and DeFillippi 1996, 92). The film industry's network organization "is constantly being created and re-created. Firms and subcontractors combine for a specific project, disband when the project is finished, and then combine for new projects" (Jones 1996, 58). Since there is basically no formal authority overseeing this field structuration, emerging processes and outcomes are tightly linked to network structure and properties (White 1970). In other words, the structure of the network is highly informative on its members' choices and decisions.

Second, through network resources, industry participants may gather different kind of relevant information. For instance, network resources may provide information about *talent*. As Sorensen and Waguespack (2004, 7) suggest: "Industry participants clearly believe that certain actors draw audiences, that some directors usually deliver good movies and that a few producers have good instincts for producing winning pictures. And they act on this belief." Thus, while a production company may know some key players (writers, directors, composers, cinematographers), a cinematographer may know other critical talents (designers, camera operators). Since, as Nahapiet and Goshal (1998, 252) argue, "who you know affects what you know," industry participants with a good position in this range of connections are more likely to match the right talents for the project (Jones et al. 1997). This matching process is part of the filmmaking process: once the script is done – explains director Talor Hackford – "... you're ready to go out and try to match the unique talents of other artists with the qualities of the script" (Brouwer and Wright 1991, 34).

Networks also permit the companies to gather information about *ideas* or project *opportunities* such as promising scripts, books or stories, new technologies, whose copyrights have not yet been secured. While a film project must bring together a complex assortment of creative and business-oriented people, the most important element in this mix is the creative idea itself (Litman 1998). Good ideas, however, are scarce and hard to locate, a point compellingly made by Kathleen Kennedy, president of Spielberg's Amblin Entertainment: "This company always looks for a good story. That sound very simplistic... but you'd be surprised – it is the most difficult thing to find" (Brouwer and Wright 1991, 17). Network resources thus enhance the ability to gain control of valuable assets (talents and ideas), pick the right project, and thus pursue winning ventures.

The previous discussion suggests that fitness in creative industries – and in the Hollywood film industry in particular – results from properly balancing artistic and commercial performance. Because for the reasons illustrated before a firm's human capital and network resources are expected to influence both artistic and commercial performance, and by implication firm fitness, establishing the relative importance of their influence might help properly strike that balance.

4 Methods

4.1 Data

To provide empirical grounding to our discussion, we collected data on all feature length movies distributed in the United States by the eight Majors (listed earlier) and the two largest independent distributors (Miramax and New Line) over the twelve-year-period 1992-2003. This time period was one of relative technological and market stability, for which comprehensive data are available. In the last decade, these distribution companies accounted for approximately 90% of the entire U.S. box-office. Our dataset does not include documentaries, foreign-made films, short films, and compilation screen classics. Figure 1 shows the yearly distribution of the sampled movies.

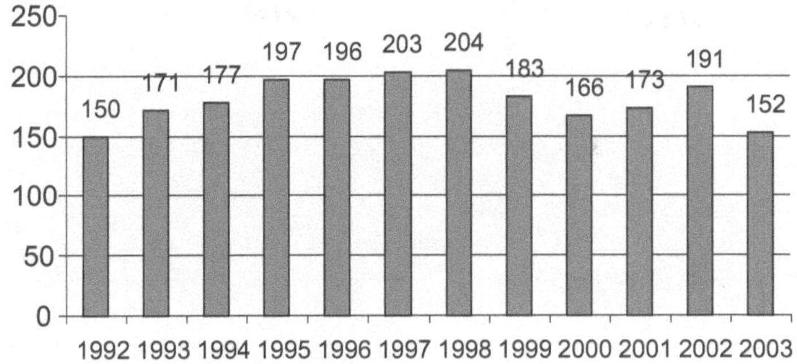

Fig. 1 Frequency distribution of movies by year of release

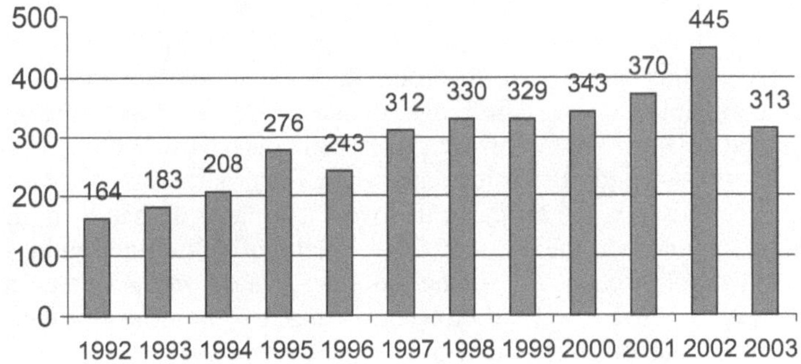

Fig. 2 Frequency distribution of production companies by year

Our dataset contains 2,146 movies. For each movie, we identified the corresponding production companies, resulting in a population of 1,912 firms (Figures 2 and 3). It is worth noting that about 85% of all the movies involve co-productions. Our data source was the Internet Movie Database (IMDB) – an online source owned by Amzon.com and largely supported through advertising. Most of the data provided by IMDB are submitted on a voluntary basis and validated by an in-house staff of 70 members. Whenever possible, we crosschecked the reliability of the data with the Alan Goble Film Index (Goble 2003).

Using the same data sources, we also identified the 12, 244 professionals involved in these companies. Given the great variety of professional profiles that operate in filmmaking, the analysis was restricted to the following set of roles: producer, director, writer, cinematographer, editor, production designer and actors (the four main actors based on the order of appearance in the movie credits). These roles, taken together, are the key representative of the so-called "core team" – i.e., the group of professionals involved during the actual shooting whose contributions are considered to be paramount for the realization of the film (Goldman 1983). This team represents the organizational heart of the film-project during its shooting phase, as it includes all the key roles from each of the "departments" (art department, camera crew, lighting crew, etc.) in which the production crew is divided (Ascher and Pincus 1999, 210).

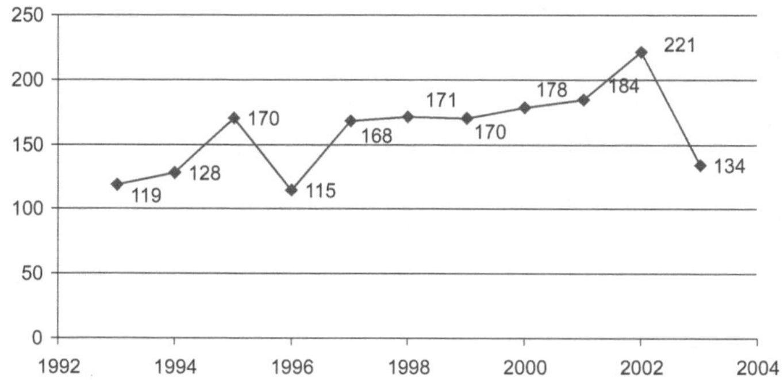

Fig. 3 Production companies entries (1992-2003)

We analyzed the network of co-productions or alliances established by the production companies in our sample over the study period by creating a set of 12 squared valued adjacency matrices (one per each year). The data for this analysis are derived from two-mode affiliation data, where the production companies are the actors and each movie is the event. In this way, a connection between production companies is assumed on the basis of their collaborative production activity. We set elements ij and ji equal to 1 if the database treated production companies i and j as movie co-producers. Furthermore, because two or more companies could co-produce more than one movie in a given year, the element ij (and the symmetric ji) may also assume values greater than 1. The 12 resulting matrices were used to compute all network measures at the production and the professional levels.

It is worth noting that, while some production companies were active in the industry before the beginning of the observation period, others started during the study period. As a result, firms differ with respect to their histories and past experiences. Similar considerations hold true also for the professionals involved in the movies produced by the firms in our sample. In the analysis we accounted for differences in initial conditions by adopting a window of three years and running the analysis for the period 1995-2003.

4.2 Measures

4.2.1 Dependent Variables

We measured commercial and artistic performance – the two key fitness dimensions in our setting – by looking at movie box office receipts and critical reception.

Commercial Performance. We measured commercial performance in terms of box office receipts, following the lead of many other studies that have investigated issues of performance in the film industry (Faulkner and Anderson 1987; Baker and Faulkner 1991; Sawhney and Eliashberg 1996). While the advent of new technologies – television, VCR, cable and DVD – has clearly expanded the number of viable

Table 1 Measure of Critical Reception

Dimensions	METASCORE RANGE
General Meaning of Score	Movies & Music
Universal Acclaim	81-100
Generally Favorable Reviews	61-80
Mixed or Average Reviews	40-60
Generally Unfavorable Reviews	20-39
Overwhelming Dislike	0-19

Source: metacritic.com

revenue sources, box office remains the "the most important benchmark when considering a film, as these ancillary revenues tend to correlate highly to the movie's performance during its theatrical exhibition period" (Sorensen and Waguespack 2004, 14). Given the uncertainty surrounding the movie business, we evaluated commercial performance based on the "annual slate of productions rather than on the performance of individual films" (Miller and Shamsie 2001, 731) – i.e., summing up the box office receipts of all movies each production company produced in a given year. Box office receipts were adjusted by a price deflator based on the consumer price index (CPI) per year (with 2003 as our base year).

We used revenues instead of profits as our dependent variable mainly for two reasons. First, budgets often include costs that are tied to the overall success of the movie (e.g., bonuses in actors' contracts). The upshot is that such costs vary endogenously with our dependent variable (on this point, see also Sorensen and Waguespack 2004). Second, our choice was made for practical reasons: only a small portion of the movies in the sample reports budget data.

Artistic Performance. As an indicator of artistic performance, we used an aggregate measure of critical reception. Our data came from a public online source ("www.metacritic.com") that rates movies distributed in the U.S. using an original algorithm. The meta-score is a weighted average of up to 30 reviews from national critics and publications for a given movie. For each review, the score given by the critic is converted to a 0-100 point scale – and for those critics who do not provide a score, a score from 0-100 based on the general impression given by the review is assigned (see Table 1). The choice of weights varies: some critics are weighted more heavily reflecting the fact that they consistently write better (more detailed, insightful and articulate) reviews than others. Moreover, some publications typically have more prestige and carry more weight in the industry. We cross checked the reliability of metacritic scores by using alternatives measures such as the number of awards and nominations (with and without Oscar awards/nominations) received by the movies each production company produced in a given year, but found no significant difference.

4.2.2 Independent Variables

Human Capital. We measured the quality of human capital – *Human Capital* – by calculating the cumulative number of awards won and the nominations received up to time $t-1$ by the key professionals (i.e., producer, writer, director, leading and supporting actors, editor and cinematographer) involved in all movies realized by the focal production company at time t. We then scaled the resulting measure by the total number of movies produced by the focal company at time t. Our goal was to have a measure capturing the average quality of the human capital a production company employed in a given year.

Our human capital measure is fine-grained in comparison to that used in other studies because we do not focus exclusively on Academy Awards – as is often the case with research in the film industry (Faulkner and Anderson 1987; Miller and Shamsie 1996, 1999). Our data was gathered on the six most prestigious awards assigned by the film industry: Academy Awards, Golden Globes, Guilds Awards (Directors Guild of America, Writers Guild of America, Producers Guild of America), National Board of Review Awards (NBR), Los Angeles Film Critics Awards (LAFC) and New York Film Critics Awards (NYFC). These awards are voted either by peer experts from the industry (Academy Awards & Guilds Awards), foreign press representatives (Globes), or film critics (NBR, LAFC and NYFC). Besides being assigned to individuals of exceptional ability, these awards reflect how successful a production company was at recruiting, combining and supporting talents (Miller and Shamsie 1999). The primary source for these data was a complete list of movie awards published by O'Neil (2000). We referred to the Internet Movie Database for the years not covered by the guide.

Network Resources. We measured a production company's network resources – *Network Resources* – by looking at its structural position in the network, using a network centrality measure. The level of centrality in a network indicates the extent to which the firm (actor in network language) occupies a strategic position by virtue of being involved in many significant ties (Wasserman and Faust 1994). High centrality enhances the informational values of network resources mainly through two dimensions: access and power. Being at the point of convergence of multiple sources of information conveyed by their ties, central actors are more likely to discover new opportunities (Valente 1995) and thereby have timely access to promising new ventures (Powell et al. 1996). Since central connectedness affects reputation and generates visibility, firms that are more centrally located enjoy status benefits (Brass and Burkhardt 1992).

We operationalized production companies' centrality using the *Eigenvector Centrality* approach (Bonacich 1972). The eigenvector approach builds on the notion of closeness/distance and is an effort to find the most central actors (i.e., those with the smallest "farness" from others) in terms of the "global" or "overall" structure of the network, and to pay less attention to patterns that are more "local" (Hanneman 2001). Given the adjacency matrices A, the eigencentrality of vertex i (denoted c_i), is given by $c_i = \alpha \sum A_{ij} c_j$ where α is a parameter. The centrality of each vertex is therefore determined by the centrality of the vertices to which it is connected.[1] We computed all measures using UCINET VI (Borgatti et al. 2002).

[1] The parameter α is required to give the equations a non-trivial solution and is therefore the reciprocal of an eigenvalue. It follows that the centralities will be the elements of the corresponding eigenvector. The normalized eigenvector centrality is the scaled eigenvector centrality divided by the maximum difference possible expressed as a percentage (Bonacich 1972).

4.2.3 Controls

Given the twofold nature of the outcome variable (artistic and commercial performance), we included several control variables to rule out alternative explanations for our results.

Major Co-production. The production of a new movie critically depends on the availability of financial resources. Not surprisingly, it is quite common for the distributor to be also a co-producer of the movies that it releases. A production company's involvement with a major distributor typically "includes a production loan . . . and ensures preferential access to the best theaters during the most favorable times" (Litman 1998, 183). For each production company, we thus created the variable *Major Co-production* – i.e., the ratio of movies co-produced by a major distributor to the total number of movies produced in a given year by a given production company.

Sequel. The oligopolistic structure of the distribution market, with a few Majors controlling the entire channel, might lead to an overemphasis on the bottom line at the expense of truly creative outcomes. Firms might then focus on formulaic content, such as sequels, which reduce the risk of a new production (Shamsie 2003), and hence forestall the search for artistic novelty. Several movies that do well at the box office are indeed sequels. Accordingly, we computed the variable *Sequel* as the ratio of movies that are sequels to the total number of movies each company produced in a given year.

Non-original script. Similarly, movies that are adaptations of, or are based on, a previously known story (e.g., books, novels, comic strips, or TV shows) are more likely to appeal to the audience than movies that rely on an entirely new script because the public is already familiar with the story (Litman and Kohl 1989). To account for this effect, we created the variable *Non-Original Script* as the percentage of all movies produced by a company in a given year that is based on prior material.

Family movie. Another important factor on the creative side is the rating assigned by the Motion Picture Association of America (MPAA). Ratings signal the degree of graphic sequences, violence and harsh language in a movie. Prior research suggests that features produced for mature audiences (R and NC-17) perform worse at the box office (Ravid 1999). Moreover, since movies rated G, PG and PG-13 have greater audience potential, and mall owners sometimes by contract require theaters not to show NC-17 films, quite often studios exert some pressure on producers and directors to ensure their films receive a rating aligned with their market aspirations. This practice can obviously constrain creativity such as, for example, in the treatment of controversial material or the choice of scenes to edit. We accounted for this possible source of interference by creating the variable *Family Rating* as the percentage of all movies produced by a company in a given year that fall in P, G or PG-13 category.

Opening Theatres. Following prior studies, we calculated the average number of opening screens – *Opening Theatres* – on which each firm's movies are initially released in a given year. This measure partly reflects the distribution and marketing strategy of each production company.

Release Date. The release dates of a motion picture also provide some indications on its box-office potential. Since moviegoers tend to crowd during certain periods (e.g., for Christmas and in the summer), high caliber movies are released only on those dates. We thus created a variable – *Release Date* – measuring the percentage of movies shown during these two peak periods.

Year. Since we have no a priori expectations about the existence of a time trend over the study period, we ran the analysis by using firm-year fixed effects. We also entered the variable year into the model as a continuous variable, but found the results to be qualitatively similar to those reported here.

4.3 Analysis I: A visual inspection of the data

We use Pajek (de Nooy et al. 2003) to offer a visual framework in which one can simultaneously appreciate the dynamics between human capital and network resources at the firm level with respect to the two fitness dimensions of interest (i.e., commercial and artistic performance). We expect these dimensions to be associated with different network structures and resources endowments. Pajek is a software package that permits the visualization and inspection of large size networks and has been used in diverse disciplines from physics (Albert et al. 2000) to management (Owen-Smith et al. 2002) to sociology (Powell et al. 2005).

The approach we follow has two major attractions. First, Pajek generates visualizations of networks that position nodes (firms) in a two-dimensional Euclidean space such that their nearness is a function of their *level* and *intensity* (in case of valued graphs) of connectedness. This is accomplished through a "spring embedded" network, drawing algorithms that simulate the network of inter-organizational connections as a system of interacting particles, where organizational nodes (production companies) repel one another unless network ties act as springs to draw connected nodes closer together. In our network, all companies that co-produced the same movie are assumed to have varying intensity of reciprocal attraction. The attraction between companies is stronger when: a) the collaboration is repeated multiple times over time; and b) the collaborators have third party ties in common. The network layout so generated is an iterative representation that minimizes the variation in length of the lines. As a result, highly connected nodes are placed at the core and weakly connected nodes at the periphery, while the relative distance among nodes is a function of their shortest network paths.

The second major attraction of this approach is that, while we are able meaningfully to represent the topology of the field based on the relational data, we can also combine this analysis with non-relational attribute data at the firm level. In practice, we can discern among firms based not only on their structural position, but also on the level of human capital possessed as well as performance achieved. This multidimensional representation of the data allows for a rich and immediate

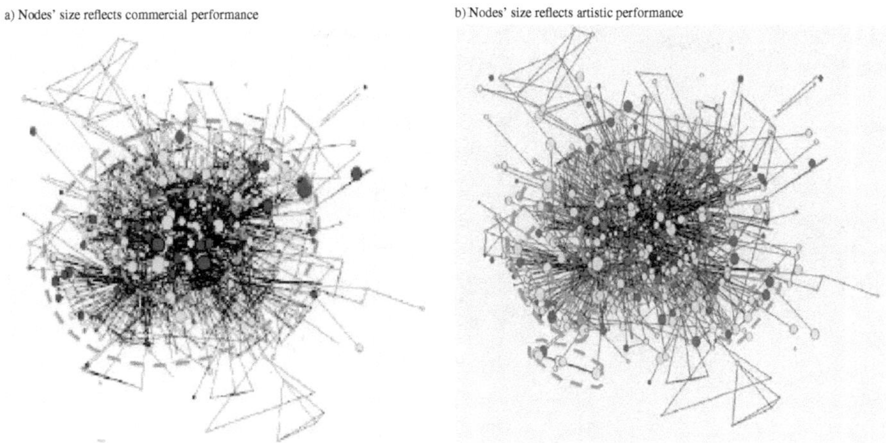

Fig. 4 1995 Production companies' network topology by commercial (a) and artistic performance (b) and human capital endowment

appreciation of the theoretical issues of interest. It can thus be thought of as a "visual goodness of fit" (Powell et al. 2005) for our conjectures.

The analysis is presented in Figures 4 to 6, where each drawing is a snapshot at time t of the network of connections among the production companies still active up to that point in time. We obtained the drawings by sequentially running two algorithms. Following Powell et al. (2005), we first drew on the Fruchterman-Reingold algorithm (1991), which optimizes network images without reference to the graph theoretic distance among nodes. This provides a reference topology for all nodes in the network. We then turned to the Kamada-Kawai (1989) algorithm to obtain a representation of substantive relational interest with nodes' Euclidean distance proportional to their graph theoretic one.

Nodes represent individual production companies and lines represent co-production agreements. The size of the nodes is proportional to the firms' commercial (Figures 4a-6a) and artistic (Figures 4b-6b) performance, respectively.

To show how the traits of human capital influence artistic and commercial performance, we assign each node the value 1 (dark grey nodes) or 0 (light grey nodes) based on whether the value of "human capital" at the firm level is greater or lesser than the average value of human capital for all companies in the sample. We opted for this simple dichotomization as it helps maintain clarity in the appreciation of the network drawings without subtracting from their illustrative intent. These visualizations provide a multi-faceted appreciation of the industry network topology, in which nodes' structural characteristics and attributes can be appreciated simultaneously. Due to space constraints, we only present a selection of drawings at 3 points in time, i.e., 1995-1999-2003.

Several interesting features stand out in the Figures. First, starting from the 1995 data on commercial performance shown in Figure 4a, we see that nodes of bigger size are highly clustered towards the centre (inner ellipse) of the network field, while small size nodes are 'pushed' to the periphery (outer ellipse). The nodes within the inner ellipse are, on average, 0.4 times bigger than the nodes that lie

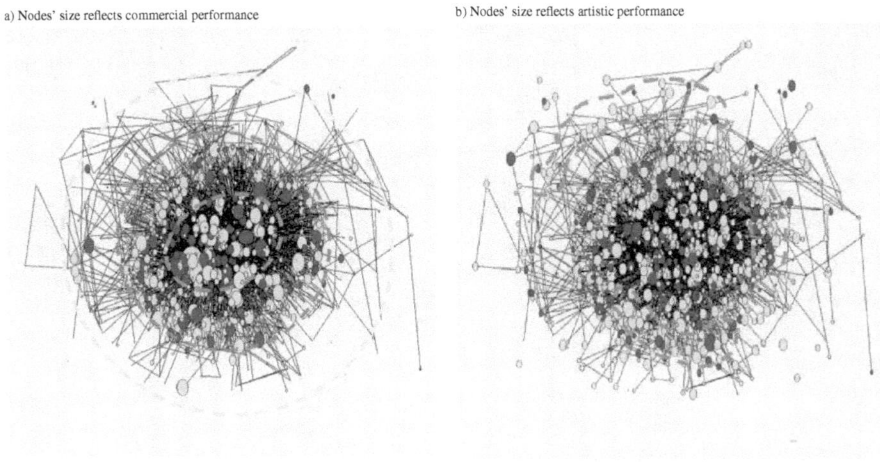

Fig. 5 1999 Production companies' network topology by commercial (a) and artistic performance (b) and human capital endowment

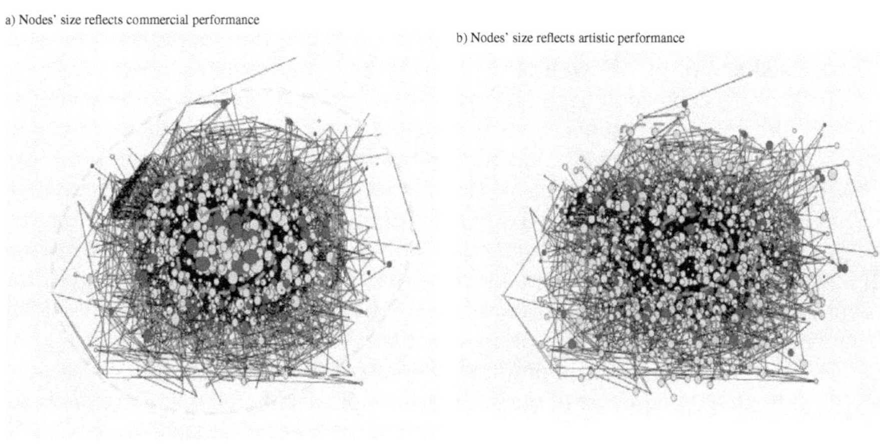

Fig. 6 2003 Production companies' network topology by commercial (a) and artistic performance (b) and human capital endowment

outside, suggesting a positive association between commercial performance and central network topology.

This structural stratification is not only consistent, but also more evident over time. In fact, the 1999 and the 2003 network drawings suggest the emergence of a second periphery (outermost ellipse) as the performance gap between the relational core and the periphery gets larger. In 1999 (Figure 5a), the nodes in the core are, on the average, 0.5 times bigger than those in the first periphery (band of nodes between inner and outer ellipse) and 0.9 times bigger than those in the outermost periphery (group of nodes between 2nd and 3rd outer ellipses). The gap is roughly the same in 2003 (Figure 6a). Furthermore, the dominant color within the clustering core is always dark grey. This means that greater commercial performance appears to be associated not only with richer relational positions, but also with richer human capital endowments. On the average, dark grey nodes are 0.37 bigger than blue nodes in 1995, 0.50 in 1999 and 0.47 in 2003. Network and human resources do seem to play a critical role when looking at commercial performance.

We get quite a different picture when we turn to our second fitness dimension: artistic performance. Starting from artistic performance in 1995, shown in Figure 4b, it is apparent that the relational core of the system is not where artistic work is actually produced. The drawings suggest quite the opposite. While the average size of the nodes in the relational core is only slightly lower than the size of peripheral nodes, the periphery is now scattered and displays a few pockets of high artistic performers (we highlighted them by tracing dashed line around some of these groups). This reversal in the role of the network finds confirmation in subsequent drawings. As we move through time, the results do not change. Again, in 1999 the nodes in the network core do not outperform those in the periphery: several 'islands' of peripheral nodes (dashed lines) do better than the core. Year 2003 exhibits a very similar pattern, suggesting that art may just as well thrive on the fringes of the network.

The saliency of human capital for artistic performance is less clear than in the other case. In 1995, the 'dark grey' appears uniformly scattered across nodes of varying size. On the average, dark grey nodes – i.e., firms with high level of human capital – are only 0.02 bigger than light grey nodes, indicating that companies employing highly talented professionals perform just as well as companies employing less talented professionals. In 1999 (Figure 5b), the size gap between dark and light grey nodes increases slightly, rising from 0.02 to 0.04, but it moves back to 0.01 in 2003. Once again, this suggests a very weak association, if any, between the two attributes.

Although these visualizations provide many suggestive insights into the shifting role of key organizational traits in the pursuit of organizational fitness, they cannot support in any statistical sense our conjecture nor can we infer any direction of causality by merely inspecting the drawings. Is a central position in the network of production alliances conducive to greater commercial performance? Is this effect consistent across fitness dimensions? Are firm with high levels of human capital more likely to achieve organizational fitness?

To address these questions and test the insights derived from the visualizations, we now turn to a multivariate analysis in which we account for our traits of interests as well as many other control variables.

4.4 Analysis II: Estimation model

To test the previous hypotheses, we estimated a random intercept effects model. The model has the following basic form:

$$y_{it} = \mu_t + \beta x_{it} + \gamma z_i + \alpha_i + \varepsilon_{it} \quad (1)$$

Instead of assuming that α_i represents a set of fixed parameters as in the fixed-effects model, in the random-effects model each α_i is a random variable with a specified probability distribution. Typically, it is assumed that α_i has a normal distribution with a mean of 0 and constant variance, and that it is independent of x_{it}, z_i and ε_{it}. Although our focus is mainly on the role of *human capital* and *network resources*, we included in the model several other variables that previous studies found to be important fitness determinants.[2] To correct for the skeweness in the dependent variables, we took the log of the two measures.[3] The two finals models are:

$$\begin{aligned}\log(\textit{Box Office}_{it}) &= \alpha_i + \beta_1(\textit{Firm Movies}) + \beta_2(\textit{Major Co-Production}_{it})\\ &+ \beta_3(\textit{Sequel}_{it}) + \beta_4(\textit{Non-Original Script}_{it}) + \beta_5(\textit{Opening Theatres}_{it})\\ &+ \beta_6(\textit{Family Rating}_{it}) + \beta_7(\textit{Release Date}_{it}) + \beta_8(\textit{Human Capital}_{it-1})\\ &+ \beta_9(\textit{Network Resources}_{it-1}) + \textit{Year Dummies} + \epsilon_{it}\end{aligned} \tag{2}$$

and

$$\begin{aligned}\log(\textit{Meta-score}_{it}) &= \alpha_i + \beta_1(\textit{Firm Movies}) + \beta_2(\textit{Major Co-Production}_{it})\\ &+ \beta_3(\textit{Sequel}_{it}) + \beta_4(\textit{Non-Original Script}_{it}) + \beta_5(\textit{Opening Theatres}_{it})\\ &+ \beta_6(\textit{Family Rating}_{it}) + \beta_7(\textit{Release Date}_{it}) + \beta_8(\textit{Human Capital}_{it-1})\\ &+ \beta_9(\textit{Network Resources}_{it-1}) + \textit{Year Dummies} + \epsilon_{it}\end{aligned} \tag{3}$$

We obtained our estimates using PROC MIXED and PROC GENMOD in SAS (version 9.1).

5 Results

Descriptive statistics is presented in Tables 2 and 3. We checked for multicollinearity by calculating the tolerance factor, which is measured as the difference between 1 and the R^2 value from the model. Usually statisticians suggest 0.3 as a threshold for the tolerance factor below which multicollinearity might become an issue (Allison 1999). We found no variable to violate such level.

Random-effects estimates for the model where commercial performance (i.e., box office receipts) is the dependent variable are displayed in Table 4, while those for the model where artistic performance (i.e., critical reception) is the dependent variable are displayed in Table 5. Both tables report the estimates for the baseline model (Model 1), the model including the controls and the measure of human capital (Model 2), the controls and measure of network resources (Model 3), and the full model (Model 4), respectively. In each model, we entered year dummies. In the last column of each table, we computed standardized coefficients for the full model to estimate the relative importance of each variable in affecting the dependent variable.

The baseline model (Model 1) in Table 4 presents the results for the controls that turned out to be significant and in the expected direction, with the exception of the variable controlling for the number of movies produced by a firm in a given year (*Firm Movies*) and the variable that accounts for whether a Major is a co-producer

[2] Our approach is then somehow consistent with the *systems* approach to fitness (see Drazin and Van den Ven 1985; Van de Ven and Drazin 1985; Galunic and Eisenhardt 1994).

[3] We should expect the dependent variables to be skewed because most of the Hollywood movie productions perform poorly and just a few are very successful.

Table 2 Descriptive Statistics

Variables	N	Mean	Std Dev	Minimum	Maximum
1. Box Office (log)	5096	16.39	2.01	7.69	20.32
2. Critical Reception	5253	5.53	1.46	0.7	9.5
3. Rated Family	5182	0.49	0.5	0	1
4. Sequel	5269	0.08	0.27	0	1
5. Non-Original Script	5224	0.39	0.55	0	3
6. Major Co-production	5280	0.29	0.45	0	1
7. Opening Theatres	4552	6.31	2.33	0	8.25
8. Release Date	5262	0.31	0.46	0	1
9. Firm Movies (log)	5280	0.9	1.01	0	3
10. Human Capital	5280	60	88	0	904
11. Network Resources	5280	0.04	0.09	0	0.55

(*Major Co-production*). Movies that are based on an existing story (*Non-Original Script*) appeal to the audience more than movies that rely on a new script. Similarly, a movie's commercial success is enhanced when the movie release dates coincide with Christmas or the summer, and the movie addresses a broader audience (*Family Rating*), is a sequel, and is distributed on a large scale as captured by the number of opening theatres.

The baseline model (Model 1) in Table 5 offers quite a different picture. Artistic performance declines when production companies produce movies that address a broader audience (*Family Rating*), are sequels, are distributed on a large-scale distribution – i.e., movies for which box office considerations are typically more salient – and a production company distributes a larger number of firm in that year. In contrast, movies that are based on an existing story (*Non-Original Script*) and are co-produced with a Major on the average attain higher level of artistic performance.

In Model 2 of Tables 4 and 5, we entered our measure of human capital, i.e., the average number of awards and nominations that the core members staffed by the production company received until the year prior to the focal one. The variable is statistically significant and in the expected direction in both models, suggesting that the quality of the professionals' human capital enhances production companies' commercial and artistic performance. Also, the inclusion of the human capital variable improves the overall fit of the model relative to the baseline model, as indicated by variation in the value of the -2 Log Likelihood statistics.

In Model 3 of Tables 4 and 5, we entered our measure of network resources, based on the Eigenvector approach, i.e., the relational position of the companies with respect to the overall structure of the network. The variable is statistically significant and in the expected direction for the model where commercial success is the dependent variable, suggesting that production companies that are part of rich inter-organizational network of co-production alliances are more likely to pick the right projects and talent and hit the market with high grossing films. The inclusion of the variable improves the overall fit of the model. Production companies' network resources, on the contrary, seem to have no effect on movies' artistic performance.

The results for the full model (Model 4) are consistent with those of the previous models. While the quality of professionals' human capital employed in a movie is significant and positively affects both commercial success and artistic perfor-

Table 3 Pearson Correlation Coefficients

Variables	1	2	3	4	5	6	7	8	9	10	11
1. Critical Reception	1										
2. Box Office (log)	0.126	1									
3. Rated Family	-0.134	0.262	1								
4. Sequel	-0.102	0.147	0.061	1							
5. Non-Original Script	0.115	0.102	0.096	0.015	1						
6. Major Co-production	0.008	0.132	0.07	0.055	0.009	1					
7. Opening Theatres	-0.286	0.675	0.288	0.174	0.025	0.126	1				
8. Release Date	0.141	0.164	0.151	0.083	0.081	0.011	0.039	1			
9. Firm Movies (log)	-0.016	0.173	0.068	0.04	0.019	0.804	0.188	0.006	1		
10. Human Capital (1-year lag)	0.25	0.224	0	0.044	0.11	0.003	0.049	0.104	0.008	1	
11. Network Resources	-0.018	0.11	-0.009	0.045	0.017	0.298	0.121	-0.001	0.448	0.028	1

Table 4 Determinants of *Commercial Performance*. Random Intercept Regression

Dependent Variable = Movie Box Office Receipts, 5280 Observations

Variables	Model 1	Model 2	Model 3	Model 4	Standardized Coefficients
Intercept	12.75***	12.41***	12.72***	12.39***	
	(0.099)	(36.167)	(36.176)	(36.198)	
Year (dummies)	yes	yes	yes	yes	
Rated Family	0.114**	0.164***	0.120**	0.169***	0.042
	(0.046)	(0.045)	(0.046)	(0.045)	
Sequel	0.223**	0.214**	0.215**	0.207**	0.028
	(0.077)	(0.075)	(0.077)	(0.075)	
Non-Original Script	0.318***	0.251***	0.317***	0.251**	0.069
	(0.038)	(0.038)	(0.038)	(0.038)	
Major Co-Production	0.182	0.149	0.200*	0.164	0.037
	(0.121)	(0.116)	(0.120)	(0.115)	
Opening Theatres	0.525***	0.519***	0.525***	0.518***	0.6
	(0.010)	(0.010)	(0.010)	(0.010)	
Release Date	0.621***	0.548***	0.621***	0.548***	0.125
	(0.046)	(0.045)	(0.046)	(0.045)	
Firm Movies (log)	0.055	0.048	0.006	0.008	0.004
	(0.045)	(0.044)	(0.047)	(0.045)	
Human Capital (1-year lag)		0.003***		0.003***	0.131
		(0.0003)		(0.0003)	
Network Resources			1.290***	1.071**	0.048
			(0.351)	(0.343)	
-2 Log Likelihood	15278	15090	15264	15081	
Likelihood ratio test (vs. baseline)		376***	28***	394***	

$^{*}p < 0.1, ^{**}p < 0.05, ^{***}p < 0.001$
Two-tailed tests for all variables

mance, network resources have a positive and statistically significant effect only in the case of commercial success. The last column in Tables 4 and 5 reports the standardized coefficients to establish which variable is more important in affecting the dependent variable. Since we concern ourselves primarily with the human capital and network resources variables, we restrict the comparison to them. Interestingly, professionals' human capital has a stronger effect – almost 3 times more – than production companies' network resources on movie box office. In conclusion, the results suggest that the quality of the human capital employed is important for movie commercial and artistic performance, while production companies' network resources (i.e., relations to other companies) have a significant positive effect only on movie box office.

Robustness Tests. We tested the robustness of the results to alternative model specifications. First, we estimated the random effects model using the Generalized Estimating Equations (GEE). In the case of linear models, GEE is equivalent to generalized least squares (Allison 2005). The GEE method makes no explicit assumptions about random components in the regression model, but simply allows for correlation in the dependent variable across observations over time. We assumed this correlation to be equal across time using an exchangeable correlation

Table 5 Determinants of *Artistic Performance*. Random Intercept Regression

Dependent Variable = Critical Reception, 5280 Observations

Variables	Model 1	Model 2	Model 3	Model 4	Standardized Coefficients
Intercept	6.384***	5.964***	6.386***	5.970***	
	(0.095)	(0.094)	(0.095)	(0.094)	
Year	yes	yes	yes	yes	
Rated Family	-0.307***	-0.244***	-0.308***	-0.246***	-0.061
	(0.044)	(0.043)	(0.044)	(0.043)	
Sequel	-0.351***	-0.360***	-0.350***	-0.357***	-0.048
	(0.074)	(0.071)	(0.074)	(0.071)	
Non-Original Script	0.309***	0.224***	0.309***	0.224***	0.061
	(0.037)	(0.036)	(0.037)	(0.036)	
Major Co-Production	0.248**	0.207*	0.246**	0.201*	0.045
	(0.121)	(0.109)	(0.121)	(0.109)	
Opening Theatres	-0.163***	-0.172***	-0.163***	-0.172***	-0.199
	(0.010)	(0.009)	(0.010)	(0.009)	
Release Date	0.423***	0.332***	0.423***	0.332***	0.076
	(0.044)	(0.043)	(0.044)	(0.043)	
Firm Movies (log)	-0.001***	-0.017	0.004	-0.001	-0.001
	(0.044)	(0.042)	(0.046)	(0.043)	
Human Capital (1-year lag)		0.004***		0.004***	0.175
		(0.0002)		(0.0002)	
Network Resources			-0.135	-0.425	-0.019
			(0.340)	(0.325)	
-2 Log Likelihood	14971	14633	14970	14631	
Likelihood ratio test (vs. baseline)		776***	2	780***	

$^{*}p < 0.1,^{**}p < 0.05,^{***}p < 0.001$
Two-tailed tests for all variables

structure (Liang and Zeger 1986). It is worth noting that the GEE coefficients are population averaged (i.e., they describe what happens to the whole population, not a particular individual, if everyone's predictor values are increased by one unit). The results for the GEE model turned out to be qualitatively similar to those for the random intercept model and are available from the authors upon request. Second, we compared the results with those for the fixed effects model to verify whether unobserved heterogeneity might be a problem in our analysis. Again, the coefficient estimates for the fixed effects regression model are qualitatively similar to those obtained using the random intercept and the GEE models.

We also ran additional analyses using alternative measures for artistic performance, such as the total number of awards and nominations (with and without Oscars) received by each production company in a given year. Though not reported here, the results did not vary appreciably from those presented in the paper.

Like in many other industries, success or failure is highly dependent on the intensity of competition in the marketplace. Previous research (for a review see Litman 1998) showed that high levels of box office concentration for the top (i.e., most successful) movies have a negative impact on both theatrical rentals and the length of run. We likewise controlled for the degree of concentration of the industry at the firm, not the movie, level using the top four production company concentration ratio – $C4$ – which we computed by taking the revenues accruing to the top four firms in a given year as the percentage of total industry revenues for the year in question. Moreover, intense competition makes it more difficult for firms to survive and grow. Organizational crowding and resource scarcity force firms to rely on inferior resources, jeopardizing their ability to spot valuable opportunities. Production companies can stimulate demand by offering more innovative products to counter declining performance as the environment becomes less munificent (Miller and Shamsie 1996). Since movies compete directly with each other for the same consumer dollars, awards and critical attention, we controlled for the degree of market competition by including the variable *Film Crowding*, which we calculated as the total number of movies released each year in the industry. Using year dummies already controls for the degree of industry concentration and film crowding. But the results did not change when we entered these two variables into the model with *year* as a continuous variable.

Furthermore, we made an assumption about the duration of sampled co-production alliances whose termination dates were not available. Following a common practice in alliance research (e.g. Bae and Gargiulo 2004; Gulati and Gargiulo 1999), we used various moving windows to account for the duration of each alliance, making the adjacency matrixes time-varying. Results were consistent across 2, 3 and 4 year-window specifications.

Finally, we re-estimated our models adopting other measures of network centrality than the Eigenvector. We opted for two measures that are well-suited to deal with centrality in valued matrices. In one set of analyses, we used a simple measure of degree centrality that counts the number of co-production partners with which the focal firm works. Next, we employed *Bonacich Power* measure (Bonacich 1987), which is a generalization of degree-based approaches to centrality and estimates an actor's centrality as a function of the number of connections that both she and the actors in her neighborhood have. In all such cases, the results were consistent with our prior findings.

6 Discussion and conclusions

We began with the observation that fitness is a multidimensional concept and its components are contest-specific. As such, the determinants of fitness can only be identified with respect to a particular context. On the premise that the nature of the selection environment determines which traits eventually affect the degree of organizational fitness, we employed multiple fitness measures and studied their relationships with specific fitness dimensions. We situated the analysis in the Hollywood motion picture industry, since, as in the case of other cultural/creative industries, firms are confronted with the challenge of striking a balance between seemingly conflicting needs such as artistic performance and commercial imperatives. These are two critical fitness dimensions in that organizational fitness in those industries stems from striking a balance between them.

Besides identifying some of the key determinants of fitness, our primary objective was to gauge the relative contribution of different organizational attributes – most notably, human capital and network resources – to movie production companies' commercial and artistic performance, respectively. In the movie industry, firm performance critically depends on the ability to replenish creative resources – which are typically embedded in individuals and groups; and to establish social networks with other companies to have easier access to such resources.

The results of the analysis suggest three main findings. Fist, the quality of human capital is valuable not only in generating a creative outcome, but also in attaining higher levels of commercial performance. Second, network resources contribute to enhancing production companies' commercial success, but seem to have no impact on artistic performance. Finally, the quality of human capital is more strongly associated than network resources with production companies' commercial success.

An important implication of these findings is that the impact of different attributes on organizational fitness varies with the particular dimension of fitness that is being considered. For instance, if one measures fitness in terms of commercial success, network resources have a positive impact. But different conclusions are drawn if one uses artistic performance as a measure of organizational fitness. Since, in our setting, artistic performance and commercial success are not independent, in that organizational fitness is the result of properly balancing between artistic and commercial performance, focusing on attributes that influence only one dimension but not the other, or attributes the influence of which is different for each dimension, might prove to be misleading. The attention should then be directed to the identification of organizational attributes with a similar effect on both dimensions. Ceteris paribus, the quality of human capital production companies employ in each new movie is especially important.

The finding that network resources – which in this paper we defined as firms' network centrality – has no effect on artistic performance deserves further attention. We surmise that one possible explanation for this somewhat unexpected result is that an increase in connectedness could engender a proclivity towards conformity. As firms become entrenched into the relational system, their incentive to deviate from the 'norm' declines, mostly because of the higher stakes they have in case of failure. Moreover, creating and maintaining many tight-knit links between different actors cost time and energy, with the effect of smothering creativity under a blanket of homogeneity.

By contrast, firms that remain at the periphery of the system could be in a better position to contribute some 'freshness' to the dominant practices of the system. Peripheral firms are in fact less strongly embedded in a given network as compared to more central ones. Not only are they more likely to have a distinctive perspective resulting from divergent ideas sparked by isolates or outsiders; they can also pursue these ideas without facing the constraints of breaking established norms, or worrying about deviating from accepted tenets of the network (Perry-Smith and Shalley 2003). As we noted before, the visual inspection of the data further suggests that firms are more likely to achieve higher levels of artistic performance when they are located on the fringes of the network.

This line of reasoning is consistent with evidence from research on the conditions favoring the introduction of breakthrough inventions. In the telecommunications industry, for instance, corporations interested in point-to-point wireless telegraphy (e.g., RCA, Westinghouse, AT&T) viewed the scattering of radio waves as a nuisance, whereas amateur radio enthusiasts saw the potential of point-to-multipoint broadcasting (Sawhney and Lee 2003). Also, recent findings in the movie industry have shown how an intermediate position between the core and the periphery of the social network enhances individual creativity (Cattani and Ferriani 2005). Likewise, peripheral firms – which are not too far from the core – are more likely to retain the ability to think outside the box because a peripheral position "may suggest connections outside the network that facilitate creativity" (Perry-Smith and Shalley 2003, 98).

The need to strike a balance between opposing polarities is not unique to cultural/creative industries, but is common to a wide range of industries characterized by high levels of ambiguity and dynamism (Lampel et al. 2000). The trade-off between exploration and exploitation in research on learning bears some similarity with the issue under investigation (March 1991; Levinthal and March 1993). According to this research, for instance, firms have "to cope with confusing experience and the complicated problem of balancing the competing goals of developing new knowledge (i.e., exploration) and exploiting current competencies in the face of dynamic tendencies to emphasize one or the other" (Levinthal and March 1993, 95). There is an intuitive similarity between the concept of exploration and the quest for a creative outcome; and the pursuit of commercial performance, which appears well matched to the idea of exploitation. Although the distinction between *creative protectionist* and *balanced producers* reflects the different emphasis firms place on one fitness dimension or the other, organizational performance and survival depends on the ability to reconcile the aforementioned polarities. A study on whether traits which are typically referred to as important fitness determinants have a similar or a different effect depending on the particular fitness dimension being considered represents a step towards a sharper understanding of how that reconciliation can be obtained.

There are some obvious limitations to the study. First, the unique nature of the industry raises questions about the generalizability of our findings that can only be answered by examining other contexts. Industries such as fashion, architecture, design, publishing, and the life science in general, are all intriguing candidates due to the friction between innovative endeavors and budgetary constrains that lies at the very core of their business models. Second, while network and human resources are both important traits in the movie industry, a finer understanding of their effect on movie performance would probably require a deeper examination of their relationship. For instance, many studies suggest that, while relational ties and connections critically improve the value of human capital within an organization (Seibert et al. 2001), superior human resources likewise enhance an organization's networking capability (Florin et al. 2003). Accounting for the interaction effect between human and network resources in shaping artistic and financial performance, therefore, would probably represent a viable extension of this study. In addition, a deeper analysis of the relationship between human capital and network resources, and other organizational fitness traits should be investigated more deeply and might thus represent an interesting avenue for future research.

Finally, though we drew from prior studies to create measures of human and network resources, our composite measures are indirect proxies for largely unobservable phenomena. As such, they might lack the required precision. For instance, if one takes the perspective of a movie production company, an individual's web of relations is an integral dimension of the human capital resource. The overall quality of each individual's human capital can in fact be more correctly conceptualized as resulting from talents and social capital. We thus need a more fine-grained measure of human capital to estimate its actual impact on firm fitness. In a similar vein, it would be interesting to account for network resources resulting not only from alliances (co-productions) between production companies, but also from the relationships between production companies and distributors. Finally, while we opted for a linear model linking different fitness measures to its determinants, a viable alternative might be the use of structural equation modelling to more effectively capture the trade-off between artistic and financial performance.

Of course, this paper is a first attempt to explore determinants of fitness in the context of cultural/creative industries where the complex nature of the relation between commercial and artistic performance critically affects firm performance and survival. The results are, therefore, still preliminary and do not lend themselves to any generalization. Nevertheless, the pattern shown in the analysis corroborates our initial intuition and is consistent with the nature of the industry under investigation.

References

Ahuja G (2000) Collaboration networks, structural holes, and innovation: A longitudinal study. Admin Sci Quart, 45: 425–455

Albert R, Jeong H, Barabási AL (2000) Error and attack tolerance in complex networks. Nature, 406: 378–382

Allison PD (1999) Logistic Regression Using the SAS System: Theory and Application. SAS Institute, Cary, NC

Allison PD (2005) Fixed Effects Regression Methods for Longitudinal Data. SAS Institute, Cary, NC

Alvarez JL, Svejenova S (2002) Symbiotic careers in movie making: Pedro and Agustin Almodovar. In: Peiperl M, Arthur M, Anand N (eds.) Career Creativity: Explorations in the Remaking of Work, Oxford University Press, Oxford

Argote L, Ophir R (2002) Inter-organizational learning. In: Baum JAC (ed.) Companion to Organizations, 181–207, Blackwell, Oxford

Ascher S, Pincus E (1999) The Filmmaker's Handbook: A Comprehensive Guide for the Digital Age. Plume

Bae J, Gargiulo M (2004) Partner substitutability, alliance network structure, and firm profitability in the telecommunications industry. Acad Manage J, 47, 6: 843–859

Bailey EE, Helfat CE (2003) External management succession, human capital, and firm performance: An integrative analysis. Managerial Dec Econ, 24, 4: 347–369

Baker WE, Faulkner RR (1991) Role as resource in the Hollywood film industry. Amer J Sociol, 97, 2: 279–309

Barley SR (1990) The alignment of technology and structure through roles and networks. Admin Sci Quart, 35: 61–103

Baty GB, Evan WM, Rothermel TW (1971) Personnel flows as inter-organizational relations. Admin Sci Quart, 16: 430–443

Becker G (1993) Human Capital: A Theoretical and Empirical Analysis with Special Reference to Education. University of Chicago Press, Chicago, IL, 3rd edn.

Becker H (1982) Art Worlds. University of California Press, Berkeley

Bonacich P (1972) Factoring and weighting approaches to status scores and clique identification. J Math Sociol, 2: 113–120

Bonacich P (1987) Power and centrality: A family of measures. Amer J Sociol, 92: 1170–1182

Borgatti SP, Everett MG, Freeman LC (2002) Ucinet 6 for Windows. Analytic Technologies, Harvard

Brass DJ, Burkhardt ME (1992) Centrality and power in organizations. In: Nohria N, Eccles RG (eds.) Networks and Organizations, HBS Press, Boston

Bröcheler V, Maijoor S, van Witteloostuijn A (2004) Auditor human capital and audit firm survival: The Dutch audit industry in 1930-1992. Account Org Soc, 29, 7: 627–646

Brouwer A, Wright T (1991) Working in Hollywood. Reissue Edition. Avon Books, NY

Burns T, Stalker GM (1961) The Management of Innovation. Tavistock, London

Castanias R, Helfat C (1991) Managerial resources and rents. J Manage, 17: 155–171
Cattani G, Ferriani S (2005) A Social network perspective on individual creative performance: Evidence from the Hollywood film-making industry, 1992-2003. Paper presented at the 21st EGOS Conference, June 30 - July 2, Berlin
Caves ER (2000) Creative Industries. Harvard University Press, Cambridge, MA
Chandler A (1962) Strategy and Structure. MIT Press, Cambridge, MA
Child J (1972) Organizational structure, environment and performance: The role of strategic choice. Sociology, 6: 1–22
Coff R (1997) Human assets and management dilemmas: Coping with hazards on the road to resource-based theory. Acad Manage Rev, 22: 374–402
Corman R, Jerome J (1990) How I Made a Hundred Movies in Hollywood and Never Lost a Dime. Random House, New York
De Jong G (1994) The fitness concepts and the description of natural selection. Q Rev Biol, 69, 1: 3–29
de Nooy W, Mrvar A, Batagelj V (2003) Exploratory Social Network Analysis with Pajek. Cambridge University Press, New York
Doty DH, Glick WH, Huber GP (1993) Fit, equifinality, and organizational effectiveness: A test of two configurational theories. Acad Manage J, 36: 1196–1250
Drazin R, Van den Ven AH (1985) Alternative forms of fit in contingency theory. Admin Sci Quart, 30: 514–539
Faulkner RR, Anderson AB (1987) Short-term projects and emergent careers: Evidence from Hollywood. Amer J Sociol, 92: 879–909
Flides RA (1990) Strategic challenges in commercializing biotechnology. Calif Manage Rev, 63–72
Florin JM, Lubatkin M, Schulze W (2003) A social capital model of new venture performance. Acad Manage J, 46, 3: 374–384
Fruchterman T, Reingold E (1991) Graph drawing by force-directed replacement. Software Pract Exper, 21: 1129–1164
Galunic DC, Eisenhardt KM (1994) Renewing the strategy-structure-performance paradigm. In: Staw BM, Cummings LL (eds.) Research in Organizational Behavior, vol. 16, 215–255, JAI Press, Greenwich, CT
Gimeno JF, Folta TB, Cooper AC, Woo CY (1997) Survival of the fittest? Entrepreneurial human capital and the persistence of underperforming firms. Admin Sci Quart, 42: 750–783
Glynn MA (2000) When cymbals become symbols: Conflict over organizational identity within a symphony orchestra. Organ Sci, 11, 3: 285–298
Goble A (2003) The Complete Index to World Film. Valan Publishing
Goldman W (1983) Adventures in the Screen Trade. Scribner, New York
Grant RM (1996) Toward a knowledge-based theory of the firm. Strategic Manage J, 17: 109–122
Gulati R (1999) Network location and learning: The influence of network resources and firm capabilities on alliance formation. Strategic Manage J, 20, 5: 397–420
Gulati R, Gargiulo M (1999) Where do inter-organizational networks come from? Amer J Sociol, 104, 5: 1439–1493
Hannan MT, Freeman JH (1989) Organizational Ecology? Harvard University Press, Cambridge, MA
Hanneman R (2001) Introduction to Social Network Methods. University of California, Riverside
Hawley AH (1950) Human Ecology. A Theory of Community Structure. The Ronald Press, New York
Hirsch P (1972) Processing fads and fashions: An organization-set analysis of cultural industry system. Amer J Sociol, 77, 4: 639–659
Hitt MA, Bierman L, Shimizu K, Kochhar R (2001) Direct and moderating effects of human capital on the strategy and performance in professional service firms: A resource-based perspective. Acad Manage J, 44: 13–28
Hodgson GM (1996) Economics and Evolution. The University of Michigan Press, Ann Arbor
Holbrook MB (1999) Popular appeal versus expert judgments of motion picture. J Consum Res, 26: 144–155
Honthaner EL (2001) The Complete Film Production Handbook. Focal Press, Burlington, MA, 3rd edn.
Jones C (1996) Careers project-networks: The case of film industry. In: Arthur MB, Russeau DM (eds.) The Boundaryless Career, Oxford University Press, New York

Jones C, DeFillippi RJ (1996) Back to the future in film: Combining industry and self knowledge to meet career challenges of the 21st Century. Acad Manage Exec, 10, 4: 89–104
Jones C, Hesterly WS, Lichtenstein B, Borgatti SP, Tallman SP (1997) Intangible assets of teams: How human, social end team capital influence project performance in the film industry. Working Paper, Organization Studies Department, Boston College
Kamada T, Kawai S (1989) An algorithm for drawing general undirected graphs. Inform Process Lett, 31: 7–15
Knudsen T (2004) General selection theory and economic evolution. J Econ Methodology, 11, 2: 147–173
Lampel J, Lant T, Shamsie J (2000) Balancing act: Learning from organizing practices in cultural industries. Organ Sci, 11, 3: 263–269
Lawrence P, Lorsch J (1967) Organization and Environment. Harvard Business School Press, Boston, MA
Learned E, Christensen C, Andrews K, Guth W (1965) Business Policy: Text and Cases. Irwin, Homewood, IL
Lee JJ (2000) The Producer's Business Handbook. Focal Press, Burlington, MA
Levinthal DA (1997) Adaptation on rugged landscapes. Manage Sci, 43, 7: 934–950
Levinthal DA, March JG (1993) The myopia of learning. Strategic Manage J, 14: 95–113
Liang KY, Zeger SL (1986) Longitudinal data-analysis using Generalized Linear-Models. Biometrika, 73, 1: 13–22
Litman BR (1998) The Motion Picture Mega-Industry. Allyn and Bacon, Boston
Litman BR, Kohl LS (1989) Predicting financial success of motion pictures: The '80s experience. J Media Econ, 2: 35–50
March JG (1991) Exploration and exploitation in organizational learning. Organ Sci, 2, 1: 71–87
McEvily B, Zaheer A (1999) Bridging ties: A source of firm heterogeneity in competitive capabilities. Strategic Manage J, 20: 1133–1156
Metcalfe SJ (1998) Evolutionary Economics and Creative Destruction. Routledge, London, UK
Meyer MW, Gupta V (1994) The performance paradox. In: Cummings LL, Staw BM (eds.) Research in Organizational Behavior, 309–369, JAI Press, Greenwich CT
Miles RE, Snow CC (1978) Organizational Strategy, Structure and Process. McGraw-Hill, New York
Miller D, Shamsie J (1996) The resource-based view of the firm in two environments: The Hollywood film studios from 1936 to 1965. Acad Manage J, 39, 3: 519–543
Miller D, Shamsie J (1999) Strategic responses to three kinds of uncertainty: Product line simplicity at the Hollywood film studio. J Manage, 25, 1: 97–116
Miller D, Shamsie J (2001) Learning across the life cycle: Experimentation and performance among the Hollywood studio heads. Strategic Manage J, 22: 725–745
Morley E, Silver A (1977) A film director's approach to managing creativity. Harvard Bus Rev, 55, 2: 59–68
Nahapiet J, Goshal S (1998) Social capital, intellectual capital and the organizational advantage. Acad Manage Rev, 23: 242–266
Nonaka I, Takeuchi H (1995) The Knowledge-creating Company. Oxford University Press
O'Neil T (2000) Movie Awards. Variety Books
Owen-Smith J, Riccaboni M, Pammolli F, Powell WW (2002) A comparison of U.S. and European university-industry relations in the life sciences. Manage Sci, 48, 1: 24–43
Pennings JM (1987) Structural contingency theory: A multivariate test. Organ Stud, 8: 223–240
Pennings JM, Lee K, van Witteloostuijn A (1998) Human capital, social capital and firm dissolution. Acad Manage J, 41, 4: 425–440
Penrose ET (1959) The Theory of the Growth of the Firm. Blackwell, Oxford
Perry-Smith JE, Shalley CE (2003) The social side of creativity: A static and dynamic social network perspective. Acad Manage Rev, 28, 1: 89–106
Pfeffer J (1994) Competitive Advantage Through People. Harvard Business School Press, Boston
Porter ME (1980) Competitive Strategy. Free Press, New York
Powell W, Kenneth W, Koput W, Smith-Doerr L, Owen-Smith J (1999) Network position and firm performance: Organizational returns to collaboration in the biotechnology industry. Research in the Sociology of Organizations, 16: 129–159
Powell W, Koput KW, Smith-Doerr L (1996) Inter-organizational collaboration and the locus of innovation: Networks of learning in biotechnology. Admin Sci Quart, 41: 116–145

Powell W, White DR, Koput KW, Owen-Smith J (2005) Network dynamics and field evolution: The growth of inter-organizational collaboration in the life sciences. Amer J Sociol, 110, 4: 1132–1205
Rao H, Drazin R (2002) Overcoming resource constraints on product innovation by recruiting talent from rivals: A study of the mutual fund industry, 1986-1994. Acad Manage J, 45, 3: 491–507
Ravid SA (1999) Information, blockbusters, and stars: A study of the film industry. J Bus, 72, 4: 463–492
Reed R, DeFilippi RJ (1990) Causal ambiguity, barriers to imitation, and sustainable competitive advantage. Acad Manage Rev, 15, 1: 88–102
Rivkin J (2000) Imitation of complex strategies. Manage Sci, 46: 824–844
Robins JA (1993) Organization as strategy: Restructuring production in the film industry. Strategic Manage J, 14: 103–118
Saundry S (1998) The limits of flexibility: The case of UK television. Brit J Manage, 9: 151–162
Sawhney H, Lee S (2003) Arenas of Innovation: Fringe Groups and Discovery of New Liberties of Action. Working Paper of the Center for Social Informatics SLIS, Indiana University Bloomington
Sawhney MS, Eliashberg J (1996) A parsimonious model for forecasting gross box-office revenues of motion pictures. Marketing Sci, 15, 2: 113–131
Scott AJ (2002) A new map of Hollywood: The production and distribution of American motion pictures. Reg Stud, 36: 957–975
Scott WR (1975) Organizational structure. Annu Rev Sociol, 1: 1–20
Seibert SM, Kraimer ML, Liden RC (2001) A social capital theory on career success. Acad Manage J, 2: 219–237
Shamsie J (2003) Exploiting the tie that binds: Dealing with issues of resource mobility in the U.S. motion picture industry. Working Paper of the Michigan State University
Siggelkow N (2001) Change in the presence of fit: The rise, the fall, and the renaissance of Liz Claiborne. Acad Manage J, 44, 4: 838–857
Siggelkow N (2002) Evolution toward fit. Admin Sci Quart, 47: 125–159
Sober E (2000) Philosophy of Biology. Westview Press, Boulder, CO
Sorensen O, Waguespack D (2004) Social Networks and Exchange: Self-Confirming Dynamics in Hollywood. Working Paper presented at the Wharton School, University of Pennsylvania
Squire JE (1992) The Movie Business Book. Fireside Book, New York, 2nd edn.
Stearns SC (1976) Life-history tactics: A review of the ideas. Q Rev Biol, 51: 3–47
Stearns T, Hoffman A, Heide J (1987) Performance of commercial television stations as an outcome of inter-organizational linkages and environmental conditions. Acad Manage J, 30: 71–90
Svejenova S, Alvarez L (2002) Imprints of creative style on project network governance. Working Paper of the IESE School of Business
Valente TW (1995) Network Models of the Diffusion of Innovations. Hampton Press, Cresskill, NJ
Van de Ven AH, Drazin R (1985) The concept of fit in contingency theory. In: Staw BM, Cummings LL (eds.) Research in Organizational Behavior, 333–365, JAI Press, Greenwich, CT
Venkatraman N (1989) The concept of fit in strategy research: Toward verbal and statistical correspondence. Acad Manage Rev, 14, 3: 423–444
Vogel HL (1998) Entertainment Industry Economics. Cambridge University Press, Boston, MA
Wasserman S, Faust K (1994) Social Network Analysis: Methods and Applications. Cambridge University Press, Cambridge, Boston, MA
White H (1970) Careers and Creativity: Social Forces in the Arts. Westview Press, San Francisco
Zajac EJ, Kraatz MS, Bresser RKF (2000) Modeling the dynamics of strategic fit: A normative approach to strategic change. Strategic Manage J, 21: 429–453

José Vitor Bomtempo · Edmar Luiz Fagundes de Almeida · Ronaldo Goulard Bicalho

The selection environment for gas to liquids technology and technological strategies: Challenging the natural trajectory

Abstract The Gas to Liquids technology (GTL) consists of a chemical conversion of natural gas into a stable liquid by means of the Fischer-Tropsch (FT) synthesis. This conversion makes it possible to obtain products that can be consumed directly as a fuel (for example, Diesel) or special products such as lubricants. The original version of the FT process, using coal as the raw material, was commercially developed in Germany in the 1940s, during the Second World War. Nevertheless, with the exception of Sasol, a South African company, the FT process was abandoned by chemical companies after the war.

The decade of the 1990s witnessed the return of the FT process to the center of attention in the world petroleum and natural gas industry. A radical transformation in the selection environment of this technology has opened the door for its reemergence. This paper analyzes the way in which the GTL's selection environment is challenging its natural trajectory. We show that this selection environment is driving the technology into directions that are not in balance with the technology's natural trajectory, as defined by Nelson and Winter (1977). As a consequence, at the level of their R&D strategies, firms have difficulties in defining a clear research program. The analysis of firms' technological strategies gives evidence of a fragmented R&D effort. Some companies concentrate their efforts in the search for radical innovations trying to develop small-scale competitive processes. Others, such as big oil companies, tend to emphasize heavily economies of scale. As a result, the emergence and diffusion of the GTL technology tend to be locked in by a high level of uncertainty and a research agenda not clearly defined, which means that players tend to work with a dispersion of efforts.

Keywords Gas-to-liquids technology · Selection environment · Natural trajectory
JEL Classification O32 · O33 · L71

J. V. Bomtempo
Escola de Química, UFRJ, Rio de Janeiro, E-mail: vitor@eq.ufrj.br

E. L. Fagundes de Almeida · R. G. Bicalho
Instituto de Economia, UFRJ, Rio de Janeiro, E-mail: edmar@ie.ufrj.br · bicalho@ie.ufrj.br

1 Introduction

The evolutionary theory of technical change has inaugurated a research program that places substantial emphasis on the learning process to explain technological progress (for example: Nelson and Winter 1982; Dosi 1982). The innovation process is seen as a result of knowledge accumulation under specific evolutionary patterns. Knowledge that generates technological innovation is assumed to be local, tacit to some extent, and path dependent. The evolutionary theory claims that we can explain the rhythm and direction of technical change, analyzing the history of the learning process. We can even identify the determinants of technical change if we recognize the technical, scientific and economic factors that drive the learning process.

This approach has identified clear trends for the technological progress in different industries that can be grasped as "natural trajectories" (Nelson and Winter 1977). The trend identified for Process Industries – such as Chemical, Refining, and Petrochemicals – is to innovate by exploiting scale economies. However, critical changes in the economic environment of these industries are making more difficult the adoption of this technological strategy. This paper discusses the discrepancy between the natural trajectories and the technological selection environment for the case of Gas to Liquids technology.

Gas to Liquids technology (GTL) consists of a chemical conversion of natural gas into stable liquids by means of FT synthesis. This conversion makes it possible to obtain products that can be consumed directly as fuel (for example, diesel) or special products such as lubricants. The technological features of GTL are quite similar to other process technologies. Gas conversion uses reactors and catalysts to recombine the hydrocarbon molecules. Therefore, we would expect that the development of larger scale processes would be the "natural trajectory" for its development. Nevertheless, the development of smaller scale processes in order to allow the exploitation of smaller gas reserves is one of the main driving-forces for the technological development of GTL option today.

The second section of this paper discusses in more detail the notions of "natural trajectory" and selection environment as proposed by Nelson and Winter (1977) and reinterpreted by other authors.

The third section describes the potential economies of scale in the GTL plants operating in the world, and most technological options have only been tested in small-scale pilot plants. The technology commercially available today does not allow the construction of large-scale trains (more than 20,000 barrels/day).

The fourth section presents the current selection environment for GTL technology and discusses the contradictions between its dimensions and the search of economies of scale in GTL technology. A combination of a series of factors prompted the transformation in the environment of GTL technology: i) the great increase in the gas reserves (especially associated and remote reserves); ii) the renewed activity of environmental legislation creating market niches for clean fuels; iii) finally, the impacts of liberalization of the natural gas and electricity industries in the main world markets, resulting in higher transaction costs for traditional technological options for gas monetization (LNG and pipelines).

Finally, the fifth section analyzes the technological strategies at the level of firms, and presents evidence of the impact of the contradiction between natural trajectory and the current selection environment. It is shown that firms have difficulties in defining their technological strategies for GTL technology. The R&D effort is mostly fragmented. Some companies concentrate their efforts in the development of radical innovations trying to make small-scale processes competitive. Other companies focus on scaling-up traditional processes, given their complementary assets (large gas reserves, market position), pursuing the natural trajectory of economies of scale. Even though GTL technology started from a knowledge base developed in the 1920s, players in the energy market seem to believe that this technology still has large technological risks. These technological risks affect the feasibility of GTL projects, resulting in lots of smoke and practically no fire in the GTL landscape.

2 The Natural Trajectory in Technological Development

Nelson and Winter (1977) proposed the term *natural trajectories* to refer to technological development paths that proceed relatively independently from changes in market conditions. Natural trajectories concern "heuristics that apply when a technology is advanced in a certain direction, and payoffs from advancing in that direction that exist under a wide range of demand conditions." (Nelson and Winter 1977, 56)

Nelson and Winter further argued that a trajectory is usually "specific to a particular technology or broadly defined 'technological regime'" (Ibid. 1977: 57). The regime concept refers to the cognitive foundation underlying engineering "beliefs about what is feasible or at least worth attempting" (Ibid. 1977: 57). Related to the concept of a technological regime is the concept of technological paradigm, proposed by Dosi (1982) as the technological analog of the concept of scientific paradigms introduced by Kuhn (1970). Similar to a regime, a technological paradigm refers to a codified set of scientific, technical and design principles, which remain largely unchanged during the stage of routinized innovation. Dosi (1982, 154) suggested that a technological trajectory could be represented "by the movement of a multi-dimensional trade-off among the technological variables which the paradigm defines as relevant".

Regimes or paradigms thus differ in their underlying scientific and technological principles. As long as technological principles are not codified, selection among different sets of technological principles cannot be explained by heuristics. At this pre-paradigmatic stage of development, different sets of principles compete. Once principles become increasingly codified, the choice behavior is increasingly guided by perceived differences in the relative future development potential of different technologies. At this point, standardization becomes more likely. In many cases, the phase of standardization is marked by what has been called a "dominant design" (Abernathy and Utterback 1978; Utterback 1994), a successful product model based on a particular set of technological principles (Model T Ford, Douglas DC3, Windows OS). It works as an "exemplar" for other producers and for users and enhances the acceptance and credibility of the underlying technological paradigm.

The dominant design idea was associated to the Product Life Cycle - PLC theory. The PLC theory tried to determine the interaction between the technologies evolution and the industry organization and structure. The most influential work in this field has been the research of Willian J. Abbernathy and James M. Utterback[1]. This theory claims that there is some regularity concerning the way technologies and industry organization evolves. The basic assumption is that the locus of innovation moves progressively from product to process technology during the PLC

Klepper (1997) has shown that some patterns of development of mature industries are not captured by the PLC theory. In this stage of the PLC, industry shakeouts have often been verified as a result of the renewal of product technological competition. In addition, the impact of this technological renewal has not been the same from industry to industry. Empirical studies have shown that, for some cases, the incumbent firms have been the survivors of the industry's shakeouts, while in other cases, later entrants have been the survivors. Therefore, the empirical studies on the PLC call for further clarifications concerning the co-evolution of pattern of innovations and industry structure. Therefore, Klepper (1997) shows radical innovation in mature technologies can take place, challenging the "common sense" in the industry with surprising effects on the industry structure.

As far as the process industries are concerned, the exploitation of economies of scale has historically been the main direction for technological development. Levin (1977) pointed out how the technical change in industries such as petroleum refining and chemicals is deeply related to the search of optimal scales. This direction can be seen as a "natural trajectory" determined by the intrinsic characteristic of this technology that is independent of the market conditions. Process industries deal with fluid/gas transfers/manipulation and are intensive in material and energy. One of the main characteristics of these processes is the fact that the capacity of product transfers/manipulation increases more than proportionally to the escalation of material in reactors, pipes, reservoirs and other equipments. Therefore, process industries are simply very sensitive to scale economics. This fact has made the scaling-up the natural trajectory for these industries.

However, considering that innovation represents ruptures with the established technological practices, challenging a natural trajectory can be a strategy for many companies. This perspective has been suggested by Nelson (1991) when he emphasized that "where company strategies and associated capabilities differ significantly, their patterns of innovation are likely to differ significantly as well" (Nelson 1991, 69). Therefore, we can think of technological strategies as going beyond the natural trajectory.

In Nelson´s perspective, firms differ by their ability to create distinctive capabilities. When a new technology develops in a mature industry, the probability of the success of the incumbents and new entrants will depend on the extent to which their capabilities are reinforced or destroyed (Tushman and Anderson 1986). In the case in which a new capability is required, incumbent firms should implement strategic changes in order to take advantage of new business opportunities. According to Stopford and Baden Fuller (1990, 1994), some firms found subjacent change in mature industries and took the opportunity of rejuvenation. Rejuvenation in mature industries implies some degree of innovation not only in equipments and technology but also in strategy. Finally, Stopford and Baden-Fuller's case studies suggest that the decline of firms in mature industries could be "caused by failure to adjust and innovate strategy continuosly in a changing world" (Stopford and Baden Fuller 1990, 412).

[1] See Abernathy and Utterback (1978) and Utterback (1994).

3 Economy of Scale in The GTL Technology

The process of conversion of natural gas into liquid products involves a preliminary stage of transformation of natural gas into synthesis gas – syngas – (a mixture of carbon monoxide – CO and hydrogen – H2). After being produced, the syngas is converted into liquid hydrocarbon through the FT synthesis (Vosloo 2001; Wilhelm et al. 2001).

The production of hydrocarbons by FT synthesis does not directly result in products of commercial interest. An additional step is needed, hydroprocessing, in which the heavy molecular hydrocarbons are decomposed into smaller molecules (naphtha, diesel oil, lubricant oil, paraffin and others). Therefore, the conversion of natural gas into fuels can be better characterized by three distinct types: 1) generation of syngas, 2) conversion of syngas and 3) hydroprocessing.

The plant for the generation of syngas corresponds to about 50% of the capital costs of GTL plants. For this reason, a great innovative effort has been carried out by the companies in this segment (Wilhelm et al. 2001; Sundset et al. 1994; Dybkjaer 2003). At least five different types of processes are available, or in an R&D stage, for the generation of syngas[2]. Most efficient technology options involve a process with the presence of catalysts, the utilization of pure oxygen to react with natural gas at a temperature of about 1,000 degrees (Rostrup-Nielsen 2000). The result is the generation of syngas in a H2/CO ratio near to 2. This ideal proportion permits a maximum efficiency of the conversion process by FT synthesis.

There are currently two well-defined routes of FT synthesis; one at a low temperature and another at a high temperature. The first is employed in the production of waxes, which are converted into naphtha or diesel oil after the hydroprocessing. The second is employed in the production of gasoline and alfa-olefins (Voss 2002).

Hydroprocessing is employed in the treatment of wax produced by the FT synthesis at a low temperature. The wax is basically composed of linear paraffin and small quantities of olefins and oxygenates. The hydrogenation of the olefins and the oxygenated compounds, besides hydrocracking of wax, can be conducted in conditions that are not very severe, with the production of naphtha and diesel oil.

Since GTL technology is at its preliminary stage of development, there is a huge potential for technological development by scaling-up the process. There are only two commercial plants operating in the world, and most technological options have only been tested in small-scale pilot plants. The technology commercially available today does not allow the construction of large-scale trains (more than 20,000 barrels/day or 20,000 bpd).

Currently, the maximum capacity for the syngas generation trains is about 10,000 bpd. There are proposals for larger trains but they are still in a project or study stage. BP/Davy Process Technology suggests a new version of the old steam reformation that could attain a train of 17,000 bpd (Freide et al. 2003). Haldor Tpsoe, a technology engineering specialist, proposes a version of ATR (autothermal reforming) technology that could permit the construction of 24.000 bpd synthesis gas trains (Dybkjaer 2003).

[2] Steam Methane Reforming – SMR; Partial Oxidation – POX; Catalytic Partial Oxidation – CPO; Autothermal Reforming – ATR; Catalytic Membrane Reforming – CMR.

Table 1 Evolution in the Scale of the FT Reactors "Sasol Advanced Synthol – SAS"

	Diameter (meters)	Capacity (bpd)
1981	0.866	90
1989	5	3,000
1995	8	11,000
1998	10.7	20,000

Source: Sasol

One of the important obstacles to the construction of larger trains is the limited size of air-separation plant available today. Another obstacle is the capacity of syngas reactors. The syngas process is well known and largely diffused in the chemical industry. It is used in the production of commodity petrochemicals such as methanol and amonia. Even if, in the case of these products, the scale up of syngas plants would have been quite successful (current commercial plants of methanol are larger than 1 MM tonnes/year), FT technology requires larger scales comparable to current petrochemical tecnology.

Similarly, the maximum scale available for FT reactors is 20,000 bpd. Table 1 shows how the size of Sasol's FT reactors has been scaled-up[3]. Apparently, if we take in account the progress in the scale of FT reactors by Sasol in the 1990s, and consider the literature on the FT process, the restrictions to scale up are less important here than in the synthesis gas step.

Several authors have indicated the existence of potential scale-economies in GTL plants. Black & Veatch Company has estimated the scale function for the GTL technology, from the 13 plants built in the world (mostly pilot plants). This regression resulted in the following function (Rettenmaier 2002):

$$Y = 1.33X^{0.622}\text{R-square} = 0.967,$$

where $y =$ investment cost (US$ millions), $x =$ capacity (bpd).

This expression is consistent with the traditional scale up in chemical engineering which is based on the six tenth rule (Chauvel et al. 2003). According this rule, the scaling up of a piece of equipment tends to be proportionate to an impressive reduction in the capex (capital expenditure/tonne of product) due to the material, engineering and overhead economies that can be achieved by building a larger plant. On average, in order to duplicate a chemical plant, the capital expenditure can be multiplied by only 1.5, giving a considerable reduction in the capex.

Using the Black & Veatch function above, we can estimate total investment cost for GTL plants, using the cost of plants already built (Mossgas and Shell) as the departing cost. This estimation has two problems. In a certain way, it does not take into consideration the engineering progress achieved in the last 10 years, after the construction of the first two commercial plants. It is currently mentioned (Den Hartog 2001; Crouch and Cline 2003) that the learning curve in GTL technology has been steep in the last years and so it would be possible to achieve today a capex around US$ 20.000/bpd compared to the almost US$ 50.000/bpd in Shell's Bintulu plant in 1987.

[3] Sasol is a South African synthetic fuel producer and technology provider for GTL plants.

On the other hand, the potential economy of scale could be overestimated by the function above. As the available train size for syngas generation and FT conversion is limited to 10,000 bpd and 20,000 bpd, respectively, the large scale plants proposed in the industry (more than 50,000 bpd) are constrained to a modular concept (several trains of 10,000 to 15,000 bpd). As a result, part of the potential economy of scale cannot be captured by the project. It is important to mention that the central point in the scaling up of chemicals plants has been historically the construction of single train processes.

For that reason, in order to reach the potential reduction in the investment cost, an important effort of technological innovation is necessary. However, in order to innovate in this direction, it is very important to consider the potential market. The research for the scaling-up process is very expensive and requires extensive learning-by-doing. The feedback from the learning-process during the construction and the operation of new plants is very important. Therefore, the GTL technology's selection environment is an essential dimension for the implementation of the research program necessary for the scaling-up of the process.

4 The Selection Environment in GTL Technology

According to Dosi (1982), the selection environment can be defined as the economic forces and the social and institutional factors that actuate as selection mechanisms for technologies. As GTL technology is essentially a process innovation, firms, as pointed out by Nelson and Winter, could better "make an assessment of profitability by considering the effects on costs, with far less concern for consumer reaction" (Ibid. 1977: 65) than in the case of product innovation. We may expect in this case that firms should interpret more easily the selection environment drivers. However, some empirical works on selection environment dynamics (for example Van den Belt and Rip 1997; Glynn 2002) have emphasized a somewhat more complex process. In these analyses, firms and other actors are seen as playing a decisive role in the shaping of the selection environment, not only in its interpretation.

In the GTL landscape, a combination of a series of factors prompted the transformation in the environment of the GTL technology: i) the great increase in the gas reserves (especially associated); ii) the renewed activity of environmental legislation making viable the creation of niches in the market for clean fuels; iii) finally, the impacts of liberalization of the natural gas and electricity industries in the main world markets, creating higher transaction costs for traditional technological options for gas monetization (pipelines or gas liquefaction and cryogenic transportation – LNG).

4.1 Increase on the stranded reserves

The great effort in the search for alternative reserves to Arab petroleum after the petroleum shocks in the decade of the 1970s had a great impact on the increase of tested gas reserves, which jumped from 40 tcm (tera cubic meters) in 1970 to about 140 tcm in 1996. The reserves/production ratio increased from 59 years at the end of 1985 to 65 years at the end of 1996. The Natural Gas Industry is thus experiencing a period of abundant gas reserves. However, this abundance is not a reality in the regions in which the markets are concentrated. For example, while Europe and North America accounted for 56% of total world gas consumption in 2003, their gas reserves represented only 13% of the world total.

One of the most important problems for the gas industry is the higher cost of transportation as compared to liquid fuels. Natural gas has a volume of about 1.000 times the volume of oil for the same energy content. Therefore, gas transportation requires specific technological options that are very capital intensive. The transportation cost of the same amount of energy in the form of gas is 3 to 10 times more expensive than the cost of oil transportation, depending on the distance. Therefore, when associated gas reserves are far from end-market and when the amount of gas produced is low, investments in the infrastructure necessary for the monetization of the reserve are not profitable. Gas flaring and ventilation is then the best economical option. So, the stranded reserves would be monetized only if new technological options become available (Baker 2004) and GTL is probably one of the best options to consider.

4.2 Increase in environmental restrictions

The World Bank estimates that about 4.3 Tcf of gas are flared and ventilated every year in the World. This represents 12.9% of gross world gas production, or about 70% of the total gas traded via the LNG chain. Gas flaring and ventilation represent a very important environmental problem at the global level (greenhouse gas emissions). Governments are imposing restrictions on gas flaring and ventilation, since the growth of carbon emissions was declared no longer acceptable under the Kyoto protocol of 1998, which aims at mitigating global warming through the reduction of carbon emissions by industrialized countries.

The environmental restrictions on flaring are being introduced by several means. National governments and regulators are creating rules and standards making difficult the development of gas fields that requires substantial flaring. Some governments have signed agreements with oil producers in order to implement "zero burn" programs. Regulatory restrictions to gas flaring are inducing companies to invest in the monetization of an important amount of gas reserves being burned each year. The gas flared each year in the world would be enough to anchor the about 1,3 million barrels a day in GTL capacity. Since each ton of CO2 has a price of $3 to $4, the gas currently being burned has a price in the international market, which represents about US$ 1,05 billions or US$ 25 cents per million Btu. Therefore, an important way to burst GTL projects would be to capture the value of the CO2 reduction (Almeida 2003).

The environmental restrictions regarding the quality of fuel represent great incentives for GTL, too. The environmental regulations contributed to the increase in the cost of oil refining during the 1990s. The proposed standards for the next decade suggest further increases in environmental restrictions. In order to respond to these new regulations, several countries are developing markets for "environmental additives". In Europe, for example, biodiesel is produced on a large scale, to be blended with conventional diesel. Since GTL diesel can improve diesel quality when blended, it can add value to conventional diesel. Therefore, GTL diesel could capture an important share of world diesel market. Since it aggregates value to conventional diesel, it can be sold at a premium price (Almeida 2003).

4.3 Increase in demand for flexibility

Current innovative efforts in GTL technology represent a response to the increase in the requirement for asset flexibility, due to higher uncertainty in the energy markets. These uncertainties are related to the increase in price volatility with the energy market liberalization.

The selection of the GTL technology corresponds to an attempt to improve the classical trade-off between flexibility and scale (Dosi 1988). The objective of the innovation effort is to allow a flexibility to the physical assets, without incurring higher costs of operation in small-scale activities. GTL allows the reduction of the rigidity of the natural gas chain, and an escape from highly specific assets and from the contractual difficulties associated with increasing transaction costs (Williamson 1979). This technological option has the potential for reducing the amount and the life-cycle of the investments, through equipment compaction. Therefore, GTL can improve the companies' portfolio liquidity and contribute to companies' asset optimization.

Increasing asset flexibility implies higher production costs due to the sacrifice of economies of scale. R&D programs are influenced by this diversity in situations. Some efforts are focused on technology versatility and adaptability; while others explore the trajectory "standardization-scale". The outcome of these efforts will depend on the specific characteristics of countries in terms of natural resources and the composition of company portfolios.

5 Technological Strategy in the GTL Market

GTL players can be classified into three groups. The first is companies linked to oil, gas and fuel markets. Some big oil companies have been investing in GTL programs in many ways: new plants, alliances with strategic players, R&D programs, etc. The most important players in this group are: Shell, ExxonMobil, BP, ChevronTexaco and ConocoPhillips. Some of these companies have strong patent portfolios, but their most remarkable strength is their complementary assets (Teece 1986). The second group represents technology companies and other research institutions. Syntroleum and Rentech are the most notable players in this group as potential technology providers to GTL projects. In the specific and crucial step of synthesis gas, Haldor Topsoe should be mentioned as a proactive player in the industry. These players are concentrated in the area of research and development of GTL technologies. As a rule, they lack any experience in the market side of energy products. Their weakness is due to a quite poor position in terms of complementary assets. The third group consists of companies with operational experience in GTL plants. Sasol, a South-African company, is the strongest player in terms of operational capability. It has been running for decades plants for the conversion of coal to liquid hydrocarbons in South Africa and has supplied since the beginning of the1990s FT technology to PetroSa's (former Mossgas) GTL plant in Mossel Bay, South Africa. Shell is also a competitor with a high level of operational expertise. Its plant at Bintulu, Malaysia, operated successfully from 1993 until an accident in 1997.

A fourth group could also be considered, which consists of countries such as Qatar, Iran, Nigeria and others, with large non-explored reserves. These countries

Table 2 Summarized data for GTL complexes

	Implemented[1]		Under implementation[2]	
Location	South Africa (Mossel Bay)	Malaysia (Bintulu)	Nigeria (Escravos)	Qatar (Ras Laffan)
Startup	1993	1993	2007	2005
Capacity (1000 bpd)	22.5	12	34	34
FT Technology	Sasol	Shell	Sasol	Sasol

Sources: 1 - Bakhtiari (2001); 2-Alperowicz (2004)

have been strongly pushing the implementation of projects aiming to monetize their reserves. GNL and GTL are frequently mentioned as the suggested alternatives.

Table 2 shows summarized data for implemented and under implementation GTL complexes. Projects with feasibility studies not yet completed were omitted. Industry experts estimate that there are at the moment around 40 project announcements, but no more than a projected 700.000 bpd capacity could be operating by 2012 (Hydrocarbon Processing 2004), if the majors players as ExxonMobil, Shell and Sasol can handle the capital requirements and risk exposure.

One can see that design capacities of first and second generation of GTL complexes remain at around 30,000 bpd. This is in contrast with announced industry figures on GTL economy of scale. Most analysts and particularly big oil companies consider nowadays that optimal scales are at least around 60,000 to 70,000 bpd (Chang 2000; Den Hartog 2001; Rockwell 2002). In fact, it seems that attention has been shifting to large reserves where there are alternatives to GTL, such as pipelines, LNG or domestic use (Hosoe and Brown 2002). In the beginning, the plan was to improve technology aiming to monetize remote or associated gas with little or no value, and to focus on more desirable products such as ultra clean fuels. Apparently, some competitors have changed their interpretation of the selection environment and have changed their behavior. Perhaps they have shifted to large reserves because using remote gas increased the cost of building a plant, and GTL has not yet become viable. Crouch and Cline, 2003, suggest that the stranded gas approach is no longer viable, given that companies have concluded that stranded gas costs are righer than previously supposed. The authors stress this opinion, mentioning what they call the "stranded gas myths". In that situation, the rewards of the natural trajectory – the traditional increasing of scale in the process industries – are, in the view of some players, the target of the technological strategies. Shell and ConocoPhillips are good examples of that strategy.

Conoco, now ConocoPhillips, was almost absent in the GTL competition until recently. Four years ago (Chang 2000), the firm revealed it was developing its own GTL process and stressed a very clear strategy: scale is the key in GTL business. In Conoco's words, several players in the GTL arena are proposing plants too small to be profitable. The firm thinks GTL is all about scale. The ConocoPhillips GTL group consists of 200 engineers, scientists and technicians, 16 lab and pilot syngas reactors, 25 lab and pilot FT reactors and a 400 bpd demonstration plant. The firm intended to build a commercial plant, scheduled for 2008, of at least 60,000 bpd (Rockwell 2002). However, the absence of descriptions of this project suggests that it was postponed.

Shell has also been emphasizing economies of scale in its technological strategy. In a paper by Den Hartog, 2001, the firm suggests that large-scale GTL projects are a viable alternative to LNG limitations. Shell considers that GTL products are not specialty products for niche markets. They will compete in the global market for middle distillates, where crude oil refining sets product prices. In Shell's perspective, commercial scale for a new GTL plant is now in excess of 50,000 bpd. The firm states it is now possible to build a 75,000 bpd plant using Shell GTL technology. Recent announcements by Shell indicate the intention to concentrate its investments in GTL in an impressive project in Qatar. Shell plans to spend around US$ 5 bn to build a 140.000 bpd plant, in two 70.000 bpd units (Petroleum Economist 2003). ExxonMobil has also signed an agreement with the government of Qatar, and is planning to build an 100,000 bpd plant in the country before the end of 2008 (Petroleum Economist 2004).

In contrast with Shell and ConocoPhillips, the BP strategy seems less dependent on scale. In the release "Gas and Power 2000: An Insight into Europe's Developing Gas and Power," BP maintains that "GTL technologies have the potential to release stranded gas reserves, which are too small to justify an LNG scheme and too remote to be piped to markets" and concludes that BP is working to develop new technology that could allow much smaller plants, supported by smaller fields, to be constructed economically (Weeden 2001; Thomas 2001). In line with this perspective, Freide et al. (2003) and Fleisch et al. (2003) have presented a new reforming technology, based on steam reformation that could enable converting stranded gas into synthesis gas. The technology, developed by BP and Davy Process Technology, would be sufficiently compact to allow commercial scale units to be transported to offshore platforms or remote sites.

Nevertheless, BP recently released an evaluation in which it compares GTL and LNG value chains. According to Fleisch (2002), they have similar capital expenditure exposure and both applications are projected to represent similar integrated returns. In that analysis, BP does not mention stranded gas and seems to take a perspective that emphasizes scale. One may wonder if the firm is shifting to a more traditional strategy, similar to Shell, ConocoPhillips and others. Probably yes, if we consider another recent BP presentation (Freide 2003) in which the central point is the definition of GTL technology as one the possibilities of an option game, competing against pipelines, electricity generation, and particularly LNG and methanol. At the same time, BP is sponsoring a research program at three universities, with a budget of US$ 30 millions in 10 years, in order to develop a quite revolutionary approach to GTL: the direct conversion of natural gas to liquids, without the syngas step (Voss 2002).

Technology companies, such as Syntroleum and Rentech, are clearly betting on a small-scale strategy (Agee 1999; Broyles 2000; Waddacor 2003; Bigger and Tomlinson 2004).

Rentech has been active in GTL conversion since the 1980s. Its first initiative was a small plant (253 bpd), built at Pueblo, Colorado, to convert landfill methane into liquid fuels. This GTL venture failed because gas volumes were lower than expected. Rentech has been trying to license its technology globally in a small scale perspective. Small gas fields and other sources such as lignite and steelwork off-gases has been envisaged by Rentech (Waddacor 2003) Apparently Rentech has been addressing a particular dimension of the selection environment: the very stranded and small reserves without any other alternative of utilization.

Syntroleum, founded in 1984, is an R&D company in Tulsa, Oklahoma. It has spent around US$ 100 MM from government grants, investors and shareholders in GTL research, pilot and demonstration plants. Syntroleum has announced the development of a proprietary GLT technology that the company intends to license or to offer in order to forge strategic alliances. Syntroleum licensees include Arco, the Commonwealth of Australia, KerrMcGee, Marathon Oil, Repsol YPF and Texaco (Waddacor 2003) but any decision concerning investiment in a commercial plant has been taken. The company has proposed a quite original concept of a barge-mounted plant built as a floating and mobile system to develop stranded gas fields (Bigger and Tomlinson 2004).

The strategies of technology companies are compatible with some dimensions of the selection environment and with their weakness in terms of complementary assets (natural gas reserves, engineering know how, energy market expertise). Their technological strategy is oriented to the production of technological breakthroughs, so their proprietary technology can be licensed to firms that have the complementary assets.

Finally, Sasol is possibly taking a quite original approach (Petroleum Economist 1999; Alperowicz 2004) to the problem from big oil companies. Even if it is aware that scale will play a role in GTL future, Sasol considers that its technology is still on a very steep learning curve. The company emphasizes alliances with oil firms, including Haldor Topsoe for synthesis gas technology and ChevronTexaco for refining technology. If successful, these alliances could give to Sasol the complementary competencies needed in order to lead the GTL industry, taking advantage of its own tecnological and operational knowhow. As was shown in Table 2, the projects under implementation are both controlled by Sasol. Sasol intends to invest aggressively in the GTL industry in order to consolidate a leading position in the industry (Alperowicz 2004).

In summary, two opposite factors influence the scale of plants:

- the natural trajectory of economy of scale and the rewards of firms' efforts in exploiting it. Big oil firms particularly target the GTL future in large-scale plants, which may be viable in comparison with LNG alternatives.
- the opportunities linked to small gas fields, associated or remote gas and to environmental restrictions related to gas flaring and ventilation require the development of economical GTL technologies for small-scale plants.

GTL technology needs improvements in many respects, but big oil companies tend to emphasize economies of scale in their strategies. The natural trajectory becomes thus the dominant trajectory in the industry, consistent with the search of flexibility in energy markets and with the exploitation of large remote gas reserves. By contrast, other dimensions of the selection environment are incongruent with the natural trajectory of economies of scale. As a result, GTL actors have divergent interpretations of the selection environment, and the emergence and diffusion of the GTL technology tend to be blocked by a high level of uncertainty. The research agenda is not clearly defined, which means that players tend to work with a dispersion of efforts.

6 Conclusions

This paper has shown that the GTL technology has an important potential for innovation exploiting economies of scale. The scaling-up of different segments of the technology can be considered the natural trajectory for the development of cheaper GTL processes. The R&D effort for making possible this technological development requires a scenario of probable diffusion for the new GTL plants. The R&D for GTL's scaling-up is expensive and very sensitive to learning-by-doing. Therefore, GTL's selection environment will play an important role in this technological development.

The paper has also shown that the GTL's selection environment favors small-scale plants. The monetization of small gas reserves, which are stranded today, is the most obvious potential niche markets for GTL. The large gas reserves can be monetized through other technological options (LNG and pipelines). Given the competition with other technological options in large gas reserves market, the higher technological risk becomes an important hurdle to the economic viability of GTL technology.

We have also shown that firms have difficulties in defining a clear research program for the GTL technology. The analysis of firms' technological strategy gives evidence of a fragmented R&D effort. Some companies concentrate their efforts in the development of radical innovations trying to make small-scale processes competitive. Others favor the scaling-up of developed processes, given their complementary assets (large gas reserves) and their interpretation of the selection environment. Even though GTL technology was developed in the 1920s, players in the energy market believe this technology still has large technological risks. These technological risks are affecting the feasibility of GTL projects, resulting in lots of smoke and practically no fire in the GTL industry landscape.

References

Abernathy W, Utterback J (1978) Patterns of industrial innovation. Technol Rev, 1978, June-July: 41–47

Agee M (1999) Taking GTL Conversion Offshore. 1999 Offshore Technology Conference, Houston, TX

Almeida E (2003) Increase in the Environmental Restrictions as a Leverage to GTL Projects. 26th Annual Conference of the International Association for Energy Economics New Challenges for Energy Decision Markers, Prague, June

Alperowicz N (2004) Sasol, new capacity fuels expansion. Chem Week, 2004, April 28: 17–19

Baker J (2004) Stranded gas options. Eur Chem News, 2004, 1-7 march: 19–20

Bakhtiari AMS (2001) Gas-to-liquids: To much smoke, little fire. Hydrocarb Process, 2001, December: 19–21

Bigger JM, Tomlinson HL (2004) Consider barge-mounted plant to produce ultra-clean diesel. Hydrocarb Process, 2004, July: 39–41

Broyles K (2000) Underdog Rentech furthers gas-to-liquids plans. Oil Gas J, 2000, October 9: 55–56

Chang T (2000) Conoco says economies of scale key. Oil Gas J, 2000, July 31: 46–47

Chauvel A, Bournier G, Raimbault C (2003) Manual of Process Economic Evaluation. Editions Technip, Paris

Crouch A, Cline G (2003) GTL: A new era. Petroleum Economist, 2003, special issue on Fundamentals of Gas to Liquids: 42–44

Den Hartog M (2001) Latest Developments in GTL. paper presented at the LNG-Natural Gas – GTL – Syngas Conference, September 6-7, Kuala Lumpur, printed in Hydrocarbon Processing Asia, October: 46-48
Dosi G (1982) Technological paradigms and technological trajectories: A suggested interpretation of the determinants and directions of technical change. Res Pol, 11: 147–162
Dosi G (1988) Sources, procedures and microeconomic effects of innovation. J Econ Lit, 26, 3, september: 1120–1171
Dybkjaer I (2003) Synthesis gas tecnology. Petroleum Economist, 2003, special issue on Fundamentals of Gas to Liquids: 16–19
Fleisch T (2002) BP's Gas To Products Activities Progress in GTL. CWC Conference: GTL – What's in it for the USA, November 20, Houston
Fleisch T, Sills R, Briscoe M, Freide JF (2003) GTL-FT in the emerging gas economy. Petroleum Economist, 2003, special issue on Fundamentals of Gas to Liquids: 39–41
Freide JF (2003) The Options Game: Pipeline, LNG, Gas-by-wire and Gas-to-liquids. GTL Summit, CWC Conference, May, London
Freide JF, Gamlin T, Ashley M (2003) The ultimate clean fuel – Gas to liquids products. Hydrocarb Process, 2003, February: 53–58
Glynn S (2002) Constructing a selection environment: Competing expectations for CFC alternatives. Res Pol, 31: 935–946
Hosoe T, Brown J (2002) Asian markets present little near-term opportunity for GTL fuels. Oil Gas J, 2002, February 25: 62–63
Hydrocarbon Processing (2004) GTL projects: finally moving forward? 2004, July: 24
Klepper S (1997) Industry life cycles. Ind Corp Change, 6, 1: 145–181
Kuhn TS (1970) The Structure of Scientific Revolutions. University of Chicago Press
Levin R (1977) Technical change and optimal scale: Some evidence and implications. Southern Econ J, 44: 208–221
Nelson R (1991) Why firms differ, and how does it matter? Strategic Manage J, 12: 61–74
Nelson R, Winter S (1977) In search of a useful theory of innovations. Res Pol, 6: 36–76
Nelson R, Winter S (1982) An Evolutionary Theory of Technical Change. Harvard University press, Cambridge, MA
Petroleum Economist (1999) New perspectives emerge. January: 26-27
Petroleum Economist (2003) Shell to make Qatar its GTL hubs. 2003, November: 1
Petroleum Economist (2004) World gas: LNG vs GTL: Using all the option. May: 16-25
Rettenmaier A (2002) GTL Projects from the Viewpoint of the EPC Contractor. CWC Conference: GTL- What's in it for the USA. November 20, Houston.
Rockwell J (2002) Commercialization of ConocoPhillips GTL Technology. CWC Conference: GTL – What's in it for the USA, November 20, Houston TX
Rostrup-Nielsen JR (2000) New aspects of syngas production and use. Catal Today, 63, 2-4: 159–164
Stopford J, Baden Fuller C (1990) Corporate rejuvenation. J Manage Stud, 27, 4: 399–415
Stopford J, Baden Fuller C (1994) Rejuvenating the mature business. Harvard Business School Press, Boston
Sundset T, Sogge J, Strøm T (1994) Evaluation of natural gas based synthesis gas production technologies. Catal Today, 21, 2-3: 269–278
Teece D (1986) Profiting from technological innovation: Implications for integration, collaboration, licensing and public policy. Res Pol, 15: 285–305
Thomas V (2001) An oil major redefines its role. Petroleum Economist, 2001, 3-4, February
Tushman M, Anderson P (1986) Technological discontinuities and organizational environments. Admin Sci Quart, 31, 3: 439–465
Utterback J (1994) Mastering the Dynamics of Innovation. Harvard Business School Press, Boston
Van den Belt H, Rip A (1997) The Nelson-Winter-Dosi model and synthetic dye chemistry. In: Bijker W, Hughes T, Pinch T (eds.) The Social Construction of Technological Systems, MIT Press
Vosloo AC (2001) Fischer-Tropsch: A futuristic view. Fuel Process Technol, 71: 149–155
Voss D (2002) Hitting the natural gas jackpot. Technol Rev, 2002, January
Waddacor M (2003) Modern-day gas alchemy yields cleaner fuels. Petroleum Economist, special issue on Fundamentals of Gas to Liquids, 5–7

Weeden SL (2001) Progress and prospects: Financial commitments brighten 2001 GTL prospects. Oil Gas J, 2001, Mars 12: 58–63
Wilhelm DJ, Simbeck DR, Karp AD, Dickenson RL (2001) Syngas production for gas-to-liquids aplications: Technologies, issues and a outlook. Fuel Process Technol, 71: 139–148
Williamson O (1979) Transaction cost economics: The governance of contractual relations. J Law Econ, 22, 2

Tommaso Antonucci

Innovation and employment in Europe: A sectoral perspective

Abstract Following the Neo-Schumpeterian and Evolutionary stream of analytical and empirical literature on the relationship between technological change and employment, the paper provides an econometric model through which the complex web of relationships among employment growth, heterogeneous innovation strategies and structural factors such as demand and labor cost dynamics, is tested. The results seem to confirm the relevance of the product and process dichotomy to disentangle the innovation-employment relationship, and the role of demand growth either in pushing employment directly or in sustaining the reabsorption of technological unemployment.

Keywords Product innovation · Process Innovation · Employment · Demand
JEL Classification O3 · J6

1 Introduction

Following the Neo-Schumpeterian and Evolutionary stream of analytical and empirical literature on the relationship between technological change and employment[1], the paper focuses on testing the hypothesis of the contrasting effects of product and process innovation strategies on employment at the sectoral level.

The attempt to account for these effects relies on the use of a unique sectoral database, described in section 5, which combines a rich set of qualitative and quantitative data on innovation, provided at 22 manufacturing sector breakdown and for 10 major European countries, with STAN OECD economic and structural data, over the period 1994-2001.

T. Antonucci
University of Cassino, Faculty of Economics
E-mail: tantonucci@regione.lazio.it

[1] See, among others, Vivarelli (1995); Vivarelli and Pianta (2000); Freeman and Soete (1987, 1994); Petit (1995); Edquist et al. (2001).

In section 2, the methodological and analytical set into which the effects of the variety of innovation processes on employment growth have been assessed, is defined.

Building on previous empirical analyses, the paper provides an econometric model through which the complex web of relationships among employment growth, heterogeneous innovation strategies and structural factors such as demand and labor cost dynamics, is tested. The hypotheses set on these relationships are drawn in section 3 and 4.

Econometric results, presented in section 6 and critically discussed in the conclusion – section 7 – seem to confirm much of the model's predictions, concerning mainly: (a) the *relevance* of the product and process dichotomy to disentangle the innovation-employment relationship; (b) the role of demand growth either in pushing employment directly or in sustaining the reabsorption of technological unemployment.

2 Theory and concepts

2.1 The variety of innovation processes

Since the time of the classical economists, technical change has been considered a possible cause of job displacement, although an unconditioned belief in its beneficial economic effects has often prevented a more rigorous analysis of its net employment consequences.[2]

In particular, the neoclassical view of technological change as an exogenous and undistinguished process which, as 'manna from heaven' pushes economic growth leading to full employment, has limited the investigation of the heterogeneity in innovation processes and of their different effects on employment.

It is only in the last few decades that the concept of variety in innovation has become the object of the analyses of economists of technological change, especially within the 'evolutionary stream' arising from the seminal contributions of Nelson and Winter (1977, 1982); Dosi (1988); Dosi and Nelson (1994).

Furthermore, since recently, data constraints in innovation studies have been severe and have limited the possibility of exploring variety in technology and innovation behaviors in a systematic way[3]. The technological activities covered by official statistics (both at the national and international level) has been limited and mainly confined to R&D and patent activities (Sirilli 1997), which, though highly reliable, have been able to grasp only a limited part of innovation activities.[4]

Also, in the early neo-Schumpeterian literature, innovation activities have been conceptualized as linear processes with the scientific and technological advancement stemming from formalized R&D (Freeman 1997). Stressing the role of radical innovations in driving economic growth, this literature privileged the 'creative destruction' force in place of the smooth process of incremental innovations. The

[2] See, on this point, Petit and Soete (2001).

[3] See, on the different sources of variety ('technological' versus 'behavioral diversity'), Dosi (1988).

[4] See Evangelista and Mastrostefano (2005, 1-2).

discontinuous nature of radical innovation, mainly driven by scientific and technological discoveries, tended to create and expand new technological sectors while displacing more traditional ones[5]. The positive effect of innovation on employment in this set was ensured by the expansion of the new and more innovative sectors.

From the eighties, the 'technological regimes' theory offered new insights to the analysis of the innovation-employment relationship by providing theoretical and empirical evidences of the existence of high innovation heterogeneity across sectors.[6] The pioneering contribution in this area is the sectoral taxonomy proposed by Pavitt (1984). This taxonomy describes and explains similarities and differences among sectors in the nature, objectives, activities and sources of innovation.

Since then, numerous contributions have underlined the presence of different and largely sector-specific 'technological regimes'. Different types of knowledge and conditions in the level of technological appropriability, opportunity and cumulativeness can explain cross-industry differences in the level of concentration of innovation activities, in the size distribution of innovative firms, and in the level of technological barriers preventing the entrance the sector of new competitors (Breschi et al. 2000; Malerba 2004).

Recent works moreover, which have exploited the great bulk of qualitative and quantitative information on innovation offered by the SIEPI database (Castellacci 2004; Evangelista and Mastrostefano 2005), have shown that sectors seem to differ mostly in terms of the overall financial commitment to innovation, in the type of innovation activities performed and in terms of overall innovation strategies performed.

In particular, the most relevant cross-industry technological differences found seem to refer to: (i) the resources devoted to innovation in terms of the percentage of personnel involved in R&D activities or the percentage of total turnover affected by innovation; (ii) the ways through which firms innovate, that is internally in R&D activities or through the adoption of technologies embodied in new machineries and equipment. Sectors differ also, other than in the propensity to patent, in the (iii) specific objectives pursued by firms and among them, the greater variance is associated with product oriented innovation objectives ('replacement of phased-out products' and 'extending the product range').[7]

Following this literature and exploiting the richness of information of the SIEPI database, the present work aims at analyzing the manner in which the variety of forms in which innovation processes take place exerts its impact on employment. In particular, following the seminal work of Pavitt (1984), the analysis focuses on the effects on employment of different *objectives, activities and sources of innovation* across European industries.

Once the undistinguished approach to innovation is abandoned, the innovation-employment relationship assumes new and more complicated features.

Different innovation objectives such as 'reducing labor cost' and 'improving product quality' pursued by firms in different sectors, may generate very different employment outcomes in sectors either in the sign or in the intensity of growth. Heterogeneous outcomes may occur also when different innovation activities such as 'R&D' rather than the 'adoption of technologically advanced capital' prevail in

[5] See, on this point, Archibugi and Santarelli (1990); Walsh (1990).

[6] See Malerba (2002); Malerba and Orsenigo (1993, 1996, 1997); Evangelista (2000).

[7] See Evangelista and Mastrostefano (2005, 19).

the sector. The employment displacement which may occur in the latter case differs radically from the likely increase in employment due to market expansion of new products driven by R&D expenditures. Various sources of innovation, finally, such as 'suppliers of equipment and materials' rather than universities or private research centers', which operate as principal transmitters of sectoral technological opportunities to be exploited, may differently inform sectoral economic and employment growth.

From a methodological point of view, the analysis implies three features:

1. The analysis of the innovation-employment relationship is misleading if it does consider innovation as an uniform process rather than as having a multiplicity of forms and intensities;
2. It is not possible defining *a priori* the effects of innovation on employment but only empirically when looking at, among others, the net effects of different characteristics and nature of innovation processes;
3. The sectoral level seems offering the appropriate analytical set when looking at the innovation heterogeneity and its effects on employment.

Following previous empirical analyses carried out at the sectoral level on this issue[8], the product/process dichotomy has been used to disentangle the effects on employment of a great heterogeneity of innovation processes included in the SIEPI database.

The contrasting effects on employment of different innovation objectives, activities and sources are assessed through the positive/negative effects framework generally associated with the introduction of product/process innovations.

In particular, a set of indicators mainly oriented towards product and process innovations has been built for each objective ('extending product range and quality' rather than 'reducing material consumption'), activity ('R&D' rather than 'new machinery') and source of innovation ('patent disclosure' rather than 'competitors'). This set of indicators has been tested against employment dynamics, as described in section 5.

2.2 The sector level of analysis

Most of the literature that has investigated the relationship between technological change and employment has adopted both an aggregate and a micro level approach to innovation[9]. Yet the theoretical and empirical advancements in recognizing the sector-specific nature of technological activities as one of the key factors, if not the principal[10], suggests the use of sectoral approaches instead of firm or macro ones.

There are other reasons sustaining the sectoral approach to the innovation-employment relationship.

Firstly, the sectoral level of analysis allows levelling off employment gains and losses within the sector. This is a major limitation to firm level analyses, since the

[8] See Antonucci and Pianta (2002); Vivarelli and Pianta (2000); Mastrostefano and Pianta (2004).

[9] For a thoughtful review, see Pianta (2004); Spezia and Vivarelli (2002).

[10] See the results of the ANOVA analysis on 'the source of variety of innovation' in Evangelista and Mastrostefano (2005, 15).

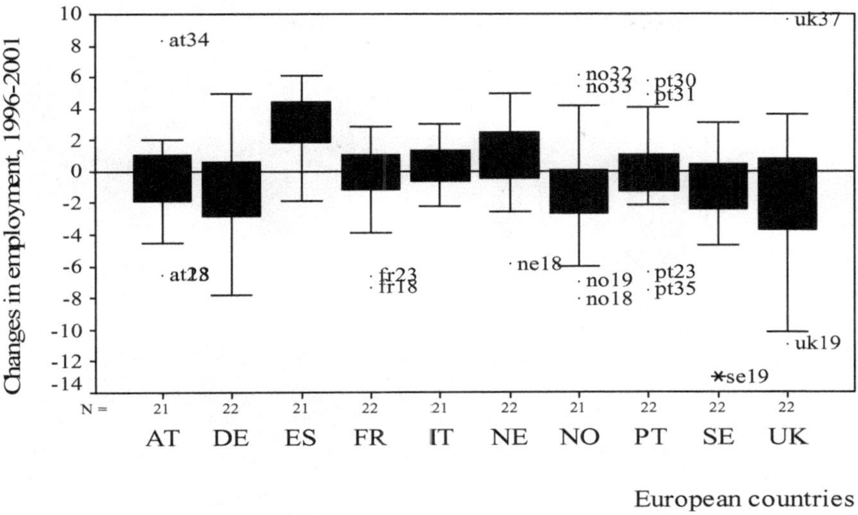

Source: OECD STAN database

Fig. 1 Employment changes. 1996-2001. Country patterns

gains of the most innovating firms may be obtained at the expense of less innovative competitors, undermining the net employment performance of innovative behaviors.

Moreover, sectoral studies may better account for structural changes occurring in European economies. Since changes in the sectoral composition of economies would determine the decline or the consolidation of sector-specific innovative behaviors and strategies, the consequences on employment may vary greatly.[11]

Thirdly, when looking at European employment dynamics, the variety of sectoral patterns seems also to be high.

Figs. 1 and 2 show for the manufacturing sector (disaggregated in 22 sectors), and for 10 major European countries, respectively, country and sectoral patterns of employment changes[12] over the period 1996-2001.

In Fig. 1, sectoral values are included within a minimum and a maximum, whose distance expresses the sectoral variance within the country. All countries but Spain, Italy and the Netherlands seem to reveal employment loss in the majority of industries.

The exceptions of Spain, Italy the Netherlands seem to confirm the existence of systemic interactions and contextual factors, largely assessed by the National Systems of Innovation literature, able to influence employment dynamics[13]. The empirical analysis carried out in section 6 will control for country effects.

Country outliers are also provided in Fig. 1. It is notable that lower bound outliers are concentrated in few sectors such as Wearing Apparel and Dressing (18), Leather, Leather products and Footwear (19), Coke and Refined Petroleum

---

[11] See Perez (1983); Evangelista (1999); Marsili (2001); Marsili and Verspagen (2002).

[12] Employment changes are expressed as compound annual averages changes.

[13] See, on this point, the seminal works of Freeman (1995); Lundvall (1992); Nelson and Rosenberg (1993).

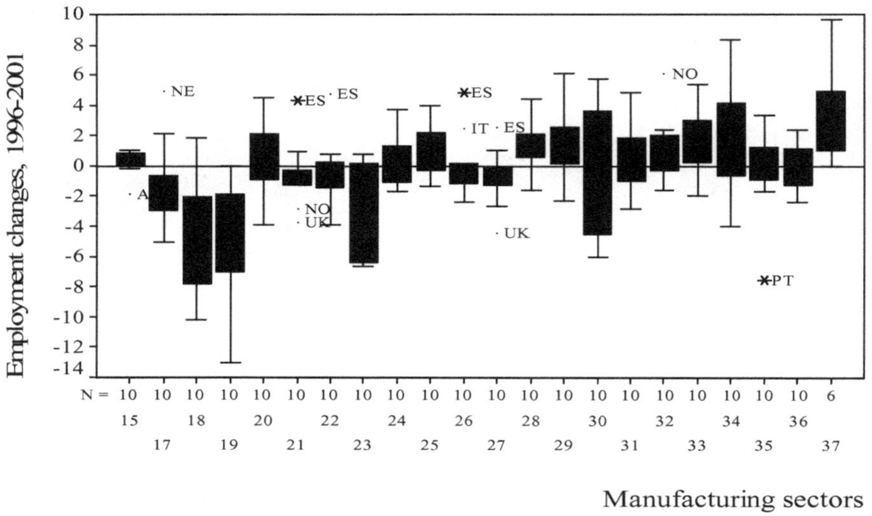

Source: OECD STAN database

Fig. 2 Employment changes, 1996-2001. Sectoral patterns

products (23) classified by Eurostat[14] as low and low-medium tech manufacturing industries. By contrast, ,upper bound outliers are mainly high-tech manufacturing industries, including Office, Accounting and Computing Machinery (30), Electrical Machinery and Apparatus (31), Radio, Television and Communication Equipment (32) and Medical, Precision and Optical Instruments (33). This fact, further investigated later and empirically tested in section 6, seems, at this stage, to suggest that sectoral intensity of innovation and innovation performance (stylized in section 4.1 as 'commitment to innovation') are generally positive correlated with employment dynamics.

When looking at Fig. 2, similar indications are straightforward. Furthermore, the figure shows, as expected, that a much greater variety of patterns emerges across sectors than across countries. At first glance and excepting few sectors, the picture shows that, for manufacturing, countries tend to behave, although with different intensities, similarly within the same sector, supporting the existence of sectoral employment patterns. Whether these patterns are influenced by different innovation strategies mainly oriented towards product or process innovations, emerging as dominant at the sectoral level, is the objective of the next sections.

3 Product and process-oriented innovation strategies

Conceptualizing innovation strategy and its determinants is a difficult task. Starting from the seminal work of Pavitt (1984), innovation strategies are defined here as the combination of three elements: *objectives* pursued by firms, *activities* undertaken to

[14] This classification is provided by Eurostat Statistics on Innovation in Europe (2000), Part 4, Methodological Note: 114.

achieve the objectives set and *sources* which provide with the existing technological opportunities to be exploited.

The conceptualization of innovation strategies mainly oriented towards product or process innovations, moreover, takes its lead from the classification developed in Antonucci and Pianta (2002)[15]. In that paper, three different technological strategies pursued by firms are assumed. The first strategy, searching for technological competitiveness, is based on *product innovation* and productivity growth rooted in quality advantages. The second, a strategy of active price competitiveness, is based on *process innovations* and productivity growth rooted in restructurings and specialization. The third, the passive price competitiveness strategy, is pursued by non-innovators relying on cost-cutting efforts.

Such strategies have substantially different outcomes on growth and employment, in relation to changes in demand, as argued more in detail in sections 4.2 and 4.3.

There is a large consensus in the literature in recognizing the *direct* positive/negative effects on employment of product and process innovations.

The general assumption is that sectors in which disembodied innovation activities are carried out (R&D, design, engineering) seem to have better employment performance than sectors that are more committed to embodied innovation activities (machinery and equipment). This is typical, respectively, of firms at the technological frontier, leaders in their market segments/sectors or entering new fields of activity rather than firms competing on price reductions in established markets.

The empirical work carried out in the paper, following the conceptualization of innovation strategies made, will focus on assessing more systematically the employment effects of different product or process-oriented innovation objectives, activities and sources chosen by firms, and on analyzing whether there exists a coherence of effects within the strategy.

In particular, the set of hypotheses that will be tested in section 6 is summarized below:

H1 The direct effects of product and process-oriented innovation objectives, activities and sources are expected to be, respectively, positive and negative;

H2 The oriented product or process objectives, activities and sources are expected to be *strategically coherent* when looking at their effects.

3.1 Product and process-oriented innovation indicators

Based on the hypotheses made above, a set of product/process-oriented innovation indicators have been created.

The general criteria followed to select the appropriate variables for the empirical analysis among the several available in the SIEPI Database is as follows:

(i) Avoiding investigation of the effects of all indicators at once through, for example, a principal component analysis which would have not allowed identifying the effects of single objectives, activities and sources.

---

[15] See Antonucci and Pianta (2002, 15).

(ii) Distinguishing as clearly as possible between product or process-oriented objectives, activities and sources in order to avoid an ambiguous interpretation of the estimation results.

The following choices have been made:

Innovation objectives: incremental innovation objectives have been preferred to radical ones in order to reduce the displacement effects due to products substitutability, since the functionality seems to vary less in the former than in the latter case.[16]

In particular, the product-oriented objective indicator has been created as the mean of two variables – the share of firms improving product quality (OIMP) and extending product range (OEXT) – on total population of firms. The process innovations indicator concerns the objective of cost reductions – the share of firms reducing material consumption (OMAT) on total population of firms – which usually are pursued along with the introduction of new machineries.

Innovation activities: the traditional distinction between the expenditure in R&D and in new machinery and equipment has been adopted. In particular, the following indicators have been created: total (internal and external) R&D expenditure (RDX) on sales and machinery and equipment expenditure (RMACX) on total innovation expenditure.

Innovation sources: the indicator of product innovation sources includes those which usually make use of R&D activities, such as universities, higher education institutes, government or private no-profit research institutes, patent disclosures. In particular, the product-oriented source indicator has been created as the mean of three variables: the share of firms having universities and higher education institutes (SUNI), government or private no-profit research institutes (SGMT) and patent disclosures (SPAT) – all divided by the total population of firms. In the case of process innovation sources, the indicator selected is the opportunity offered by competitors – share of firms having competitors as source of information (SCOM) on total population of firms. Both choices on product and process innovation sources indicators reflect Pavitt's intuitions and his taxonomy (Pavitt 1984).

3.2 The indirect effects of product and process-oriented innovations

The existence of several indirect effects prevents the possibility of determining *a priori* net employment outcomes of heterogeneous innovation processes. The attempt is to find significantly different relations between, respectively, product and process innovation strategies and employment growth requires, when carrying out the econometric tests, to control for their indirect effects.

From a theoretical point of view, indirect effects are caused by the existence of:

1. several mechanisms which may compensate the negative effects of *process innovations* as well as of the limitations to their functioning;
2. displacement effects which may diminish the positive effects of *product innovations*;
3. complementarities and overlaps between *product and process innovations*.

[16] See Edquist et al. (2001, 19).

1. Controlling **through demand** for the negative effects of process innovations
   Far from reviewing the effectiveness and limitations of the 'compensation theory', the attention here is focused on the role of demand in determining the full functioning of the mechanisms.
   Following the seminal work of Vivarelli (1995), the employment effects of the compensation mechanisms (via decreases in wages, via decreases in prices, via new investments, via increases in income[17]), with the exemption of 'via new machines', depends largely on demand dynamics, since the redistribution of productivity gains, due to process innovations, among wages, profits and prices leads to sustained increases in consumption and investments and in turn full employment only if demand fully satisfies supply (Say's law)[18].
   If consumption and investments do not grow enough when prices go down and/or wages and profits go up, or when increases in technological unemployment, by decreasing wages, lead to use only partially labor-intensive techniques, then the direct (negative) effects on employment associated to process innovations are not fully reabsorbed.
   To synthesize, if demand grows less than productivity, then the net employment outcome of process innovations is always negative, independent of the way productivity gains are distributed.
   If this is the case, the more innovative firms will take over the market shares of the non/ less innovative ones, given the fact that market has not grown enough to absorb the increased supply of all firms.
   Following the considerations made above, the negative effects of process-oriented innovation strategies on employment are expected to be only partially reabsorbed, where the extent of the reabsorbtion is mainly determined by demand growth. A focused analysis and a more formal hypothesis on this point are provided in section 4.3.
   It is important to underline, at last, that this approach may underestimate the intensity of the reabsorption driven by demand if, as it occurs in this work, manufacturing only is considered, while new jobs may be created also through increased demand for services.
2. Controlling **through the inclusion of all manufacturing sectors** for displacement effects of product innovations
   Turning to product innovation, its employment impact strongly depends on the degree of substitutability among different goods.[19] In Katsoulacos' work (1984), it emerges that the extension in the product range increases consumer's utility function and requires additional employment to produce the new goods (welfare effect). On the other hand, the expansion effect on employment may diminish if the new products replace the old ones (displacement effect).
   On the extent of the displacement effect there is a lack of investigation in the literature, although, in theory, this effect should be larger when 'vertical integration' (improved quality products) occurs and smaller when vertical integrations

[17] In the work of Vivarelli (1995), the compensation mechanism 'via new income' is considered as a "complementary force" (p. 61) and not as a mechanism *strictu senso*. The 'income effect on demand' has been also treated in Edquist et al. (2001, ch.5: 3).

[18] Which in turn implies that price elasticity of demand for the commodity and income elasticity of demand are, respectively, greater than one. See, on this point, Edquist et al. (2001, ch. 5).

[19] See Katsoulacos (1984, 1986).

allows greater competition, decreases in prices and greater demand for the new products.[20]
The coexistence of welfare and displacement effects makes uncertain and only empirically determinable the net employment effects of product innovations.
Here, the displacement effects are controlled by considering all sectors in Manufacturing. The net effect on employment due to the displacement of a product by a new one – either in the case of displacement being intra-sector or inter-sector – is well captured by controlling for all the 22 sectors composing Manufacturing at the two-digit level.[21]

3. Controlling **through the inclusion of all manufacturing sectors** for the existence of complementarities and overlaps between product and process innovations
   The inclusion in the empirical analysis of all manufacturing sectors over a period of three years allows us, finally, not to overestimate the positive role of product innovations, when used as process ones, or underestimate the negative role of process innovations, when implemented to create new products, since time seems to be long enough to capture these changes in the nature of the innovation process.[22]

4 The role of structural factors

Following the results coming from previous studies on this subject[23], the inclusion of other structural factors in the analysis, other than innovation variety, allows us to better disentangle the net effects of heterogeneous innovation strategies on employment.

In particular, the structural factors considered as relevant dimensions when investigating the innovation-employment relationship and indeed included in the econometric specifications, are:

(i) Commitment to innovation;
(ii) Demand and cost of labor dynamics;
(iii) The technology-demand growth circle.

A synthetic description of the relationship among these factors and employment is provided in the next paragraphs, and hypotheses on their functioning are formulated.

4.1 Commitment to innovation

The question whether innovation commitment affects employment dynamics is an important one. Following the Neo-Schumpeterian studies on sectoral patterns of

---

[20] Vivarelli (1995, 50).

[21] However, also in this case, the eventual displacement effects between manufacturing and services (purchasing of a new car versus renting, for example) are not taken into account, underestimating eventual employment reabsorption.

[22] However, also in this case, the overlap effects between manufacturing and services are not taken into account, underestimating in this way the potentiality of this reabsorption mechanism.

[23] See footnote 8.

innovation, the work extends the idea of a 'critical mass' of resources necessary to be successful innovators to its employment effects and tests this hypothesis empirically.

The hypothesis take its lead from the argument, discussed in section 2, that resources are devoted to innovation activities coherently with the dominant strategy in the sector, which in turn affects employment.

Going back to the study on *Schumpeter Mark 1-2 innovation patterns* (Breschi et al. 2000), the product-oriented innovation strategy, for example, seems to dominate the Mark 2 pattern, which is, in turn, characterized by large innovating firms aiming at patenting new scientific discoveries and/or introducing new products in the market by intensive R&D activities and research collaborations with universities and other research institutes.

The presumption here is that the higher the amount of resources devoted to innovation, the greater the possibility that product-oriented innovation strategies will be dominant in the sector, which, in turn, creates better opportunities for employment growth. However, since it is not possible, at this stage, to presume the positiveness of the innovation intensity-employment relationship, the hypothesis has been restricted to shed light on the relevance of innovation commitment for employment growth, regardless of sign.

This hypothesis is tested making use of two different common indicators of innovation intensity. The first indicator is the ratio of total innovation expenditure on sales (*intensity indicator*), while the second is built as the percentage of innovative firms on total population of firms (*diffusion indicator*).

The comparison of the econometric results associated with the two indicators is expected to reveal the uninformative nature the diffusion indicator when looking at employment performances. This indicator, which only expresses whether the firm is innovator, gives evidence neither of the commitment to innovation of the firm nor of the fact that the firm has introduced product or process innovations or pursued prices rather than quality-based competitiveness.

In particular, following the considerations made above, the hypothesis tested in section 6 is set as follows:

H3 The *intensity indicator*, once controlling for country effects, is expected to be significantly correlated with employment dynamics, while the *diffusion indicator* is not. The sign of the coefficients of both indicators is uncertain, although, when significant, is (only) presumed to be positive.

Fig. 3 shows the relationship between total innovation expenditure on sales and the share of innovating firms on the total population of firms for the 10 European countries and the 22 manufacturing sectors considered, over the period 1994-1996.

The figure shows the absence of a clear relationship between the two indicators confirming that, among the innovating firms, there are several with scarce commitments to it.

4.2 The role of demand and labour compensation

Demand represents the engine and a powerful constraint to employment growth. In the following empirical analysis, the effects of changes in demand are controlled to

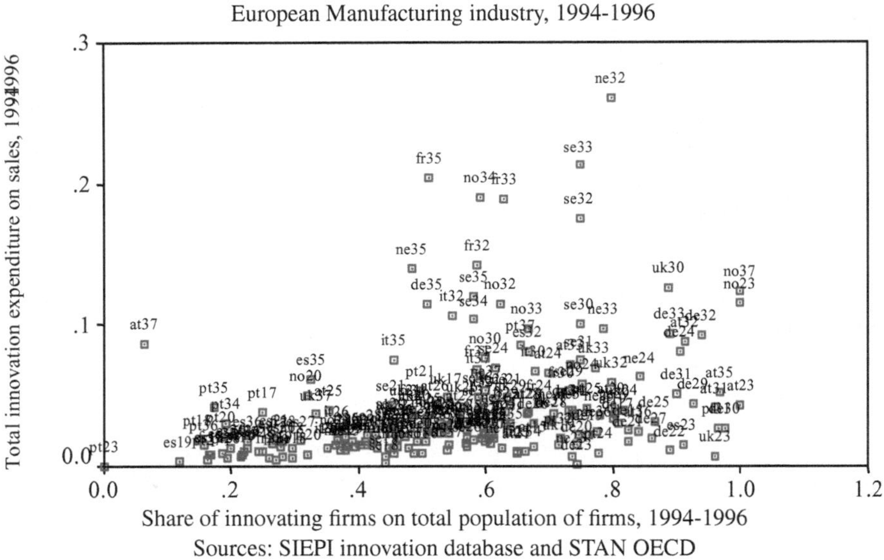

Fig. 3 Innovating firms and innovation intensity

avoid catching the effects of innovation on employment due to exogenous aggregate economic growth or endogenous demand growth driven by technical change.

Demand dynamics are controlled at the sectoral level in order to account for structural changes in European economies which deeply affect employment growth. An increase in the international competition or the emergence of exogenous shocks can determine restructuring in some sectors and the expansion of others. Controlling for changes at the sectoral level allows us to account for these events.

Fig. 4 shows the relationship between changes in employment and value added for 22 manufacturing sectors and the 10 European countries considered over the period 1996-2001.

Three elements seem to emerge. The first concerns the strong and positive relationship between demand and employment growth for most of the sectors and countries considered, confirming the need for controlling the co-movement of the two variables. Secondly, changes in value added are followed by second order (lower) employment changes. Some of the sectors characterized by positive economic growth rates have had no or little employment growth (‘*jobless growth*’ phenomenon). The third element consists in the presence of three clusters: low tech sectors (traditional ones) on the bottom-left side of the figure, a great heterogeneity of performances of middle tech industries, and a small number of highly innovative sectors (upper right in the figure), which seems to confirm, at first approximation, the role of commitment to innovation in stimulating either economic or employment growth.[24]

[24] Note that other four low-tech industries (sector 19 for the UK and Sweden, sector 18 for the UK and Norway) and three high or medium-high tech industries (sector 32 and 30 for France and sector 34 for Austria), which were, respectively, on the bottom left side and on the upper right side of the figure, have not been included in the figure.

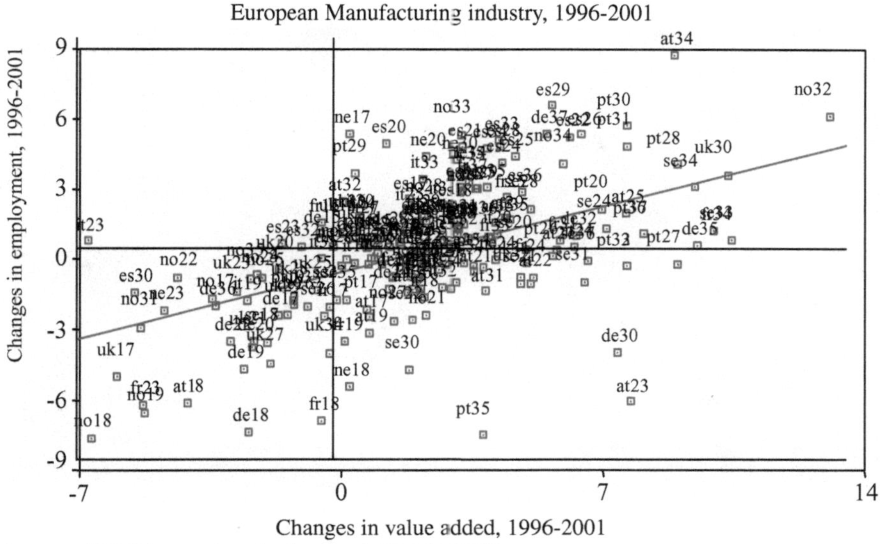

Sources: SIEPI innovation database and STAN OECD

Fig. 4 Changes in employment and value added

As far as the relationship between employment growth and labor compensation dynamics is concerned, the attempt is to control for the effects that increases in labor cost may have on employment.

Fig. 5 shows a negative relationship, although not very significant, between employment growth and labor compensation dynamics over the period 1996-2001 for the 10 European countries and the 22 manufacturing sectors considered. Same sectors across different countries have completely different employment patterns and no clear country specificities or cluster of sectors/countries seem to emerge. The relationship is expected to be more significant when controlling for country effects, as carried out in section 6.

4.3 The technology-growth effect

The technology-growth effect is investigated in order to capture the positive impact that process-oriented innovation strategies may have on employment when demand grows.

These effects are indirect effects *strictu sensu* in the case of demand growth being driven by technological change, and are exogenous shocks when demand growth is driven by global upturns.

The technology-growth effect is investigated here as combining both effects regardless of their being endogenously or exogenously determined. As a result, the new variable sheds light on the relevance of the demand dynamics in determining employment outcomes and on the effectiveness of employment compensations due to increases in aggregate demand.

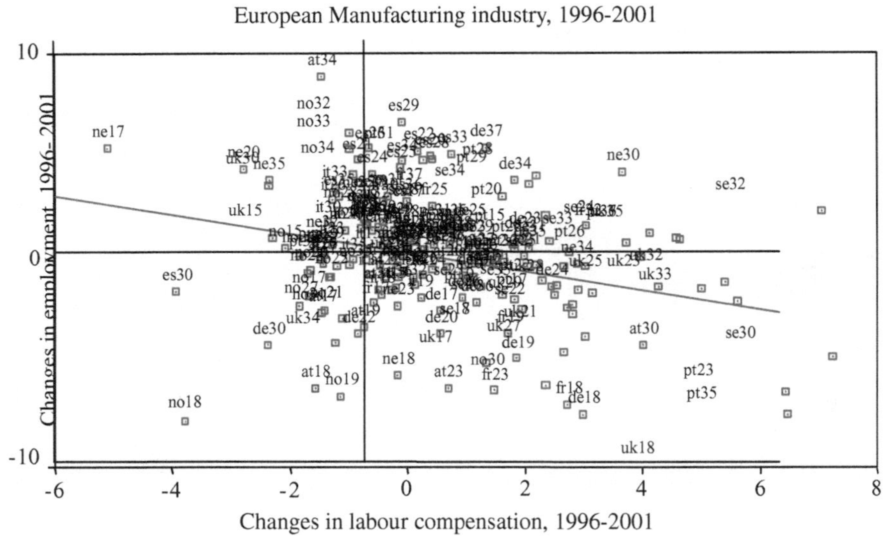

Sources: SIEPI innovation database and STAN OECD

Fig. 5 Changes in employment and labor compensation

The empirical analysis of the 'technology-growth effect' is carried out through the application of a multiple regression analysis[25] in which value added changes (proxy for demand changes) and process innovation variables (objectives, activities and sources) are simultaneously entered in the regressions.

In particular the following hypothesis is set:

H4 The multiplicative variable (value added changes times process innovation variables) is expected to be positively correlated to employment dynamics when demand grows, partially offsetting the negative effects of process innovations.

5 Data and Methodology

5.1 The integrated database

The empirical analysis makes use of an integrated dataset called 'The SIEPI Database'[26] which merges CIS2 data (1994-1996) on different quantitative and qualitative dimensions of innovation processes – objectives, activities, performance, co-operation, hampering factors, sources – with economic and structural data drawn

[25] See, on this subject, Greene (1997, 235-236).

[26] The SIEPI database is a research outcome of the European research project SIEPI – The Structure of Innovation and Economic Performance Indicators (2002-2004) – financed within the Fifth FP, "Improving the Human Research Potential and the Socio-Economic Knowledge Base", DG Research. Further information on the project and its outcomes are available at the SIEPI Project website: www.econ.uniurb.it/siepi

from the OECD STAN database, for 22 manufacturing sectors – Nace Rev.1[27] and 10 European countries[28] over the period 1996-2001.

In particular, as far as the OECD STAN database is concerned, the set of variables taken into consideration are: Total Employment (Number engaged); Value added; Labor Compensation of employees at constant prices. Labor compensation of employees has been divided by the total number of employees to account for the sectoral dimension.

Finally, value added data have been deflated with sectoral deflators (elaborated from the OECD STAN database), while GDP deflators have been used to deflate data on Labor Compensation.

Detailed work to standardize OECD STAN with innovation data (CIS2) has been carried out in order to have a consistent dataset at two-digit level of sectoral disaggregation – Nace Rev. 1. Data coverage of the database is over 90% and data quality has been widely checked and improved under the work conducted for the SIEPI European research project.

5.2 The econometric model

Building on previous empirical analyses carried out at the sectoral level[29], the paper provides an econometric model through which the complex web of relationships, either direct or indirect, among employment growth, heterogeneous innovation strategies and structural factors is empirically evaluated. In particular, the set of hypotheses presented in the previous paragraphs is investigated through a cross section analysis carried out on 220 observations (22 manufacturing sectors and 10 European countries) over the period 1994-2001, using the OLS estimator.

The econometric equation used for the analysis is the following:

$$\Delta E_{ij} = \alpha + \beta \Delta D_{ij} + \gamma \Delta W_{ij} + \delta I_{ij} + \lambda INNOV_{ij} + \theta INNOV_{ij} * \Delta D_{ij} + CD_j + \varepsilon_{ij} \quad (1)$$

where, for sectors i and countries j:

α is a constant,

[27] The 22 manufacturing sectors, which cover the whole Manufacturing industry, are: Food and Beverages (15); Textiles (17); Wearing Apparel, Dressing (18); Leather, Leather products and Footwear (19); Manufacture of Wood and Products of wood (20); Pulp, Paper and Paper products (21); Publishing and Printing (22); Coke and Refined Petroleum products (23); Chemicals and Chemical products (24); Rubber and Plastics products (25); Other Non-Metallic Mineral products (26); Basic Metals (27); Fabricated Metal products, except machinery and equipment (28); Machinery and Equipment (29); Office, Accounting and Computing Machinery (30); Electrical Machinery and Apparatus (31); Radio, Television and Communication Equipment (32); Medical, Precision and Optical Instruments (33); Motor Vehicles, Trailers and Semi-Trailers (34); Other Transport Equipment (35); Manufacturing NEC (36); Recycling (37).

[28] The countries are: Austria, Germany, France, Italy, Norway, Portugal, Spain, Sweden, The Netherlands and The United Kingdom.

[29] See footnote 27.

ΔE , the endogenous variable, *is the compound annual rate of change of Total Employment – Number Engaged (EMPN) 1996-2001.*[30]

ΔD *is the dummy of value added growth.* This variable is built on the compound annual rate of growth of value added 1996-2001, taking value '1' when the growth rate is over the mean of all manufacturing industries and value '0' when is below the mean;[31]

ΔW *is the dummy of labor compensation per employee growth.* This variable is built on the compound annual rate of change of labor compensation per employee 1996-2001 taking value '1' when the growth rate is over the mean of all manufacturing industries and value '0' when is below the mean;[32]

The use of dummies of real value added and real labor compensation per employee growth, instead of their actual values, is a result of the need to reduce endogeneity from the regression. The endogeneity arising from the influence of employment dynamics on value added and labor costs growth is expected to be less when exogenous variables are expressed as dummies. Moreover, the endogeneity, which may exist at the aggregate level, should be less strong when sectoral data are considered.

As a matter of fact, the value added growth generated by positive employment dynamics in one sector may not lead automatically to changes in value added at the same rate in the sector, since the new wage-based demand will spread over all manufacturing and service activities through the complex web of sectoral interdependencies.[33]

I *is the intensity of total innovative efforts* measured by:

i) total innovation expenditure on sales over the period 1994-1996;

ii) innovating firms on total population of firms over the period 1994-1996.

$INNOV$ represents the set of product and process-oriented innovation objectives, activities and sources indicators defined in section 3;

$INNOV * \Delta D$ is the *technology-growth effect* which is captured by the multiplicative regression variable defined in section 4.3, that is, the product between value added growth (the dummy of) and process-oriented innovation variables.

CD is a country dummy;

ε is the error term.

6 Testing the model: the econometric results

Tables 1 and 2 show the results of equation (1). Tables 1 and 2 are organized in columns corresponding to different econometric specifications.

---

[30] Data cover 22 manufacturing sectors and all European countries considered but the United Kingdom whose employment data are not expressed in number Engaged but in number of Employees. Changes in employment refer to the 1996-2001 period for all countries considered except Norway and Sweden, for which data arrive up to 2000 and for Portugal whose data coverage is up to 1999.

[31] Data cover 22 manufacturing sectors and all countries considered except Norway, whose data arrive to 2000, and Portugal, whose data coverage is up to 1999.

[32] Data cover 22 manufacturing sectors and all countries considered except Spain, the United Kingdom and Norway, whose data coverage is up to 2000, and Portugal, whose data are available only until 1999.

[33] Further developments of this work will focus on the use of instrumental variables, whose application has not given satisfactory results yet.

Table 1 The determinants of employment change, 1996-2001

Dependent variable: Average annual compound growth rate of employment
Pool of 22 manufacturing industries in: AT, DE, ES, FR, IT, NE, NO, PT, SE, UK

	Structural factors without fixed effects 1	Structural factors with fixed effects 2	Innovation intensity 3
Structural factors			
Changes in value added (dummy variable)	2.53 (6.02)***	2.05 (4.84)***	2.05 (4.84)***
Changes in labour compensation per employee (dummy variable)	-0.97 (-2.17)**	-1.05 (-1.99)**	-1.08 (-2.02)**
Total innovation expenditure on sales	3.55 (0.68)	10.73 (2.08)**	10.06 (1.83)*
Share of innovating firms on total population of firms			0.43 (0.35)
Country effects			
Austria		0.45 (0.53)	0.36 (0.41)
Germany		0.70 (0.86)	0.59 (0.67)
Spain		3.93 (4.65)***	3.97 (4.64)***
France		0.70 (0.86)	0.73 (0.89)
Italy		1.96 (2.30)**	1.97 (2.29)**
The Netherlands		1.81 (2.19)**	1.76 (2.10)**
Norway		0.20 (0.20)	0.18 (0.18)
Portugal		1.56 (1.83)*	1.61 (1.86)*
The United Kingdom		0.31 (0.38)	0.26 (0.32)
Constant	-0.70 (-1.59)	-1.87 (-2.23)**	-2.04 (-2.09)**
Adjusted R-square	0.17	0.26	0.26
F	15.57***	7.33***	6.74***
Number of cases	204	195	194

T-statistics between brackets. Significance levels: *90%, **95%, ***99%.

By adding variables to the first specification – column 1, table 1 – the general relationship between innovation and employment and its robustness at changes in specification is tested. In particular, table 1 captures the correlations among employment growth and the dynamics of the structural factors (value added, labor cost, innovation intensity) with and without fixed effects (country dummies for all countries, except Sweden, have been included).

Table 2 The determinants of employment change, 1996-2001

Dependent variable: Average annual compound growth rate of employment
Pool of 22 manufacturing industries in: AT, DE, ES, FR, IT, NE, NO, PT, SE, UK

	Innovation activities 1	Objectives of innovation 2	Sources of innovation 3
Structural factors			
Changes in labour compensation per employee (dummy variable)	-0.76 (-1.44)	-1.08 (-1.87)*	-1.40 (-2.51)**
Innovation activities			
Total R&D expenditure on sales	28.36 (3.52)***		
Machinery and eq. expenditure on total innovation expenditure	-0.7 (-0.73)		
Technology-Growth effect	2.82 (3.40)***		
Innovation objectives			
Share of firms improving product quality and extending product range on tot. pop. of firms		5.58 (2.47)**	
Share of firms reducing material consumption on tot. pop. of firms		-5.66 (-2.16)**	
Technology-Growth effect		5.66 (4.81)***	
Innovation sources			
Share of firms having science based sources of information on tot. pop. of firms			11.50 (2.67)***
Share of firms having competitors as sources of information on tot. pop. of firms			-7.19 (-3.10)***
Technology-Growth effect			5.99 (5.04)***
Country effects			
Austria	1.09 (1.35)	0.11 (0.12)	0.53 (0.55)
Germany	0.58 (0.75)	0.02 (0.03)	0.34 (0.36)
Spain	4.33 (5.31)***	5.57 (4.71)***	3.43 (3.37)***
France	0.64 (0.83)		0.11 (0.13)
Italy	2.04 (2.48)**	1.68 (1.81)*	1.04 (1.16)
The Netherlands	2.32 (2.91)***	1.31 (1.49)	1.71 (1.76)*
Norway	0.46 (0.49)	0.32 (0.31)	-0.58 (-0.58)
Portugal	1.84 (2.16)**		
The United Kingdom	0.41 (0.47)	-0.16 (-0.18)	-0.00 (-0.04)
Constant	-1.63 (-1.89)*	-1.93 (-1.65)	0.57 (0.57)
Adjusted R-square	0.23	0.29	0.32
F	5.60***	7.16***	8.06***
Number of cases	180	153	163

T-statistics between brackets. Significance levels: *90%, **95%, ***99%.

Starting from the specification of table 1 the general innovation-employment relationship is investigated in table 2 by looking at the innovation heterogeneity of activities, objectives, and sources, along the product and process innovation dichotomy. In column 1 of table 2, the general innovation intensity indicator is substituted with an indicator of the type of innovation activity carried out in the sector through R&D expenditure and the purchase of new machinery. Two further specifications (column 1 and 2) look at the objectives (improve product quality and reducing material consumption) and at the innovation sources selected at the sectoral level (science-based source and competitors).

The three econometric specifications of table 2 include value added growth, which is strongly correlated with employment dynamics all through table 1, and fixed effects, the control of which allows cross-sectors heterogeneity to emerge. In table 2, demand effects are tested by the technology-growth effect to account for the positive effects of process innovations when demand grows.

The econometric results of table 1 and 2 are summarized below, following the scheme of hypotheses sketched in the previous sections.

6.1 Product-oriented vs. process- oriented strategies

The regression estimates presented in table 2 seem clearly to confirm that product and process-oriented innovation strategies have contrasting effects on employment. As expected, the sign of the former elements (objective, activities and sources) are always positive and those of the latter are always negative, with coefficient estimates significant at the conventional level. Findings are stable when different variables are introduced one by one (estimates are not provided in tables) or in couple (column 1, 2 and 3).

In particular, the product-oriented strategy has a clear positive effect on employment, whether we look at innovation objectives pursued by firms (5.28** the coefficient estimate of its indicator) or consider the innovation activity (28.36***) and the set of sources (11.50***) exploited by them.[34] Process-oriented strategy coefficients, on the other hand, are always negative and significant (except in column 1) confirming the labor substituting nature of process innovations.

The results confirm the *relevance* of the product/process distinction when investigating employment effects of innovation. The trade-off between the need to account for the heterogeneity of innovation processes carried out by firms and the need of synthesizing the complex relationship between innovation and employment in few relevant dimensions has a powerful compromise in the product-process impact differentiation.[35]

The stability of the signs of coefficients in all the specifications of table 2 highlights the robustness of the results when alternative indicators of the same strategy are considered, which confirms the expectations on the *strategic coherence* hypothesis made in section 3.

---

[34] Asterisks on coefficient estimates reveal the significance levels: *90%; **95%; ***99%.

[35] It must be noted, in this regard, that these results are robust to changes in process-oriented innovations variables, such as, among the objectives, OLBR (reduce labor cost) and OPDT (improve production flexibility) and among sources, SSUP (Suppliers of equipment, materials, components or software) and SEXB (fairs, exhibitions).

The fact that employment grows when, for example, R&D expenditure increases is a relevant result in itself. However, this result becomes even more informative when compared to the negative effects of machinery and equipment expenditure, especially for policy makers. Finally, the fact that R&D expenditure is an activity within a strategy which implies other product-oriented objectives, performance, sources and collaborations undertaken by firms, reinforces the overall positive/negative effects on the employment growth of different strategies.

6.2 The role of innovation commitment

Table 1 column 3 shows the different correlation estimates between the two innovation intensity indicators and employment growth. The role of total innovation expenditure on sales in shaping employment is significant (10.06*), while the number of innovators on total population of firms is not (0.43), confirming the expectations and the hypotheses made in section 4. In particular, the corrected R-sq does not change (0.26) from column 2 to 3 when the *diffusion indicator* is included, confirming the uninformative nature of it.

Commitment to innovation – proxied by the amount of resources devoted to innovation on sales – other than being significant, has a positive impact on employment. As said in section 4, there was high uncertainty as to the sign of the intensity coefficients due to the existence of heterogeneous innovation strategies and of several indirect effects. The positiveness of both coefficients (although the other innovation intensity indicator is not significant) suggests, however, an overall positive effect of technological change on employment growth.

Starting from this broad result, the existence of a 'critical mass' of resources needed in order to be a successful innovator and net employment creator, seems to be confirmed, offering an empirical baseline for employment friendly policies.

Regressions estimates shown in column 2 and 3 answer in the negative to the question whether being an innovator is a sufficient condition to sustain employment and set the groundwork for further research in the direction of qualifying the type and the intensity of innovation strategies in order to predict employment growth. It is worth to underlining at this point that the positive sign of the innovation commitment coefficient contrasts with the one found in Antonucci and Pianta (2002), where the overall effect of technological change on employment seemed to be negative (and significant). The possible reasons for this discrepancy are, first, the different period of time covered by the two analyses (1994-1999 instead of 1996-2001), second, the different number of countries considered (8 instead of 10), and third, the different number of sectors included (10 instead of 22).

Among these differences, the increased level of sectoral disaggregation used in the present work is likely to be the most important explanation of the different results, since the higher number of sectors considered (more than doubled) allowed the heterogeneity of innovation strategies and their contrasting effects on employment to emerge.

6.3 The role of demand and labor compensation

The impact of demand dynamics on employment growth is positive and highly significant across all countries and sectors, while the impact of labor compensation per employee is negative and significant (although less) across all countries and sectors.

These results seem to be highly robust to changes in the econometric specifications and through all the regressions carried out.

The picture that emerges from tables 1 and 2 seems to confirm the expectation expressed in section 4 about employment growth being higher in those industries and countries in which demand grows faster and labor costs increase slowly.

The role of demand growth in sustaining employment has two important implications. First, the beneficial effect of technological change on employment, by increasing productivity, occurs only if demand grows at least as much as productivity, requiring to control for its determinants. In this regard, further research on the mechanisms and the extent demand may reabsorb technological unemployment needs to be carried out, replicating, for example, the seminal work of Vivarelli (1995) at a more disaggregated level and looking at regularities and differences across sectors.

The second implication concerns the role of the sectoral composition and specialization of the economy within the international division of labor in driving employment growth. Given that demand seems to have high and positive effects on employment dynamics, it seems that there would be less technological unemployment for those countries whose economies are structurally based on growing sectors. This fact should stimulate industrial policies in moving and concentrating resources towards the most technological advanced sectors.

6.4 The role of country effects

The effect of total innovation expenditure on sales on employment growth seems not to be significant when fixed effects are excluded from regressions, confirming the existence of a great heterogeneity of innovation patterns across countries.

Looking at table 1, column 2, where fixed effects are included, the coefficient of the total innovation expenditure on sales becomes significant and increases greatly along with the corrected R-square, which goes from 0.17 to 0.26.

Looking at the country dummies in all regressions of table 1, Spain, Italy and the Netherlands have a positive and significant coefficient showing better employment patterns relatively to Sweden (the intercept). This picture is confirmed also in table 2.

The persistence of positive country effects for Spain, Italy and the Netherlands through different econometric specification suggests the need to carry out deeper investigations on their recently good employment performance, starting from the labor market reforms introduced since ninety's in Italy and the Netherlands and the sustained economic growth which has characterized Spain in the last decade.

6.5 The role of technology-growth effect

As far as the 'technology-growth effect' estimates are concerned, table 2 shows that the negative effects of process-oriented innovation objectives, activities and sources are compensated when demand increases. In particular, estimates of the 'technology-growth effect' are always positive and highly significant in all the regressions carried out.

Following the way the 'technology-growth effect' variable has been built, these results seem to suggest that, in those sectors in which value added has grown over the mean of manufacturing, process innovations have, with respect to employment, a positive sign. Concerning the net effect on employment, however, it is worthy to notice that, when looking at the size of the coefficients, the direct effects seem to be only partially compensated by indirect ones, leading to overall net employment losses.

7 Conclusions

The very first issue to be commented on, in conclusion, regards the methodological approach undertaken to disentangle the relationship between technological change and employment dynamics.

An empirical approach has been mainly followed in this work, based on the belief that it is impossible to define *a priori* whether innovation has a positive or negative impact on employment due to the existence of a variety of innovation processes which may have contrasting effects on employment. The heterogeneity of innovation processes being largely recognized as sector specific, although variance arises also from different national innovation systems, has required the analysis to be, once controlling for country fixed effects, methodologically focused on sectoral data.

A great amount of data on innovation characteristics, objectives, activities, performances, sources at work in European manufacturing industries were necessary to test whether employment performance were differently influenced by various innovation strategies emerging as dominant in sectors. In particular, the existence of two innovation strategies mainly oriented, respectively, towards product or process innovations, has been the interpretative tool for comprehending at once the variety of innovation, on the one hand, and its multiple effects on employment, on the other.

The richness of qualitative and quantitative information provided by the SIEPI innovation database at a high level of sectoral disaggregation and across a large number of European countries has allowed testing this relationship under these methodological achievements. When looking at the results comprehensively, econometric estimations seem to confirm much of the hypotheses and expectations coming from the model, concerning: (a) the contrasting effects of product and process-oriented strategies; (b) the strategic coherence of effects of heterogeneous innovation objectives, activities and sources of innovation within the innovation strategy associated; (c) the relevance of the 'commitment to innovation' in affecting employment growth versus 'being an innovator'; (d) the role of demand growth either in pushing employment directly and in sustaining the reabsorption of technological unemployment, (d) the negative role of labor cost increases, and (e) the importance of country effects.

The employment sensitiveness to heterogeneous innovation strategies, as mainly induced by the sector in which firms are located, reveals the importance of the sectoral specialization of the economies as well as of their structural changes in affecting the sign and the speed of employment growth. These considerations become clearer when looking at the relationship between employment dynamics and R&D expenditure. The estimate of the (internal plus external) R&D activity coefficient shows the strongest positive relationship with employment among those considered, shedding new light on the employment opportunities provided by this activity relative to others and on the risks undertaken when the average size of firms falls in the economy, as small and very small size firms are generally recognized as being less committed to R&D.

The strong commitment to innovation associated with R&D activities and research collaborations, moreover, seems to be confirmed by the positive (and significant) relationship between the intensity of innovation and employment growth, although empirical tests have been limited, at this point, to enlighten the uninformative nature of 'being an innovator' compared to its commitment.

When turning to the net employment impact of process innovations strategies, things are complicated by the co-existence of direct and indirect effects. Indirect effects occur in the case of demand growing exogenously, allowing for compensation mechanisms to exert their full functioning, and/or when demand is endogenously increased by these processes. Productivity gains due the introduction of new vintages of capital may be used to reduce prices and/or rising wages or profits, which may lead to stimulate consumption and investments and consequently employment reabsorption.

In this regard, the econometric estimates of the '*technology-growth effect*' variable, which has been built to capture the co-effects of process-oriented innovations and demand propulsions, show a positive, although only partial, compensation of direct effects.

The results of the '*technology-growth effect*' strengthen the role of demand in fostering employment growth, as distinct from the persistent positive and highly significant estimates of the direct relationship between demand and employment growth.

It is worth mentioning, finally, that the results concerning the contrasting effects of product and process innovations as well as of demand and labor cost dynamics on employment are consistent with those found in previous works on this issue (Vivarelli and Pianta 2000; Antonucci and Pianta 2002) making use of different databases – respectively, CIS I (1992-1994), New Cronos (1994-1996) and OECD (1992-1999) datasets.

The similarity of results, which are robust to changes in data and econometric specifications, strengthens the conclusions sketched above and, once heterogeneity is taken into account, sets the groundwork for overwhelming the uncertainty characterizing the net effects of technological change on employment.

References

Antonucci T, Pianta M (2002) Employment effects of product and process innovations in Europe. Int Rev Appl Econ, 16, 3: 295–308
Archibugi D, Santarelli E (1990) Cambiamento Tecnologico e Sviluppo Industriale. FrancoAngeli, Milano
Breschi S, Malerba F, Orsenigo L (2000) Technological regimes and Schumpeterian patterns of innovation. Econ J, 110: 388–410
Castellacci F (2004) How does innovation differ across sectors in Europe. Evidence from the CIS-SIEPI database. Tech. rep.
Dosi G (1988) Sources, Procedures and Microeconomic Effects of Innovation. J Econ Lit, 26: 1120–1171
Dosi G, Nelson R (1994) An Introduction to Evolutionary Theories in Economics. J Evolutionary Econ, 4: 153–172
Edquist C, Hommen L, McKelvey M (2001) Innovation and employment: product versus process innovation. Elgar, Cheltenham
Evangelista R (1999) Knowledge and Investment. The sources of innovation in industry. Elgar, Aldershot
Evangelista R (2000) Sectoral patterns of technological change in Services. Econ Innovation New Tech, 9: 183–221
Evangelista R, Mastrostefano V (2005) Firm Size, Sectors and Countries as Sources of Variety of Innovation. Econ Innovation New Tech, 15, 3: 247–270
Freeman C (1995) The National System of Innovation in Historical Perspective. Cambridge J Econ, 19, 1: 15–24
Freeman C (1997) The 'national system of innovation' in historical perspective. In: Archibugi D, Michi J (eds.) Technology, Globalization and Economic Performance, Cambridge University Press, Cambridge
Freeman C, Soete L (1987) Technical Change and Full employment. Oxford, Blackwell
Freeman C, Soete L (1994) Work for All or Mass Unemployment. Pinter, London
Greene HW (1997) Econometric analysis. Prentice Hall
Katsoulacos YS (1984) Product innovation and employment. Europ Econ Rev, 26: 83–108
Katsoulacos YS (1986) The Employment Effect of Technical Change. Wheatsheaf, Brighton
Lundvall B (1992) National Systems of Innovation. Pinter, London
Malerba F (2002) Economia dell'Innovazione. Carocci, Roma
Malerba F (2004) How innovation differ across Sectors and Industries. In: Fagerberg J, Mowery D, Nelson R (eds.) Handbook of innovation, Oxford University Press, Oxford
Malerba F, Orsenigo L (1993) Technological Regimes and Firm Behaviour. Ind Corp Change, 2: 45–71
Malerba F, Orsenigo L (1996) Schumpeterian Patterns of Innovation Are Technology-Specific. Res Pol, 25: 451–478
Malerba F, Orsenigo L (1997) Technological Regimes and Sectoral Patterns of Innovative Activities. Ind Corp Change, 6, 1: 83–117
Marsili O (2001) The Anatomy and Evolution of Industries: Technological Change and Industrial Dynamics. Elgar, Cheltenham
Marsili O, Verspagen B (2002) Technology and the dynamics of industrial structure: an empirical mapping of Dutch manufacturing. Ind Corp Change, 11, 4: 791–815
Mastrostefano V, Pianta M (2004) The dynamics of innovation and its employment effects. An analysis of innovation surveys in European industries. Tech. rep.
Nelson R, Rosenberg N (1993) Technical Innovation and National System. In: Nelson R (ed.) National Innovation Systems. A Comparative Analysis, Oxford University Press, New York
Nelson R, Winter S (1977) In Search of a Useful Theory of Innovation. Res Pol, 6: 36–76
Nelson R, Winter S (1982) An Evolutionary Theory of Economic Change. Harvard University Press, Cambridge
Pavitt K (1984) Sectoral Patterns of Technical Change: towards a Taxonomy and a Theory. Res Pol, 13: 343–375
Perez C (1983) Structural change and the assimilation of new technologies in the economic and social system. Futures, 15: 357–375
Petit P (1995) Employment and technical change. In: Stoneman P (ed.) The Economics of Innovation and Technical Change, Blackwell, Oxford

Petit P, Soete L (2001) Technology and the Future of European Employment. Elgar, Cheltenham
Pianta M (2004) Innovation and employment. In: Fagerberg J, Mowery D, Nelson R (eds.) Handbook of innovation, Oxford University Press, Oxford
Sirilli G (1997) Science and Technology Indicators: the State of the Art and Prospects for the Future. In: Antonelli G, Liso ND (eds.) Economics of Structural and Technological Change, Routledge, London
Spezia V, Vivarelli M (2002) Innovation and employment. In: Greenan N, L'Horty Y, Mairesse J (eds.) Productivity, inequality and the digital economy, MIT Press, Cambridge
Vivarelli M (1995) The Economics of Technology and Employment. Elgar, Aldershot
Vivarelli M, Pianta M (2000) The Employment Impact of Innovation. Evidence and policy. Routledge, London
Walsh V (1990) Invenzione e innovazione nell'industria chimica: 'demand-pull' o 'discovery-push'? In: Archibugi D, Santarelli E (eds.) Cambiamento Tecnologico e Sviluppo Industriale, FrancoAngeli, Milano

Part 5:

Knowledge diffusion, spillovers and firm strategies

Andrea Morone · Piergiuseppe Morone · Richard Taylor

A laboratory experiment of knowledge diffusion dynamics

Abstract This paper aims to study, by means of a laboratory experiment and a simulation model, some of the mechanisms that dominate the phenomenon of knowledge diffusion in the process that is called 'interactive learning'. We examine how knowledge spreads in different networks in which agents interact by word of mouth. We define a regular network, a randomly generated network and a small world network structured as graphs consisting of agents (vertices) and connections (edges), situated on a wrapped grid forming a lattice. The target of the paper is to identify the key factors that affect the speed and the distribution of knowledge diffusion. We will show how these factors can be classified as follows: (1) learning strategies adopted by heterogeneous agents; (2) network architecture within which the interaction takes place; (3) geographical distribution of agents and their relative initial levels of knowledge; (4) network size. We shall also attempt to single out the relative effect of each of the above factors.

Keywords Knowledge diffusion · Interactive learning · Heterogeneous agents · Network architecture
JEL Classification D83 · Z13 · C9

1 Introduction

Several authors have recently attempted to study the way in which knowledge diffuses by means of informal interactions. Both theoretical and empirical works

A. Morone
University of Bari, Department of Economics, Via Camillo Rosalba 54, Bari, Italy
E-mail: a.morone@dse.uniba.it

P. Morone
University of Napoli "L'Orientale", Department of Social Science, Italy
E-mail: p.morone@gmail.com

R. Taylor
Centre for Policy Modelling, Manchester Metropolitan University, Manchester, U.K.
E-mail: richard@cfpm.org

have addressed this topic with the aim of shedding some light on the complex mechanisms that regulate informal learning (Ellison and Fudenberg 1993; Rauch 1993; Bala and Goyal 1995, 1998; Acemoglu and Angrist 1999; Chwe 2000). The growing interest in this field of research is due to the increasing importance of knowledge for economic growth and development, coupled with the recognition of the dominant role of informal interactions in producing and diffusing this knowledge. In fact, the modern economy has often been described as knowledge-based, or as a learning economy, due to the central role that knowledge and learning play for economic development (OECD 1996).

The aim of this paper is to present an original contribution to the debate on informal learning using a laboratory experiment designed to reproduce the complex learning dynamics that take place when people exchange knowledge by means of face-to-face interaction. The experiment is based upon the conceptual model of knowledge diffusion specified in our earlier work (Morone and Taylor 2004) and presented in section 3.1. In section four, we present the results of the experiment. Through analysis of these data we identify the learning strategies followed by the experimental subjects, and use this knowledge to inform agent design in the simulation model[1]. We then compare the experimental findings with the outputs of simulations and we analyse the performance of the agents' artificial learning strategies in terms of the efficiency and equality of knowledge diffusion. The final part of this paper then further explores the behaviour of the model by simulating over a much larger range of parameter settings than would be possible with laboratory methods, thereby extending the analysis of the influence of network factors upon knowledge diffusion patterns and contributing to the debate on informal knowledge diffusion.

2 Literature Review

The mechanisms that dominate informal processes of knowledge diffusion have been investigated by evolutionary economists as well as by game theorists and applied economists. Mechanisms of innovation diffusion (Clark 1984; Rogers 1995) are often viewed as good examples of informal learning processes because they tend to occur through interaction within geographical and other informal networks, involving social externalities. Several researchers have investigated the patterns through which different agents adopt new technologies by means of theoretical as well as simulation models (Ellison and Fudenberg 1993, 1995; Bala and Goyal 1995, 1998). Another common way of modelling the mechanisms of social learning and technology diffusion makes use of evolutionary game theory (Chwe 2000; Ellison 1993, 2000; Anderlini and Ianni 1996; Berningaus and Schwalbe 1996; Goyal 1996; Akerlof 1997; Watts 2001).

Along with the speed of new technology diffusion, several researchers have focused on the impact of peer behaviour upon individual decisions in areas such as propensity to crime, use of drugs, school dropout and school attainment (Brock and Durlauf 1995; Bénabou 1993; Durlauf 1996; Glaeser et al. 1996). What all the studies considered so far have in common is their reliance on the idea that learning

[1] We are not in any way claiming that the underlying agent motivations are the same as in the experimental case.

from neighbours occurs and that under certain conditions it leads to the desirable stable equilibrium. However, none of these studies go beyond a binary definition of learning.

Jovanovic and Rob (1989) proposed for the first time a model in which incremental improvements in knowledge were defined as a complex process of assembling different ideas by means of information exchange among heterogeneous agents. The new insight brought by the authors is that knowledge was defined as something more complex than a binary variable and that, therefore, growth of knowledge could be defined as an interactive process tightly linked to its diffusion.

Cowan and Jonard (2004) made a subsequent attempt to study the effects of incremental innovations and their diffusion within a network of heterogeneous agents. Knowledge in their model was considered as a vector of values and was exchanged via a simple process of barter exchange. Depending on the network structure, the authors found a trade-off between the speed of knowledge diffusion and the variance of knowledge. In other words, there is a spectrum of states of the world, varying from a situation of high knowledge inequality and fast knowledge diffusion to the opposite situation, more equal in terms of knowledge variance but less efficient in terms of knowledge diffusion.

Along the lines of these works, Morone and Taylor (2005) defined a model in which agents exchange knowledge exclusively by means of face-to-face interactions. The network structure was endogenous to the model and could vary over time. The authors showed how small-world networks emerged and coexisted with both a very unequal and a very equal diffusion of knowledge, different outcomes depending upon the initial conditions.

3 The Model

The model presented in this paper is based on the above revised literature and aims at bringing, within this intellectual framework, new insights on the features that shape learning patterns. We shall argue that there are three fundamental factors which influence the speed and the distribution of knowledge diffusion within any closed network. Namely these factors are: (1) the learning strategies adopted by heterogeneous agents; (2) the network architecture within which the interaction takes place; (3) the geographical distribution of agents and their relative initial levels of knowledge. We shall also investigate how system size might affect knowledge flows.

In what follows we will single out the impact of each of these factors on learning dynamics by testing the model by means of a laboratory experiment. Subsequently, the experimental results will be compared with the outcomes of simulation modelling, using the same network sizes, architectures, geographical distributions, and initial levels of knowledge. In doing so we aim at replicating the experimental learning dynamics by means of simple behavioural rules followed by artificial agents.

The first objective of the simulation modelling, therefore, is to test a number of artificial learning strategies to see how they compare with the experiments in terms of the efficiency and equality of diffusion. The second objective is to further explore the model by considering much larger multi-agent systems, and to analyse these over hundreds of repetitions to remove any unwanted artefacts associated

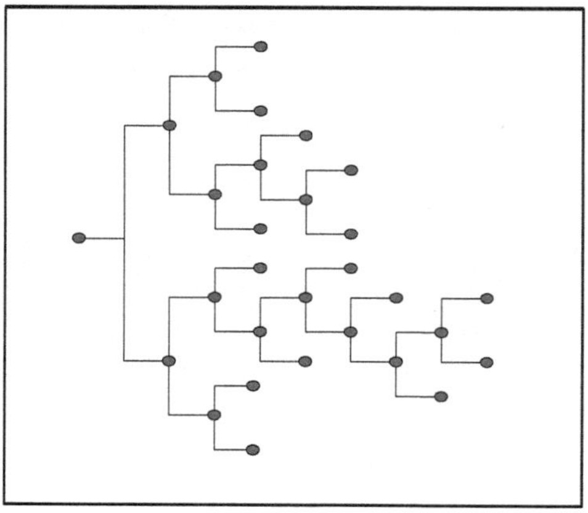

Fig. 1 Cognitive Map

with particular configurations or initialisation routines. We will now discuss more in detail the overall framework of the model, the experimental setting and the simulation methodology.

3.1 Defining Complex Learning Dynamics

This paper is grounded on the idea that learning is a complex process (Jovanovic and Rob 1989; Ancori et al. 2000). More precisely, the theoretical foundation of the experiment relies on the works of Morone and Taylor (2004, 2005), where agents exchange knowledge by means of face-to-face interactions, and every time there is a knowledge transfer, the knowledge is mastered through a backward process by which it is confronted and articulated with previous knowledge. This learning process is achieved through the introduction of a nonlinear complex cognitive structure. Basically, each agent is endowed with a cognitive map (CM) that resembles the structure of a tree in which each vertex (node) represents a piece of crude knowledge, and each edge (link) represents knowledge that agents have already mastered and learned how to use (see Figure 1).

Moving from left to right in the cognitive map, we advance from less to more specialised knowledge. This observation justifies the assumption that new nodes can only be activated if they are directly connected to active nodes. In other words, new knowledge can only be acquired if it can be integrated with the existing, accumulated knowledge in order to be used.

The model is stepped forward through simulated time, where each discrete time-step is known as a 'cycle'. From one cycle to the next, agent states may change according to rules specified in the model. These rules are executed simultaneously for each member of the agent population. Interaction amongst agents is based on the transmission of knowledge. Every cycle each agent has got the opportunity to initiate a learning activity by contacting one of her/his acquaintances in the social network (see section 3.3 for description of the network upon which interactions take place) as follows. An agent, whom we shall call A, contacts an acquaintance, B, from her/his list of acquaintances. She/he then selects a node of her/his cognitive map which she/he would like to learn (of course this node has to be pegged to already active nodes). If the contacted agent possesses the bit of knowledge required, the player that initiated the contact acquires it and activates it in her/his cognitive map. Otherwise, he/she will remain at the former level of knowledge.

3.2 Initialising the Cognitive Maps

While constructing agents' cognitive maps, it was assumed that everybody knew the first node (i.e. was endowed with the most elementary level of knowledge). The nodes in the second column were assigned subject to a specialisation process (i.e. each agent could specialise either in scientific knowledge, knowing the upper part of the tree, or in technical knowledge, knowing the lower part of the tree). The initial CM configuration was constructed as follows. Firstly, an integer indicating the depth of knowledge was selected at random from the interval between 1 and N_{cols}, where N_{cols} is the total number of columns in the CM. In all experiments reported here, we used a value of 6 for the integer N_{cols}. Subsequently, another integer was chosen to define the kind (i.e. specialisation) of knowledge possessed by each agent. This number was either a 1, indicating scientific knowledge, or a 2, indicating technical knowledge.

If the first integer took the maximum value N_{cols}, then the whole map was activated in the corresponding part of the tree. Otherwise, the nodes in all the columns up to and including the column indexed by the first integer value were activated in the region of specialisation defined by the second integer.

3.3 Setting of the Experiment

The aim of the "game" for each player is to increase his/her own level of knowledge. The cognitive map assigned to players resembles the graphical representation reported in Figure 1. In Figure 2 we reproduce a demonstrative cognitive map that each player might see displayed on his/her screen. Each tab is labelled between 1 and 63 and represents a node that can be acquired. Yes/No values indicate whether or not the tab is activated, i.e. whether that particular piece of knowledge is at present in the player's cognitive map. As already discussed, in order to activate a new tab (i.e. to learn a new bit of knowledge), the acquired knowledge has to be linked to an already activated tab in the cognitive map.

The game is simultaneous and composed of N periods. The networks upon which interactions take place are constructed following the method first used by Watts and Strogatz (1998). Each agent is initially assigned a random, unique position in a one-dimensional wrapped grid (i.e. a ring). The social network is then created by connecting an agent with all other agents located within her/his neighbourhood. Social neighbourhoods are defined as the region on the grid that includes the adjacent cells falling within the agent's visible range. We therefore specify a regular network shaped as a ring lattice within which each agent interacts only with its n nearest neighbours.

Departing from this initial network structure we then constructed two other networks following the *re-wiring procedure* first introduced by Watts and Strogatz (1998). With a certain probability p, we allowed any individual to disconnect

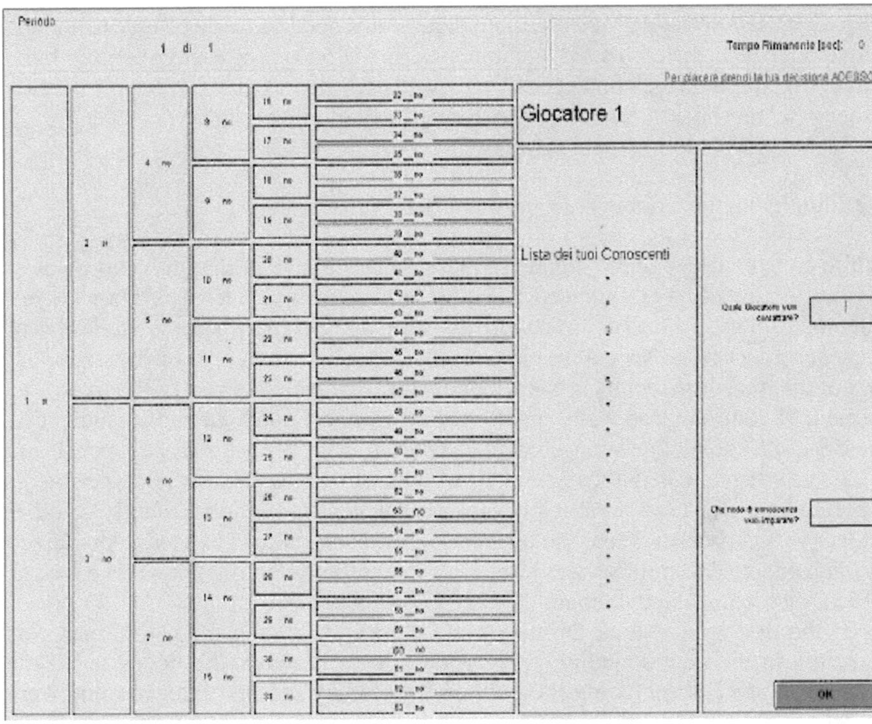

Fig. 2 Experimental definition of the cognitive map

him/herself from the local connection and to re-connect that edge to a vertex chosen uniform randomly over the entire lattice. As the value of p changes, the structure of the network changes. By choosing a fairly small value of p, the lattice remains almost regular and highly clustered (i.e. with high degree of cliquishness), but because each non-local connection is a potential short-cut between two vertices the graph has the low average path length of an almost-random graph (Cowan and Jonard 2004). The concept of average path length introduced here is a measure of the efficiency of the model, giving the average number of steps required to connect each pair of vertices in the lattice. We defined our two new networks architectures by setting the p parameter to 0.1[2] and 1. The experimental networks were then defined as follow: regular network ($p = 0$), small world network ($p = 0.1$) and random network ($p = 1$).

The huge appeal of small-world networks lies in the impact they are said to have on dynamical systems. Watts and Strogatz (1998) maintain the point that "models of dynamical systems with small-world coupling display enhanced signal propagation speed, computational power, and synchronizability." Furthermore, the authors showed, using simplified model for the spread of an infectious disease, that "infectious diseases are predicted to spread much more easily and quickly in

[2] Given the small number of agents composing our network (this constraint was due to the experimental nature of the work) we chose the highest possible value to the rewiring probability to generate a small world network. We tested the network structure following the Watts and Strogatz (1998) criteria for identifying the network as small world.

a small world" than in regular lattices (1998: 442). These findings have profound implications for many manmade and natural systems. For instance, in a transportation network, small-world architecture could improve the flow of people or goods through the network. In a network within which people are exchanging knowledge by means of face-to-face interactions, small-world connectivity might improve the ease with which ideas and knowledge diffuse through the system.

Small world models were intended by the above authors to be used only in the case of investigating large, sparse networks. However, under experimental conditions it is not practical to undertake studies involving hundreds of participants. In order to take into account this practical limitation of our experimental analysis we shall, in section 4.3, corroborate our experimental findings with simulation findings obtained over a much wider network of agents. By combining the experimental approach with the simulation approach and by grounding the simulation model upon experimental findings, it becomes possible to extend the scope of our investigation being nonetheless loyal to the empirical evidence. In this sense, we want to promote the use of agent-based models as a 'wider laboratory' for experimental economics.

4 Empirical Findings: Comparing Experimental and Simulation Results

In this section, we will present our experimental findings. We will then compare those results with simulation outcomes in order to investigate artificial learning strategies. By basing the agent strategies upon those identified through analysis of the experimental data, we shall develop a sort of history friendly model (which we should call experimentally-friendly model), following the methodological approach first developed by evolutionary economists such as Nelson and Winter (1982), Silverberg et al. (1988), and Dosi et al. (1995).

After investigating the impact of different learning strategies upon knowledge flows, we will turn our attention to the specific role played by the network architecture. By means of simulation analysis we shall single out the impact of different network structures upon the learning dynamics and knowledge dispersion, comparing the performances of small world networks, regular networks and random networks. We will finally turn our attention to the role played by the geographical distribution of knowledge, pointing out the importance of 'access to knowledge' and 'equal learning opportunities' as an inequality-decreasing device.

4.1 Studying Learning Strategies

The experiment was set up with $A_N = 14$ agents[3], and with an average number of acquaintances (n) of four. The number of time periods of the experiment was fixed as $N = 100$. As mentioned earlier, in every cycle each player had the opportunity to contact another individual, by choosing from her/his acquaintance list, and to choose one bit of knowledge to request from that individual. Independently of the network structure the experimental agents always reached the steady state within the time frame of the experiment. In Figure 3 we present the learning dynamics (both

[3] We are in negotiation to scale up this investigation.

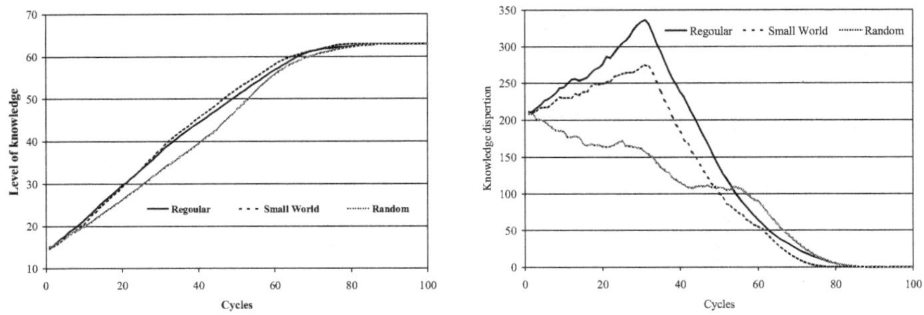

Fig. 3 Small world network, learning dynamic

in terms of learning speed and knowledge dispersion) for the three experiments carried out.

As a first step we shall focus our analysis on the overall picture. As we can see all 14 players converge to the steady state in a time frame of less than 90 cycles, gaining on average 40 nodes in 90 cycles (hence learning almost at an average speed of half node per cycle). Moreover, the knowledge variance grows in the short term in two of the three network structures (i.e. in the small world and in the regular network), showing an initially unequal learning process. After the first 35 cycles it starts converging towards the zero-variance steady state. A possible interpretation of this behaviour is that after 35 cycles the most knowledgeable agents have reached the highest possible level of knowledge (i.e. they have saturated their cognitive map) and therefore the system variance starts decreasing (we will come back to this point in section 4.2).

Another interesting result (and somehow counterintuitive) is that the random network is the worst performer in terms of speed of learning. However, we shall leave the interpretation of this result to the next section when we will tackle the issue of the impact of different network architectures over knowledge diffusion, looking more closely at experimental and simulation results as well as comparing small scale systems with much larger systems.

In order to elicit the learning strategies followed by experimental agents we studied the cognitive maps of each player as well as used questionnaires asking them how they decided the node to be learned and the acquaintance to contact. For the former we allowed a choice of three possibilities: random strategy, width-first strategy or depth-first strategy (we also provided an 'other' checkbox). The depth-first strategy would entail a specialised learning process (i.e. horizontal learning), whereas a width-first strategy would entail the acquisition of less specialised knowledge first (i.e. vertical learning).

An example may be useful to clarify the different learning strategies. Let us assume that a subject has the following cognitive map: $\{1, 2, 3, 4, 5\}$ (see Figure 4). In the first cycle she/he can choose the new bit of knowledge to be learned from a set of six possible nodes: $\{6, 7, 8, 9, 10, 11\}$. We shall maintain that if she/he asks one of the following nodes $\{6, 7\}$, then she/he adopts a vertical strategy (hence, learning the least specialised knowledge first). On the contrary, if she/he tries to learn $\{8, 9, 10, 11\}$, then she/he adopts a horizontal strategy (hence, she/he is specialising). Finally, we shall note that if our subject's strategy shows some kind

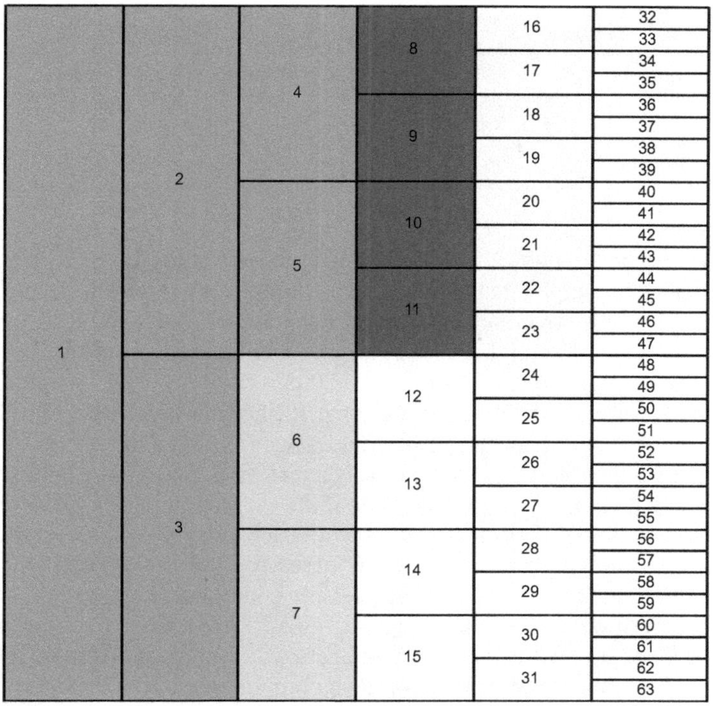

Fig. 4 Cognitive Map and learning strategies

Table 1 Different node selection strategies

	Questionnaires		CM investigation	
	share	percentage	share	percentage
width-first strategy	10	27.0%	24	57.1%
depth-first strategy	2	5.4%	5	11.9%
random strategy	25	67.6%	13	31.0%

of 'time inconsistency' (i.e. it changes over time), then we will classify her/him as a random strategy follower.

It is interesting to compare the results obtained studying the learning strategy actually adopted by experimental players with the results obtained through a questionnaire in which players were asked to state the learning strategy they adopted.

The figures obtained through direct investigation of players' CMs seem to support rather strongly the hypothesis that subjects try to learn in a vertical way: 57% of the subjects always adopted a horizontal strategy, whereas only 12% of the players consistently adopted a vertical one. Confronting these results with those obtained through our questionnaire shows that most of the players who adopted a width-first strategy were unable to categorise their behaviours as such. Hence, we can conclude that they were unconsciously following a particular strategy.

In part based on this analysis, we designed the artificial learning strategies for both node and acquaintance selection. These are summarised in Table 2. Compar-

Table 2 Possible Learning Strategies

Node selection strategy	Acquaintance selection strategy
random strategy	random strategy
width first strategy	sequential strategy
depth first strategy	preferential model strategy

ing the results of the questionnaires and the analytic study of CMs, we defined three strategies for acquaintance selection: "random" or zero-intelligence (ZI), intelligent "sequential" and one based on the adaptation of the relative "strength" of connections, the preferential model strategy, which was shown to be efficient in previous works (Morone and Taylor 2004, 2005).

The "sequential" strategy requires some further clarification. Following this strategy, in the first cycle the agent would select an acquaintance to contact at random. Depending on the outcome of this interaction, the agent would select the same acquaintance in the following cycle if the interaction was successful. If the previous interaction was unsuccessful however, the agent would revert to random selection from among acquaintances. We therefore have three possible strategies for each option, making a total of nine possible combined strategies.

A first step towards understanding the impact of learning strategies involved initialising the model with all agents homogenously following the same combined strategy, and tabling the outcomes for each combination under different network configurations. In doing so, it was found that in all cases the "width-first" strategy was the most efficient in terms of generating the highest average level of knowledge. Further, it was found that in the case of "sequential" acquaintance selection, and "width-first" node selection, the artificial agents outperformed the experimental players by some margin.

In contrast, the "random" acquaintance selection strategy performed very poorly. Some of the results are less easy to explain. For example "depth-first" node selection performed far better upon the random network than upon the other two networks. Maybe (as we discuss later) some of these peculiarities can be explained by the particular instantiation of the experimental network.

None of the nine simulations replicated closely the experiments under all three network configurations. However the "width-first"/ "sequential" combination did match the experiment quite closely under the small-world configuration and the regular configuration. There was greater discrepancy under the random network configuration: here the diffusion pattern of the experiment looked quite different to any of our simulation results.

In Figure 5 we present a comparison of the experimental learning dynamics over a "width-first"/ "sequential" combined strategy. The simulation results were obtained by averaging batches of 100 runs, all initiated with the same parameters and the same configuration. As already discussed, the simulation result replicated quite closely the experimental results, both in the case of the regular network and in the case of the small world network.

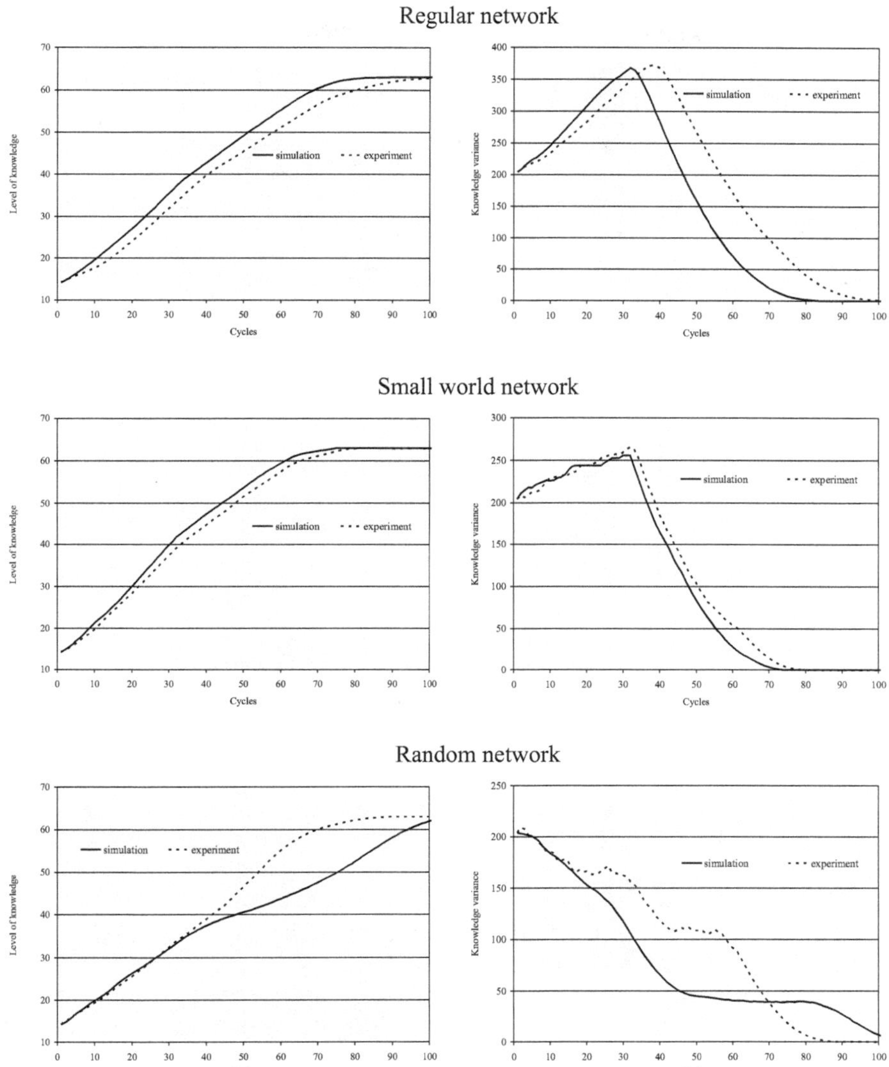

Fig. 5 Learning dynamic comparison (experimental results vs. simulation results)

4.2 Studying Network Structure and Geographical Distribution

We can now start investigating the role played by different network architectures in the learning process. In order to properly study the network effect, we shall both look at experimental and simulation results obtained using small networks (i.e. 14 agents), as well as look at new sets of results obtained by running much bigger simulations. This is to avoid the aforementioned problem related to the system's size.

However, we shall first go back to the experimental results in order to get further insights on the different learning dynamics. In section 4.1 we observed that

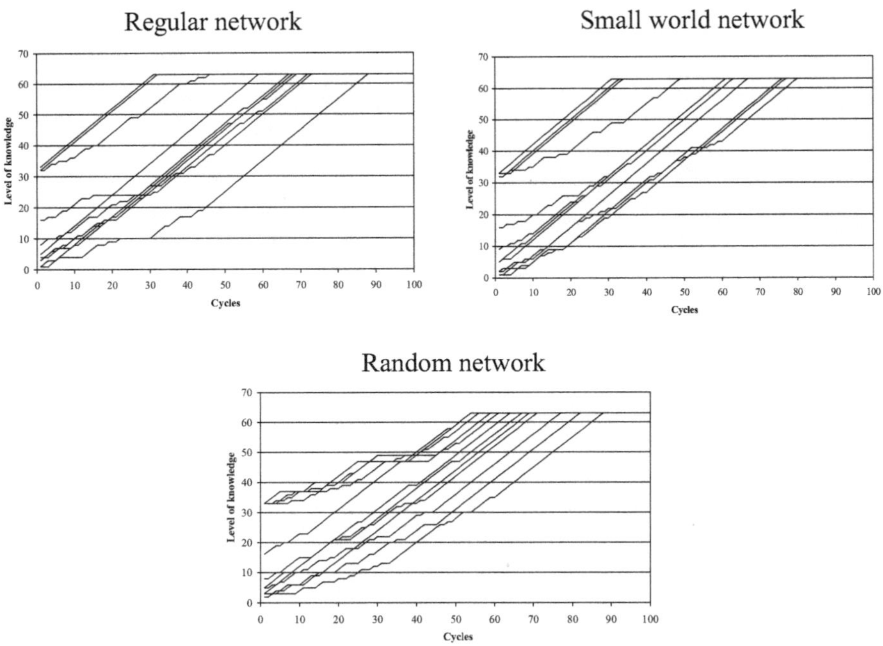

Fig. 6 Experimental individuals learning dynamics

the system always reached the long-run steady state converging, within 90 cycles, towards a stable equilibrium. We also observed that after 35 cycles the variance of the system started a descending trajectory in two of the cases examined (i.e. small world and regular network), whereas in the third network analysed the variance was constantly decreasing. In order to shed some light on this peculiar pattern we will analyse the individual learning trajectory for each of the 14 agents.

Our earlier prediction is confirmed by this analysis: the decrease in system variance of regular and small world networks after 35 cycles is imputable to the fact that after this time frame the most knowledgeable agents (i.e. those who are initially endowed with a higher level of knowledge) reach the steady state and, therefore less knowledgeable agents start closing the gap. The fact that in the short-run transition the system variance is increasing in the regular and small world networks, is due to the faster learning dynamic of knowledgeable agents clearly observable in Figure 6 above.

With respect to different network architectures and overall system behaviours another interesting result (already pointed out in section 4.1) was that the random network underperformed (in terms of average speed of knowledge diffusion) if compared with the other two network architectures (see Figure 3). This result appears counterintuitive, as the random network is also the system with the shortest average path length and therefore is expected to perform better in terms of speed

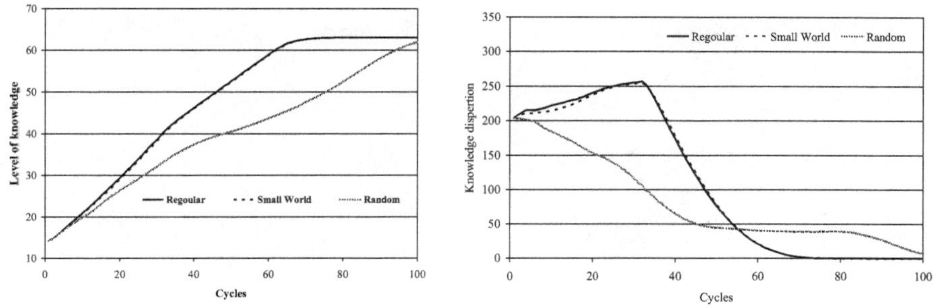

Fig. 7 Comparing network performances using simulation results (average over batches of 100 simulations)

of knowledge diffusion. However statistically significant[4], this result comes from a single experiment and therefore cannot be considered robust.

In order to corroborate this finding we then ran 100 repetitions of each calibrated simulation, using the same initial parameters, network configuration and agent distribution that were used in the experiments.

As shown in Figure 7, the simulation findings agree with experimental results: the random network always underperforms compared with regular and small world networks.

The answer to this apparently odd result lies in the different geographical distribution of agents obtained in the random network. Looking at the three network's structures, we clearly see how the most knowledgeable agents tend to be clustered together in the small world and in the regular network, whereas the same agents are quite disconnected in the random network (Figure 8). In light of these observations we could expect to see knowledgeable agents learning faster in the first two networks as opposed to the random one. This observation was corroborated when looking at individual learning patterns: the learning ability of the most knowledgeable agents in the random network was far less efficient than that of the same agents in the other two network structures. This gap in the performance was due to the fewer learning opportunities available to knowledgeable agents in the random network.

This result leads us to conclude that learning dynamics are not just affected by the network structure and the learning strategy, but also (and perhaps mainly) by the learning opportunities provided to different agents in the network.

In order to test independently the effect upon learning dynamics of the network structure and of the geographical distribution of agents we ran batches of 100 simulations for each network, always reallocating the agents in different ways. Then, we computed the average performance of each network hence clearing out the geographical effect. The results obtained in this way show, in fact, a different picture: the random network structure no longer performs the worst in terms of speed of learning, but actually appears to be the most efficient network. On the other hand, the relative performance of small world and regular network is very similar in terms of speed of convergence towards the long-run steady state.

[4] We calculated t tests under the alternative hypothesis of the random network knowledge mean value being greater than the knowledge mean value of the other two networks. In both cases the alternative hypothesis was rejected with a significance level smaller than 1%.

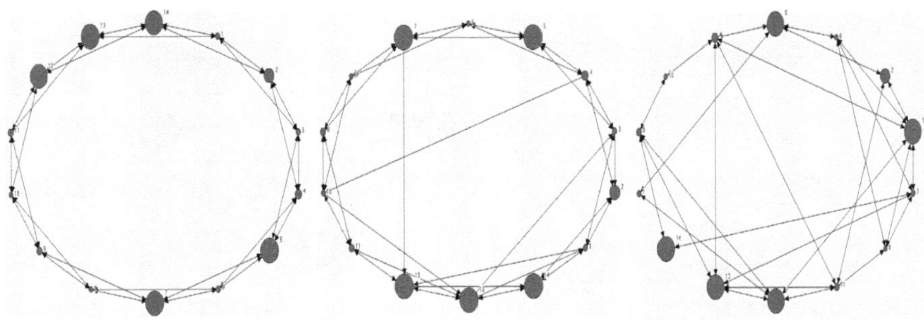

Fig. 8 Experiment network structures: Regular network, Small world network and Random network

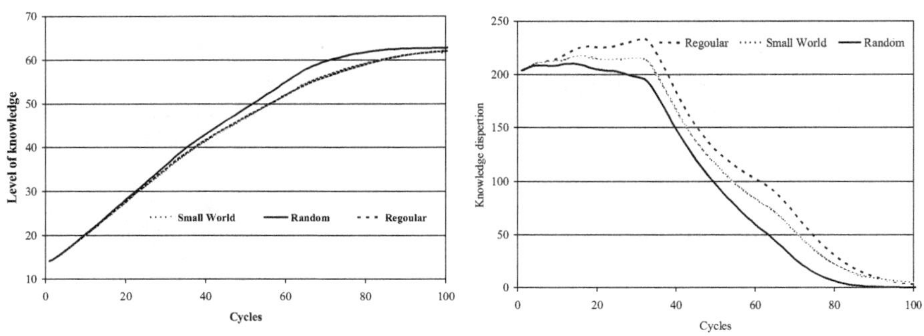

Fig. 9 Different networks, learning dynamic comparison (average over batches of 100 simulations)

From a distributional point of view the random structure shows almost a constant decrease in the knowledge variance in the short-term transition towards the natural equilibrium of the model. This contrasts with small world and regular architectures, which display a general inequality increase over the short run (Figure 9).

What emerges from this analysis is that *the small world is not the most efficient nor the most unequal network*. This is a predictable result as in diffusion models there are obvious benefits to short path length, whereas cliquishness is not directly related to the speed of diffusion. However, this finding contrasts with earlier findings in the literature on knowledge diffusion. In fact, as anticipated in section 2, Cowan and Jonard (2004) found opposite results. Their model (in which knowledge flows by means of barter exchange) shows that the long-run average level of knowledge is a non-monotonic function of p, with a clear peak in the small world region. More precisely, their system performs best at $p = 0.09$.

In the authors' words, "a clear implication of this result is that the average path length of a network of relationships is actually not an unequivocal measure of the performance of this structure. Diminishing the distance between members of an organization or economic system by reallocating links does not always improve performance: the architecture of links matters or, put another way, there is value to cliquishness" (2004: 1564).

A possible explanation for these contrasting results could be the different network size used in different simulations. In fact, our experiment, and subsequently

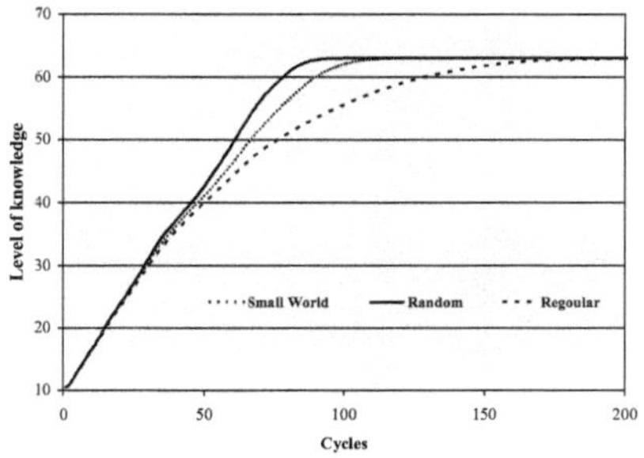

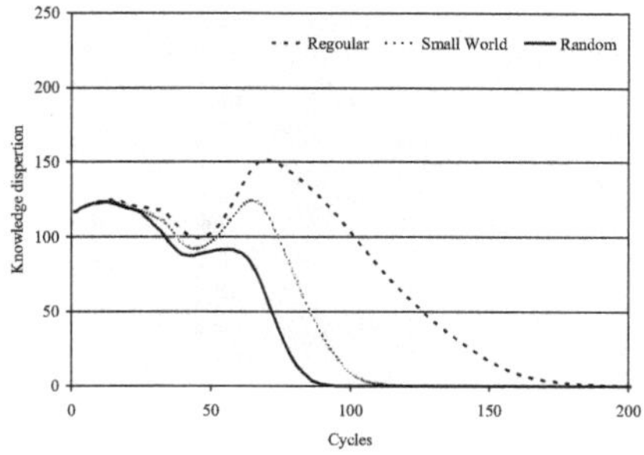

Fig. 10 Different networks, learning dynamic comparison (500 agents' networks, 10 agents' neighbourhoods, average over batches of 100 simulations)

our simulations, were based on a rather small network (i.e. 14 agents) compared to that of Cowan and Jonard (500 agents). According to Watts (1999), small world phenomena can only be observed in large and sparse systems. We can certainly say that, in such small networks, the regular and the random would tend to look more similar in terms of sharing small L and high C values.

Since we do not know how the scaling issue might be affecting the result, and in light of the small-networks limitation upon the experimental setting, we used new batches of simulations with subsequent increases in the size of the network. This allowed us to extend our analysis to much larger networks.

4.3 Studying Scale in Small World Systems

The final stage in the investigation of the model therefore focused upon the impact of system scale upon the result that small world networks are neither the most efficient nor the most unequal. While increasing the size of the network, we also investigated the way in which results are sensitive to neighbourhood size. We ran batches of simulation experiments, significantly increasing the size of the network. In the first set of simulations, we increased the number of nearest neighbours $n \in \{4, 6, 8, 10\}$ with 500 agents. This latter simulation set up replicates exactly the Cowan and Jonard case. In the second set we varied the number of agents $A_N \in \{50, 100, 500\}$ using the number of acquaintances as before.

As a first result, we observe that much bigger networks displayed similar results to those obtained with 14 agents: random networks being consistently the best performing in terms of speed of knowledge diffusion as well as in terms of knowledge distribution, followed by small world networks and finally by regular networks. This result is confirmed regardless of the neighbourhood size. As already mentioned, by setting the neighbourhood size equal to 10 we replicate exactly the network structure employed by Cowan and Jonard (2004); however, as showed in Figure 10, the random network over performs the small world and regular networks. In terms of the variance of the distribution, we confirm again our earlier results with the random network being constantly more equal than small world and regular systems. We can now confidently state that *the small world is not the most efficient nor the most unequal network.*

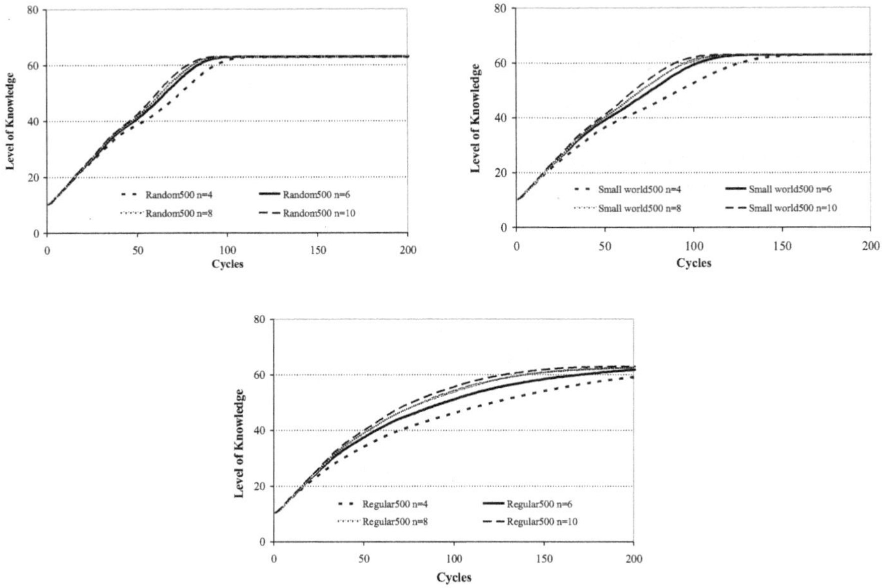

Fig. 11 Comparing neighbourhood size and speed of convergence (4, 6, 8 and 10 agents' neighbourhoods, average over batches of 100 simulations)

A second, not surprising, result is that in increasing the neighbourhood size, the speed of convergence increases.

Finally, we investigate possible scale effects. Looking at Figure 12, an interesting observation can be drawn: larger networks show a faster converging process (i.e. knowledge flows faster). This is indeed an interesting result as it shows that, other things being equal, the system scale is a key determinant of the relative speed of knowledge diffusion, with larger networks being more efficient. This is true in two cases out of the three analysed here: we observe a scale effect both for small world and random architectures. From this result we can draw some interesting conclusions. First, there is no scale effect in regular networks. This is probably due to the fact that, in a static regular network, agents act only locally and are therefore not influenced by the network size. Second, small world and random networks are characterised by short average path length and are hence more affected by network size. This could be due to the fact that combining short average path length with large sized networks might generate higher *learning opportunities* for involved agents.

It is also interesting to observe that increasing the network size does not necessarily imply a constant increase in network efficiency. In fact, shifting from a network with 50 agents to a network with 100 agents produces an increase in efficiency; however, shifting from 100 agents to 500 agents has virtually no effect on the speed of diffusion. This suggests that there is a critical network size above which there are no scale effects, whereas shifting from relatively small networks to bigger ones generates (for small world and random architectures) an improvement in the diffusion performance.

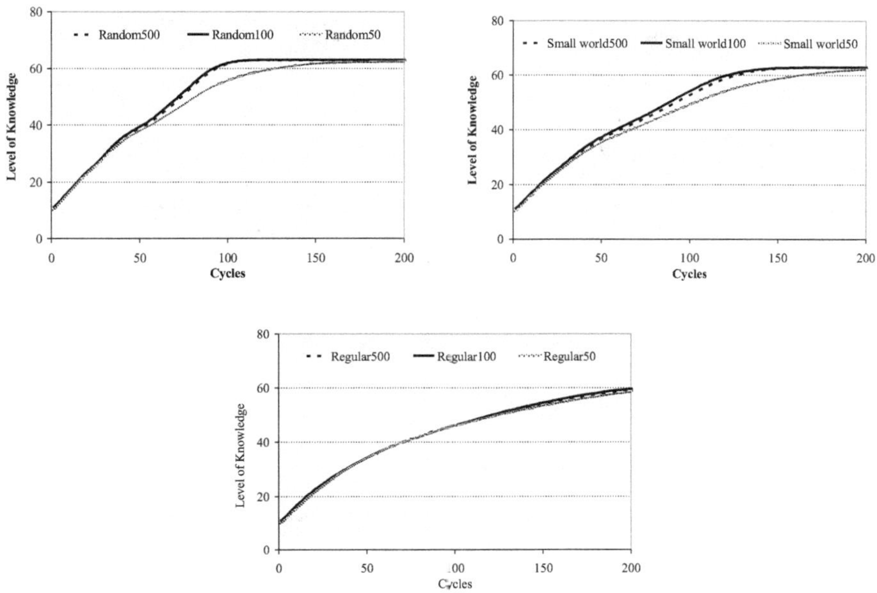

Fig. 12 Comparing size of networks and speed of convergence (500, 100 and 50 agents' networks, average over batches of 100 simulations)

5 Concluding Remarks

In this paper we investigated the dynamics of knowledge diffusion in a small population of agents. As stated in the introduction, the aim of the paper was to present an original contribution to the debate on informal learning, combining a laboratory experiment designed to reproduce complex learning dynamics with simulation experiments to investigate learning strategies. In this sense, we share the view that "experimental economics and agent based modelling jointly are the 'laboratory' of the economic science" (Pajares et al. 2004). Departing from this analysis, we further explored the behaviour of the model by simulating over a much larger range of parameter settings than would be possible with laboratory methods, thereby extending the analysis of the influence of network factors upon knowledge diffusion patterns. Employing the experimental methodology constrained us to use initially small size networks. This represented a problem as with small systems it was arguable the presence of small world characteristics. We overcame the problem by contrasting always experimental results with simulation results obtained with much larger systems. However, a clear advantage of studying small networks in the laboratory was that it allowed us to easily trace the paths of agents' interaction patterns and learning processes.

While conducting this investigation, several elements arose as key determinants of flow dynamics within a closed network. Namely these factors were: (1) the learning strategies adopted by heterogeneous agents; (2) the network architecture within which the interaction took place; (3) the geographical distribution of agents and their relative initial levels of knowledge; (4) the network size.

Concerning the learning strategy, we were able to identify a set of nine possible *combined learning strategies* which allowed us to investigate the two actions that each player had to undertake while attempting to acquire knowledge. Any time an agent tried to learn a new bit of knowledge, she/he had to make two choices: first, decide what to learn (this decision was constrained to the set of possible nodes which could be articulated with previous knowledge); second, decide with whom to interact (also this decision was bounded to the set of acquaintances in the agent's social network). One conclusion was that when agents pursued the strategy of "width-first" learning (i.e. non-specialisation) knowledge diffusion was much faster. Secondly, our artificial agents following this simple strategy were able to outperform the experimental agents: clearly many of our subjects were playing the "game" in a very sub-optimal way.

The second step in our investigation was examining those factors affecting knowledge flows. A preliminary result of our investigation showed that learning dynamics is heavily affected by the *learning opportunities* provided to each agent in the network. By learning opportunities we mean the chances each agent has got to interact with more knowledgeable agents. In other words, a particular geographical distribution of agents (endowed with different knowledge) could substantially affect learning dynamics. In order to test independently the effect upon learning dynamics of the network structure and of the geographical distribution of agents, we ran batches of simulations for each network's structure, while reallocating the agents in different ways. Then, we computed the average performance of each network, hence clearing out the geographical effect. Once corrected for any possible geographical bias we could conclude that small world networks do perform better than regular networks, but consistently underperformed when compared with random networks. This finding contrasts with previous literature (Cowan and Jonard 2004), which maintained that the small world network is the most efficient (as well as the most unequal) system. In our research we have found a clear counterexample to Cowan and Jonard's claim. Moreover, this finding proved to be consistent over system sizes of different orders of scale. Eliminating the question of network scale and neighbourhood size, we infer that the different outcomes are due to the way the agents and their interaction processes are defined. It leads us to conclude that it is very difficult to derive and generalise optimal configurations when the modelling environment changes.

A further result obtained in this investigation, which is to our knowledge an original finding, was that bigger networks over perform smaller ones. However, the size-performance relation (which applies only to small world and random networks) is not monotonic, being true only when shifting from small networks to medium/large networks.

In conclusion, investigating the relative effect of each and every of the four factors, we can maintain that studying the nexus between structure and flows is a rather complex task which involves a large number of aspects that concur to define flow dynamics. As a suggestion for further research we propose to make an attempt to produce a clear taxonomy of all the factors that might affect knowledge flows that occur in social networks. It may be interesting to divide these factors into some broad categories. As a pure exercise we could suggest *structural* and *individual*, the first one including all those factors directly referring to the network architecture and size, and the second one referring to individual decisions such as strategy decisions.

References

Acemoglu D, Angrist J (1999) How Large are the Social Returns to Education? Evidence from Compulsory Schooling Law. NBER, Working Paper
Akerlof G (1997) Social distance and social decisions. Econometrica, 65: 1005–1027
Ancori B, Bureth A, Cohendet P (2000) The economics of knowledge: The debate about codification and tacit knowledge. Ind Corp Change, 9, 2: 255–287
Anderlini L, Ianni A (1996) Path dependence and learning from neighbours. Games Econ Behav, 13: 141–177
Bala V, Goyal S (1995) A theory of learning with heterogeneous agents. Int Econ Rev, 36, 2: 303–323
Bala V, Goyal S (1998) Learning from neighbours. Rev Econ Stud, 65: 595–621
Bénabou R (1993) Working of a city: Location education and production. Quart J Econ, 58: 619–652
Berningaus S, Schwalbe U (1996) Evolution, interaction, and Nash equilibria. J Econ Behav Organ, 29: 57–85
Brock W, Durlauf S (1995) Discrete Choice with Social Interaction I: Theory. NBER Working Paper 5291
Chwe MSY (2000) Communication and coordination in social networks. Rev Econ Stud, 67: 1–16
Clark G (1984) Innovation Diffusion: Contemporary Geographical Approaches. Geo Books, Norwich
Cowan R, Jonard N (2004) Network structure and the diffusion of knowledge. J Econ Dynam Control, 28: 1557–1575
Dosi G, Marsili O, Orsenigo l, Salvatore R (1995) Technological regimes, selection and market structures. Small Bus Econ, 411–436
Durlauf S (1996) Neighborhood feedbacks, endogenous stratification, and income inequality. In: Barnett W, Gandolfo G, Hillinger C (eds.) Dynamic Disequilibrium Modelling, Cambridge University Press, Cambridge
Ellison G (1993) Learning, local interaction, and coordination. Econometrica, 61: 1047–1071
Ellison G (2000) Basins and attraction, long run stochastic ability and the speed of step-by-step evolution. Rev Econ Stud, 67: 17–45
Ellison G, Fudenberg D (1993) Rules of thumb of social learning. J Polit Economy, 101, 4: 612–43
Ellison G, Fudenberg D (1995) Word-of-mouth communication and social learning. Quart J Econ, 110, 1: 93–125
Glaeser EL, Sacerdote B, Scheinkman P (1996) Crime and social interactions. Quart J Econ, 111: 507–548
Goyal S (1996) Interaction structure and social change. J Inst Theoretical Econ, 152: 472–494
Jovanovic B, Rob R (1989) The growth and diffusion of knowledge. Rev Econ Stud, 56, 198-9: 569–82
Morone P, Taylor R (2004) Small world dynamics and the process of knowledge diffusion. The case of the metropolitan area of greater Santiago De Chile. Journal of Artificial Societies and Social Simulation, 7, 2
Morone P, Taylor R (2005) Learning dynamics with complex cognition. In: Pyka A, Hanusch H (eds.) Applied Evolutionary Economics and the Knowledge-based Economy, Edward Elgar, Cheltenham
Nelson R, Winter S (1982) An Evolutionary Theory of Economic Change. Harvard University Press
OECD (1996) Employment and Growth in the Knowledge-Based Economy. Paris
Pajares J, Hernández C, López A (2004) Agent-Based Modelling in Economics: Beyond the Neoclassical Paradigm. proceedings of the 2nd International Conference of the European Social Simulation Association, University of Valladolid - Spain, 16-19 September
Rauch J (1993) Productivity gains from geographic concentration of human capital: Evidence from the cities. J Urban Econ, 34: 380–400

Rogers EM (1995) Diffusion of Innovation. Free Press, New York, 4th edn.
Silverberg G, Dosi G, Orsenigo L (1988) Innovation, diversity and diffusion: A self-organization model. Econ J, 98: 1032–1054
Watts A (2001) A dynamic model of network formation. Games Econ Behav, 34: 331–341
Watts DJ (1999) Small Worlds. Princeton University Press, Princeton
Watts DJ, Strogatz S (1998) Collective dynamics of small-world networks. Nature, 393: 440–442

Aykut Lenger · Erol Taymaz

To innovate or to transfer?

A study on spillovers and foreign firms in Turkey

Abstract FDI has been considered by many development economists as an important channel for transfer of technology to developing countries. It is suggested that modern, advanced technologies introduced by multinational firms can diffuse to domestic firms through spillovers. In this paper, we study innovation and technology transfer activities of domestic and foreign firms in Turkish manufacturing industries, and the impact of horizontal, vertical and labor spillovers on these activities. Our analysis shows that foreign firms are more innovative than their domestic counterparts, and transfer technology from abroad (mostly from their parent companies). Horizontal spillovers from foreign firms seem to be insignificant. The effects of foreign firms on technological activities of other firms in vertically related industries are ambiguous. High-tech suppliers tend to have a high rate of innovation when the share of foreign users is high, but the opposite is true for users: high-tech users supplied mainly by foreign firms tend to have a lower rate of innovation. Labor turnover is found to be the main channel of spillovers. Our findings reiterate the importance of tacitness of knowledge, and confirm that technology cannot easily be transferred through passive mechanisms.

Keywords FDI · Innovation · Technology transfer · Spillovers · Productivity

JEL Classification O14 · O33 · O31 · F23

Earlier versions of this paper were presented at the Schumpeter 2004 Conference (Università Bocconi, 9–12 June 2004) and the Conference in Honour of Keith Pavitt (University of Sussex, 13–15 November 2003). We thank our discussants, Robin Cowan and Martin Bell, conference participants and an anonymous referee of this journal for their valuable comments. Giovanni Dosi let Aykut Lenger benefit from the excellent research environment he has created at the Sant'Anna School of Advanced Studies. We appreciate very much his hospitality and his suggestions for the paper. This paper depends on a background report prepared for the Technology Capability Project that was carried out by the Scientific and Technical Research Council of Turkey.

A. Lenger (✉) · E. Taymaz
Department of Economics, Ege University, 35040 Izmir, Turkey
E-mail: lengera@metu.edu.tr, etaymaz@metu.edu.tr

1 Introduction

There are two strands of the literature in development economics that have attracted substantial interest in the last couple of decades: the importance of technological change for long term economic growth and the role of foreign direct investment (FDI) in the process of economic development. Studies on technological change emphasize the fact that innovation (the development of new products, processes and organizations) is basically an interactive process. Recent advances in science and technology have led to, on the one hand, an increase in the knowledge content of products and processes, and, on the other hand, an appreciation of the importance of generic technologies that can be used in various products and processes. These two processes, which form two sides of the same coin, have increased the need to extend the knowledge base of industrial firms. As Rosenberg suggested 20 years ago, the process of innovation cannot fit into the boundaries of a single firm. Firms can now innovate only within an intensive web of interactions with other firms (suppliers, buyers, and, even, competitors), consumers, research institutions, etc, i.e., they can be innovative, and, thus, competitive, only if they can form and be part of innovation networks (for a small group of studies, see Lundvall 1988; Nelson and Rosenberg 1993; Smith 1995; OECD 1999, 2000).

FDI has been considered by many development economists as an important channel for the transfer of technology to developing countries. It is suggested that modern, advanced technologies introduced by multinational firms can diffuse to domestic firms through spillovers (imitation, demonstration effects, training local labor, vertical technology transfers, etc.). However, empirical studies show that the net benefits the host country can enjoy from FDI depend on host country characteristics, such as industry and the policy environment (Blomström and Kokko 1998), the level of human capital stock (Borensztein et al. 1998; Noorbakhsh et al. 2001), and the absorptive capacity of domestic firms (Kinoshita 2001).

This paper contributes to the existing literature by presenting new evidence on the interactions between domestic and foreign firms engaged in technological activities in Turkish manufacturing industries. The aim of this paper is to analyze technology acquisition decisions and to test the impact of various types of spillovers on technological activities (innovation and technology transfer activities).

First, we model and estimate the determinants of two types of technology acquisition, innovation and technology transfer, and test whether foreign ownership matters for technology decisions. We study the determinants of technology transfer because policy-makers in Turkey since the early 1980s have consistently claimed that FDI is an important channel for transfer of technology from abroad, and have introduced various measures to attract foreign capital. However, there is no comprehensive study that analyzes the contribution of foreign firms in transferring technology from abroad, and its subsequent diffusion within the manufacturing industry through spillovers. In this paper, we test whether foreign firms are more likely to transfer technology from abroad, and whether they have any impact on the technology transfer decisions of domestic firms. Since earlier studies have shown that innovative activities by domestic firms are essential to building technological capabilities and to becoming competitive in international markets (for the Turkish case, see Özçelik and Taymaz 2004), we also look at the determinants of innovative activities, and test the hypothesis that foreign firms tend to be more innovative than domestic firms.

Second, we identify the effectiveness of different types of spillovers in enhancing domestic firms' innovative and absorptive (technology transfer) capability (see also, Reger 1998; Smith 1995). We focus on three types of spillovers:

- *Horizontal spillovers* (spillovers from foreign firms to others operating in the same industry or in the same region)
- *Vertical spillovers* (spillovers from foreign firms to others operating in vertically related industries, i.e., from foreign suppliers to domestic users, and from foreign users to domestic suppliers, all located in Turkey)
- *Labor spillovers* (spillovers through labor turnover, i.e., employment by domestic firms of workers who worked for foreign firms)

Following Pavitt's warning on the importance of inter-sectoral differences in technological activities (see Pavitt and Patel 1999), we analyze the effects of spillovers for low-technology and medium- and high-technology industries separately.

Finally, we estimate productivity equations for low-technology and medium/high-technology industries to observe the impact of innovation and technology transfer activities on productivity in these two types of industries.

The paper is organized as follows: the second section provides background information on FDI in Turkey, and presents the data on productivity differentials between foreign and domestic firms in low-technology and medium/high-technology industries. The differences in innovativeness between foreign and domestic firms are analyzed as a possible factor behind productivity differentials. The data source, variables and models used to test the impact of spillovers on innovation and technology transfer activities are explained in Section 3. Estimation results are presented in Section 4. The last section summarizes basic findings and implications of our analysis.

2 Foreign direct investment and productivity differentials

Turkey introduced the first legislation governing foreign investments in the early 1950s. The Foreign Capital Law, enacted in 1954, and the related Decree of the Council of Ministers remained in force until the late 1980s. Although this early legislation provided a liberal framework designed to create a favorable environment for FDI, the cumulative FDI authorized from 1950 to 1980 reached only 229 million USD (Öniş 1994). Restrictive bureaucratic practices were blamed for the low level of FDI in Turkey in the pre-1980 period (see, for example, Erdilek 1982).

Turkey had to abandon the import substitution industrialization strategy followed in the 1960s and 1970s after the severe balance of payments crisis in the late 1970s. On January 24, 1980, the Turkish government announced a stabilization program that was implemented under the military regime after September 1980. The new program was based on an outward-oriented trade strategy and foreign trade, product, and, later, capital markets were liberalized to a large extent (for a comprehensive overview of the Turkish economy, see Kepenek and Yentürk 2000). The administrative system regulating FDI was reorganized in the early 1980s and all discriminatory treatment of foreign investors, requirements on local equity participation, and restrictions on the transfer of earnings were gradually eliminated (Erdilek 1986; Akpınar 2001).

The share of foreign firms[1] in the total number of private firms in the manufacturing industry was about 1% in 1983, but it increased continuously up to 2% in 1999, and 3.5% in 2000 through acquisitions and entry.[2] The share of foreign firms in private manufacturing employment was about 6%, with 50 thousands people employed by foreign firms in 1983. The employment share of foreign firms increased gradually, especially after 1988, and reached 11% in 2000.

Foreign firms prefer to invest in medium- and high-technology industries:[3] their share in value added increased continuously from about 25% in the mid-1980s to almost 50% in the late 1990s (see Fig. 1). On the other hand, the share of foreign firms in low-tech industries increased gradually until the mid-1990s and stabilized at around 13–15% afterwards. In other words, the increase in foreign investment in manufacturing since the mid-1980s is mainly due to the attractiveness of high-tech industries.[4]

Foreign firms are on average more productive than domestic firms, and the productivity[5] differential is much wider in low-tech industries: foreign firms in low-tech industries have been 2.5–3 times more productive than their domestic counterparts (Fig. 2). High-tech foreign firms in Turkey have been two times more productive, but the productivity differential has widened in recent years and reached 150% (2.5 times) in 2000.

We use US productivity data in order to compare the productivity of foreign firms with the "best practice" productivity. It is interesting to observe that low-tech foreign firms in Turkey are as productive as US firms, but high-tech foreign firms in Turkey lag behind their US counterparts.

Why are foreign firms more productive than domestic firms? Technological differences between foreign and domestic firms can explain a part of this productivity differential.[6] If foreign firms use superior technologies and if they are more innovative so that they maintain their technological superiority over domestic firms, persistent productivity differentials may arise.

---

[1] Following the usual convention, "foreign firms" are defined as those joint ventures in which foreign ownership is 10% or more. If the foreign share is less than 10%, it is considered to be portfolio investment. Joint ventures with more than 50% foreign ownership are "majority-owned foreign firms."

[2] The data refers to all private establishments employing ten or more people, and all public establishments. The statistical unit is the "establishment" which is the main decision-making unit.

[3] We use OECD's definition of low-, medium- and high-technology industries. Since the number of firms operating in high-technology industries is small, medium- and high-technology industries are grouped together, and defined as "high tech."

[4] After the elimination of local equity participation and minimum export requirements in 1986 (Öniş 1994: 96), majority-owned foreign firms realized a rapid growth in their valued added share. Therefore, all the expansion in value added share since the late 1980s was achieved by majority-owned foreign firms, whereas minority-owned foreign firms (with equity participation within the 10–50% range) kept their shares almost constant. Cieslik and Ryan (2002) also found a similar shift from minority-owned joint ventures in Central and Eastern Europe in favor of wholly-owned foreign firms.

[5] Productivity is defined as "labor productivity," i.e., value added per employee. Current exchange rate is used for currency conversion.

[6] Another important reason behind the productivity differential between domestic and foreign firms is the size differential. Foreign firms are, on average, 2.5 times larger than domestic ones. Note that the productivity differential between large (employing at least 150 people) and small (employing 10–149 people) firms in low-tech industries was about 60% in the late 1990s whereas the same rate was almost 150% in high-tech industries.

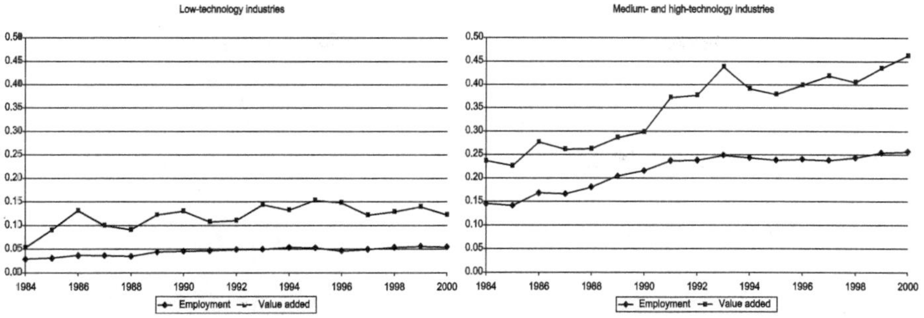

Fig. 1 Share of foreign firms in manufacturing employment and value added in Turkey, 1984–2000

Table 1 summarizes the data on the innovativeness of domestic and foreign firms in the periods 1995–1997 and 1998–2000 for low-tech and high-tech industries. It is interesting to observe that there is almost no difference in terms of product innovations between domestic and foreign firms in low-tech industries. For example, only 11.2% of domestic firms introduced any product innovation in the period 1995–1997, whereas the proportion of foreign firms that introduced product innovations in the same period is even slightly lower (9.1%). The proportion of innovative firms has increased in the second time period (1998–2000), but the difference between domestic and foreign firms is not significant. Foreign firms in low-tech industries seem to have become more successful in process innovations than their domestic counterparts in the second time period.

Firms operating in the high-tech industries are almost two times more innovative than firms operating in low-tech industries, and foreign firms in these industries are undoubtedly superior to domestic firms in innovativeness. The data provide strong evidence that support the argument that domestic firms are technologically weaker than foreign firms in high-tech industries.

The relative importance of product and process innovations differs in low-tech and high-tech industries, and the ownership of the firm matters for the type of innovation. Process innovations are more common than product innovations in low-tech industries. Moreover, foreign firms put more emphasis on process innovations than do domestic firms. Since low-tech industries tend to have "mature" product technologies, process innovations are likely to play a more important role for productivity and competitiveness, where foreign firms seem to have a com-

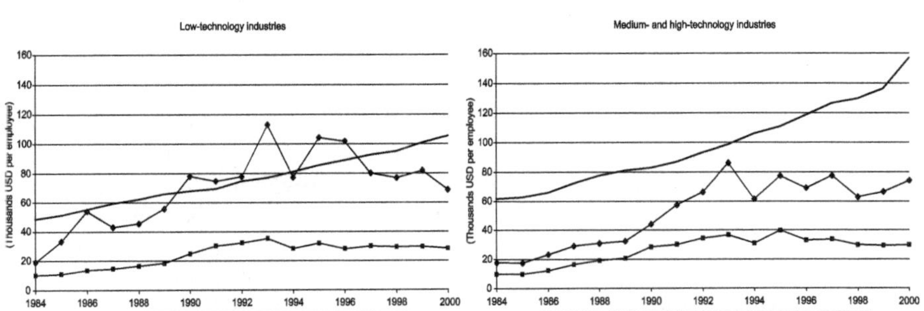

Fig. 2 Labor productivity, domestic and foreign firms in Turkey, and US firms, manufacturing industries, 1984–2000

Table 1 Innovativeness of domestic and foreign firms, 1995–1997 and 1998–2000 (proportion of innovative firms)

	1995–1997		1998–2000	
	Domestic firms	Foreign firms	Domestic firms	Foreign firms
Product innovations				
Low tech	0.112	0.091	0.143	0.162
Medium and high tech	0.278	0.526	0.325	0.601
Process innovations				
Low tech	0.159	0.163	0.193	0.387
Medium and high tech	0.280	0.453	0.279	0.483
Innovative (product and/or process innovations)				
Low tech	0.191	0.169	0.250	0.425
Medium and high tech	0.378	0.563	0.419	0.680
Product/process innovators ratio				
Low tech	0.704	0.558	0.741	0.419
Medium and high tech	0.993	1.161	1.165	1.244
n				
Low tech	1301	68	1391	83
Medium and high tech	646	79	770	94

Source: SIS, Innovation Surveys, 1995–1997 and 1998–2000

petitive advantage over domestic ones. The product/process innovator ratio is much higher in high-tech industries than in low-tech industries, and foreign firms have an even higher ratio of product-to-process innovations. This finding supports the perception that high-tech industries play a leading role in developing new products.

Since foreign firms, on average, are more innovative than domestic firms and they are more productive (possibly due to their superior technologies), there could be spillovers from foreign to domestic firms. We identify three types of (or mechanisms for) spillovers: horizontal, vertical and labor.

Horizontal spillovers flow between firms operating in the same industry or in the same region. Horizontal spillovers arise as a result of imitation (demonstration effects, reverse engineering, etc) or competitive pressure exerted by foreign firms. Horizontal spillovers can be observed in the same industry (for example, in the case of industry-specific technologies), or in the same region if geographical proximity is important.

Vertical spillovers refer to the transfer of technology through the transfer of embodied technology, information exchange (disembodied technology) or imitation between vertically related firms (suppliers/users). They can flow from suppliers to users when the supplier provides new or improved machinery, equipment or intermediate goods that enable product and/or process changes. However, competent, demanding users can also provide valuable technological information to suppliers, and can force them to improve technologically (for an early study on information flows between users and suppliers, see Lundvall 1988).

Labor spillovers take place when technology is transferred from one firm to another through the employment relationship. If a worker, employed in a technologically superior firm, moves to another one, he can transfer, at least, a part of that technology. Labor turnover could be an important mechanism for spillovers especially when the technology is tacit, so that it is difficult to be imitated and transferred through other means.

3 Technological activities and spillovers: the model

As noted in the previous section, foreign firms in Turkish manufacturing industries are more productive and innovative than their domestic counterparts. Differences in foreign and domestic firms may lead to spillovers in various forms. In this section, a model for estimating the impact of horizontal, vertical and labor spillovers on technological activities is presented. The analysis takes into account two types of technological activities: in-house innovative activities and technology transfer (from abroad).

The choice to be innovative and to transfer technology depends on a number of firm- and sector-specific factors:

$$INNO_{it} = \alpha_0 + \sum \alpha_j x_{ijt} \quad (1)$$

$$TECHNO_{it} = \beta_0 + \sum \beta_j x_{ijt} \quad (2)$$

$$i = 1, \ldots, n, \ j = 1, \ldots, m$$

where x's are m variables that determine innovativeness (INNO) and technology transfer (TECHNO); t denotes the time period (1995–1997 or 1998–2000); INNO and TECHNO are dummy variables that take the value 1 if the firm is innovative and transferred a technology through license/know how agreement, respectively.

The data on innovativeness were collected by the State Institute of Statistics through two Innovation Surveys following the methodology set by the Oslo Manual (OECD 1997), and the Community Innovation Survey of the European Union. The first survey conducted in 1998 covers the period 1995–1997 and the second one conducted in 2002 covers the period 1998–2000. "Technological innovation" is defined in the questionnaire as "technologically new products and processes or significant technological improvements in products and processes." Innovation is explicitly defined at the firm level, i.e., "innovation occurs when a firm implements a new or improved product or process, which is technologically novel for the firm, not for the market." The response rates were more than 50% in both surveys. The SIS performed a non-response analysis and estimated sample weights for each respondent.

The data on technology transfer come from the Annual Survey of Manufacturing Industries, collected by the SIS. The TECHNO variable is defined as a binary variable, and takes the value 1 if the firm transferred any technology through license or know-how agreements in the period under consideration (1995–1997 or 1998–2000).

A number of variables are defined as proxy for horizontal, vertical and labor spillovers.

Horizontal spillovers We use three variables to capture the effects of horizontal spillovers from foreign firms: the first variable, SFDI, measures the market share[7] of foreign (FDI) firms (for variable definitions and data sources, see Table 2). If there are (sectoral) horizontal spillovers from foreign firms in the form of demonstration effects, imitation, etc., other firms in the same industry may invest in innovative activities to benefit from these spillovers. In a similar way, informational spillovers may make technology transfer more likely. If competition from foreign firms forces other firms to adopt better technologies through innovative activities and/or transferring technologies, the SFDI variable will have a positive coefficient as well. Although foreign firms are, on average, more productive than domestic firms, and, therefore, are likely to generate spillovers for domestic firms, R&D intensive foreign firms are likely to be the main source of spillovers. Therefore, we use two additional variables, SREGRD and SSECTRD, to estimate the impact of spillovers from R&D activities of foreign firms. The SREGRD variable is defined as the ratio of foreign firms' R&D expenditures to total output in the province in which the firm operates, whereas the SSECTRD variable is measured similarly at the (4-digit) industry level. Thus, the SREGRD variable captures regional foreign R&D spillovers, and the SSECTRD variable sectoral foreign R&D spillovers.

Vertical spillovers There are two variables used as proxies for the extent of vertical spillovers: SSUP and SBUY measure the weighted average of foreign market share in supplier and user industries, respectively. These variables are defined as follows:

$$SSUP_i = \sum \omega_{ij} s_j$$

$$SBUY_i = \sum \omega'_{ij} s_j$$

where s_j is the market share of foreign firms in market j, ω_{ij} the jth sector's share in inputs used by the ith sector, and ω'_{ij} the share of jth sector in total consumption of the ith sector's output. Thus, in a sense, SSUP measures the proportion of firm's inputs produced by foreign firms, and SBUY measures the proportion of firm's output used by foreign firms. ω and ω' variables are calculated from the 1996 Input–Output Table. If vertical relations are used to transfer knowledge from foreign firms, these two variables are expected to have a positive impact on technological activities.

Labor spillovers There is an extensive literature that emphasizes the importance of tacit knowledge in technological activities (for a recent review and extensions, see Cowan et al. 2001). Technology is, at least partly, tacit and embodied in people who develop and use it. Therefore, the transfer of workers, formerly employed by for-

[7] The "market" and "industry" are defined at the ISIC (Revision 2) four-digit level.

Table 2 Descriptive statistics (average values for the period 1995–2000)

Label	Variable definition	Unit	Low-tech industries	Medium- and high-tech industries
Innovation and technology transfer				
INNOVAT	Innovativeness	Binary (0/1)	0.25	0.44
TECHNO	Technology transfer	Binary (0/1)	0.01	0.07
FDI spillover variables				
SLABOR	Labor turnover in foreign firms	Percentage	0.01	0.02
SFDI	Market share of foreign firms	Percentage	0.10	0.31
SREGRD	Regional foreign R&D intensity	Percentage (*100)	0.06	0.06
SSECTRD	Sectoral foreign R&D intensity	Percentage (*100)	0.00	0.09
SSUP	Market share of foreign firms in supplier ind	Percentage	0.07	0.12
SBUY	Market share of foreign firms in user ind	Percentage	0.03	0.07
Foreign ownership				
FDI	Foreign-owned firm	Binary (0/1)	0.03	0.06
R&D and R&D spillovers				
RDINT	R&D intensity	Percentage (*100)	0.02	0.22
DREGRD	Regional domestic R&D intensity	Percentage (*100)	0.04	0.05
DSECTRD	Sectoral domestic R&D intensity	Percentage (*100)	0.02	0.09
Other variables				
LTURN	Labor turnover ratio	Percentage	0.14	0.15
INTERNET	Internet intensity	Percentage	0.54	0.71
GROUP	Member of a business group	Binary (0/1)	0.07	0.08
SUBIN	Share of subcontracted inputs	Percentage	0.05	0.03
SUBOUT	Share of subcontracted outputs	Percentage	0.07	0.01
SKILLED	Proportion of skilled employees	Percentage	0.16	0.20
LQ	Output (log)	Million 1997 TL	10.95	10.90
LL	Employment (log)		3.90	3.75
LM	Inputs (log)	Million 1997 TL	10.50	10.28
LE	Electricity consumption (log)	Thousands kWh	12.74	12.39

Table 2 (continued)

Label	Variable definition	Unit	Low-tech industries	Medium- and high-tech industries
LK	Depreciation allowances	Million 1997 TL	7.17	7.17
LRW	Real product wage (log)	Million 1997 TL	4.86	5.35
n	Number of observations		1978	1043

Sources: Innovation and Internet intensity variables: SIS, Innovation Surveys, 1995–1997 and 1998–200
R&D variables, SIS, Annual R&D Surveys, 1995–2000. All other variables: SIS, Annual Surveys of Manufacturing Industries, 1995–2000

eign firms, could constitute an important channel for spillovers. Since there is no data about the flow of workers between firms, we use a proxy variable, SLABOR, to measure the extent of spillovers through labor flows. The SLABOR variable is defined as the ratio of the number of separations (quits and fires) from foreign firms to the total number of employees in a given industry. Therefore, the higher the value of the SLABOR variable, the higher the probability that former employees of foreign firms would be employed by other firms operating in the same industry. If there are spillovers through labor turnover, the coefficient of the SLABOR variable will be positive.

Other variables used to explain innovativeness and technology transfer are as follows:

FDI is a dummy variable that takes the value 1 for joint ventures in which the share of foreign ownership is 10% or more. This dummy variable is used to test whether foreign firms are more innovative and/or whether foreign firms are more likely to transfer technology from abroad, possibly from their parents.[8]

The main input for innovation process is investment in R&D activities. The R&D intensity (RDINT, R&D expenditures/sales ratio) is used to determine the effect of R&D activities on innovation. Since there could be a complementarity between in-house R&D and technology transfer, it is also included in the technology transfer model. Moreover, the effects of regional and sectoral knowledge spillovers from domestic firms are captured by the R&D intensity of firms operating in the same province (DREGRD) and in the same sector (DSECTRD), respectively.

The size of the firm is considered to be one of the main determinants of innovativeness. Thus, we include the (log) number of employees (LL) to test the impact of firm size on technological activities. Moreover, the proportion of skilled employees, SKILLED, is used to test the contribution of skilled employees on innovation and technology transfer activities.

The effect of subcontracting relations on technological activities is tested by using two variables, SUBIN (the share of subcontracted inputs in total inputs) and

[8] We also experimented with a dummy for majority-owned foreign firms. Since most of the foreign firms in the sample are majority-owned foreign firms, there was not any major change in our results.

SUBOUT (the share of output subcontracted by other firms in total output). These variables are used to check whether subcontract-receiving (SUBOUT) and subcontract-offering (SUBIN) firms are more innovative/more likely to transfer technology from abroad.

Finally, there are three additional firm-specific variables: GROUP is a dummy variable that takes the value 1 if the firm belongs to a business group. This variable is used to test whether membership in a business group yields any benefit for technological activities. The variable Internet is defined by the proportion of employees who have direct access to the Internet on the job. If technological activities require extensive exchange of information (and, of course, if the Internet provides the basis for information exchange), this variable is expected to have a positive coefficient in both innovation and technology transfer models. The third variable, LTURN, is the ratio of the number of separations in a year to the average number of employees (average employment *plus* the number of separations). This variable is used to measure labor flexibility that is likely to have a negative impact on innovative activities (see Kleinknecht 1998; Michie and Sheehan 2003).

In order to test the productivity effects of innovation and technology transfer activities, we estimate a simple production function, defined as follows:

$$Q_{it} = f(A_{it}, K_{it}, L_{it}, E_{it}, M_{it}, SKILLED_{it}, LRW_{it}) \qquad (3)$$

$$A_{it} = A_0 e^{\delta INNOit + \gamma TECHNOit + \lambda t} \qquad (4)$$

where Q is (real) output, K, L, E and M are (real) capital, labor, energy and materials inputs. SKILLED and LRW are the share of skilled employees and real product wages, respectively. These variables are used to control for labor quality. Subscripts i and t denote firm and time period, respectively. A_0 is the base-line productivity level, and δ and γ are the effects of innovation and transferred technologies, respectively, on productivity.

Since the innovation and technology transfer variables are endogenous in the output model (Eq. 3), we first estimate Eqs. (1) and (2), and then estimate the output Eq. (3) by adding the inverse-Mills ratios (obtained from the estimation of Eqs. 1 and 2) to have unbiased estimation. Since the Innovation Surveys are available for two time periods, 1995–1997 and 1998–2000, the data for these two time periods are pooled together in the regression analysis, and a dummy variable for the second period is used to capture exogenous changes in the dependent variables over time. Moreover, dummy variables for 2-digit industries are added into all models to control for unobserved sector-specific factors.

4 Determinants of innovation and technology transfer: estimation results

Table 2 presents descriptive statistics on all variables used in the regression analysis. As noted in the previous sections, the share of innovative firms is much higher in high-tech industries than in low-tech industries. Moreover, the share of technology transferring firms is also higher in high-tech industries. Firms in high-tech industries are somewhat smaller than firms in low-tech industries, but spend proportionately much more on R&D activities. (However, note that the average R&D intensity is only 0.22% for firms in high-tech industries.) Regional R&D intensity is almost the same for both groups of firms, i.e., high-tech firms do not cluster in specific regions (provinces), but, as expected, sectoral foreign and domestic R&D intensities (SSECTRD and DESCTRD) are much higher in high-tech industries. The average market share of foreign firms is higher in high-tech industries, and foreign firms have a larger market share in supplier industries than in user industries. Finally, the possibility that a firm will employ a former employee of a foreign firm is much higher in high-tech industries than in low-tech industries because of the larger share of labor turnover in high-tech foreign firms.

Regression results summarized in Table 3 show that *labor spillovers* from foreign firms contribute significantly to the innovativeness of Turkish manufacturing firms. However, as may be expected, labor spillovers do not have any impact on the probability of technology transfer from abroad. This finding indicates that tacit knowledge, embodied in people, plays an important role for innovativeness.[9] Moreover, in supporting this argument, the skilled variable (the proportion of skilled employees) is also found to be one of the main determinants of innovativeness.

As for *horizontal spillovers*, neither the foreign presence in the industry, nor R&D spillovers, make any contribution to technological activities (neither innovation nor technology transfer). There seem to be some regional spillovers from the R&D activities of foreign firms in low-tech industries, but the coefficient of the variable SREGRD is statistically significant only at the 5% level. Economic significance of regional spillovers in low-tech industries is also low because of low level of R&D intensity.

Vertical spillovers are significant only for innovativeness in high-tech industries, but with a mixed outcome. User firms operating in industries supplied mainly by foreign firms tend to be less innovative, i.e., the higher the share of foreign firms in supplier industries, the lower the innovativeness of firms in user, high-tech industries. On the other hand, high-tech firms supplying their output mainly to foreign firms tend to be more innovative. This finding may point to the importance of users in technological activities (for the importance of "learning from users," see Carlsson and Jacobsson 1991).

Since most of the spillover variables have statistically insignificant coefficients, we estimated a number of additional models to test the robustness of our results. First, we estimated our models by excluding all but one spillover variable to eliminate possible multicollinearity among spillover variables. The results for the models including only one spillover variable are qualitatively the same as those obtained by including all spillover variables together. The only difference is that the presence of foreign firms (the SFDI variable) in high-tech industries becomes significant at the 5% level in the technology transfer model. Second, it is suggested in the literature that domestic firms can benefit from spillovers only if they are equipped with necessary technological capabilities/absorptive capacity. We used two variables, the firm size (dummy) and the share of skilled employees as mea-

[9] In a recent study on Dutch manufacturing, Brusoni et al. (2005, p. 230) found no link between the availability of codified knowledge and the level of innovativeness. Thus, they suggest that "... attempts to improve the distribution power of the innovation system by supporting codification exercises, such as computer-information networks, will have a limited impact on overall rates of innovation."

Table 3 Determinants of innovativeness, technology and productivity

	Low-tech industries						Medium- and high-tech industries					
	Innovativeness		Technology transfer		Production		Innovativeness		Technology transfer		Production	
	Coeff	Std dev	Coeff	Std dev	Coeff	Std dev	Coeff	Std dev	Coeff	Std dev	Coeff	Std dev
Labor spillovers												
SLABOR	11.45	2.88**	−6.91	22.61			7.78	3.22*	5.55	6.86		
Horizontal spillovers												
SFDI	0.22	0.33	−0.89	1.57			−0.09	0.26	0.61	0.52		
SREGRD	0.73	0.35*	0.47	3.79			−0.63	0.59	0.82	1.55		
SSECTRD	20.07	29.73	−7.03	275.59			−0.15	0.50	−0.44	1.04		
Vertical spillovers												
SSUUP	−1.64	1.09	1.98	5.97			−2.33	0.72**	0.61	1.55		
SBUY	0.61	0.94	−1.57	4.03			2.19	0.87*	3.27	2.81		
Foreign ownership												
FDI	0.05	0.19	1.81	0.48**	0.11	0.03**	0.47	0.19*	1.17	0.26**	0.05	0.03
R&D and R&D spillovers												
RDINT	25.38	9.63**	6.97	34.71			25.44	5.73**	7.31	4.06		
DREGRD	−0.40	0.46	−3.44	8.88			0.53	0.52	0.41	1.14		
DSECTRD	5.35	1.44**	1.69	7.16			−0.14	0.49	−0.09	1.10		
Other explanatory variables												
LTURN	−0.66	0.30*	−0.05	2.55			−0.46	0.37	0.27	1.07		
INTERNET	0.84	0.08**	0.35	0.49			0.99	0.13**	0.56	0.41		
GROUP	0.04	0.13	0.38	0.60			−0.03	0.20	0.58	0.27*		
SUBIN	−0.64	0.41	−0.08	2.15			−0.67	0.76	1.62	1.42		
SUBOUT	0.33	0.19	−5.23	13.61			−0.56	0.71	−0.46	2.00		
SKILLED	0.69	0.23**	−0.61	1.94	0.00	0.02	1.06	0.31**	0.61	0.77	0.23	0.05**
LL	0.17	0.03**	0.30	0.18	0.13	0.01**	0.06	0.05	0.60	0.10**	0.22	0.01**

Table 3 (continued)

	Low-tech industries						Medium- and high-tech industries					
	Innovativeness		Technology transfer		Production		Innovativeness		Technology transfer		Production	
	Coeff	Std dev	Coeff	Std dev	Coeff	Std dev	Coeff	Std dev	Coeff	Std dev	Coeff	Std dev
LM					0.68	0.00**					0.67	0.01**
LE					0.06	0.00**					0.03	0.01**
LK					0.08	0.00**					0.07	0.01**
LRW					0.00	0.00**					0.00	0.00**
INNOVAT					0.25	0.03**					0.31	0.01**
TECHNO					−0.04	0.11					0.00	0.00**
λ_{inno}					−0.12	0.02**					−0.17	0.03**
λ_{techno}					0.01	0.05					−0.02	0.04
ρINNNO TECHNO	0.072	0.274					0.065	0.141				
n	1978				1978		1042				1042	
Log-likelihood	−1007				−793		−756.25	−1595			−433	
Spillovers test	48.4**						41.3**					
Adj. R^2					0.945						0.948	

All models include sector dummies for ISIC 2-digit industries, a dummy for year 2000, and a constant term
**(*) means statistically significant at the 1% (5%) level, two-tailed test

sures for domestic firms' capabilities, and used interactions of these variables with spillover variables to test whether larger firms or firms that employ proportionately more skilled people benefit more from spillovers. Among 24 coefficients estimated for the innovation models, only one turned out to be statistically significant at the 5% level.[10] In other words, interactions with size and skill levels did not change our results.

Among other explanatory variables, foreign ownership is found to matter for innovativeness in high-tech but not for low-tech industries, i.e., foreign firms in high-tech industries tend to be more innovative than domestic firms even after controlling for all other factors. However, foreign ownership is one of the main determinants of technology transfer in both sectors: foreign firms tend to transfer technology from abroad.

Firm size has a positive impact on innovative activities in low-tech industries, and on technology transfer in high-tech industries. It seems that small firms are as innovative as large firms in high-tech industries, where innovativeness is essential for competitiveness. High-tech firms belonging to business groups tend to transfer technology, but, apparently, business groups fail to improve the innovativeness of their members. As expected, R&D intensity is correlated with innovativeness. Sectoral knowledge spillovers from domestic firms are correlated with innovativeness in only low-tech industries, but regional spillovers do not have any impact on technological activities. The access to the Internet has a positive impact on innovativeness in both sectors, and labor turnover, i.e., labor flexibility, has a negative impact on innovativeness in low-tech and high-tech industries, although the latter is not statistically significant. Subcontracting relationships do not play any role in technological activities.

Estimation results for production equations reveal a difference between low-tech and high-tech industries. Although the coefficients of input variables (except labor input variables, LL and SKILLED) are almost the same for low-tech and high-tech industries, the innovativeness variable has a somewhat larger coefficient in the high-tech industries model. It seems that what matters for productivity in high-tech industries is internally generated technological capability. Technology transferred from abroad improves productivity neither in low-tech nor in high-tech industries after controlling for all other variables. It is interesting to observe that although foreign ownership makes low-tech firms more productive, it does not have any impact on productivity in high-tech industries, where foreign firms are assumed to be in a more advantageous position. Foreign firms in high-tech industries could be more productive only if they were to become more innovative.

---

[10] The interaction between the SFDI and skilled variables has a statistically significant (at the 1% level) negative coefficient in the innovation model for low-tech industries, i.e., those low-tech firms that employ more skilled people benefit less form foreign presence in their sectors. In the case of technology transfer models, only four out of 24 interaction variables had coefficients statistically significant at the 5% (size-SFDI, size-SSECTRD, and size-SBUY interactions had positive coefficients, and skilled-SSECTRD interaction had negative coefficient in high-tech industries).

5 Conclusions

FDI has been considered by many development economists as an important channel for the transfer of technology to developing countries. It has been suggested that modern, advanced technologies introduced by multinational firms can also diffuse to domestic firms through spillovers. Our analysis shows that foreign firms in Turkey are more innovative than their domestic counterparts in medium- and high-tech industries, but not in low-tech industries. However, in both industries, foreign firms tend to transfer technology from abroad (mostly from their parent companies).

The type of spillovers seems to matter: in terms of horizontal spillovers, neither the foreign presence in the industry, nor foreign R&D spillovers, make any significant contribution to technological activities (neither innovation nor technology transfer) in low- and high-tech industries. There are no vertical spillovers in low-tech industries and their effect on innovativeness of medium- and high-tech firms is ambiguous. High-tech suppliers tend to have a high rate of innovation when the share of foreign users is high, but the opposite is true for users: high-tech users supplied mainly by foreign firms tend to have a lower rate of innovation. The main channel of spillovers is labor turnover. Our findings reiterate the importance of tacitness of knowledge, and confirm that technology cannot easily be transferred through passive mechanisms (demonstration effects, reverse engineering, etc.). Therefore, the policy aimed at encouraging innovativeness should pay due attention to in-house technological activities.

References

Akpınar A (2001) Turkey's Experience with Bilateral Investment Treaties, paper presented at the OECD Investment Compact Regional Roundtable, May 28–29, Dubrovnik [http://www.oecd.org/daf/conference/Dubrovnik/akpinar.pdf]

Blomström M, Kokko A (1998) The impact of foreign investment on host countries: a review of the empirical evidence. J Econ Surv 12:247–277

Borensztein E, De Gregorio J, Lee JW (1998) How does foreign direct investment affect economic growth? J Int Econ 45(1):115–135

Brusoni S, Marsili O, Salter A (2005) The role of codified sources of knowledge in innovation: empirical evidence from Dutch manufacturing. J Evol Econ 5(2):211–231

Carlsson B, Jacobsson S (1991) What makes the automation industry strategic? Econ Innov New Technol 1(4):257–269

Cieslik A, Ryan M (2002) Characterising Japanese direct investment in Central and Eastern Europe: a firm level investigation of stylised facts and investment characteristics. Post-Communist Econ 14:509–527

Cowan R, David PA, Foray D (2000) The explicit economics of knowledge codification and tacitness. Ind Corp Change 9(2):211–253

Erdilek A (1982) Direct foreign investment in Turkish manufacturing. J.C.B. Mohr, Tübingen

Erdilek A (1986) Turkey's new open-door policy of direct foreign investment: a critical analysis of problems and prospects. METU Studies in Development 13:171–191

Kepenek Y, Yentürk N (2000) Türkiye ekonomisi, 10th edn. Remzi Kitapevi, Istanbul

Kinoshita Y (2001) R&D and technology spillovers via FDI: innovation and absorptive capacity, University of Michigan Business School Working Paper No. 349a

Kleinknecht A (1998) Is labour market flexibility harmful to innovation? Camb J Econ 22: 387–396

Lundvall B-Å (1988) Innovation as an interactive process: from user–producer interaction to national systems of innovation. In: Dosi G, Freeman C, Nelson R, Silverberg G, Soete L (ed) Technological change and economic theory. Pinter, London

Michie J, Sheehan M (2003) Labour market deregulation, 'flexibility,' and innovation. Camb J Econ 27:123–143
Nelson RR, Nathan R (1993) Technical innovation and national systems. In: Nelson RR (ed) National innovation systems: a comparative analysis. Oxford University Press, New York
Noorbakhsh F, Paloni A, Youssef A (2001) Human capital and FDI flows into developing countries: new empirical evidence. World Dev 29:1593–1610
OECD (1997) Oslo manual: the measurement of scientific and technological activities proposed guidelines for collecting and interpreting technological innovation data. OECD, EC and Eurostat, Paris
OECD (1999) Boosting innovation: the cluster approach, OECD Proceedings. OECD, Paris
OECD (2000) Innovation and economic performance. OECD, Paris
Öniş Z (1994) Liberalization, transnational corporations and foreign direct investment in Turkey: the experience of the 1980s. In: Fikret Ş (ed) Recent industrialization experience of Turkey in a global context. Greenwood, Westport, Connecticut, pp 91–109
Özçelik E, Taymaz E (2004) Does technology matter for international competitiveness in developing countries? The case of Turkish manufacturing industries. Res Policy (33):409–424
Pavitt K, Parimal P (1999) Global corporations and national systems of innovation: who dominates whom? In: Archibugi D, Howells J, Michie J (ed) Innovation policy in a global economy. Cambridge University Press, Cambridge, pp 94–119
Reger G (1998) Changes in the R&D strategies of transnational firms: challenges for national technology and innovation policy. Sci Technol Ind (22):243–276
Smith K (1995) Interactions in knowledge systems: foundations, policy implications and empirical methods. Science, Technology, Industry (16):69–102

Giovanni Battista Dagnino ·
Marcello M. Mariani

Dynamic gap bridging and realized gap set development: The strategic role of the firm in the coevolution of capability space and opportunity space

Abstract Building on the hypothesis that firm strategy is fundamentally a dynamic process of gap bridging between capability space and opportunity space, this paper investigates the strategic role of the firm in the coevolution of the capability space and the opportunity space. In more detail, it contributes to the strategy literature by introducing and discussing a typological distinction of a few dynamic gap sets (i.e., potential gap set, realized gap set, deliberate gap set, emergent gap set), which are deployed in a comprehensive conceptual framework underscoring the causes and consequences of the gap sets evolution. Additionally, the framework proposed is able to shed new light on the received distinction between deliberate and emergent strategies, and to reconnect them to the objectives of *innovation* and *execution* they aim to achieve. This intellectual effort allows us to synthesize the dichotomy, already familiar to strategy investigation, between strategy formulation and strategy implementation and to enhance the hermeneutic and interpretive capacity of the conceptual backbone of this work. In an integrative fashion, whereas strategy formulation refers to the firm's mindful predisposition of the capabilities required to bridge the strategy gap between capabilities and opportunities, strategy implementation is concerned with the 'real' operational closure of the strategy gap.

Keywords Capability space · Opportunity space · Dynamic gap bridging · Co-evolutionary framework · Typology · Innovation
JEL Classification O31 · O32 · D83

The authors presented earlier versions of this study at the 10th International Schumpeter Society Conference, held in Milan's Bocconi University, June 9-12, 2004, and at the 24th International Conference of the Strategic Management Society, convened in Puerto Rico, October 31-November 3, 2004. For insightful comments, we thank participants to the sessions of these venues where the paper was discussed. We acknowledge Franco Malerba for his kind support.

G. B. Dagnino
University of Catania, E-mail: dagnino@unict.it

M. M. Mariani
University of Bologna, E-mail: mmariani@unibo.it

1 Introduction

Moving from the assumption that firm strategy is a fundamentally dynamic process of gap bridging between capability space and opportunity space, this paper emphasizes the strategic importance of the firm in influencing the coevolution of the capability space and the opportunity space. More specifically, it contributes to the strategy literature by introducing and discussing a typological distinction of a few dynamic gap sets (i.e., potential gap set, realized gap set, deliberate gap set, emergent gap set), which are deployed in a comprehensive conceptual framework relating to the interactions among them and underscoring the causes and consequences of the gap sets evolution. In addition, the framework is able to shed light on the distinction between deliberate and emergent strategies, and to reconnect them to the objectives of strategic *innovation* and *execution* they aim to achieve. This intellectual effort proffers a way of synthesizing the stark dichotomy, already familiar to strategy analysis, between strategy formulation and strategy implementation. In an integrative fashion, whereas strategy formulation refers to the firm's mindful predisposition of the capabilities required to bridge the strategy gap between capabilities and opportunities, strategy implementation is concerned with the 'real' operational closure of the strategy gap.

In order to illustrate the multiple links between the dynamic process of gap bridging and the realized gap set development, the chapter is structured in seven sections. In section 2, the main building blocks and the strategic cornerstones of the work (i.e., the capability lifecycle, the coevolution of capability space and opportunity space, and Mintzberg's distinction between deliberate strategies and emergent strategies) are discussed and the research questions are introduced. In section 3, in order to confer hermeneutical vigor to the analytical framework, a typology of various strategy gaps and gap sets is introduced, illustrated and discussed. Additionally, we scrutinize the interactions and connections among these original notions. In section 4, the influx of the firm dynamic gap bridging on the coevolution of capability space and opportunity space is investigated by means of examining the fundamental drivers of the realized gap set (RGS) evolution and of their composed effects: entrepreneurial innovation and the reduction of environmental uncertainty. Section 5 explores the role and features of the coevolutionary process that takes place along capability space and opportunity space. On the basis of this analysis, the stark dichotomy existing between the formulation and implementation of strategy is stressed and analyzed. Section 6 attempts to characterize the dynamic bridging process according to its objectives of innovation and execution. In the seventh and concluding section, we elucidate a few theoretical implications and suggest novel itineraries for further coevolutionary research in strategy.

2 Theoretical background and motivating questions

In this chapter, we start our investigation path with a few motivating questions that are epitomized as follows. How can firm strategies addressing either innovation or execution objectives be characterized as emergent versus deliberate? Which is the fundamental connection among capabilities, opportunities and innovation?

Investigation into these issues requires a deeper exploration of the strategic processes through which strategic opportunities are seen and seized. Yet, it also implies a fine-grained understanding of the crucial role firm capabilities play in seizing strategic opportunities. Accordingly, it appears that a thorough and rigorous conceptualization of the capability space (CS) and the opportunity space (OS) is at the very basis of firm strategy, which is fundamentally directed to bridge the strategy gap between them. More precisely, firm capabilities (Langlois 1995; Teece et al. 1997; Eisenhardt and Martin 2000) are undertaken and deployed to seize strategic opportunities (Kirzner 1997; Winter et al. 2003) that a firm is able to perceive and see (Penrose 1959; Miller 2003) in the opportunity space. Consequently, firm strategy is intended as a means to bridge the strategy gap between the capability space and the opportunity space over time. Being strategically interrelated, capability and opportunity spaces display two additional inherent properties:

1. they coevolve, as they are coevolutionary in nature (Lewin and Volberda 1999; Murmann 2003);
2. they influence each other in a recursive process, or they display the property of feedforward (Levinthal and Myatt 1994; Dagnino 2003).

Three main theoretical antecedents in the strategic management literature contribute to provide a well thought-out axes suitable to develop the initial research questions: (a) the notion of capability lifecycle; (b) the concepts of capability space (CS) and opportunity space (OS), conceptualized as dynamic systems that display the property of feed-forward; (c) the Mintzberg-like distinction between deliberate strategies and emergent strategies.

As we intend to construct our framework in a piecemeal fashion, we first analyze each of the three basic conceptual foundations independently. Rooted in the quest for a dynamic resource-based view of the firm (Peteraf 1993; Rumelt 1987, 1991) and in the evolutionary perspective (Nelson and Winter 1982, 2002), the notion of capability lifecycle (Helfat and Peteraf 2003) is targeted to highlight the main evolutionary phases of capabilities (i.e., founding, development, and maturity) in a way that helps explain the sources of heterogeneity in organizational capabilities. This notion contributes to both refining and enhancing the concept of capability space, allowing us to characterize its evolution (expansion or contraction) according to a few recognized patters and to firm strategic intervention. Correspondingly, we maintain that the opportunity space may evolve as opportunities (and their entrenched Schumpeterian environments; Schumpeter 1934, 1935) may show instability and uncertainty and may expand or contract over time.

The capability space and the opportunity space are both understood as dynamic systems (Boulding 1956; Anderson 1999). This appreciation implies that multiple interactions are likely to occur among different interconnected elements of the same space: each capability can interact with one or more other capabilities displaying different directionality of the influence flux (i.e., capability ‘A’ can influence capability ‘B’ and vice versa), and each opportunity can intermingle with one or more opportunities with different directionality of the influence flux (i.e., opportunity ‘C’ can influence opportunity ‘D’ and vice versa). As we will show, these interactions, respectfully, between capabilities and between opportunities are significantly influential in bridging CS and OS.

Instead of treating the strategy process as an hyperrational analytic process meant to establish long-range goals and action plans as concerns firm organization, according to Mintzberg and Waters (1985) we consider strategy as “a pattern in a stream of decisions" (see also Mintzberg 1972, 1978; Mintzberg and Waters 1982, 1984; Mintzberg et al. 1988; Mintzberg and McHugh 1985). We stress two basic nuances of this concept: the concept of *deliberate* strategy (according to which strategies are realized as previously intended), and the concept of *emergent* strategy (according to which strategies are realized differently from, and somewhat despite, the ex-ante intentions). Since the interrelation and coevolution between capability space and opportunity space is governed by differentiated firm strategies, we draw upon and expand the received distinction between deliberate and emergent strategies (Mintzberg et al. 1976; Mintzberg and Waters 1985).

The nuances heretofore discussed allow us to distinguish between two kinds of mirroring strategic gaps: the *emergent* strategy gap (ESG) and the *deliberate* strategy gap (DSG). The two kinds of strategy gaps are seemingly different, but nonetheless interrelated, and proceed together in the strategy process. Their inner nature and major features will be discussed in detail in the next section. The framework we develop henceforth rests on the three above mentioned main strategic cornerstones of capability lifecycle, dynamic systems, and the distinction between deliberate and emergent strategies, which will be instrumentally and conveniently recombined and integrated.

3 Toward a typology of dynamic strategy gaps and gap sets

In order to establish a lexicon adequate to our purpose, we have elaborated a typology of strategy gaps and gap sets. The same way we have accepted the hypothesis that concerns the existence of the capability space and the opportunity space, we introduce a variety of strategy gap bridging processes and strategy gap sets. Because it is sufficiently 'short and snappy' and pretty effective, this concise dictionary comes to enhance the hermeneutic and interpretive capacity of the conceptual framework depicted above. Subsequently, we shall observe the main interactions among these original notions in strategy and assess their evolutionary potential.

A Concise Dictionary of Dynamic Strategy Gaps and Gap Sets. Here follows a detailed glossary of the original concepts that we intend to initiate and introduce in the strategic management literature:

(a) the strategy gap (SG) bridging process is a coevolutionary process through which it is possible to join one or more elements that compose the opportunity space with one or more elements that compose the capability space;
(b) the potential gap bridging (PGB) refers to the strategy set (of capabilities) *potentially* capable of reducing the strategy gap to zero;
(c) the realized gap bridging (RGB) concerns the strategy set actually capable of reducing the strategy gap to zero or, in other words, the set of implemented strategies which can qualify as intentional or non-intentional. They in turn originate, respectively, the deliberate gap bridging (DGB) and the emergent gap bridging (EGB);
(d) the potential gap set (PGS) is seen as the set of all strategy gaps that may potentially be bridged through firm strategies; i.e., through firm strategies that are even hypothetically possible to elaborate. It is *de facto* the *universe* containing all the strategy gaps;

(e) the realized gap set (RGS) is viewed as the set of all strategy gaps that are really bridged through firm strategies that are actually put into practice;
(f) the deliberate gap set (DGS) is the set of all the strategy gaps that are really bridged through firm strategies *actually* put into practice in an *intentional* and deliberate fashion;
(g) the emergent gap set (EGS) is the set of all the strategy gaps that are actually bridged through firm strategies realized in a *non-intentional* fashion. It is thus a subset of RGS and the natural complement to DGS.

As they are comparatively more important in the strategic process of capability-opportunity gap bridging, we focus our attention on the latter two strategy gaps. The distinction advanced above between two kinds of mirroring strategic gaps (i.e., the *emergent* strategy gap (or ESG) and the *deliberate* strategy gap (or DSG)) suggests that they are dissimilar, but interconnected, and evolve concurrently in the strategy process. The emergent strategy gap uncovers itself when a firm is driven unintentionally to deploy new capabilities – reformulating its strategy or capability space – in order for it to seize the same opportunity that evolves over time. The deliberate gap is displayed when firm capabilities are deployed to seize an opportunity that is intentionally and deliberately perceived and discovered. Thus, the gap bridging process is elucidated as a fundamentally coevolutionary process, which shows three basic properties: (1) it is probabilistic and somewhat unpredictable ex ante[1]; (2) it exhibits intense feedback-feedforward processes; (3) it is amenable to encompass both the relevant processes of strategy formulation and strategy implementation.

According to Mintzberg and Waters (1985, 258), for a strategy to be perfectly deliberate, at least three conditions seem to be required. First, "there must have existed *precise intentions* in the organizations, articulated in a relatively concrete level of detail", so that what was desired before any action is clear *a priori*. Second, such a priori intentions should be shared at an organizational level, thus being common to virtually all the actors. Third, these "collective intentions must have been realized *exactly* as intended, which means that no external force could have interfered with them". This implies that the environment is perfectly predictable, totally benign or under the complete control of the organization.

For a strategy to be perfectly emergent, there must be order in the *absence* of intention about it. The assumption of total absence of intention which leads to action is quite extreme, thus suggesting that both the purely emergent strategy and the merely deliberate one, are actually rare or, more simply, accommodating and helpful abstract extremes. Yet, some patterns that firms originate and develop come rather close to one of them and the two shape the countervailing poles of a continuum along which real-world strategies typically fall.

A variety of eight strategies types fall along this continuum. Consistent with Mintzberg and Waters (1985, 270), in Table 1 we identify the following strategies: planned strategy, entrepreneurial strategy, ideological strategy, umbrella strategy, process strategy, unconnected strategy, consensus strategy, and eventually imposed strategy.

[1] Namely, gap bridging processes occur with a time lag not necessarily equal to 1. An opportunity that changes at time t_0, does not necessarily imply that this will affect change in capability space in time t_1. In fact, the capability space may change afterwards at times t_2, and/or t_3.

Table 1 A Comprehensive Description of Strategy Types

Strategy	Major features
Planned	Strategies originate in formal plans: precise intentions exist, formulated and articulated by central leadership, backed up by formal controls to ensure surprise-free implementation in benign, controllable or predictable environment; strategy is mostly deliberate
Entrepreneurial	Strategies originate in a central vision: intentions exist as personal, unarticulated vision of single leaders, and so that are adaptable to new opportunities; organizations are under the personal control of the leaders and located in protected niche environments; strategy is relatively deliberate but can be emergent
Ideological	Strategies originate in shared beliefs: intentions exist as collective vision of all actors, in inspirational form and are relatively immutable, controlled normatively through indoctrination and/or socialization: organizations are often proactive vis-à-vis the environment; strategy is rather deliberate
Umbrella	Strategies originate under constraints: leadership is in partial control of organizational actions, defines strategic boundaries or targets within which the other actors respond to their own forces or to complex and perhaps also unpredictable environments; strategy is partly deliberate, partly emergent and partly deliberately emergent
Process	Strategies originate in process: leadership controls process aspects of strategy (hiring, structure, et cetera), leaving the content aspects to the other actors; strategy is partly deliberate, partly emergent (and, again, partly deliberately emergent)
Unconnected	Strategies originate in enclaves: actor(s) are loosely coupled to the rest of the organization, produce(s) patterns of their own actions in absence of, or in direct contradiction to, the central or common intentions; strategy is organizationally emergent whether or not deliberate for actor(s)
Consensus	Strategies originate in consensus: by means of mutual adjustment, actors converge on patterns that become pervasive in absence of central or common intentions; strategy is rather emergent
Imposed	Strategies originate in the environment: the environment dictates the patterns in actions, either through direct imposition, or through implicitly preempting or bounding organizational choices; strategy is mostly emergent, although may be internalized by organizations and rendered deliberate

Source: Mintzberg and Waters (1985, 270)

Drawing on the established typological distinction indicated above, we can dig more deeply the concepts of realized gap bridging and realized gap set as follows:

(a) the *planned* gap bridging (PlanGB) is a rational coevolutionary process through which one or more elements included in the opportunity space are linked to one or more elements that compose the capability space, after a firm's leaders have articulated their intentions through a detailed plan (that takes the form of budgets, schedules and so on) in order to minimize the confusion and discretion that could impede its realization. The set of strategy gaps which are bridged through planned strategies can be defined Planned Gap Set (PlanGS);
(b) the *entrepreneurial* gap bridging (EntrGB) process refers to the strategy set undertaken by one individual in personal control of an organization and successful in imposing his vision on it. This set of strategies (EntrGS) implies the existence of intentions deriving from one individual who, nonetheless, does not need to articulate or elaborate them (because he is the 'one and only' and is not subject of any other external interference);

(c) the *ideological* gap bridging (IdeolGB) concerns the process through which an organization, whose members share a vision and pursue it as an ideology, is able to reduce the strategy gap to zero. At the very basis of this process, we can identify the ideological gap set (IdeolGS), including all the strategies deployed in an intentional fashion, being intentions organizational and not individual as in the case of PlanGS and EntrGS;
(d) the *umbrella* gap bridging (UmbrGB) involves all strategies realized through the intervention of leaders who only have partial control over other actors of a firm, designing general guidelines and boundaries of behavior and letting other actors maneuver within them. The set of strategies these processes are based upon can be defined Umbrella Gap Set (UmbrGS);
(e) the *process* gap bridging (ProcGB) is viewed as the process through which the strategy gap is reduced to zero in such situations as the ones in which the firm's leadership controls the *process* of strategy making, and therefore of strategy bridging, while leaving the *content* of strategy to other actors. We can identify a Process gap set (ProcGS) as the set of strategies designed by the central leadership allowing other actors the flexibility to evolve strategy patterns within it;
(f) the *unconnected* gap bridging (UnconGB) uncovers the strategy gap when one part of the organization with considerable discretion – as, for instance, a subunit, or even a single individual – because only loosely coupled to the rest of the organization, is able to realize its own strategic patterns. The set of strategies on which this bridging process rests can be depicted as unconnected gap set (UnconGS);
(g) the *consensus* gap bridging (ConsGB) allows the reduction of the gap between the CS and the OS through the intervention of different firm's actors who naturally converge on the same theme, or pattern, so that it becomes pervasive in the organization, without the need for any central direction or control. The set of strategies that is undertaken in this case is defined the consensus gap set (ConsGS);
(h) the *imposed* gap bridging (ImpGB) is generally driven by the external environment, forcing the organization into a pattern of its stream of actions, regardless of the presence of central controls. This process of gap bridging takes place on the basis of a gap set identifiable as ImpGS.

As much as the accommodating typological distinction carried out at the level of the realized gap set (RGS) is concerned, we focus especially on the five key concepts of EntrGS, UmbrGS, ProcGS, UnconGS, and ImpGS. We have decided to do so because we consider the five mentioned gap sets as the most essential and critical for framing the firms' evolutionary paths and the most significant as regards the coevolutionary matching of capability space and opportunity space. In the following subsection, we shall take into consideration in more detail the crucial interrelations among the various strategic gap sets.

Strategic Interactions Among Strategy Gap Sets. The interrelations among the strategy gap sets in the typological distinction reported above may be epitomized by a virtual image of two concentric circles, including two semicircles, and is graphically represented as reported (see Figure 1).

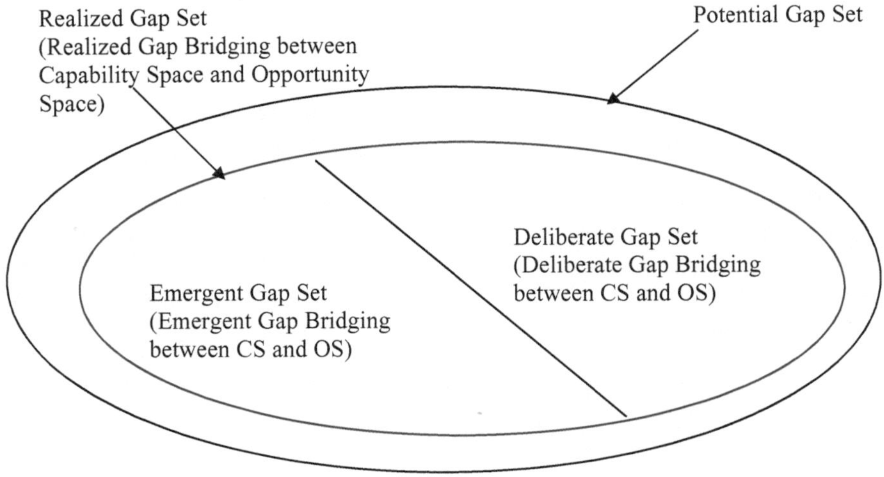

Fig. 1 A Representation of the Relevant Interactions Among Strategy Gap Sets

The potential gap set (PGS) is a straightforward container which includes the set of all strategic choices potentially imaginable and conceivable (i.e., strategic conjectures). In our view, the most interesting evolutionary dynamics concern, however, the realized gap set (RGS), whose composition and development depend on the recombination of its various subsets (i.e., PlanGS, EntrGS, IdeolGS, UmbrGS, ProcGS, UnconGS, ConsGS and ImpGS). Each of these subsets contains some real-world strategic choices that can be associated with the more abstract distinction between deliberate and emergent gaps.

Undeniably, a variety of interactions between the different kinds of RGS subsets are likely to occur, and circumscribed regions of overlap between them may exist. For instance, UmbrGB and ProcGB processes are perfectly compatible one to another, if we look at a firm whose main means to reduce its strategic gap to zero is dependent on the intervention of leaders who have only a partial control over the other firm actors, or who only have control of the *process* of strategy making, while they allow other actors the formulation of strategy or strategy *content*.

In the next section, starting from the conceptualization of the gap sets heretofore depicted, we illustrate the key drivers which play a strategic role in the evolution of the gap sets.

4 Drivers of realized gap set (RGS) evolution

This section is articulated into two basic subsections: the former deals with the developmental processes of the strategy gap sets as regards organizations in general, where we consider 'the organization' as a general archetype, and a conveyor representative of all types of human-driven organized action. The latter examines the same developmental processes by focusing more specifically the attention on the business firm.

Development Processes of Strategy Gap Sets: An Application to Organizations as General Archetypes. The realized gap set (RGS) is far from being a static set of elements. Conversely, it may distinctively modify its extension, range and scope parallel to the coevolution of capability space (CS) and opportunity space (OS). This process also affects the realized gap set (RGS) composition, leading to an increase or reduction of its actual subsets (e.g., the PlanGS or the EntrGS). Whereas these subsets do not in fact necessarily follow the expansion/restriction of the realized gap set, nonetheless they are key factors contributing to shape the RGS net final development process. As a consequence, deliberate gap set (DGS) and emergent gap set (EGS), as combinations of the real-world subsets, amplify or reduce their borders and scope, according to the changes taking place in the real-world RGS subsets.

Several underlying factors drive the processes of development (i.e., expansion and contraction) of the RGS. These factors display various key characteristics that, by influencing the interaction between the CS and the OS, can/cannot support the actual closure of the potential strategy gaps. More specifically, it is possible to epitomize these process factors as follows:

1. the introduction/exploitation of an entrepreneurial innovation[2];
2. the increase/reduction in the degree of environmental uncertainty;
3. the affirmation of an ideology at the organizational level;
4. the mechanisms of intra/inter-organizational power consolidation;
5. the change in the processes of definition of organizational responsibility;
6. the creation of organizational consensus about organizational processes, standards, routines[3].

Each of the above mentioned six process factors may affect a specific real-world subset. For example, the introduction/exploitation of an entrepreneurial innovation could definitely influence the extension of the EntrGS, but it could also influence the UnconGS if we, for instance, consider the degree of discretion the R&D function of a pharmaceutical firm might be allowed. In addition, the UmbrGS may be influenced if that pharmaceutical firm has subsidiaries that could experiment and benefit from spillover effects from its R&D activity (Henderson and Cockburn 1994; Doz et al. 2000).

The increase/reduction in the degree of environmental uncertainty could in turn induce direct changes in the ImpGS, but it could also generate variation in the extension of other different subsets (for example the UmbrGS). Consider the case of the periodical reduction in the crude oil quotas operated by OPEC, which typically affects the oil prices and price revision strategies determined by both the oil firms depending on OPEC (UmbrGS) and by almost all the other oil and oil-related firms embedded in the global economy.

The affirmation of an ideology at the organizational level can modify the IdeolGS, by having an impact on the shared vision of the organization's members.

[2] This process factor is increasingly important should we consider that organizations have innovation objectives besides execution ones. According to this line of reasoning, it is exceedingly probable that the EntrGS is the most relevant gap set to be taken into consideration.

[3] This process factor appears particularly relevant should we consider that organizations have day-to-day execution objectives apart from those related to innovation processes. It is thus highly probable that the ImpGS is the most relevant GS to be taken into account.

Since ideology can be the by-product of the institutionalization of the vision of a charismatic leader (possibly an entrepreneur), it does certainly generate an adaptation in the EntrGS. For instance, in the first part of the last century, Henry Ford I was able to generate not only a lucrative entrepreneurial vision, but also an industrial and production ideology, which may be considered to be the most successful practical expression of the "scientific management movement" earlier pioneered by Fredrick W. Taylor. This socio-economic ideological movement is widely known in industrial applications as 'Fordism' and has become a popular antonym of mass production[4]. Henry Ford's strategic choices generated different fashion waves in managerial practice as they were the baselines through which Ford's capabilities (and therefore the CS) could match strategic opportunities (and therefore the OS), thus generating changes in both the EntrGS and the IdeolGS. These changes in fact have had a thorough influence on the actual shape and composition of the RGS.

The mechanisms of intrafirm and interfirm power consolidation are important factors that influence UmbrGS, while the change in the processes of definition of organizational responsibilities generates modifications in the ProcGS and the UnconGS (and also in the ConsGS); if we consider that the process of responsibility definition implies different degrees of autonomy to the firm's actors.

Development Processes of the Strategy Gap Sets: An Application to the Business Firm as a Strategic Crux. In this subsection, we focus more specifically the analysis of the strategy gap sets as relates to the level of the business firm. This is observed as a particular kind of organization that displays certain degrees of intentional behaviors and, since it has been traditionally prominent in strategy enquiries, as a sort of 'strategy crux'. Accordingly, our scrutiny of the firm leads us to maintain that the most significant factors in the development processes of the strategy gap sets are seemingly the following ones:

1. the introduction of an entrepreneurial innovation;
2. the reduction of the degree of environmental uncertainty.

Since the two factors (1) and (2) above are not always present conjointly, we hypothesize that the RGS may expand if only one of the following circumstances is in action:

(a) we simultaneously have an entrepreneurial innovation and a reduction in environmental uncertainty;
(b) we have an entrepreneurial innovation and a simultaneous increase in environmental uncertainty, provided that the effect of the entrepreneurial innovation is greater than the effect of environmental uncertainty (or the net sum is in favor of the entrepreneurial innovation);
(c) we have zero entrepreneurial innovation and a simultaneous reduction in the degree of environmental uncertainty, provided that the effect of the reduction of environmental uncertainty is greater than the effect given by the absence of entrepreneurial innovation (or the net sum is in favor of the reduction of environmental uncertainty).

[4] We recall that, in the late twenties, Henry Ford himself was the one who received the call to write on "mass production" in the *Encyclopaedia Britannica*.

Accordingly, the evolutionary processes of realized gap set (RGS) expansion and contraction are read starting from the coevolutionary dynamics concerning the capability space and the opportunity space that reflect on entrepreneurial innovation and environmental uncertainty. In more detail, when in a definite time period CS expands more than OS (owing for instance to an important innovation, which initiates new capabilities able to seize opportunities present in the OS), then RGS also expands. If, conversely, in a definite time period, the OS expands at a rhythm/speed superior to the CS, then the RGS reduces itself[5].

5 The firm role in the coevolution of capability space and opportunity space, strategy formulation and strategy implementation

In this section, we delve more deeply into the firm role in the coevolutionary process of capability-opportunity matching; i.e., in reducing the strategy gap. Building on earlier work, we first show that, in any time context, a firm interacts with a capability space and an opportunity space, which coevolve jointly. Second, we discuss the effects that strategy formulation and strategy implementation have on the matching of capability space and opportunity space. We start by discussing the definition of the mutual coevolutionary relationships between the capability space and the capability lifecycle.

Capability Space and Capability Lifecycle. Drawing on previous evolutionary research in strategy and economics, Dosi et al. (2000, 1) provided us with an advanced definition of firm capability as the "know how that enables organization to perform these activities"[6]. The recognition that this definition has received suggests that the concept of capability is far from being a static one. In fact, firm capabilities are based on routines and knowledge which are embedded in individuals and organizations (Argote and Darr 2000) as a sort of organizational gene, but they also undergo an *aging* process, being often bounded in time and space, and potentially fitting only one or more parts of an organization at a time (Dagnino 2003).

Pretty much in the same vein, Helfat and Peteraf (2003) emphasized especially this latter feature of firm capabilities and initiated the concept of capability lifecycle (CLC) as a strategic notion that includes three basic specific focal stages of capabilities: the founding stage, the development stage and the maturity stage. According to Helfat and Peteraf, once a capability has reached the maturity stage, it may branch into one of at least six additional stages of the capability lifecycle: retirement (*death*), retrenchment, renewal, replication, redeployment, and recombination. These six additional post-maturity stages are influenced either by internal factors (i.e., managerial decisions) or by external factors (i.e., changes in demand, science and technology, availability of raw materials, government policy).

For the purpose of the present work, we emphasize the fact that each firm may display (and deploy) a certain number of capabilities at time t, and that each of

[5] A medical metaphor of this process is that related to the immune reaction system (Abbas et al. 2000). This metaphor was explained and illustrated in this context by Dagnino and Mariani (2004).

[6] By the word 'activities', we refer essentially to the creation of material products, the provision of a traditional service, and the development of new services.

these capabilities may be at a particular stage of its capability life cycle. This issue implies that a firm could seize one or more opportunities, deploying either a specific capability in a specific stage (e.g., founding, maturity, etc.) or a set of capabilities which, if individually taken, are seemingly at different stages of their lifecycle. For example, a firm 'Z' could seize the opportunity 'X' deploying the capabilities 'A', 'B' and 'C' which could be, respectively, in their maturity, founding and replication stages.

Additionally, the particular stage of the capability lifecycle (CLC) in which a capability is found can influence the final actual operational closure of the Strategy Gap (SG). This basic rationale is further complicated if we call to mind that the CS is a dynamic system in which capabilities interact with each other, generating either an expansion or a shrinking of the CS. Consequently, recalling the example we put forth earlier, the seizing of a certain opportunity 'X' may be influenced by both the stages in which the individual capability 'A', 'B' or 'C' is, and the net effect is thus given by the interaction between different capabilities seemingly at different stages of their lifecycles. Understandably, the way that capabilities interact among each other (and with opportunities), as well as the particular stage of their CLC, both have considerable impact on the degree of the closure intensity of the realized strategy gaps (SG)[7].

Opportunity Space. As far as opportunities are concerned, we concentrate on a subset of opportunities which are present in the opportunity space: business opportunities. Business opportunities are entrepreneurial opportunities or profit opportunities, or opportunities to make money and thrive in some way: i.e., incentives to start-up a new business, a newly exploitable technological discovery, a newly available source of convenient energy, and so on. The concept of opportunities has an extensive tradition in non-mainstream economics: Schumpeter (1934), as well as Penrose (1959) and Kirzner (1979, 1997), have emphasized that a firm's action is undertaken in order to seize profit opportunities. The opportunity space may be characterized, according to Dagnino (2003), either as in part (1) sufficiently predictable and stable, or as in part (2) complex, dynamic, time-pressured and rapidly evolving. Whereas in the first case, we pinpoint the presence of "durable opportunities", in the second we speak of "dissolving opportunities" (2003, 11)[8].

Coevolution of Capability Space and Opportunity Space. As Dagnino (2003) has previously clarified, the degree of alignment between CS (the firm's capability base with the perceived/targeted opportunities) and OS (the perceived/targeted opportunities) generates different possibilities of strategic gap (SG) bridging or closure and thus shows its inherent coevolutionary nature. More specifically, we can single out two counterpoised circumstances: (i) the perfect correspondence between CS and OS; (ii) their absolute dissonance. If there is perfect correspondence between CS and OS, then SG = 0, whereas if there is perfect dissonance, SG = MAX. Moreover,

[7] In order better to characterize the degree of closure of the strategy gap, it is beneficial to introduce a concept more comprehensive than the one of capability lifecycle. This concept is labelled the Modal Capability LifeCycle or MCLC (Mariani and Dagnino 2006) as it encompasses the set of all the capabilities of an individual firm at a certain moment 't'.

[8] Durable opportunities can be seized by means of organizational capabilities, whereas quick response (Richardson 1996) and dynamic capabilities (Teece et al. 1997) seem to be more appropriate for dissolving opportunities.

the capability-opportunity alignment is also relevant for CS and OS subspaces that are separated by what is called the local strategy gap (LSG). In the latter case, a perfect correspondence between CS subspace and OS subspace implies LSG = 0, whereas a perfect dissonance between CS subspace and OS subspace implies LSG = MAX.

Strategy Formulation and Strategy Implementation. As we have elsewhere underscored and recalled in section two, the concepts of capability space (CS) and opportunity space (OS) were originally characterized as displaying three main properties: strategic interrelatedness, coevolutionary nature, and the property of feed-forward. We have chosen to focus on the latter property (i.e., the property of feed-forward) in order better to characterize the received dichotomy between strategy formulation and strategy implementation. Feed-forward or positive feedback implies that the CS systematically influences the OS, which in its turn influences the CS. A coevolutionary process of circular causality takes place: each capability influencing the other and in its turn being influenced by the other; each opportunity influencing the other and in its turn being influenced by the other; each capability influencing each opportunity and in its turn being influenced by each opportunity.

According to Dagnino (2003), as regards the coevolution of CS and OS, the feed-forward property – together with the consideration that the firm's intervention is mainly carried out through three driving metacapabilities (foresight, judgment and mindfulness) – generates the so called *intentionality of action puzzle*. This firm-level puzzle is virtually consistent with the evolutionary economic stance that firms differ partly by choice, partly by history and partly by chance, given "an intriguing mix of intentional decisions and chance whose outcome is usually impossible to predict entirely ex ante and may turn out unintentional and unexpected consequences" (2003, 27).

Drawing on this received notion, it is our intention neither to solve the puzzle nor to overshadow it, but actually to try elucidating the terms in which the puzzle can solved by unveiling a novel connection between the coevolutionary interplay of CS and OS and the strategy formulation/implementation process.

Strategy formulation refers to the firm's identification and deployment of the capabilities essential to bridge the strategy gap between the CS and the OS and it specifically involves the PlanGS. Strategy formulation is hence a preliminary stage that typically takes place before the strategy gap is actually reduced to zero. The unfolding of this process occurs since not only do capabilities interact with opportunities, but they continuously evolve and interact with each other in a comprehensive dynamic system (the CS) guided by the feed-forward property. Correspondingly, this is nothing but the same property that characterizes the interactions among different opportunities in the OS. Strategy implementation is conversely related more closely to the operational strategy gap matching as it may involve both the DGS and the EGS, implying modifications in a few strategic gap sets such as the EntrGS and the PlanGS. Strategy implementation concerns the operational closure of the strategy gap and is achieved when one or more capabilities seize(s) one or more opportunities.

Formulation and implementation play distinctive roles in the reduction to zero of the deliberate/emergent strategy gaps (or the local strategy gaps). The reduction to zero of the deliberate strategy gaps (and the subsequent formation of DGS) implies a two-stage process which involves: 1) strategy formulation, or the firm's identification, predisposition and deployment of the capabilities essential to bridge the strategy gap between the CS and the OS; 2) strategy implementation, involving the operational closure (reduction to zero) of the strategy gap.

In other words, the DGS is the set of strategy gaps (SG) actually reduced to zero as a consequence of a firm's intentional predisposition of capabilities, through which it is capable of putting them together and deploying them: (a) in a *rational way*, and (b) in a *relational way*.

The order by which capabilities are arranged is *rational* as the firm utilizes its bounded rationality with the aim of recognizing the capabilities (or the set of capabilities) that are suitable to be matched with opportunities (or a set of opportunities). Firm rationality is bounded for, in Simonian terms (Simon 1983a,b), it cannot accomplish a perfect knowledge of the composition of CS and OS, and it is impossible for it to grasp all the possible connections between capabilities (or set of capabilities) and opportunities (or set of opportunities).

Likewise, the order by which capabilities are deployed is *relational* as it is possible to attain not only a connection between one or more capabilities and one or more opportunities, but also a connection between different capabilities in the capability space. In fact, the CS is a dynamic system in which capabilities interact with each other, influencing the way opportunities can be sensed and seized. Therefore, a firm could realize that, in order to seize a certain opportunity 'X', a set of three different but related capabilities ('A', 'B' and 'C') is required. Consequently, under bounded rationality the firm should choose the best way to predispose relationally the three capabilities 'A', 'B' and 'C' in order to seize the opportunity 'X'. This means that a set of capabilities in the capability space may be the condition necessary and sufficient for a firm to seize a given opportunity that it has sensed in the opportunity space. As they are incidentally the outgrowth of a series of interorganizational efforts, these capabilities may be termed as relational capabilities (Lorenzoni and Lipparini 1999).

Alternatively, the EGS is the set of strategy gaps (SG) actually reduced to zero lacking a two-stage process involving both formulation and implementation. It seems that, in this case of no intentionality, only *implementation* takes place in the predisposition and deployment of capabilities according to a process which is neither rational nor relational. Nonetheless, whenever one or more opportunities are seized through strategy implementation, a certain degree of rationality and relationality in the firm's way of acting and managing may be recognized ex-post. Yet, not infrequently, this may occur primarily in order to justify *au rebours* the closure of the SG.

6 Discussion

In this section, we try to depict the processes driving the firm, on the one hand, to the achievement of its execution objectives and, on the other hand, to the accomplishment of its innovation objectives, in accord with the conception of firm strategy as a process intended to bridge the strategy gap between capability space and opportunity space over time. Consequently, the section is aimed at offering a preliminary, but nonetheless consistent, answer to the fundamental research questions already presented at the beginning of section two.

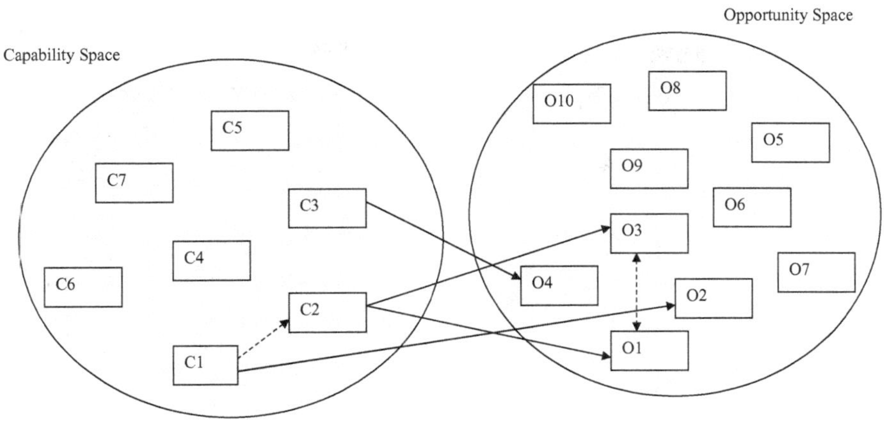

Fig. 2 Interaction between a Firm's Capabilities and Opportunities at Time t

Strategic Innovation Objectives. In the terms we have established heretofore, strategic innovation occurs whenever one or more novel *connections* between (one or more) capabilities and (one or more) new opportunities are achieved. Usually the new opportunities we deal with are dissolving opportunities, generated in complex, dynamic, time-pressured, and rapidly evolving environments. In particular, we can describe the process of strategic innovation as follows. Assume that, at time t, the capability space (CS) and the opportunity Space (OS) are as depicted in Figure 2. In this Figure, at time t the firm uses only a small subset of the entire collection of its available capabilities in the capability space (CS). We can label this subset as the subset of UsedCapabilities and formalize it at time t in the terms of the mathematical set theory as follows:

$$\text{UsedCapabilities}_t = UC_t = \{C1; C2; C3\}$$

The UsedCapabilities set is deployed in order to seize a number of opportunities, which in turn form a subset in the opportunity space (OS) and is formalizable as follows:

$$\text{SeizedOpportunities}_t = SO_t = \{O1; O2; O3; O4\}$$

As is apparent in Figure 2, the interaction between the elements of the CS and the elements of the OS is indicated with a continuous line, while the interactions within each of the two spaces are indicated with a dashed line, thus separating out the interspace linkages from the intraspace connections. We hypothesize that the firm has innovation objectives and intends to seize the dissolving opportunities 'O7' and 'O9'. It could do so once it has deployed a capability set such as the one that activates various specific capabilities that are more fit to seize specific opportunities (e.g., by means of significant R&D investments, the firm could be able to seize and exploit the new opportunities 'O7' and 'O9').

Yet, strategic innovation can take place even if there is no intentional redeployment of capabilities: in this case, we could have *emergent* or unintentional strategic innovation, or an innovation which has not been driven intentionally by the firm. Nonetheless, an innovation could emerge in the presence of a certain degree of intentionality. We can distinguish between two extreme cases: (a) the case in which

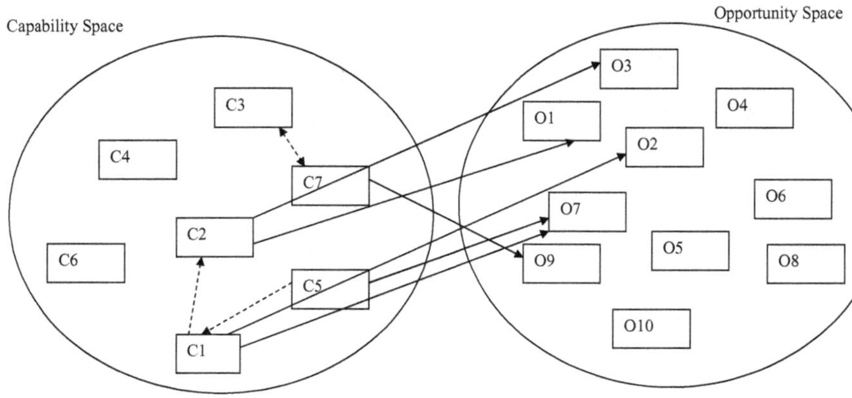

Fig. 3 The Interaction between a Firm's Capabilities and Opportunities at Time $t+1$

an emergent innovation can occur if one or more new seized dissolving opportunities are different from those the firm originally attempted to seize intentionally; (b) the case in which an emergent innovation can occur if the targeted dissolving opportunities are seized through the activation (or reactivation) of capabilities significantly *different* from those the firm has intentionally pre-deployed. In the former case, the CS and the OS at time $t+1$ can be represented as reported in Figure 3.

The Used Capabilities subset will be the following:

$$\text{UsedCapabilities}_{t+1} = UC_{t+1} = \{C1; C2; C3; C5; C7\},$$

and its opportunities counterpart is to be the following:

$$\text{SeizedOpportunities}_{t+1} = SO_{t+1} = \{O1; O2; O3; O5; O10\}.$$

Execution Objectives. Execution occurs whenever one or more old connections between (one or more) capabilities and (one or more) durable opportunities are in fact (re)activated. This is a *routinization* of the already deployed gap bridging processes. Differently from the case of strategic innovation, in the case of execution, the opportunities we deal with are durable opportunities, which are usually generated in sufficiently predictable and relatively stable environments and do not allow the handling of entirely new problems and the embarking on new experiences. Operational efficiency and refinement of the Carnegie-like routine-based or routinized behavior (March and Simon 1958; Nelson and Winter 1982) according to well-known or semi-automatic processes appears to be the most relevant element to take care of in this case. This state of affairs should not drive us to undervalue execution or to disregard it as a non-strategic activity. It is in fact never unproblematic or trouble-free for a firm to maintain its operational standards and performance, by performing its usual tasks with both efficiency and effectiveness navigating through time and space.

Explaining the Formulation/Implementation Strategy Dichotomy According to the Firm's Innovation/Execution Objectives. As is widely known, the relationship between strategic formulation and implementation has been traditionally addressed in strategy studies. Some early scholars in strategy (Andrews 1971; Chandler 1962; Learned et al. 1965) have conceived formulation and implementation as two different, separate and pretty rigid consecutive stages, with strategy ultimately determining the structure of the firm in a single period of time (wide practical support to this contention was initially found by Alfred Chandler in some large US corporations of the first half of the 20th century; e.g., GM, DuPont, Sears, Standard Oil New Jersey, etc.).

Other scholars (Mintzberg et al. 1976; Mintzberg 1979; Quinn 1980; Barney and Zajac 1994; Schendel 1994) have in turn questioned this stark dichotomic characterization of the strategy-structure relationship, supporting *de facto* the idea that this is more probably a two-way causal relationship, which presents a reciprocal influence of strategy over structure, and introducing the crucial distinction between deliberate and emergent strategies (both of which are implemented but, while the first is intentional, the second may not be so). The dichotomy between formulation and implementation is based on two related arguments:

1. the Simonian hypothesis of full or Olympic actor rationality (i.e., the strategic decision maker has complete information or in any case holds a quantity of information considerably superior to strategy implementers);
2. the hypothesis of stable contexts (i.e., the environment is so stable or predictable that no strategy reformulation and adaptation is required).

The sole absence of one of the two conditions above leads to the collapse of the traditional strategic dichotomy. Consequently, the simple linear sequence strategy formulation-strategy implementation turns into a much less predictable learning process, generating multiple feedbacks and feedforwards of implementation on formulation and vice versa, which have relevant effects on the dynamic gap bridging processes of the capability space and the opportunity space.

7 Implications for strategy theory and research agenda

Implications for strategy theory. A quick review of the arguments put forth in the previous section underscores that we do not have so far a clear distinction between formulation and implementation, especially when we have to deal with strategies addressing: (1) firm innovation; (2) firm execution and performance maintenance.

In case (1) reported above, innovation could play a role in both (1a) the stage of start-up of a brand-new firm, and (1b) the stage when factors, such as the saturation of traditional markets, impose – in order to keep the pace of a certain development rate – a diversification strategy and/or the requisites to building a more formalized organizational structure with precise responsibilities and an increased hierarchical distance along the power or command chain.

In case (1a), entrepreneurial gap bridging will be the main strategic activity of the firm. In this case, organizational learning (Argyris and Schön 1978; March 1988) is, in a nutshell, a direct consequence of a centralized process of trials and errors through which a firm's entrepreneur succeeds in figuring out the current environmental opportunities trying to match them to the firm's capabilities. Here there is an unambiguous sequential prevalence of formulation over implementation; but the opposite condition (i.e., the prevalence of implementation over formulation) is likely to prevail as well, with lower original impact and with the occurrence of an emergent innovation unrelated to the main business idea (Normann 1977) or business model.

In case (1b), it seems that the strategies actually realized are like Unconnected, Imposed and Process strategies and, therefore, the dynamic gap bridging would involve the realized gap set (RGS) subparts like UnconGS, ImpGS and ProcGS. In the latter case, organizational learning is a consequence of a *decentralized* process of trials and errors through which a firm (and especially firm's components such as divisions, units, departments) manages to sense the opportunities present in an unstable/unpredictable environmental context and seizeable on the basis of its own current capabilities.

In case (2), it seems that the realized strategy is likely to be closer to Entrepreneurial, Planned and Ideological strategies; thus the bridging of the dynamic gap should involve some realized gap set (RGS) subsections, such as EntrGS, PlanGS and IdeolGS. In this situation, the process of organizational learning is a consequence of a *centralized* process of trials and errors through which the firm (and especially the firm's composing parts, e.g., divisions, units, and departments) manages to figure out the opportunities available to seize its environment. If the environment is unstable/unpredictable, a firm will probably undertake entrepreneurial gap bridging processes, while if the environment is stable/predictable, it will undertake planned gap bridging processes (prevalence of formulation over implementation).

Following this line of reasoning, we may conclude that the two strategies of "driving innovation" and "maintaining performance", so crucially relevant to all firms operating in the competitive arena, are all but mutually exclusive. If we consider, for instance, the entrepreneurial gap bridging strategies (EntrGS), it is apparent that they do match with both strategies of addressing innovation and execution and performance maintenance. Additionally, if – on the path already traced by Henry Mintzberg (1982) – we consider that firm development is pithily a succession of subsequent sprints and pauses, we recognize that sprints are usually characterized by innovation and modification of the EntrGS (see Tushman and Romanelli 1985; Mocciaro Li Destri and Dagnino 2005). On the other hand, maintaining performance strategy is characterized by pauses which do not necessarily mean absence of EntrG bridging.

A specific condition worthwhile recalling in this vein is the one that characterizes the Chandlerian diversified or multibusiness firm (or the so-called M-form in Williamson's familiar terminology) as it customarily displays in unison a tension to ignite strategic innovation and sustain performance. Strategy planning and planning attitude (recently reappraised in Grant 2003, in relation to the empirical investigation of eight of the world's largest oil companies since the late 1970s) perform a role which proves extremely important equally in the case of strategic innovation and in that of performance maintenance. In the former, planning ordinarily supplant innovation in established firms; in the latter, planning is seen (as earlier noticed in Mintzberg and Waters 1982) as a way to line up and fine-tune a given strategy rather than the formulation of a brand new strategy.

A Proposed Research Agenda. Stemming from the arguments advanced earlier, we are able to point up the initial sketch of a vivid agenda for prospective research that is laden with a variety of intriguing issues, including at least the three following elements: (1) the introduction of an additional cognitive and learning dimension to the framework we have put forth, (2) the introduction of the concept of modal capability lifecycle as an extension of the notion of capability lifecycle, and (3) the application of the capability-opportunity coevolutionary framework at different levels of analysis vis-à-vis the firm level.

As far as point (1) is concerned, since the decision-maker rationality is necessary limited, and by no means can it be Olympic (in Simonian terms), and since the environment is potentially turbulent and highly unstable, the strategy gap *perceived* by the entrepreneur/manager may not necessarily coincide with the *actual* strategy gap. Accordingly, as there exist actual and perceived opportunities, there will be actual and perceived capabilities. An intriguing example in this vein which emphasizes the role of perception for capabilities is the one related to a firm's functional area (for example the R&D area), which could well recognize its own capabilities without knowing so well the capabilities of another area (for example, the marketing are or the manufacturing area). The same line of reasoning applies to a multinational firm in relation to its subsidiaries dispersed in different geographical regions. This logic entails in turn a lack of overlapping between the perceived gap and the actual gap, with a consequential partial closure of the actual gap, endangering an elevated feedback from implementation to formulation and a significant impact on both the expansion/reduction of the RGS, and the underlying organizational learning processes.

As much as point (2) is concerned, the intellectual efforts may concentrate on the linkages between the particular lifecycle stage of an individual capability and its degree of 'ability' to seize opportunities. Additionally, a characterization of the concept of capability lifecycle (Helfat and Peteraf 2003) consistent with the statistically distributed prevalence of capabilities in a specific stage of their lifecycle and available for a firm at a certain time '*t*', could become relevant in order to understand the effectiveness of the strategies implemented by that firm (Mariani and Dagnino 2006). Ultimately, both the strategies that tackle strategic innovation and those that address performance maintenance can be connected to the capability lifecycle (CLC) and to the modal capability lifecycle, in order to understand the conditions under which strategic innovation and/or execution are more likely to occur.

As far as point (3) is concerned, we confirm that, until this point, we have taken into account the level of the single firm. Yet, in the footsteps of Schumpeterian (Schumpeter 1934) and post-Schumpeterian inquiry (Nelson and Winter 1982; Dosi et al. 2000), an entire industry and, similarly, the economic system taken as a whole, may be elucidated in terms of a coevolutionary gap bridging process between capabilities and opportunities. We underscore that, when an entire industry is under scrutiny, the opportunity space will be somewhat similar for all the firms belonging to that industry, while the capability space will be different in relation to each of the firms considered. When instead the whole economic system is under examination, the OS will differ significantly with respect to the industry to which each firm belongs and, comparatively, the CS will vary according to the individual firm taken into account.

References

Abbas AK, Lichtman AH, Pober JS (2000) Cellular and molecular immunology. Sunders, Philadelphia (PA)

Anderson P (1999) Complexity theory and organization science. Organ Sci, 10, 3: 216–232

Andrews KR (1971) The concept of corporate strategy. Dow Jones-Irwin, Homewood (IL)

Argote L, Darr E (2000) Repositories of Knowledge in franchise organizations: Individual, structural and Technological. In: Dosi G, Nelson RR, Winter SD (eds.) The nature and dynamics of organizational capabilities, Oxford University Press, New York

Argyris C, Schön DA (1978) Organization learning. Addison Wesley, Reading (MA)

Barney JB, Zajac EJ (1994) Competitive organizational behavior: Toward an organizationally-based theory of competitive advantage. Strategic Manage J, 15, Winter Special Issue: 5–9

Boulding KE (1956) General system theory: The skeleton of science. Manage Sci, 2

Chandler A (1962) Strategy and structure: Chapters in the history of the American industrial enterprise. MIT Press, Cambridge (MA)

Dagnino GB (2003) Bridging the strategy gap: Firm strategy and coevolution of capability space and opportunity space. The Wharton School, Sol C. Snider Center for Entrepreneurial Research, Working Paper n.306

Dagnino GB, Mariani MM (2004) Multilevel learning and dynamic gap bridging: The strategic role of the firm in the coevolution of capability space and opportunity space. Presented at the Strategic Management Society Conference. San Juan, Puerto Rico: Oct. 31-Nov. 3

Dosi G, Nelson RR, Winter SD (2000) The nature and dynamics of organizational capabilities. Oxford University Press, New York

Doz YL, Olk PM, Ring PS (2000) Formation processes of R&D consortia: Which path to take? Where does it lead? Strategic Manage J, 21, 3: 239–266

Eisenhardt KM, Martin JA (2000) Dynamic capabilities: What are they? Strategic Manage J, 21, Special Issue 10-11: 1105–1121

Grant RM (2003) Strategic planning in a turbulent environment: Evidence from the oil majors. Strategic Manage J, 24, 6: 491–517

Helfat CE, Peteraf MA (2003) The dynamic resource-based view: Capability lifecycles. Strategic Manage J, 24, 10: 997–1010

Henderson R, Cockburn I (1994) Measuring Competence? Exploring Firm Effects in Pharmaceutical Research. Strategic Manage J, 15, Winter Special Issue: 63–84

Kirzner IM (1979) Perception, opportunity and profit. Studies in the theory of entrepreneurship. University of Chicago Press, Chicago

Kirzner IM (1997) Entrepreneurial discovery and the competitive market process: An Austrian approach. J Econ Lit, 35: 60–85

Langlois RN (1995) Capabilities and coherence in firms and markets. In: Montgomery CA (ed.) Resource-based and evolutionary theories of the firm: Towards a synthesis, 71–100, Kluwer, Boston

Learned EP, Christensen CR, Andrews K, Guth WD (1965) Business policy: Text and cases. Richard Irwin, Homewood (IL)

Levinthal D, Myatt J (1994) Co-evolution of capabilities and industry: The evolution of mutual fund processing. Strategic Manage J, 15, Winter Special Issue: 45–62

Lewin AY, Volberda HW (1999) Prolegomena on coevolution: A framework for research on strategy and new organizational forms. Organ Sci, 10, 5: 519–534

Lorenzoni G, Lipparini A (1999) The leveraging of interfirm relationships as a distinctive organizational capability: A longitudinal study. Strategic Manage J, 20, 4: 317–338

March JG (1988) Decisions and organizations. Basil Blackwell, Oxford

March JG, Simon HA (1958) Organizations. John Wiley, New York

Mariani MM, Dagnino GB (2006) Unveiling the modal capability lifecycle: The coevolutionary foundations of the firm's effectiveness in capturing strategic opportunities. Forthcoming: International Journal of Learning and Intellectual Capital, 3

Miller DA (2003) An asymmetry-based view of advantage: Towards an attainable sustainability. Strategic Manage J, 24, 10: 961–976

Mintzberg H (1972) Research on strategy-making. Proceedings of the 32nd Annual Meeting of the Academy of Management

Mintzberg H (1978) Patterns in strategy formation. Manage Sci, 24, 9: 934–948

Mintzberg H (1979) An emerging strategy of 'direct' research. Admin Sci Quart, 24, 4: 580–589

Mintzberg H, McHugh H (1985) Strategy formation in adhocracy. Admin Sci Quart, 30, 2: 160–197

Mintzberg H, Otis S, Shamsie J, Waters WJ (1988) Strategy of design: A study of architects in copartnerships. In: Grant J (ed.) Strategic Management frontiers, JAI Press, Greenwich (CT)

Mintzberg H, Raisinghani D, Theoret A (1976) The structure of unstructured decision processes. Admin Sci Quart, 21, 2: 246–275

Mintzberg H, Waters WJ (1982) Tracking strategy in an entrepreneurial firm. Acad Manage J, 25, 3: 465–499

Mintzberg H, Waters WJ (1984) Researching the formation of strategies: The history of Canadian Lady, 1939-1976. In: Lamb R (ed.) Competitive Strategic Management, Prentice-Hall, Englewood Cliffs (NJ)

Mintzberg H, Waters WJ (1985) Of strategies, deliberate and emergent. Strategic Manage J, 6, 3: 257–272

Mocciaro Li Destri A, Dagnino GB (2005) The Resource-Based Firm Between Value Creation and Value Appropriation. Adv Strateg Manage, 22: 153–188

Murmann JP (2003) Knowledge and competitive advantage. The coevolution of firms technology and national institutions in the synthetic dye industry, 1850-1914. Cambridge University Press, Cambridge

Nelson RR, Winter S (1982) An evolutionary theory of economic change. Belknap Press, Cambridge (MA)

Nelson RR, Winter SG (2002) Evolutionary theorizing in economics. J Econ Perspect, 16, 2: 23–46

Normann R (1977) Management for growth. John Wiley, New York

Penrose ET (1959) The theory of the growth of the firm. John Wiley, New York

Peteraf MA (1993) The cornerstone of competitive advantage. Strategic Manage J, 14: 179–191

Quinn JB (1980) Strategies for change. Logical incrementalism. Richard D. Irwin, Georgetown (ON)

Richardson J (1996) Vertical integration and rapid response in fashion apparel. Organ Sci, 7, 4: 400–412

Rumelt R (1987) Theory, strategy and entrepreneurship. In: J TD (ed.) The Competitive Challenge, Ballinger, Cambridge (MA)

Rumelt RP (1991) How much does industry matter? Strategic Manage J, 12: 167–185

Schendel D (1994) 'Introduction' to Competitive organizational behavior: Toward an organizationally-based theory of competitive advantage. Strategic Manage J, 15, 8 (Winter Special Issue): 1–4

Schumpeter J (1934) The theory of economic development. An inquiry into profits, capital, credit, interest, and the business cycle. Harvard Business Press, Cambridge (MA)

Schumpeter JA (1935) The analysis of economic change. Rev Econ Statist, 17, 4: 2–10

Simon HA (1983a) Models of bounded rationality. Vol. 2: Behavioral economics and business organization. MIT Press, Cambridge (MA)

Simon HA (1983b) Reason in human affair. Basil Blackwell, Oxford

Teece DJ, Pisano G, Shuen A (1997) Dynamic capabilities and strategic management. Strategic Manage J, 18, 7: 509–533

Tushman ML, Romanelli ER (1985) Organizational evolution: A metamorphosis model of convergence and reorientation. In: Cummings LL, Staw B (eds.) Research in Organizational Behavior, JAI Press, Greenwich (CT)

Winter SD, Fang C, Denrell J (2003) The economics of strategic opportunity. Strategic Manage J, 24, 10: 977–990

Part 6:

Patents – patenting strategies and impacts of patents

Carine Peeters · Bruno van Pottelsberghe de la Potterie

Innovation strategy and the patenting behavior of firms

Abstract This paper investigates whether firms' innovation strategies affect their patenting behavior, as measured by both the probability of having a patent portfolio and the number of active patents held. Three main dimensions of an innovation strategy are taken into account: the relative importance of basic research, applied research and development work in total R&D activities, the product or process orientation of innovation efforts, and the extent to which firms enter into collaborative R&D with other institutions. The major findings can be summarized as follows: *(1)* taking into account the various dimensions of an innovation strategy turns out to approximate the patenting behavior of firms better than the traditional Schumpeterian hypotheses related to firm size and market power; *(2)* there is a positive relationship between the patent portfolio of firms and an outward-oriented innovation strategy characterized by R&D partnerships with external organizations — scientific institutions and competitors in particular; *(3)* process-oriented innovators patent less than product-oriented innovators; *(4)* a stronger focus on basic and applied research is associated with a more active patenting behavior; *(5)* firms that perceive high barriers to innovation (internal, risk-related or external

The authors thank the participants to the 2004 International Schumpeter Society Conference, the 2004 Association d'Econometrie Appliquee Conference on Innovation and Intellectual Property, and the 2005 DRUID Summer Conference, and an anonymous reviewer for their insightful comments. This research was partly performed when Bruno van Pottelsberghe was Visiting Professor at the Institute of Innovation Research (IIR), Hitotsubashi University. Carine Peeters greatly acknowledges a post-doctoral fellowship from the Belgian American Educational Foundation (BAEF).

C. Peeters (✉)
The Fuqua School of Business, Duke University, Box 90120,
Durham, NC 27708-0120, USA
E-mail: carine@duke.edu

B. van Pottelsberghe de la Potterie
Solvay Business School, Universite Libre de Bruxelles (ULB),
Centre Emile Bernheim, DULBEA and CEPR, CP 145-01, av. F.D. Roosevelt 50,
1050 Brussels, Belgium
E-mail: bruno.vanpottelsberghe@ulb.ac.be

barriers) have smaller patent portfolios; *(6)* the perceived limitations of the patent system do not significantly influence the patenting behavior, suggesting that firms patent for other strategic reasons than merely protecting innovation rents.

Keywords Product innovation · Innovation strategy · Collaborative R&D · Patent portfolio · Patent system

JEL Classification O31 · O32 · O34 · L25

1 Introduction

The literature attempting to understand the patenting behavior of firms has mainly focused on 'traditional' determinants related to firm and sector characteristics. These determinants are generally borrowed from the literature on innovation performance. They relate to the well known Schumpeterian hypotheses (i.e., large firms and firms with a strong market power innovate more) and to indicators of technological opportunity and research efforts. The present paper adds to the existing body of knowledge on patenting determinants by investigating the influence an innovation strategy may exert on the likelihood that a firm develops a patent portfolio and on the number of patents it holds.

A formal firm-level measurement of innovation strategy enables us to account for three main dimensions of an innovation strategy: the relative importance of basic and applied research in the total R&D activities of a firm, the product or process orientation of innovation efforts, and the extent to which R&D is performed jointly with other institutions (competing firms, vertical partners, universities, public labs...). The theoretical framework further suggests that a firm's patenting behavior might be influenced by its perception of the limitations and inefficiencies of the patent system. Furthermore, it is argued that the patenting behavior might correlate with the perceived barriers to the innovation process (internal barriers, risk and cost-related barriers and external barriers to innovation).

This paper departs from existing empirical studies on patent determinants by going beyond the traditional factors related to firm size, market power, technological opportunity, and research efforts. The main contribution consists in explicitly taking into account differences in innovation strategies and in the perception of barriers to innovation and to patenting. It also differs from previous studies in the indicator of patenting behavior used. While the existing literature on patent determinants has mainly focused on the number of yearly patent applications made by firms, this research uses data on the number of patents firms have accumulated over time in their patent portfolio, and for which they still pay renewal fees. This approach enables us to control for contextual effects that may affect the patenting behavior of a firm in a given period without necessarily reflecting its general attitude towards patenting.

The main finding of the present paper concerns the use of R&D partnerships in a firm's innovation strategy. An outward-oriented innovation strategy that relies on collaborations with external organizations is associated with a more active patenting behavior than an inward-oriented innovation strategy that relies on in-house R&D exclusively. The type of partners firms work with is also important in determining their attitude vis-à-vis patenting. Firms that do R&D with scientific institutions and competitors patent more than firms that collaborate with their customers, their suppliers or with consultants. The positive effect of partnerships with scientific institutions probably results from the basic nature of the research activities involved, which are more likely to result in patentable knowledge. This effect is called the 'novelty' effect. Undertaking R&D activities with external organizations implies, at least to some extent, a mutual access to the partners' knowledge bases. Such partnership induces a higher need for intellectual property (IP) protection, especially in the case of collaborations with competing firms. This effect is referred to as the 'need' effect. More generally, a patent also helps clarifying issues of ownership over co-developed inventions and prior knowledge of each partner, which further explains the positive relationship observed between R&D partnerships and patenting behavior.

The econometric results also show that an innovation strategy that targets the development of new processes is associated with lower patenting behavior, while product-oriented innovators patent more than other firms. The importance given to research activities, as opposed to development work, also positively affects the probability of having a patent portfolio. The perceived barriers to innovation turn out to be a good predictor of firms' patenting behavior, as opposed to the perceived barriers to the patenting system, which do not have any significant impact.

The paper is structured as follows. The next section summarizes the main findings of the existing empirical literature on determinants of patenting activities. Section 3 presents the theoretical model underlying this study with a particular emphasis on three hypotheses related to the role played by innovation strategy variables. The empirical implementation is developed in Section 4. Two econometric models are used to evaluate the impact of several potential determinants of patenting behavior. The first model estimates the probability for a firm to have a patent portfolio. The second model estimates the number of patents possessed. Section 5 discusses the empirical results. Section 6 concludes.

2 Traditional determinants of patenting

Pioneer work in the field of patent economics probably started with the contributions of Schmookler (1957), Nelson (1959), Arrow (1962), and Scherer (1965). Since then, academic research has increasingly tackled various aspects of the patenting behavior of firms, from the theoretical analysis of patent systems (e.g., Baumol 2002), to the use of patent data to measure innovation performance and knowledge spillover indicators (e.g., Griliches 1990). With the development of extensive and accessible patent databases, several authors have analyzed the micro-determinants of innovation using patent indicators as a measure of innovative output (e.g., Crépon et al. 1996, 1998; Duguet and Kabla 1998; Cohen et al. 2000).[1] Most of these studies have focused on traditional determinants of patenting

[1] Patent-based indicators are only one measure of innovation output. They are imperfect for three main reasons (Griliches, 1990). First, not all innovations are patentable, since the three conditions of non-obviousness, inventive step and industrial application must be satisfied in order to get a patent application granted. Second, the propensity to patent 'patentable' inventions varies considerably across firms, time, and industry (see for instance Scherer 1983; Hall et al. 1986; Arora 1997). Third, in some sectors, patent protection is relatively inefficient and secrecy is favored as a mechanism to secure the rents due to an invention. The importance of the various protection mechanisms varies across industries and patents are important for only a few of them, mainly chemicals and pharmaceuticals (Mansfield 1986; Levin et al. 1987).

behavior, such as firm size, market power, market and technological opportunities, and R&D efforts.

The effect of firm size on patenting is systematically taken into account in the extant literature. This derives from the well-known Schumpeterian hypothesis that large firms are more innovative than smaller ones (Schumpeter 1942). The advantage of being large comes from three main factors summarized by Cohen and Levin (1989). First, large firms benefit from economies of scale and scope that make them more competitive in comparison to their smaller competitors. Second, they can benefit from complementarities and spillovers between different departments. Third, large firms are favored by capital markets for the financing of risky innovation projects. Although the empirical evidence seems to tilt towards a validation of the Schumpeterian hypothesis, some authors argue that the relationship is not straightforward. For instance, Baldwin et al. (2002) find that the effect of firm size depends on the innovation indicator used, with a weaker relationship when relying on patent data than when relying on the percentage of innovative sales, i.e. an indicator of innovation output. According to van Ophem et al. (2001) the effect of firm size on patent applications is debatable. Large firms can more easily rely on market lead to secure innovation rents, and hence are less likely to need patent protection. However, they are better able to set up a patent department and to face potential litigations. The econometric analysis of van Ophem et al. (2001) shows a positive effect of firm size on the number of patent applications. However, some authors find no significant impact of the size variable when it is controlled for other factors such as industry effects, differences in access to external know-how, and appropriability conditions (Duguet and Kabla 1998; Crépon et al. 1998; Cassiman et al. 2001). Brouwer and Kleinknecht (1999) shed some light on this debate by using two different patent indicators. They find that the probability of having at least one patent application increases more than proportionately with the number of employees, while the number of patent applications increases less than proportionately. This means that small firms that do apply for patents do it proportionately more, probably to compensate for disadvantages in terms of market share and brand name.

Another determinant of innovation that is quite controversial relates to the level of competition. This is usually measured by an indicator of market share or by an index of industry concentration. The debate originates from Schumpeter's hypothesis that firms with a higher market power are more innovative than firms with weak market power (Schumpeter 1942). This hypothesis has been challenged by several authors. Two effects work indeed in opposite directions. On the one hand, there is the replacement effect, or cannibalization effect, implying that firms with more market power invest less in innovation because the potential gains they would get would only replace current gains (Arrow 1962). On the other hand, there is the efficiency effect, following which firms with a high market power invest more in innovation because they do not face competition for the exploitation of their inventions (Gilbert and Newberry 1982). The impact of this variable varies quite importantly according to the indicator of innovation used. Results of studies using the same innovation indicator are often contradictory as well, as illustrated by Cohen and Levin (1989) in their review of the literature on the relationship between R&D and market power. Concerning the number of patent applications, Duguet and Kabla (1998) and Nielsen (2001) find a positive impact of firm's market power, i.e. the efficiency effect would dominate the replacement effect.

Two variables that are also traditionally included in patent equations are market and technological opportunities. Market opportunities reflect the existence of some market in demand of novelty. Technological opportunities are generally measured at the industry level and were defined by Levin et al. (1987) as the extent to which an industry relies on science-based research. Demand-pull variables are expected to have a stronger impact on innovation output indicators than on R&D investments because output measures are more directly linked to the market. In this respect, the status of patent indicators is ambiguous. Patents are a kind of intermediate indicator reflecting the output of research activities but not necessarily implying the commercialization of an invention. Firms in high technological opportunity sectors are found to patent more than other firms (Crépon et al. 1996, 1998; Brouwer and Kleinknecht 1999) but the difference is not always significant (Duguet and Kabla 1998; Baldwin et al. 2002). As regards market opportunity variables, Crépon et al. (1996) find that they have a positive and significant impact on patent applications, but other authors find no significant effect (Duguet and Kabla 1998; Crépon et al. 1998; Cassiman et al. 2001).

Another issue that has largely been studied in the literature is the relationship between R&D and patents. Scherer (1965) considers patents as an indicator of R&D success. In this perspective, R&D precedes patent applications and the causality goes from R&D to patents. More recently, Hall et al. (1986) argued that there is a strong contemporaneous effect between R&D and patenting and that it is difficult to find the adequate lag structure between them. Most studies that include an R&D indicator in patent equations find a positive and significant relationship (e.g., Duguet and Kabla 1998; Crépon et al. 1998; Brouwer and Kleinknecht 1999). Actually, the relationship between R&D and patents can be seen as a virtuous cycle - the former induces the latter, which in turn requires further development costs in order to reach the market.

3 Theoretical framework and hypotheses

Beside the traditional determinants related to firm and sector characteristics widely discussed in the literature, this paper explicitly takes into account the effect that different innovation strategies may have on the patenting behavior of firms. Variables that reflect the perception firms have of certain barriers to innovation and to the use of the patent system are introduced as well. Different firms pursuing similar innovation strategies may indeed have different attitudes as regards patenting because they differ in their perception of the limitations of the patent system. Differences in observed patent portfolios may also reflect a lower innovativeness resulting from higher perceived barriers to innovations. Both effects will be controlled for in the empirical models.

As far as the innovation strategy is concerned, three dimensions are tackled in this paper: the kind of innovation a firm seeks to develop (product or process), the type of R&D activities it undertakes (basic and applied research or mainly development work), and the extent and type of R&D collaborations with external organizations (outward- versus inward-oriented innovation strategy).

Hypothesis 1a The importance of the development of new products in a firm's innovation strategy is associated with a higher probability to have at least one patent, and with a larger patent portfolio.

Hypothesis 1b The importance of the development of new processes in a firm's innovation strategy is associated with a weaker probability to have at least one patent, and with a smaller patent portfolio.

A firm's innovation strategy can focus on product innovation, process innovation, or both. It is traditionally found that new processes are less likely to be patented (Arundel and Kabla 1998; Brouwer and Kleinknecht 1999) as secrecy is a more appropriate protection mechanism for this type of innovation (Cohen et al. 2000). It is indeed more difficult to track down imitation of processes than imitation of products. Therefore, the publication of technical information a patent requires may be acceptable in the case of a product innovation, for which infringement is easier to detect, but not for a process innovation. Moreover, imitating a process innovation might be more difficult than a product innovation because a lot of specific know-how is generally needed in order to make use of a new process, specific expertise that imitators lack. Firms might therefore opt for a non-legal protection mechanism such as secrecy. At the opposite end of the spectrum, product innovations might be easier to imitate through reverse-engineering and legally enforceable protection might be needed.

Hypothesis 2 A higher proportion of basic and applied research in total R&D budget leads to a higher probability of patenting and to larger patent portfolios.

If the positive relationship between R&D efforts and patenting has been widely illustrated, there is no evidence so far about the content of R&D. R&D is traditionally composed of basic research, applied research and development. Since patents are by definition a codification of an invention, they might rather be the outcome of basic and applied research as opposed to development activities. The latter would surely be associated with patenting (development of inventions), provided a sufficient share of total R&D is devoted to basic and applied research.

Hypothesis 3 Taking part in research partnerships leads to a higher probability of having at least one patent, and to larger patent portfolios.

In order to develop new products and processes, certain firms rely on in-house R&D exclusively. Other firms are more outward-oriented and enter into R&D collaboration agreements in order to access external knowledge, share the risks and costs of innovation with other organizations, and accelerate the innovation process. Organizations with which firms can collaborate to implement R&D projects are numerous. They include competitors, customers, suppliers, universities, research institutes, consultants, etc. Launching R&D partnerships is likely to increase the need for patent protection because it implies, at least to some extent, a sharing of one's knowledge with external organizations. A legally enforceable protection mechanism such as a patent is also helpful to clarify issues of ownership over co-developed knowledge. Both arguments stand in favor of a positive effect of R&D partnerships on patenting activity. Moreover, when patented, a firm's knowledge-base and innovation output become tradable assets that can be very useful when negotiating future collaboration agreements. Using a dummy variable, Brouwer and Kleinknecht (1999) and van Ophem et al. (2001) find that firms participating in research partnerships apply for more patents than firms that focus more on internal research.

The theoretical model underlying this study suggests that, in addition to the traditional determinants discussed in extant literature and the role played by the type of innovation strategy pursued, the patenting behavior of firms may be influenced by the limitations of the patent system they recognize and by the barriers that perceive as hindering their innovation efforts.

The advantage for a firm to patent an invention is not always clear, since a patent offers protection to its holder at the high indirect cost of revealing important technical information. Applying for a patent does not seem to be the most popular protection mechanism for manufacturing firms, which often favor secrecy and lead time over competition (Levin et al. 1987; Brouwer and Kleinknecht 1999; Cohen et al. 2000; Arundel 2001). The risk to having competitors "inventing around" and the disclosure of critical information are the most important reasons patents are not always considered as an efficient protection mechanism of innovation rents (Levin et al. 1987; Scotchmer and Green 1990; Cohen et al. 2000). This is illustrated by Mansfield et al. (1981), who find that patent protection does not increase imitation time and costs dramatically. Firms that perceive a higher "ineffectiveness" of the patent system and a higher cost of patenting are therefore likely to patent less than other firms.

Some authors use innovation survey data to test the effect of potential barriers to innovation on firms' innovation activities (e.g., Lööf and Heshmati 2002; Veugelers and Cassiman 1999). They generally find that a lack of interest from customers, a lack of technological information and a lack of qualified personnel have a negative impact on innovation performance. Cassiman et al. (2001) find that high innovation costs and lack of financing have a positive effect on firms' propensity to patent. This counter-intuitive positive effect highlights a recurrent problem of measurement of the barriers to innovation. Indeed, it is often difficult to discern firms' perception of barriers from the barriers that actually hinder their innovation efforts and patenting activities. Moreover, at similar levels of innovation output, firms that encountered difficulties affording the high development costs of their innovations may be more likely to patent than other firms. Nevertheless, it seems reasonable to expect firms that perceive higher barriers to innovation to patent less than other firms, as a result of lower innovativeness.

4 Empirical implementation

4.1 The model

Two econometric models are used in order to identify the determinants of firms' patenting behavior. The first one focuses on whether firms have a patent portfolio or not, i.e. the probability of a firm having at least one active patent. The second one is intended to explain the size of the patent portfolio, i.e. the number of active patents a firm possesses. Crépon et al. (1996) and Brouwer and Kleinknecht (1999) have already adopted this dual approach. However, they used information about the number of yearly patent applications, while the present data concern active patents in the patent portfolio of firms.

Three reasons justify the choice of patent portfolios as indicators of patenting behavior. First, patents in a firm's portfolio are 'active' in the sense that renewal fees have been paid. In the case of older patents, this guarantees that the innovation rents they are supposed to protect are actually still protected. Second, these patents bear in themselves an element of 'quality' that patent applications lack. These patents have indeed been granted, which means that the inventive step and industrial applicability of the underlying technology have been recognized by a competent institution. As a result, studying patent portfolios instead of patent applications guarantees that patents that are looked at can actually be used both as a protection mechanism and as strategic tools to build a firm's technological and competitive position. Third, a patent portfolio is built over time. It is therefore a more adequate indicator of patenting behavior since it is less subject to particular events that may affect the number of patents a firm applies for in a given period of time. A firm may be a recurrent innovator with a strong preference for patent protection but no patent application in the particular time frame in which a study is interested. Another firm may happen to apply for one or more patents because of a significant technological invention that would be made during the studied period. This may, however, be highly contextual and not reflect a general active patenting behavior. Looking at the patent portfolio a firm builds over time enables us to lower these potential contextual biases and provides a more stable indicator of patenting behavior.

The empirical methodology is illustrated in Fig. 1. The explanatory variables are listed in the left-hand side box. The econometric models and corresponding dependent variables are listed in the middle and right-hand side boxes, respectively. A binary logit model is used to estimate the probability of a firm having at least one patent. A count model with a negative binomial specification is used to estimate the number of active patents firms possess.[2] Since the dependent variable is not especially determined by independent stochastic processes, in order to check the robustness of the results a tobit model is also used to estimate the size of the patent portfolio.

4.2 The data

The data set used in this study comes from an original survey on firms' innovation competencies and performance undertaken in Belgium in 2001. The questionnaire was sent to the CEO's of 1,301 large firms. The selection criteria depended on the main sector of activity of firms. The OECD terminology was used to classify the sectors.[3] For lower-technology firms (ML and LT) and service companies, the criteria were a number of employees greater than or equal to 201 and a total turnover

[2] Since the distribution of the patent variable is skewed towards low values, a negative binomial model is more appropriate than a Poisson model. It allows the conditional mean and variance of the dependent variable to differ.

[3] The OECD classifies manufacturing sectors into four categories: high-tech (HT), medium-high-tech (MH), medium-low-tech (ML), and low-tech (LT). HT=aeronautic construction, desks and computing machines, pharmaceuticals products, radio, TV and telecommunication machines; MH=professional equipment, motorcar vehicles, electric machines, chemical industries, other transport equipment, non-electric machines; ML=rubber and plastic materials, naval construction, other industrial sectors, non-iron metals, non-metallic mineral products, metallic works, petroleum and coal, steel industry; LT=paper, printing and editing, textile industry, clothing and leather, food, drinks and tobacco, wood and furniture. A category was added for all service companies: commerce, hotels and restaurants, transports, posts and telecommunications, insurances, financial services, real estate activities, computer activities.

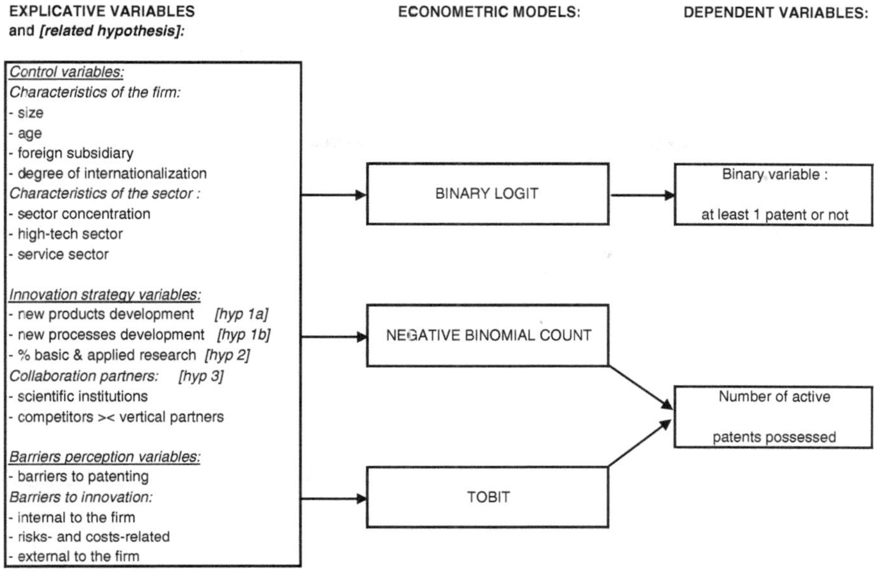

Fig. 1 Econometric framework

greater than USD 2.5 million. The criteria for higher-technology firms (HT and MH) were less restrictive, since all firms with at least 51 employees and a turnover above USD 1.25 million were selected. Data on the total number of employees and turnover of firms came from the Dun & Bradstreet database. Given Belgium's language disparity, CEO's were proposed to answer the questionnaire in Dutch or English and in French or English, depending on the region in which they were based.

Respondents were asked to rate their use of several routines and practices related to four main aspects of the innovation process: the development of a corporate culture of innovation, the generation of innovative ideas, the implementation of innovation projects, and the protection of intellectual property. For most of these questions, respondents had to answer on a Likert scale, ranging from 5 to 0 (systematically, often, sometimes, seldom, not at all, irrelevant). A few questions required yes/no answers or broad percentages. Firms were also asked to answer general descriptive questions and to provide information on the barriers to innovation they perceived and on their innovation output as regards the last three years preceding the survey.

The survey builds on existing surveys on innovation such as the Yale Survey and Carnegie-Mellon Survey on Industrial Research and Development in the U.S., the Community Innovation Survey launched in several European countries jointly by EUROSTAT and OECD, the Growing Small- and Medium-Sized Enterprise Survey in Canada, and the survey on innovation competencies launched in France by the SESSI (Industrial Statistics Service at the French Ministry of Industry). However, it distinguishes from previous surveys by combining six main peculiarities. Firstly, it adopts a systematic approach of the competencies along the entire innovation process, from ideas and knowledge generation to commercialization of new products and processes. Secondly, firms are asked to give a relatively precise evaluation of their use of several routines and practices, since most questions have to be answered on a 5-point Likert scale and not on a 'yes or no' basis. Thirdly, all firms have to answer all questions, not only those that have introduced an innovation over the last few years preceding the survey. This aims at reducing the risk of bias towards innovative firms. Fourthly, the survey is not restricted to manufacturing firms, since service companies are included in the sample as well. Fifthly, the different competencies are assessed using a large range of questions on specific routines and practices. This enables the gathering of detailed information while causing a relatively low response rate, mainly because of the length of the questionnaire. Finally, this is the first survey dedicated to innovation competencies of firms in Belgium.

4.3 Basic statistics

A total of 148 questionnaires were filed and sent back. The comparison of response rates according to firm size, sector of activity, age, region of activity in Belgium, and language of the CEO does not highlight significant differences. This limits the risk of selection bias that may result from a low response rate. Only large firms with more than 500 employees have a slightly higher response rate that should be taken into consideration when interpreting the empirical results.[4]

Large firms account for 33% of respondents (see Appendix 1). Medium and small firms represent 36% and 31%, respectively. Service companies are the most represented category with 28% of the sample, followed by medium high-tech firms that account for 26%. High-tech firms represent about 10% of responding firms. Medium-low-tech and low-tech firms represent 20% and 16%, respectively. In terms of geographical repartition, 54% of firms are based in the Flemish-speaking part of Belgium, 26% in Brussels, and 18% in the French-speaking part of Belgium. Belgian subsidiaries of foreign groups account for 52% of the sample. This repartition reflects quite well the composition of the Belgian economy, with an important percentage of major firms operating in the North of the country (Flemish-speaking) and being foreign-owned subsidiaries.

An extensive statistical analysis of the survey results can be found in Peeters and van Pottelsberghe de la Potterie (2003a). Among the participants to the survey, about 80% of firms declared some kind of R&D activity and a similar percentage claimed to have a positive share of their turnover coming from new or significantly improved products or processes in 2000. However, only about one out of two firms have applied for at least one patent between 1990 and 2000. This percentage is significantly higher for large firms and firms operating in sectors of relatively high technological opportunity (see Fig. 2).

Possible reasons as to why firms do not systematically patent their inventions are summarized in Fig. 3. The highest rated barriers that prevent firms from using the patent system to protect their innovation rents relate to the lack of effectiveness of the system. Many firms consider that market lead is more efficient than relying on the protection of a patent. More than half of firms feel unable to prevent competitors from copying their technology even if a patent protects it, and 44% of

[4] In the remainder of this section, large firms refer to firms with 500 employees or more, medium firms have between 200 and 499 employees, and small firms do not exceed 199 employees.

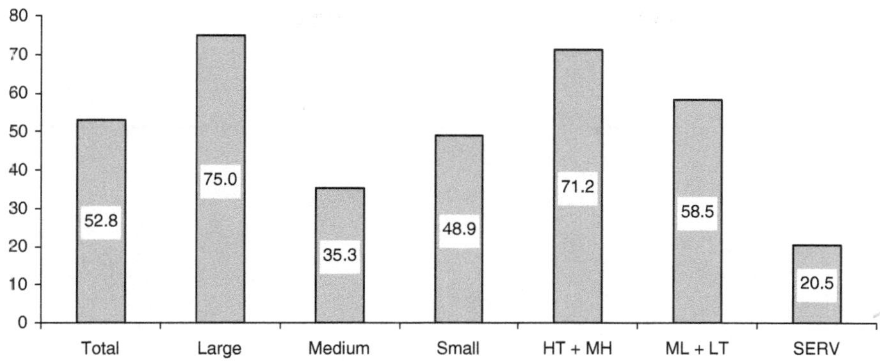

Fig. 2 Percentage of firms with at least one patent application between 1990 and 2000

firms consider secrecy as more efficient than patenting. The next category of factors that prevent firms from using the patent system relates to the cost of patents. The cost of protection in case of litigation and the cost of fees are considered to significantly limit the use of the patent system by about 40% of firms. Finally, only one firm out of five does not patent because it lacks information on the process.

Due to missing data, only 97 questionnaires could be used for the multivariate analysis. The composition of this final sample does not significantly differ from the larger sample of 148 responding firms. Although small samples limit the possibility of generalizing empirical findings, the actual size of the database should provide interesting insights into the innovation-patenting nexus, especially for R&D-active firms.

Summary statistics relative to the final sample are provided in Table 1. A little less than half the firms claim to have at least one active patent. Nearly half the firms of the sample are foreign firms active in Belgium and 37% belong to a high-tech or medium high-tech industry. The average firm size is about 600 employees. Concerning the R&D activities, 93% of firms claim to undertake R&D activities. On average, the sample allocates 35% of its R&D budget to basic and applied research, as opposed to development activities. Finally, 42% of firms give a high importance to process innovation (answer of 4 or 5 on a 5-point Likert scale for the importance of process innovation). This is higher than the percentage of firms that give a high importance to product innovation (32%).

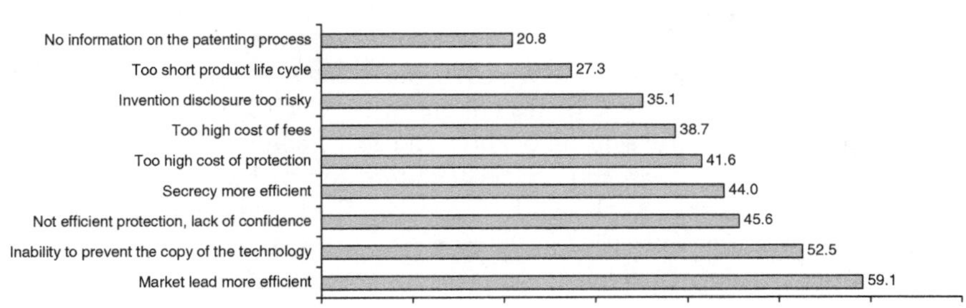

Fig. 3 Percentage of 4 or 5 answers on 5-point Likert scale to nine potential barriers to patenting

Table 1 Final sample summary statistics

Variables	Type	Percentage (%) of yes	Mean	Standard deviation
At least 1 patent?	0/1	45%		0.50
Number of patents in the portfolio	#		33	129.83
Characteristics of firms:				
Number of employees in 2000	#		595	1,102.38
Age of the firm	#		34	29.52
Foreign subsidiary?	0/1	44%		0.50
Number of operating countries	#		25	30.32
Characteristics of sectors:				
Sector concentration (C-4 ratio)	%		64	26.04
HT/MH ?	0/1	37%		0.49
Innovation strategy:				
R&D activity?	0/1	93%		0.26
Percent (%) of basic & applied research in R&D	%		35	27.43
Product orientation	0/1	32%		0.47
Process orientation	0/1	42%		0.50

The first two rows present summary statistics for the dependent variables of the two econometric models. The following rows present summary statistics for ten explanatory variables. Column 2 shows the type of variable. Column 3 shows the percentage of firms that have answered "yes" in case of a binary variable. Column 4 displays the mean of the numerical variables. The last column gives the standard deviation. Basic statistics, own survey, 2000, 97 firms.

4.4 The variables

The dependent variable of the first equation is the probability that a firm has at least one active patent. The dependent variable of the second equation is the number of active patents possessed by firms. These indicators of patenting behavior are built using answers to the survey question relative to the '*total number of patents in force*' in the patent portfolio.

Two types of explanatory variables are used in the empirical analysis: binary and numerical variables. Some numerical variables consist of the firm coordinates on factorial axes coming from factor analyses realized on sub-sections of the survey. They represent the type of institutions with which firms enter into R&D partnerships and the barriers to innovation and to patenting they perceive. The contribution of each survey item to the determination of the factorial axes and the percentage of variance explained by these axes are provided in Appendix 2.[5] The number of factorial axes used in the empirical study is determined so that the cumulated percentage of explained variance was above 50%. The explanatory

[5] A detailed description of the factor analyses may be found in Peeters and van Pottelsberghe de la Potterie (2003b).

variables are grouped into three categories: control variables, innovation strategy variables, and indicators of perceived barriers.[6]

4.4.1 Control variables

Control variables include the firm size, age, domestic or foreign nature of ownership, degree of internationalization, and indicators of sector concentration and technological opportunity. *Firm size* is measured as the total number of employees in a firm. *Firm age* is the number of years, at the time of the survey, since the creation of the company. The square of the firm age is also introduced to check for potential non-linear relationship with the patenting behavior. The firm age could influence its patent portfolio in two opposite directions. Young firms may be more dynamic and have a less rigid structure favorable to innovation, and hence patenting. They may also patent more to compensate for lower market power. However, over time, older firms may have built a larger technological base protected by a larger number of patents. Age may, therefore, act as an indicator of experience and accumulated intangible assets. Older firms have probably also more resources to sustain an active patenting behavior. The relationship between a firm's age and its patenting behavior might therefore be U-shaped, with higher records for young and old firms as opposed to firms of intermediate age.

The *domestic or foreign nature of a firm*'s *ownership* is controlled for using a binary variable taking the value of 1 if the firm belongs to a foreign group and 0 otherwise. A common thought is that foreign firms are more innovative than local ones and might therefore patent more (Baldwin et al., 2002). Nevertheless, since foreign firms are often subsidiaries of larger companies, they could be less involved in patenting. Patents could indeed be managed at the group level, in the country of origin.

The *degree of internationalization* is measured by the number of countries in which a firm operates. A firm is considered to operate in a country if it has customer contacts in this country. Firms operating in a large number of countries can be expected to patent more for two main reasons. First, they face a larger potential market than firms operating only in their national or regional market. The number of countries in which a firm operates would therefore reflect some kind of market opportunity effect that would foster innovation efforts. Second, these firms face higher international competition, what increases the need for innovation rents protection because the number of potential imitators increases and infringement is more difficult to detect. Both aspects are likely to translate into a higher patenting activity.

The *sector concentration* is proxied by the C4 concentration ratio—i.e. the total sales of the four largest firms (in terms of sales) in a firm's main sector of operation divided by the total sales of the sector. The sector of activity is determined using the four digits Nace-bel code. This is an imperfect variable since it is measured at the Belgian level, while many firms face international competition. Moreover, it is based on firms operating in the same kind of activity and does not necessarily reflect the impact of direct competitors. This is, however, the best approximation that could be made using available data.

[6] Appendix 3 provides a synthetic table with the definition of all variables introduced in the regressions.

The *sector technological opportunity* is proxied by three dummy variables based on the OECD classification of sectors of activity (see footnote 4). The first one takes the value of 1 if the firm belongs to a high-tech or medium-high-tech sector and 0 otherwise. The second one takes the value of 1 if the firm belongs to a medium-low-tech or low-tech sector and 0 otherwise. The third one takes the value of 1 if the firm is a service company and 0 otherwise. Since the three dummies sum up to 1, the second modality is removed and considered the reference group. As shown by authors such as Crépon et al. (1996, 1998) and Brouwer and Kleinknecht (1999), technological opportunities are expected to positively influence patenting behavior.

4.4.2 Innovation strategy variables

Innovation strategy variables include the relative importance of the development of new products and new processes in innovation strategy, the share of basic and applied research in total R&D budget, and the extent and type research partnerships.

The product or process orientation of a firm's innovation strategy is measured by two dummy variables. The first one takes the value of 1 if the firm answered 4 or 5 (on a Likert scale ranging from 0 to 5) to the question on the importance of product innovation. The second dummy equals 1 if the firm answered 4 or 5 to the question on the importance of process innovation. A firm's strategy can therefore be oriented towards both types of innovations or none. These two dummy variables are not mutually exclusive. They enable to test the first hypothesis discussed in Section 3, i.e. a product-oriented innovation strategy positively relates to a firm's patenting behavior and a process-oriented innovation strategy negatively relates to a firm's patenting behavior.

The relative importance of research versus development activities is taken into account in the regressions using the percentage of the total R&D budget firms allocate to basic and applied research, as opposed to development work. This variable is used to test hypothesis 2, that the higher the ratio of research activities on development work, the higher the patenting activity.

The involvement in *research partnerships* is accounted for by two variables. They are based on a factor analysis of the questions relative to the existence of R&D collaboration agreements with different types of partners (see Appendix 2). The first variable (factorial axis) distinguishes firms that do collaborate from those that do not. It is particularly correlated with the collaboration of firms with universities, research institutes and public labs. The second axis positively correlates with the collaborations with competitors and negatively correlates with the use of consultants and vertical partners as research partners. These variables enable to test hypothesis 3, that firms that do not exclusively rely on in-house R&D and collaborate with external organizations to develop new knowledge patent more than other firms.

4.4.3 Barriers perception variables

The last type of variables is derived from the assessment made by firms of certain barriers that might prevent them from innovating and from patenting. Firms had to rate, on a 5-point Likert scale, 15 potential barriers to innovation and nine potential barriers to patenting. For each type of barriers a factor analysis of the scores attributed to the various potential barriers was performed (see Appendix 2). The resulting factorial axes were used as explanatory variables to control for the fact that high perceived barriers to patenting and to innovation may negatively affect the patenting behavior of firms, beyond the effect of firm and sector characteristics and beyond the effect of the adopted innovation strategy.

The first factor analysis relates to the *perceived limitations of the patent system*. A high correlation among the different perceived limitations enables a single factorial axis to summarize effectively all barriers to the use of the patent system assessed in the survey. The coordinates of firms on this factorial axis can therefore be used as a variable that accounts for the perception of high cost of fees and protection associated with patents, the lack of effectiveness of the patent protection and preference for secrecy and market lead to secure innovation rents, the disclosure of important information, and the lack of information on the patenting process.

The second factorial analysis concerns the *perceived barriers to innovation*. Three variables are built using firms' coordinates on the factorial axes that represent three categories of potential barriers to innovation. The first category (factorial axis) accounts for the internal barriers that may hinder innovation: organizational rigidities, employee resistance to change, lack of relevant competencies, time constraints, lack of communication, and lack of leadership. The second category stands for the risk- and cost-related barriers: high costs and high economic risks associated with innovation projects, and lack of financial resources. The third category comprises the external barriers to innovation resulting from customers' organizational rigidities, customers' lack of reaction to new products, and inappropriate public regulations. As a result of their negative effect on firms' innovativeness, these barriers are likely to negatively influence the development of patent portfolios.

5 Empirical results

Estimation results are presented in Table 2. The first column relates to the probability of having a patent portfolio (binary logit model). The second column relates to the number of patents firms possess (negative binomial model). In order to check the robustness of the estimates, the size of the patent portfolio is also estimated through a tobit model. Both models yield similar results in terms of the sign and significance of the estimated parameters.[7]

[7] The estimated parameters of the logit model are not readily interpretable in terms of variation in the probability of observing a patent portfolio. However, a simple transformation enables to interpret them as variations in the odds ratios: $(e^{\hat{\beta}} - 1) = \Delta\left(\frac{P(Y=1)}{P(Y=0)}\right)$. The value of coefficients estimated through the negative binomial and tobit models cannot be interpreted directly either, as they depend on the value of the explanatory variables. The advantage of the tobit model is that it formally differentiates firms that have a patent portfolio from those that do not. However, following Hausman et al. (1984) the negative binomial estimates will be used to interpret the results. As a count model, it explicitly takes into account the non-negativity and discreteness of the data. Moreover it enables to deal with distributions that are skewed towards low values, which is the case in this study where a relatively high proportion of firms have no or only a small number of patents. The discussion in the remainder of this section will rely on average elasticities computed for a hypothetic firm characterized by all explanatory variables equal to their average value: $\lambda * \widehat{\beta}$ with $\lambda = e^{\hat{\beta}'\overline{X}}$. This hypothetic firm is referred to as the 'average firm'.

Table 2 Econometric results

	Binary logit	Negative binomial	Tobit
Control variables:			
Characteristics of the firm:			
- *Size*	0.0002 *(0.0006)*	0.0005*** *(0.0002)*	0.0282** *(0.0141)*
- *Age*	−0.0972*** *(0.0352)*	−0.0699*** *(0.0197)*	−3.5718** *(1.7028)*
- *Age square*	0.0010*** *(0.0003)*	0.0007*** *(0.0001)*	0.0385*** *(0.0134)*
- *Foreign subsidiary*	0.1390 *(1.0289)*	0.3281 *(0.4539)*	3.9264 *(41.3408)*
- *Degree of internationalization*	0.0228 *(0.0209)*	0.0068 *(0.0065)*	−0.1375 *(0.6404)*
Characteristics of the sector:			
- *Sector concentration*	0.0304 *(0.0289)*	0.0247** *(0.0122)*	1.1548 *(1.0630)*
- *High-tech*	−0.3221 *(0.9583)*	1.1814 *(0.7441)*	65.2059 *(47.0637)*
- *Service*	−1.8424 *(1.3879)*	−1.4548 *(0.9506)*	−60.4581 *(74.5017)*
Innovation strategy variables:			
New products development	1.7252 *(1.5993)*	3.1157*** *(0.7002)*	201.2267*** *(58.2937)*
New processes development	−3.1781*** *(1.0360)*	−1.5871*** *(0.4137)*	−199.6456*** *(47.9000)*
Percent (%) basic & applied research	0.0480*** *(0.0142)*	0.0186* *(0.0096)*	0.7837 *(0.8317)*
Collaboration partners:			
- *Scientific institutions*	4.6208*** *(0.8388)*	2.7775*** *(0.4954)*	217.8098*** *(48.9936)*
- *Competitors >< vertical partners & consultants*	4.1403** *(1.8146)*	1.3686** *(0.6979)*	176.5967*** *(59.9951)*
Barriers perception variables:			
Barriers to patenting	0.3359 *(0.2482)*	0.1362 *(0.1121)*	−4.3532 *(10.3145)*
Barriers to innovation:			
- *Internal*	−0.0392 *(0.3135)*	−0.3853*** *(0.1476)*	−26.3472** *(12.3453)*
- *Risks and costs*	−1.0553** *(0.5042)*	−0.5563*** *(0.2028)*	−34.5365** *(16.8960)*
- *External*	−1.1497*** *(0.3857)*	−0.2177 *(0.1774)*	−26.9095 *(17.0926)*
Constant	−2.5619 *(1.8957)*	−3.5088*** *(1.1171)*	−221.5423** *(104.5934)*
Log-likelihood	−20.5410	−192.5528	−285.3077
Mc Fadden R-squared	0.6926		
Pseudo R-squared	0.5608	0.5608	

Own survey, Belgium, 2001. 97 firms. S-E into parentheses. Significativity levels: * 10%, ** 5%, *** 1%

Not surprisingly, large firms appear to have more patents than smaller firms. However, firm size is not a significant determinant of the probability of having a patent portfolio. As regards the age of firms a U-shaped relationship is found with both the probability of having a patent portfolio and the number of patents possessed. Two effects work indeed in opposite directions. On the one hand, the need to protect the inventions may be more important for younger firms because they have no market power. On the other hand, older firms may have a larger technological base protected by more patents. The foreign ownership and degree of internationalization of firms do not seem to influence their patenting behavior, as the coefficients are not significantly different from zero. Firms operating in concentrated sectors have on average more patents than firms operating in competitive sectors, but they do not have a significantly higher probability of having a patent portfolio. The sector technological opportunity variables do not turn out to be significant determinants of patenting behavior. This may be due to the heterogeneity of firms included in the three broad sector categories, or to the difficulty of assigning one specific sector of activity to multi-business firms operating in more than one market.[8]

The innovation strategy variables are the central concern of this paper. Their effect on firms' patenting behavior appears to be much more important than the effect of firm and sector characteristics traditionally discussed in the literature. The first two variables are a measure of the orientation of the innovation strategy towards product innovations (new growth potentials) or process innovations (cost cutting). A strong focus towards process innovations negatively impacts the probability of having at least one patent. It reduces the odds ratio, i.e. the probability of having a patent portfolio divided by the probability of not having any patent, by 96% $\left(\Delta\left(\frac{P(Y=1)}{P(Y=0)}\right) = \left(e^{\widehat{\beta}} - 1\right) = e^{-3.18} - 1 = -0.96\right)$. On the contrary, giving a high importance to the development of new products in a firm's innovation strategy does not significantly impact the probability of having a patent portfolio. The negative effect of being a process innovator is significant in terms of number of patents as well, and the development of new products positively affects the size of the patent portfolio. Actually, if the 'average firm' is a process innovator its expected number of patents is reduced by six. Conversely, if it is a product innovator, its expected number of patents increases by 11 (λ=3.67 and the elasticities are 3.67*(−1.59)=−5.84 and 3.67*3.12=11.45 respectively).[9] These results validate the first hypothesis and corroborate Arundel's (2001) finding that process oriented R&D is associated with a greater importance given to secrecy as a protective means. The disclosure of a process innovation might indeed lead other firms to use the underlying technology, as infringements is difficult to track down and proving the paternity of a new process is uneasy.

The second innovation strategy variable concerns the relative involvement of firms in basic and applied research, on the one hand, and development work, on the other hand. The results clearly indicate that firms that allocate a large portion of

[8] Sector dummies would have been another way to control for sector effects. However, this would have multiplied the number of dummies in the model and eventually lead to a lack of degrees of freedom given the small size of the sample. Moreover, the problem of assigning a particular sector of activity to multi-business firms would have been accentuated.

[9] It can actually be shown that the largest patent portfolios are those of firms pursuing an innovation strategy exclusively targeted at new product development, followed by firms pursuing a mixed strategy aiming at both product and process innovations. The patent portfolio of exclusive process innovators is not significantly different from firms that do not seek high innovation goals of any kind.

their R&D budget to basic and applied research have a higher probability of having a patent portfolio. Increasing the share of basic and applied research in total R&D budget by 1% increases the odds ratio by 5%. The share of basic and applied research also positively influences the number of patents, but the significance is low (10%) and the parameter is small. The elasticity computed for the 'average firm' is 0.07, suggesting an average effect much lower than one additional patent when the share of basic and applied research in total R&D budget increases by one percent. The use of the share of basic and applied research in total R&D and not the level of these expenditures may partly explain this result.[10] In a nutshell, hypothesis 2, according to which a higher share of basic and applied research (as opposed to development), would be associated with a more active patenting behavior is confirmed for the probability to have a patent portfolio and, to a much lower extent, for the number of active patents in the portfolio. However, for the size of the portfolio the positive effect is very small. Moreover, the 'research' variable is not significant in the tobit model.

The third set of results and major findings of this study concern the strategic decisions related to the extent to which firms enter into collaborative R&D and the type of partner they choose. The two explanatory variables introduced in the model are the firms' coordinates on two factorial axes (see Appendix 2). Firms that do not exclusively rely on their in-house R&D to develop new knowledge and enter into R&D collaborations with one or more external organizations have positive coordinates on the first axis. Appendix 2 also reveals that the first axis is strongly determined by the opposition between firms that collaborate with scientific institutions (positive values on the axis) and firms that do not (negative values on the axis). The second axis shows an opposition between firms that favor their competitors as opposed to consultants and vertical partners for collaborative R&D projects. Hypothesis 3, which suggests that firms that collaborate with external organizations patent more than other firms, is empirically validated, as shown by the positive and highly significant coefficient of the first collaboration variable, i.e. the first factorial axis. A one point increase in the variable is actually associated with a 100% increase in the odds ratio and about ten additional patents for the 'average firm'.[11] Since this first factorial axis is strongly associated with scientific partnerships, it can be concluded that the positive effect of R&D partnerships on patenting is particularly significant for collaborations with universities, research institutes and public labs. Another interesting finding is that firms that enter into research partnerships with competitors, patent more than firms that prefer to partner with consultants, suppliers or customers. A one point increase of this variable increases the odds ratio by 62% and the expected number of patents of the 'average' firm by 5.

---

[10] The effect of firms' total R&D intensity (percentage of sales allocated to R&D) has been tested as well. This variable proved significant only in the count model at the 15% level and not in the binary model. In other words, the firms' R&D intensity has some determining influence for the number of patents firms hold but not for the probability of patenting. Moreover, the introduction of this variable induced a sharp decrease in the number of observations due to a low response rate (firms were more willing to provide information on the composition of their R&D activities than their total budget for R&D). Therefore the R&D intensity was not included in the final regressions.

[11] The drawback of working with factorial axes as explanatory variables is that it is difficult to interpret what a one point increase in a firm's coordinates on an axis represents in reality.

Two main factors may explain the positive and significant coefficients of the R&D collaboration variables: a 'need' effect and a 'novelty' effect. The 'need' effect refers to a higher need for patent protection resulting from the mutual access to the partners' knowledge bases. Everything else being equal otherwise, the propensity to patent an invention is indeed higher for an invention made under a collaborative framework than for an invention exclusively made in-house. The 'novelty' effect refers to potentially more fundamental and breakthrough knowledge generated by R&D collaborations compared to in-house R&D, which would result in more patents. Both effects explain the positive influence that collaborating with external organizations has on the patenting behavior of firms. It can however be argued that, in case of partnerships with scientific institutions, the 'novelty' effect would dominate, and in case of partnerships with competitors, the 'need' effect would dominate. Firms usually turn to universities, research institutes and public labs when they seek to launch science-related innovation projects that they may not be able or willing to implement alone, because of a lack of skills and capabilities, a high uncertainty associated with high R&D costs, or an inadequacy with their current portfolio of R&D projects. These 'scientific' partnerships, if successful, are likely to result in fundamental knowledge that is more likely to be patented than applied knowledge. This 'novelty' effect associated with scientific partnerships probably also captures the effect on the patenting behavior of firms of the dichotomy between research activities, on the one hand, and development work, on the other hand, which was only poorly captured by the percentage of basic and applied research in the total R&D budget.

Since R&D collaboration implies sharing knowledge with partners, the need for patent protection is likely to be higher than when innovation projects rely exclusively on a firm's own R&D activities. In that respect, partnering with competitors is particularly tricky, as any knowledge leakage could be exploited to directly compete in the firm's end-market. Even though the research teams of competing firms that perform joint R&D projects may not intend to use the knowledge they access through the partnership in a harmful way, the knowledge eventually diffuses into the respective organizations and might end up being exploited by other teams. For similar reasons, issues of ownership over the knowledge developed in partnership are of particular relevance when firms directly compete in the same end-market.

The last type of factors whose influence on the patenting behavior of firms is tested relates to perceived barriers that may prevent them from innovating and from using the patent system. Firms that perceive more barriers to innovation are likely to innovate less than other firms, which would in turn negatively affect their patenting activity. This effect is controlled for in the regressions and it appears indeed that higher perceived barriers to innovation negatively affect both the probability of patenting and the size of the patent portfolio. Whereas the perceived internal barriers negatively affect the size of the patent portfolio, the perceived external barriers influence negatively the probability to have a patent portfolio. Risk and cost related barriers to innovation affect both the probability to have a patent portfolio and its size.

A similar negative effect was expected for the perception of important limitations and high costs of the patent system. However, the empirical results do not validate this assumption, as the parameter associated with the variable that summarizes the perceived barriers to patenting is not significantly different from zero. This result may be due to the fact that firms that have a substantial experience in patenting understand the potential shortcomings (in terms of effectiveness and cost) of the patent system better than firms that do not patent. More fundamentally, this suggests that, even though the patent system has obvious limitations, firms still use it. Actually, the estimates show that firms that claim that secrecy or market lead may be more efficient than patent protection, and that patents are costly to apply for and to enforce, do develop patent portfolios anyway. This finding confirms what was already observed with the Carnegie–Mellon survey (Cohen et al. 2000), i.e. firms patent for many more reasons than merely protecting their intellectual capital from imitation. Applying for a patent is indeed a strategic decision that is not only driven by the desire to protect innovation rents (e.g., Teece 1998; Rivette and Kline 2000; Sherry and Teece 2004). A patent is also a highly valuable tool for technological negotiations with competitors or with potential collaborators, for the exclusion of rivals from a particular technological area, for licensing agreements and attraction of capital, for avoiding to be blocked by competitors' patents, and for building competitive advantage (e.g., Parr and Sullivan 1996; Teece 1998; Glazier 2000; Reitzig 2004).

6 Concluding remarks

This paper investigates the relationship between the innovation strategy adopted by firms and their patenting behavior. The patenting behavior is measured both in terms of the probability of having a patent portfolio and in terms of number of active patents held by the firm. Three main hypotheses are formulated regarding the dimensions of an innovation strategy. The first one concerns the orientation of the firm's research activities towards new products or new processes. The second one relates to the composition of R&D activities and the relative importance of basic and applied research, on the one hand, and development work, on the other hand. The third hypothesis addresses the extent to which firms enter into R&D collaborations to develop new knowledge, and the type of partners they choose. Two additional hypotheses are put forward on the potential influence of the perceived barriers to innovation and to patenting. Finally, traditional determinants related to firm and sector characteristics are introduced in the regressions as control variables.

Patent portfolios are considered a better indicator of patenting behavior than patent applications because they involve granted patents for which renewal fees are still paid. Moreover, since they are built over time, they are less subject to contextual factors that may affect the patent applications of firms in a particular period of time without reflecting their general attitude vis-à-vis patenting. A binary logit model is used to estimate the probability of having a patent portfolio, and a negative binomial model estimates the number of active patents firms hold. Empirical results reveal that innovation strategy variables play a greater role in determining firms' patenting behavior than the traditional firm and sector characteristics widely discussed in extant literature.

The key finding of the paper is the highly positive and significant influence exerted on firms' patenting behavior by an outward-oriented innovation strategy that relies on partnerships with external organizations for research projects. Two effects may explain this finding. First, a '*need*' effect results from the mutual access to the partners' knowledge base in the frame of a collaborative R&D project, which increases the need for legally enforceable protection mechanisms such as patents. The '*need*' effect would be of particular importance for R&D collaborations with competitors. Second, a '*novelty*' effect is induced by collaborative R&D oriented towards basic knowledge and risky projects. These research partnerships are likely to result into fundamental knowledge that partners may want to patent to secure innovation rents. The '*novelty*' effect is probably of higher importance for scientific partnerships with universities, research institutes and public labs. Finally, patents are useful to clarify issues of ownership over co-developed knowledge.

It is also found that process innovators patent less than product innovators and that a higher importance of basic and applied research in total R&D budget is associated with a more active patenting behavior. However, the latter effect is relatively small and probably partially captured by the '*novelty*' effect underlying the positive effect of scientific partnerships on patenting.

The survey data reveal that the most important perceived limitations of the patent system relate to its lack of effectiveness and the preference for market leadership and secrecy as means to secure innovation rents. The second category of limitations concerns the high costs of patent fees and patent protection in case of litigation. The quantitative results show that high perceived barriers to patenting do not affect the actual patenting behavior of firms. This suggests that, even though the patent system has obvious limitations, firms patent anyway because patents serve more goals than just the protection of innovation rents (e.g., patent races, blocking competition, building strong positions for technological negotiations...).

This study contributes to the debate on the relevance of patents as indicators of innovation. Existing literature argues that patent-related indicators are imperfect because their link with innovation depends on the type of firms and sectors considered. But the present findings reveal that their relevance would depend even more on the innovation strategy pursued by firms. In other words, even when considering firms of similar size, age, ownership type, market power, and market and technological opportunities, the observed patenting behavior still depends on their strategic choices regarding innovation: product versus process, the relative importance of research and development activities, and the extent and type of R&D partnerships with external organizations. Therefore, when using patent-related indicators of innovation, researchers should be aware and explicitly take into account the fact that the empirical results not only refer to a particular type of firms and sectors but also to a specific set of innovation strategies.

In terms of policy implications, it is not obvious that improving the effectiveness of the patent system and reducing the cost of patents will result in a greater level of patenting. Actually, firms seem to use the patent system for many more reasons than merely securing innovation rents, and these other more strategic reasons appear to outweigh the potentially prohibitive costs of patenting and the inefficiencies of the system.

Though the present study provides interesting insights on the patenting behavior of firms, certain limitations should be acknowledged. First, the small size of the sample of firms limits somewhat the possibility for generalizing the findings and calls for further validation of the conclusions on larger databases of companies and in other countries than Belgium. Second, the lack of panel dataset prevents from investigating more deeply the causal relationships between R&D and patents, and between collaborations and patents. It is rational to believe that R&D activities may result into inventions that firms may want to patent, but it can also be argued that patented knowledge calls for further R&D to reach the stage of a marketable innovation. Similarly, R&D collaborations may be associated with more patenting because of a 'need' effect and a 'novelty' effect. However, it could further be argued that patents facilitate R&D collaborations, since they constitute useful tools for firms to negotiate with potential partners. A dataset including a time dimension would be required to formally test these possible reverse causations.

Appendix 1: Composition of the sample of 148 surveyed firms

Percent (%) in column Percent (%) in row	HT	MH	ML	LT	SERV	Total
LARGE	21.4	42.1	50.0	29.2	19.1	*33.1*
	6.1	32.7	30.6	14.3	16.3	*100*
MEDIUM	21.4	15.8	36.7	45.8	52.4	*35.8*
	5.7	11.3	20.8	20.8	41.5	*100*
SMALL	57.1	42.1	13.3	25.0	28.6	*31.1*
	17.4	34.8	8.7	13.0	26.1	*100*
Total	*100*	*100*	*100*	*100*	*100*	*100*
	9.5	*25.7*	*20.3*	*16.2*	*28.4*	*100*

Source: Own survey, Belgium, 2001

Appendix 2: Construction of the factorial axes

The following three factor analyses relate to the type of institutions with which firms collaborate, the perceived barriers to innovation, and the perceived barriers to patenting. The asterisks mark the survey items that contribute the most to the interpretation of the factorial axes, and the percentage of total variance explained by the factorial axes used in empirical study.

Research partnerships

Partnerships	Coordinates		Contributions		Cosinus squared	
	Factor 1	Factor 2	Factor 1	Factor 2	Factor 1	Factor 2
Competitors: Yes	1.18	0.84*	8.90	12.00*	0.28	0.14*
No	−0.24	−0.17	1.80	2.40	0.28	0.14
Vertical: Yes	0.39	−0.28	4.10	5.30	0.34	0.17
No	−0.87	0.61*	9.00	11.80*	0.34	0.17*
Research instit: Yes	0.78*	0.25	11.50*	3.20	0.59*	0.06
No	−0.76*	−0.25	11.20*	3.10	0.59*	0.06

Partnerships	Coordinates		Contributions		Cosinus squared	
	Factor 1	Factor 2	Factor 1	Factor 2	Factor 1	Factor 2
Universities: Yes	0.71*	0.22	10.90*	2.70	0.65*	0.06
No	−0.92*	−0.28	14.20*	3.50	0.65*	0.06
Inside group: Yes	0.42	0.07	4.10	0.30	0.26	0.01
No	−0.62	−0.10	6.00	0.40	0.26	0.01
Consultants: Yes	0.55	−1.00*	3.90	34.00*	0.15	0.50*
No	−0.28	0.50*	2.00	17.00*	0.15	0.50*
Other firms: Yes	0.67	−0.24	7.30	2.40	0.33	0.04
No	−0.49	0.17	5.30	1.80	0.33	0

Multiple correspondences analysis, own survey, 2001, 148 firms

Factors	Eigenvalues	Percentages	Cumulated percentages
1	0.37	37.14	37.14*
2	0.14	14.01	51.15*
3	0.14	13.82	64.97
4	0.12	11.82	76.79
5	0.10	10.11	86.89
6	0.09	9.30	96.19
7	0.04	3.81	100.00

Barriers to the use of patents

Patents barriers	Coordinates on the axes		
	Factor 1	Factor 2	Factor 3
Cost of fees	0.86*	0.02	0.23
Protection cost	0.86*	0.06	0.23
Efficiency lack	0.86*	0.09	−0.08
Secrecy better	0.84*	0.27	0.13
Market lead better	0.84*	0.00	−0.25
Short PLC	0.72*	−0.34	−0.51
Disclosure risk	0.84*	0.18	0.07
Risk of copy	0.91*	0.10	−0.06
Lack of information	0.46	−0.80	0.28

Principal components analysis, own survey, 2001, 148 firms

Factors	Eigenvalues	Percentages	Cumulated percentages
1	5.87	65.26	65.26*
2	0.89	9.88	75.15
3	0.71	7.87	83.02
4	0.54	6.01	89.03

Factors	Eigenvalues	Percentages	Cumulated percentages
5	0.35	3.94	92.97
6	0.26	2.91	95.88
7	0.20	2.23	98.12
8	0.13	1.50	99.61
9	0.03	0.39	100.00

Barriers to innovation

Innovation barriers	Coordinates on the axes		
	Factor 1	Factor 2	Factor 3
Economic risk	0.28	−0.80*	0.00
High costs	0.24	−0.83*	0.14
Lack of financing	0.32	−0.58*	0.27
Internal rigidities	0.68*	0.19	0.02
Customers rigidities	0.48	0.03	−0.60*
Resistance to change	0.64*	0.17	−0.20
Lack of competencies	0.70*	0.03	0.24
Customers reaction lack	0.52	−0.02	−0.55*
Public regulations	0.26	−0.33	−0.44*
Time constraints	0.70*	−0.02	0.18
Lack of communication	0.70*	0.30	0.18
Lack of leadership	0.70*	0.30	0.36

Principal components analysis, own survey, 2001, 148 firms

Factors	Eigenvalues	Percentages	Cumulated percentages
1	3.76	31.33	31.33*
2	2.02	16.86	48.19*
3	1.24	10.33	58.52*
4	1.00	8.32	66.84
5	0.89	7.39	74.23
6	0.61	5.11	79.34
7	0.54	4.53	83.87
8	0.54	4.47	88.35
9	0.42	3.49	91.83
10	0.41	3.44	95.27
11	0.31	2.58	97.85
12	0.26	2.15	100.00

Appendix 3: Definition of variables

Name	Construction/interpretation	Type
Dependent variables:		
-Existence of a patent portfolio	The firm has at least one patent (yes/no)	0 / 1
-Size of the patent portfolio	Number of patents in the firm's patent portfolio	#
Control variables:		
Firms' characteristics:		
-Size	Number of employees in the firm in 2000	#
-Age	Number of years since the creation of the company	#
-Foreign subsidiary	The firm belongs to a foreign group (yes/no)	0 / 1
-Degree of internationalization	Number of countries in which the firm has customer contacts	#
Sector characteristics:		
-Concentration	(Sales of the four largest firms of the sector/total sales of the sector)*100	%
-High-tech	The firm is active in a high-tech or medium-high-tech sector (yes/no)	0 / 1
-Service	The firm is active in a service sector (yes/no)	0 / 1
Innovation strategy variables:		
-New products development	4 or 5 on a 5-point scale for the importance of new products development	0 / 1
-New processes development	4 or 5 on a 5-point scale for the importance of new processes development	0 / 1
-Percent (%) basic and applied in R&D	Percent (%) of the total R&D budget allocated to basic and applied research	%
Collaboration partners:		Coordinates on factorial axes
-Universities & research institutes	Scientific institutions as partners for R&D collaborations	
-Competitors><consultants & vertical partners	R&D partnerships with competitors, as opposed to consultants, suppliers and customers	
Barriers perception variables:		
-Barriers to patenting	Barriers perceived by firms to the use of the patent system	
Barriers to innovation:		
-Internal to the firm	Internal organizational barriers perceived by firms	
-Risks-and costs-related	Risks- and costs-related barriers perceived by firms	
-External to the firm	External barriers from customers and regulations perceived by firms	

References

Arora A (1997) Patents, licensing, and market structure in chemicals. Res Policy 26(4–5):391–403

Arrow K (1962) Economic welfare and the allocation of resources for invention. In: Nelson RR (ed) The rate and direction of inventive activity. Princeton University Press, Princeton

Arundel A (2001) The relative effectiveness of patents and secrecy for appropriation. Res Policy 30(4):611–624

Arundel A, Kabla I (1998) What percentage of innovations is patented? Res Policy 27(2):127–141

Baldwin JR, Hanel P, Sabourin D (2002) Determinants of innovative activity in Canadian manufacturing firms. In: Kleinknecht A, Mohnen P (eds) Innovation and firm performance: econometric explorations of survey data. Palgrave, New York

Baumol WJ (2002) The free market innovation machine. Princeton University Press, Princeton

Brouwer E, Kleinknecht A (1999) Innovative output and a firm propensity to patent. An exploration of CIS Micro data. Res Policy 28(6):615–624

Cassiman B, Pérez-Castrillo D, Veugelers R (2001) Endogenizing know-how flows through the nature of R&D investments. Int J Ind Organ 20(6):775–799

Cohen WM, Levin RC (1989) Empirical studies of innovation and market structure. In: Schmalensee R, Willig RD (eds) Handbook of industrial organization. North-Holland, Amsterdam

Cohen WM, Nelson RR, Walsh JP (2000) Protecting their intellectual assets: appropriability conditions and why US manufacturing firms patent (or not), NBER Working Paper, No. 7552

Crépon B, Duguet E, Kabla I (1996) Schumpeterian conjectures: a moderate support from various innovation measures. In: Kleinknecht A (ed) Determinants of innovation—the message from new indicators. Palgrave, New York

Crépon B, Duguet E, Mairesse J (1998) Research, innovation, and productivity: an econometric analysis at the firm level. Econ Innov New Technol 7(2):115–158

Duguet E, Kabla I (1998) Appropriation strategy and the motivations to use the patent system: an econometric analysis at the firm level in French manufacturing. Ann Écon Stat 49/50:289–327

Gilbert R, Newberry D (1982) Preemptive patenting and the persistence of monopoly. Am Econ Rev 72(3):514–526

Glazier SC (2000) Patent strategies for business, L B I Law & Business Institute

Griliches Z (1990) Patent statistics as economic indicators: a survey. J Econ Lit 28(4):1661–1707

Hall BH, Griliches Z, Hausman JA (1986) Patents and R and D: is there a lag? Int Econ Rev 27 (2):265–283

Hausman JA, Hall BH, Griliches Z (1984) Econometric models for count data with an application to the patents-R&D relationship. Econometrica 52(4):909–938

Levin RC, Klerovick AK, Nelson RR, Winter SG (1987) Appropriating the returns from industrial research and development. Brookings Pap Econ Act 1987(3):783–831

Lööf H, Heshmati A (2002) Knowledge capital and performance heterogeneity: an innovation study at firm level. Int J Prod Econ 76(1):61–85

Mansfield E (1986) Patents and innovation: an empirical study. Manag Sci 32(2):173–181

Mansfield E, Schwartz M, Wagner S (1981) Imitation costs and patents: an empirical study. Econ J 91(364):907–918

Nelson RR (1959) The simple economics of basic scientific research. J Polit Econ 67(3):297–306

Nielsen AO (2001) Patenting, R&D and market structure: manufacturing firms in Denmark. Technol Forecast Soc Change 66(1):47–58

Parr RL, Sullivan PH (1996) Technology licensing. Wiley, New York

Peeters C, van Pottelsberghe de la Potterie B (2003a) Measuring innovation competencies and performances—a survey of large firms in Belgium. IIR Working Paper No. 03-16

Peeters C, van Pottelsberghe de la Potterie B (2003b) Organizational competencies and innovation performances—the case of large firms in Belgium, IIR Working Paper No. 03-19

Reitzig M (2004) Strategic management of intellectual property. Sloan Manage Rev 45(3):35–40

Rivette KG, Kline D (2000) Discovering new value in intellectual property. Harvard Bus Rev 78:54–66

Scherer FM (1965) Firm size, market structure, opportunity, and the output of patented inventions. Am Econ Rev 55(5):1097–1125

Scherer FM (1983) The propensity to patent. Int J Ind Org 1(1):107–128
Schmookler J (1957) Inventors past and present. Rev Econ Stat 39(3):321–333
Schumpeter JA (1942) Capitalism, socialism and democracy. Harper, New York
Scotchmer S and Green J (1990) Novelty and disclosure in patent law. RAND J Econ 21(1):131–146
Sherry EF, Teece DJ (2004) Royalties, evolving patent rights, and the value of innovation. Res Policy 33(2):179–191
Teece DJ (1998) Capturing value from knowledge assets: the new economy, markets for know-how, and intangibles assets. Calif Manage Rev 40(3):55–79
van Ophem H, Brouwer E, Kleinknecht A, Mohnen P (2001) The mutual relation between patents and R&D. In: Kleinknecht A, Mohnen P (eds) Innovation and firm performance: econometric explorations of survey data. Palgrave, New York
Veugelers R, Cassiman B (1999) Make and buy in innovation strategies: evidence from Belgian manufacturing firms. Res Policy 28(1):63–80

Michaela Niefert

Effects of patenting behavior on corporate growth: A panel data analysis of German start-up firms

Abstract The present paper examines the relationship between innovations and employment growth in start-up firms. Patent applications are used as an indicator of innovative activity, using a merged data set of German enterprise and patent data. By applying fixed-effects and first-differencing panel data methods, it is shown that patenting activity has a positive effect on employment growth in the second year after application. Patenting firms do not generally exhibit higher growth rates than their non-patenting counterparts; instead, growth performance depends on their patenting activity over time.

Keywords Employment growth · Patents · Gibrat´s law · Panel data models
JEL Classification D92 · L25 · C23

1 Introduction

Innovation is universally regarded as a major source of economic growth. Not only innovation activities by established firms but also the entry of new innovative firms into the market plays an important role in structural change and the competitiveness of an economy. Innovativeness is also supposed to have a positive effect on firm performance. However, the length of time over which a competitive advantage lasts is very short in highly competitive markets, so continuous innovations are necessary to maintain a leading position. The positive relationship between innovative activities and economic performance is empirically less well established at the firm level than at the macro-level.

Quite a few studies analyze the impact of R&D and innovation on productivity, sales, and market value at the firm level. There is less microeconometric work dealing with the effect of innovation on corporate employment growth, particularly with respect to start-up firms. The sign of this effect, derived from theoretical

M. Niefert
Centre for European Economic Research (ZEW), Mannheim
E-mail: niefert@zew.de

models, is not clear: While increasing the level of demand, product innovations might replace existing products so that output and employment may decrease as well; process innovations reduce production costs but often imply a labor-saving progress. This paper seeks to illuminate the relationship between innovations and employment growth in start-up firms. It uses patent applications as an indicator of innovative activity, using a merged data set of German enterprise and patent data. Panel estimators controlling for other potential determinants of growth (particularly size and age), unobserved heterogeneity and selection bias are applied to the data.

The paper is structured as follows. The next section outlines the theoretical approaches to employment growth at the firm level while focusing on the effects of innovative activity. The methodological problems encountered when analyzing the relationship between innovative activity and corporate growth are illustrated in the third section. Section 4 surveys the relevant empirical literature. The econometric models used for the empirical analysis are explained in section 5. A description of the underlying data set and the characteristics of patenting and non-patenting firms are given in section 6. Section 7 presents the results, and section 8 concludes.

2 Theoretical Background

The theoretical literature on corporate growth has focused on the effect of firm size on corporate growth and to a discussion of Gibrat´s Law (Gibrat 1931). According to this law, which is also called the Law of Proportionate Effect (LPE), firms grow proportionally and independently of their size. Various theoretical approaches are contradictory to Gibrat´s Law. Models of optimum firm size postulate that firms converge to the minimum efficient size (MES). The smaller a firm's start-up size relative to the MES, the more urgent it is for the firm to grow. The model of "noisy selection" introduced by Jovanovic (1982) explains why most firms choose a start-up size below the optimal level. It assumes that new firms do not know their cost function in advance, but learn about their relative efficiency as soon as they enter the market. Given the information before entry, firms might be inclined to start with a suboptimal level of output to keep sunk costs low, and to expand only if subsequent performance is encouraging, leaving the market otherwise. The model implies that surviving young and small firms grow faster than older and larger ones, as older firms have already learned about their relative efficiency and have adapted their size accordingly.

More recently, the importance which the theory of endogenous growth attaches to the production of technological knowledge for the growth process has led to increased interest in the microanalysis of innovation and its consequences for firm performance. The direction of the effect of innovation on employment at the firm level is theoretically ambiguous. In addition to direct effects, indirect effects depending on parameters of the production function, the respective output and labor markets and the characteristics of the innovation itself exist (Blechinger et al. 1998). Innovations can be categorized as process or product innovations. Process innovations make it possible to produce a given amount of output with less input and change the production function of the firm. Product innovations comprise quality-improved products as well as new products and are supposed to affect the demand function a firm is facing.

The direct effects of process innovations involve an increase in productivity and a decrease in production costs. For a given amount of output, labor augmenting progress will have a negative impact on employment (displacement effect). However, the decline in marginal costs tends to reduce prices and thus increase demand and employment (compensation effect). This indirect positive effect on employment will outweigh the direct negative effect, ceteris paribus, if demand is elastic (Blechinger et al. 1998). The direct effect of product innovations is the generation of new demand and/or the conquest of market share at the expense of other firms. Consequently, firms' employment demand will rise. By offering a new or quality-improved product, a firm can obtain temporary monopolistic profits until other firms are able to imitate the product or to develop an even better one. However, the new product might replace existing products offered by the firm, so that the overall employment of the firm might decline. Katsoulacos (1986) uses a theoretical model to derive a positive net effect of product innovations on employment; conversely, he finds the net effect of process innovations to be negative.

3 Methodological Issues

There are several methodological problems associated with the empirical analysis of how innovative activity affects employment growth. Firstly, the evolution of employment size is determined by many factors. It has to be controlled for all of these factors in order to isolate the specific contribution of a certain variable. However, not all the determinant factors are observed. If these unobserved effects are correlated with the observed explanatory variables in the model, the estimated coefficients will be biased. For example, innovative firms often have unobserved comparative advantages in implementing new technologies, or may possess special strategic competencies. If employment growth in these firms is driven by these unobserved factors, the effect of innovation per se will be overestimated unless it is controlled for unobserved heterogeneity. Panel data models accounting for unobserved, time-constant individual effects may help to overcome this problem.

Secondly, the data set used might be a non-random sample of the whole population of firms, allowing the estimation to be affected by selection bias. With panel data, the problem becomes aggravated in the presence of panel attrition, i.e., if some firms drop out of the panel after a period of time. If the selection mechanism is non-random but systematically related to the response variable after conditioning on explanatory variables, the estimated coefficients might be biased. In the present case, in which only surviving firms enter the estimation procedure, such a systematic relation is very likely to exist because the growth and survival of firms can be supposed to be partially influenced by the same unobserved factors. If these unobserved factors are correlated with the observed, failure to control for them will lead to erroneous inference regarding the impact of the observables on the dependent variable. For example, it has been claimed that the negative relationship between size and growth revealed by many empirical studies is actually due to the failure to account for survival bias (Mansfield 1962). The early exit of small firms with minor growth rates leads to an overly positive picture of small firms´ growth performance and a false rejection of Gibrat´s law. If selection is correlated with the error term of the structural equation of interest, special methods correcting for selection bias have to be applied.

Further attention should be devoted to the possible endogeneity of innovative activity as a determinant of employment growth. If the innovation indicators themselves are affected by growth, econometric methods allowing for endogenous explanatory variables have to be used. Generally, one might expect a two-way relationship between R&D, innovation activities and performance at the firm level: A firm's innovativeness is an important determinant of its performance in the next period, but its current performance may also control its future innovative effort. This is plausible for performance measures such as cash flow or sales which are closely connected to the liquidity of a firm and thereby determine its ability to finance innovation activities. It may also apply to employment growth, which can be considered as a proxy for the demand expectations of a firm. In order to capture a greater part of the growing market, a firm might decide to undertake innovative efforts. However, firms can directly influence only the inputs into the innovation process. Throughput and output indicators (patents, innovations) cannot be planned exactly, since they involve R&D efforts with long gestation periods and uncertain success (Van Reenen 1997). A priori, it is therefore not clear whether one can assume innovations to be predetermined or whether one must consider them as endogenous with respect to employment growth. In their specification of an empirical model based on the innovation model of Kline and Rosenberg (1986), Klomp and Van Leeuwen (2001) preclude any influence of employment growth on innovation by allowing for a feedback loop proceeding only from sales growth to innovation output. There is no study known to the author that documents employment growth's effect on innovations. Performing a Granger causality test, Lööf and Heshmati (2004) cannot detect any significant impact of employment growth on R&D intensity.

Another problem is presented by the appropriate measurement of employment and innovation activity. Regarding employment, simply using the number of employees might be misleading. Innovations may affect various skill levels of employment very differently. There is usually a complementarity between new technology and skilled labor; this causes the demand for skilled labor to rise with technical progress, while the demand for unskilled labor declines (Blechinger et al. 1998). It is therefore desirable to have employment data distinguishing the skills required to do the job. Unfortunately, no such information was available for this study.

Different indicators have been used to measure innovative activity. There are input-oriented indicators such as share of R&D personnel in total personnel or R&D expenditures per employee, as well as output-oriented measures such as innovation counts, self-reported statements on innovations or share of turnover attributable to innovations. Measures also exist which have been referred to as an intermediate result of the production process or a throughput indicator of innovation (Licht and Zoz 1996; Blechinger et al. 1998), namely, the number of patent applications or grants. On the one hand, patents are inventions and thus the output of research activity. The application for a patent indicates that R&D efforts have been productive and have led to an invention which the enterprise considers to be worth protecting. On the other hand, patents have to be combined with information on manufacturability and user needs in order to be implemented in the production process or converted into a marketable product. They can thus be seen as an input factor for innovations, which at the same time enable firms to exert property rights and appropriate the profits from its ideas.

If the effect of innovative activity on employment is to be analyzed, output-oriented indicators incorporating economic success and thus the respective demand situation should be preferred, since a firm's employment decision depends heavily on demand (Blechinger et al. 1998). In this study, such indicators were not available for the underlying data set. For patents (which have been used instead), the link to economic success is not as strong. Like all input and throughput indicators of innovation, they affect productivity and output after a delay: the underlying inventions first have to be converted into new production techniques or marketable products. New capital equipment, training or even further R&D might be necessary. Moreover, patents can be regarded as real options guaranteeing exclusive rights which allow firms to wait on the conversion into innovations. When facing uncertain market conditions, firms might prefer to delay these investments, which are at least partly irreversible (Bloom and Van Reenen 2002). Hence, the length of time before patents affect firm performance depends on the quantity and quality of the necessary investments and on market conditions.

Moreover, the patent indicator is beset with three fundamental problems: First, not all inventions are patentable; second, not all patentable inventions are patented; and third, patented inventions differ greatly in quality (Griliches 1990). As to the first point, there are some kinds of technical progress, e.g., imitative or incremental innovations, which are too small or too applied in nature to be patentable. Still, they represent an increasingly important part of innovative activity and may affect firm performance (Licht and Zoz 1996). Referring to the second point, it is clear that patents are only one way of protecting an innovation and not always the most effective one. In some cases, other mechanisms such as secrecy, lead time or long-term employment contracts are better suited to appropriate returns on R&D. Patents disclose at least some information to competitors via patent documents and can play an important role in information diffusion (Cohen et al. 2002). The inclination to use patents for innovation protection is supposed to depend on industry and type of innovation. Patents are a more efficient protection mechanism for product than for process innovations (König and Licht 1995). For process innovations, secrecy is a more effective instrument for avoiding imitation. The last point refers to the fact that some patents reflect important inventions leading to successful innovations, while others have almost no economic significance and are not converted into innovations. Accordingly, some patents improve firm performance and others do not. This makes it difficult to estimate the average effect precisely.

4 Empirical Literature

There is a large body of empirical literature investigating the effects of firm size and age on corporate growth and survival. Size and age are used as control variables in virtually all studies on firm performance. Empirical work focusing on start-ups mostly finds that size and age are positively related to likelihood of survival, while growth rates decrease with size and age. Thus it corresponds to Jovanovic´s model and contradicts Gibrat´s law. According to Geroski´s (1995) survey article on market entry, this "stylized result" holds independent of the country, time period, and methodology employed. It has been confirmed by Audretsch and Mahmood (1994) for US manufacturing, by Mata (1994) for Portuguese manufacturing, by Almus and Nerlinger (2000) for German start-ups, and by Honjo (2004) for Japanese manufacturing. However, there is some evidence that for firms exceeding a certain size (Becchetti and Trovato 2002) and for those in specific sectors of the economy (Audretsch et al. 1999; Almus 2002), growth and size are independent of one another. Lotti et al. (2001, 2003) observe a negative effect of size on growth for firms in the Italian manufacturing and instruments industry immediately following start-up. But entrants converge to random growth rates in subsequent years as they attain the MES level of output.

Little empirical work exists on the effect of innovative activity on the post-entry performance of start-ups. Some studies compare the growth and survival chances of young firms in high- and low-tech industries without considering the innovative behavior of the individual firm. They all find the growth rates of start-ups to be higher in technology-intensive sectors of the economy (Kirchhoff and Phillips 1989; Almus and Nerlinger 1999; Audretsch 1995) but yield different results regarding survival. Almus and Nerlinger observe higher survival rates in high-tech branches, Kirchhoff and Phillips find equal survival rates in high- and low-tech industries, and Audretsch ascertains that start-ups in innovative industries have a lower survival probability. Conditional on having survived the first years, however, Audretsch observes better survival chances in these industries. He concludes that an innovative environment generally constitutes a barrier to survival, but provides exceptional growth and survival chances for those entrants able to adjust to changes in the environment and innovate successfully.

Tether (1997) derives some stylized facts regarding the mean employment creation in innovative and technology-based new and small firms: Controlling for size and age, innovative and technology-based firms significantly outperform firms from the general population in terms of the rate of job creation, but the mean rates of direct employment creation in these firms are only modest. Moreover, the distribution of the rates of job creation is highly skewed, i.e., the bulk of jobs are created by a small subset of the total population of innovative and technology-based new and small firms.

Hsueh and Tu (2004) investigate the impact of various innovation indicators on sales growth and profit rates of new Taiwanese SMEs. The cultivation of an innovative atmosphere and of the capability to innovate enhances both performance measures, especially profits. Sales growth is more strongly fostered by innovative actions such as R&D, process innovations, moving into new business areas and the use of new marketing channels. There is no study known to the author that analyzes directly the effect of innovative behavior on employment growth in start-ups at the microeconomic level. Corresponding empirical work on established firms yields mixed results. Katsoulacos' (1986) hypothesis that product innovations stimulate employment and process innovations are labor-saving has only been partly confirmed. Many studies detect a positive effect of product innovations and a negative (but often insignificant) effect of process innovations (e.g., Rottmann and Ruschinski 1997 for German manufacturing; Brouwer et al. 1993 for Dutch manufacturing; Evangelista and Savona 2003 for Italian services). But there are also several studies finding a positive effect of process innovations (Doms et al. 1995 for US manufacturing plants; Klomp and Van Leeuwen 2001 for Dutch firms) on employment growth. Peters (2004) finds that the varying motives of different types of process innovations (introduction for rationalization reasons or in order to improve product quality) may explain this contradictory empirical evidence. Other empirical work shows that the impact of product innovations is not necessarily positive but depends on firm size (Blechinger and Pfeiffer 1999) or on the character of the new product (Leo and Steiner 1995).

There are only two studies known to the author which – as with this analysis – use patents as an innovation indicator in the analysis of employment growth at the firm level. Van Reenen (1997) finds a positive but insignificant employment effect of the number of patents taken out in the US by UK manufacturing firms when we control for the number of innovations. He observes a positive relationship between the number of successful innovations[1] and the level of employment two or three periods later. Greenhalgh et al. (2001) discover that R&D intensity as well as UK patent publications have a positive impact on the employment level in British industrial and commercial companies. Instead of patent counts, they use a weighted average of patents published between two and four years prior to the employment observation, with weights reflecting the average rate of patent renewals. As with Van Reenen (1997), they are unable to find a positive impact of US patents on employment and conclude that patents in the respective domestic market rather than US patents have a significant value to UK firms.

The lack of empirical evidence on the effects of innovative activity on the success of entrants is surprising. It is widely recognized that new firms play a decisive role in the innovation process. Start-ups are often founded in order to introduce new innovations into the market. It is also well-known that innovations attract imitators, so that the competitive advantage emanating from an innovation will disappear in the long run. In order to be successful in the market, start-ups have to undertake ongoing innovations. The contribution of this paper is to investigate the impact of innovative behavior on post-entry performance over the first years of the life of a firm. Employment growth is used as performance measure, patent applications are utilized as an innovation indicator.

5 Econometric Model

The empirical analysis is based on two standard approaches for testing Gibrat's Law. The first one has been used by Evans (1987a,b). It relates employment growth of firm i to its size (S) and age (A) in a second-order expansion in a natural log specified model:

$$\begin{aligned}\ln S_{it+1} - \ln S_{it} = \beta_0 + \beta_1 \ln S_{it} + \beta_2 \ln A_{it} + \beta_3 \ln S_{it}^2 \\ + \beta_4 \ln A_{it}^2 + \beta_5 (\ln S_{it})(\ln A_{it}) + u_{i,t+1} \quad (1)\end{aligned}$$

Employment growth is independent of firm size, i.e. Gibrat´s Law holds, if the size-related coefficients are equal to zero. If panel data are available it is common to estimate equation 1 as a fixed-effects model (Das 1995; Liu et al. 1999; Heshmati 2001, e.g.). The same will be done in the present analysis after including further variables in the regression.

The second approach implies a regression of firm size on previous firm size:

[1] "Successful innovation" here means the successful commercial introduction of new or improved products or processes.

$$\ln S_{i,t+1} = \beta_0 + \beta_1 \ln S_{it} + u_{i,t+1} \qquad (2)$$

Following Cesher (1979), Gibrat´s Law will be confirmed if β_1 is equal to unity. By contrast, Gibrat´s Law is not accepted if $\beta_1 < 1$, which implies that smaller firms grow more rapidly than larger ones. The reverse holds for $\beta_1 > 1$. Equation 2 is estimated by the first-differencing approach including further explanatory variables. This econometric technique is not applied to equation 1 because that would in fact amount to an analysis of the acceleration of employment instead of employment growth. In turn, a fixed-effects estimation of equation 2 would be problematic because S_{it} is correlated with the error term in the demeaned version of this model. 2SLS instrumental variable estimation can only be used with the fixed-effects approach if strictly exogenous instruments are available for the lagged dependent variable on the right-hand side, which is not the case for the data set used in this study. By contrast, under the assumption of sequential exogeneity[2], the first-differencing model can be estimated by 2SLS using lags of S_{it} as instruments (Wooldridge 2002). This method will be applied in the empirical part.

While the fixed-effects model assumes the error term to be serially uncorrelated, the first-differencing model implies that it follows a random walk (Wooldridge 2002). In addition to the standard within estimator, a fixed-effects model, where the error term is assumed to follow a first-order autoregressive process, is estimated. The introduction of serial dependence into the disturbances makes the fixed-effects model more similar to the first-differencing model. Moreover, an extension of Heckman´s (1979) two-step selection correction procedure to the panel data context is applied in order to eliminate attrition bias.[3]

6 Description of Data

The empirical analysis is based on a sample of German firms founded in the 1990s. For its configuration, a stratified sample of 12,000 firms was drawn from two

[2] Sequential exogeneity allows the error term to be correlated with future values of the explanatory variables.

[3] The first step consists of a probit estimation of the selection equation

$$y_{i,t+1} = 1(w_{i,t+1}\delta_{t+1} + v_{i,t+1} > 0), \qquad v_{i,t+1}\,|\{w_{i,t+1}, y_{it} = 1\} \sim \text{Normal}(0,1)$$

for each t. The selection indicator $y_{i,t+1}$ takes value 1 if the firm is observed in $t+1$ and value 0 if it is missing due to permanent drop-out. $w_{i,t+1}$ should contain all explanatory variables of equation 1 to avoid exclusion restrictions on a reduced-form equation. Moreover, it should include at least one significant explanatory variable that is not part of the structural equation. Inverse Mills ratios $\hat{\lambda}_{it}$ are calculated for each of the $T-1$ probit estimations. In the second step, these are included in equation 1, yielding

$$\ln S_{i,t+1} - \ln S_{it} = \beta_0 + \beta_1 \ln S_{it} + \beta_2 \ln A_{it} + \beta_3 \ln S_{it}^2 + \beta_4 \ln A_{it}^2 + \beta_5(\ln S_{it})(\ln A_{it}) + \rho_2 d2_t\hat{\lambda}_{it} + \ldots + \rho_T dT_t\hat{\lambda}_{it} + u_{i,t+1}$$

where $d2_t$ through dT_t are time dummies. Attrition bias can be tested by a joint test of $H_0 : \rho_t = 0$ for $t = 2, \ldots, T$.

complementary firm panels maintained by the Centre for European Economic Research (ZEW), Mannheim (see Almus et al. 2000, for details). The firm data were provided by Credit-reform, the largest credit agency in Germany, which collects information on active, legally independent firms. The data contain information on variables such as industry (five-digit code), legal status, foundation date, region (district), and founding parties' human capital. They comprise virtually all Eastern and Western German firms found in the trade register. The probability of unregistered firms entering the panel depends on the scope of their credit demand and of their business relationships with other firms.

The sample drawn from the foundation panels is stratified by region: It consists of two pools of 6,000 firms each from Eastern and Western Germany, respectively. An indicator demonstrating whether each firm had possibly exited the market was applied as a further stratification criterion. Such firms were oversampled in order to counterbalance the probable positive selection encountered in enterprise panels which results from the difficulty of contacting agents of non-surviving firms and from their unwillingness to report their failure. The sample is confined to firms founded between 1990 and 1997 (more than 90% were founded between 1990 and 1993) in the manufacturing, construction, trade, transport & communication and service sectors. A large telephone survey conducted in 1999 and 2000 provided information not contained in the foundation panels, e.g., annual number of employees and exact date of firm closure. The survey ended up with 3,702 successfully interviewed firms.[4] For this study, legally dependent firms, firms that were not truly new foundations but take-overs, those that submitted a foundation year earlier than 1990 in the telephone interview, and those belonging to sectors of the economy in which patents have no relevance (communication & transporting, retail trade, and consumption-related services)[5] were not included. Furthermore, firms with an average employee base of more than 500 employees and firms for which no employment figures were obtained were excluded. Firms with implausibly high average growth rates were also dropped. In the end, 1,387 firms remained for the analysis. Annual growth rates can be calculated from the foundation year up until 1999 or the respective year of closure.

This firm data set was merged with German patent data. The patent data contain information on patent number, year of application, IPC code, an indicator of whether the application was made at the European Patent Office (EPO), year of acceptance, and number of citations. The combination of the two data sets allows analysis of the relation between patenting behavior and employment growth. In the following, some descriptive findings from examinations of the merged data set are depicted.

Only 44 of the 1,357 firms (3.2%) applied for one or more patents between 1990 and 1999 (see table 1). Altogether, the sampled firms made 128 patent applications in that period, 21 (16.4%) of which were applied for at the EPO and 56 (43.8%) of which were granted up to the year 2003.[6] The distribution by economic sector

[4] The survey is called "ZEW-Gründerstudie" and is described in detail in Almus et al. (2001).

[5] In the communication/transporting and consumption-related service sectors, not a single patent was applied for during the observation period; in the retail trade sector only one patent application was filed.

[6] The relatively low percentage of granted patents is due to the fact that the patent data are still incomplete for the year 2000 and after. The fraction of granted patents may therefore be underestimated for patent applications from the late 1990s.

Table 1 Descriptive Statistics

	All Firms	Non-Patenting Firms	Patenting Firms
Number of firms	1,387	1313 (96.8%)	44 (3.2%)
No. of patent applications	128	–	128
No. of EPO patent applications	21	–	21
No. of granted patents	56	–	56
Patents by sector (%)			
manufacturing	49.2	–	49.2
construction	1.5	–	1.5
wholesale & intermediate trade	12.9	–	12.9
business-related services	36.4	–	36.4
Firms by sector (%)			
manufacturing	22.4	21.2	59.1***
construction	34.5	35.5	4.6***
wholesale & intermediate trade	19	19.1	13.6
business-related services	24.2	24.2	22.7
Mean annual growth rate	12.7	12.7	11.8
Surviving firms (%)	73.3	72.9	84.1*
Mean employment size	16.6	15.9	37.9***
Mean firm age	3.3	3.3	3.5
Mean firm age at patent application	–	–	3
Firms by earliest legal form (%)[a]			
Ltd. liability company	58	57.3	79.6***
Civil law association	10.6	10.7	6.8
Commercial partnership	1.4	1.5	0
Sole proprietorship	29.9	30.4	13.6**
Western Germany (%)	40	39.6	52.3*
Highest lvl., founder education (%)			
doctoral level	2.9	2.8	9.1**
other academic degree	30.3	30.3	29.6
master craftsman	15.8	15.9	13.6
apprenticeship	26.2	26.1	27.3
low education	0.8	0.8	0
education unknown	24	24.1	20.4
Year of foundation (%)			
1990	26.7	26.6	31.8
1991	24.4	24.6	18.2
1992	21.8	21.8	20.5
1993	19.2	19.3	15.9
after 1993	7.9	7.7	13.6

*** (**, **) indicates a significance level of 1% (5%, 10%) in a t-test on the equality of means.
[a] The legal form of the remaining non-patenting firms is unknown. There are no stock companies in the sample.

reveals that half of the patent applications came from the manufacturing sector. This explains why the empirical literature concerning patents has focused primarily on this sector. There is, however, considerable patenting activity in business-related services as well; over a third of all patents stem from this sector. The rest come from the wholesale & intermediate trade sector and – to a very small extent – from construction.

As a comparison with the sectoral distribution of patenting firms shows, sectors obviously differ by mean number of applied-for patents. The share of manufactur-

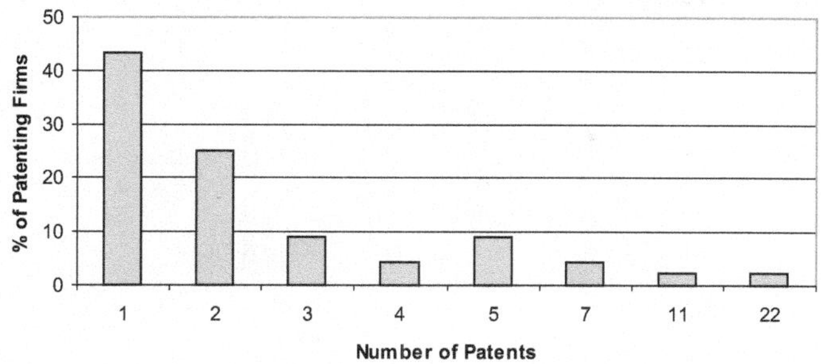

Fig. 1 Share of Patenting Firms Applying for a Specific Number of Patents, 1990-1999

ing firms in patenting firms is somewhat higher than that of manufacturing-related patent applications in all applications. The mean number of applications by patenting firm is hence lower than the average in manufacturing. By contrast, the share of business-related service firms in patenting firms is smaller than the share of applications attributable to this sector in all applications. Consequently, the mean number of patent applications per patenting firm is higher than the average in business-related services. Still, the share of patent applications from both manufacturing and business-related services exceeds by far the weight of these sectors as measured by the number of firms found in each. The opposite holds for wholesale & intermediate trade and, in particular, construction.

Overall, the distribution of patent applications across patenting firms is highly skewed. Figure 1 shows how many firms applied for a specific number of patents: 43% of all patenting firms only applied for one patent within the given period, while a quarter applied for two patents. However, only about 5% of the patenting firms applied for more than ten patents and account for more than a quarter of the total number of patent applications.

Comparing the patenting and non-patenting firms, it is apparent that average annual growth rates do not significantly differ between them. Further analysis shows that the distribution of growth rates of patenting firms exhibits less variance and is less skewed, i.e., the rates are more evenly distributed across the observed (smaller) range of growth rates. The share of firms exhibiting growth rates near the outer edge of the distribution is higher than among non-patenting firms. Following Freel (2000), figure 2 categorizes the employment trend into four groups: declining (negative growth rate), stable (growth rate equal to zero), growth (positive growth rate lower than that of the sample's upper quartile), and super growth (growth rate at least as large as that of the sample's upper quartile). It is shown that patenting firms exhibit declining employment more often than their non-patenting counterparts, but also evince much less zero growth. While both kinds of firms hardly differ with respect to growth, patenting firms show super growth somewhat more often than non-patenting firms.

These results regarding patenting firms do not entirely correspond to Tether´s (1997) findings referring to innovative, technology-based firms. Even taking into account that patenting behavior is not a perfect indicator of innovativeness, the differences with some of the stylized facts found by Tether are striking. Firstly, patent-

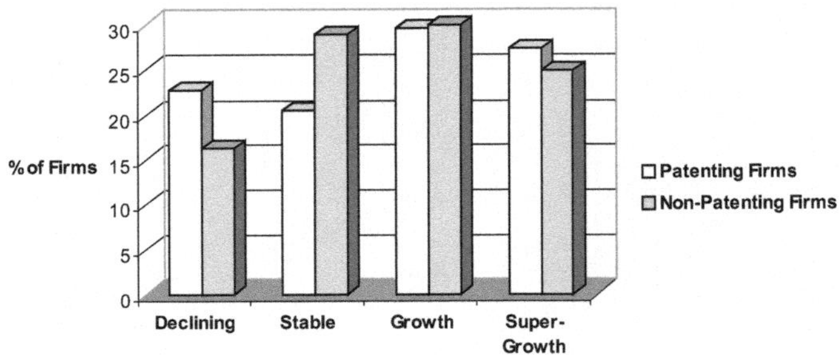

Fig. 2 Share of Firms Falling into Specific Growth Categories

ing firms do not generally outperform non-patenting firms in terms of employment growth. Secondly, it is true that the growth rate distribution of the patenting firms in the sample is skewed and exhibits a high variance, as has been generally noted of young, small firms; it is, however, less skewed and variant than the growth rate distribution of non-patenting firms.

Table 1 further indicates that patenting firms have a higher probability of survival than their non-patenting counterparts. However, the difference is only weakly significant. Average employment size (the average of the annual employment figures available for each firm) is more than twice as large for patenting firms as for non-patenting ones. The mean age of patenting firms at the time of patent application is slightly lower than their mean age over the observation period, suggesting that firms exhibit patenting activity at a relatively early stage in the life cycle.

Patenting firms are mostly founded in the legal form of limited liability companies, something which is less common among non-patenting firms; the latter are relatively more often sole proprietorships. Firms engaging in patent activity are more often situated in the western part of Germany than non-patenting ones. Comparing firm founders' highest level of education shows that the founders of patenting firms possess doctoral degrees more often than those of non-patenting companies. Somewhat surprisingly, they more often do not have other academic degrees. There is no significant difference between the two firm types concerning the distribution over the year of foundation.

7 Empirical Results

The econometric analysis is based on equation 1 and 2. The names and definitions of the explanatory variables are given in table 2. Table 3 contains the results of a standard fixed effects estimation (columns 2 and 3) and of a fixed-effects estimation with autocorrelated errors (columns 4 to 7) of equation 1, and of a first-differenced estimation of equation 2 (columns 8 and 9).[7] Next to size and age (and the second

[7] Number of observations and number of firms are lower in the fixed-effects model with autocorrelated errors than in the normal fixed-effects model because one observation per firm is needed for the estimation of the autocorrelation coefficient which cannot be used for the growth regression. Number of observations is even lower in the first-differencing model because two observations are needed to generate the instruments for the lagged dependent variable.

order expansion of these terms in case of equation 1), legal status and a measure of each firm's patenting activity are included as regressors. Patenting activity is measured either by the number of patent applications or by an indicator whether the firm applied for any patents during the year of observation; it is included with two lags in order to account for the possible delay with which patenting behavior affects employment. The patenting indicators and their respective lags are not instrumented, as they turn out to be exogenous in Granger causality tests. The test's conclusion corresponds to that of the theoretical modelling and empirical evidence concerning employment growth and innovative activity as cited in section 3: Current employment growth has no significant influence on current or future patenting activity.

In addition, either indicator variables of possible selection bias or selection correction terms are included in the regressions. Three leading selection indicators specify whether a firm is missing in the subsequent period because of permanent drop-out due to firm closure, permanent drop-out due to other reasons or temporary drop out, respectively. Using them in the regression only allows us to test for selectivity. In order to correct for it, Mills ratios are inserted as correction terms in the fixed-effects model (columns 6 and 7) as described in section 5.[8]

Gibrat´s Law can clearly be rejected on the basis of the empirical results. The highly significant negative sign of the size coefficient in the fixed-effects estimations implies that small firms grow faster than larger ones. The same conclusion can be drawn from the first-differenced estimations which yield a size coefficient smaller than one. Young firms have higher growth rates than older firms, according to the fixed-effects regressions. Depending on whether the error term is treated as serially correlated or not, the negative effect of age diminishes or increases, respectively, as the firm gets older and larger.[9] As is to be expected, there is no negative effect of age on firm size. The results from the first-differencing model reveal that the relation between the two variables it rather positive. Legal status affects employment growth as well. Firms with limited liability have significantly higher growth rates in comparison with other companies. This result is in line with other empirical work, such as Harhoff et al. (1998) and Engel (2002).

Patenting activities have a positive impact on employment growth. While the effect of both patenting indicators is largest in the period of application according to the standard fixed-effects model, it is most pronounced two years after application when serial correlation of the residuals is allowed for. Since there is presumably some state dependence in growth rates, the latter result is deemed more reliable (see footnote 9). It is also more plausible because inventions have to be converted into marketable products or implemented into the production process before they can have an impact on employment. In addition, this finding is confirmed by the first-differenced estimation. Comparing the two patenting indicators, it turns out that the number of patent applications has considerably less influence on growth

[8] Explanatory variables that are included in the probit but not in the structural regression in order to avoid multicollinearity are founders' human capital, region (Eastern or Western Germany), population density, an indicator of whether each firm has received start-up assistance, and indicators of the payment history of each firm. They all lend significant explanatory power to the selection regressions.

[9] The fixed-effects estimation allowing for autocorrelated residuals suggests a positive correlation of growth rates due to unobserved, time-varying effects. Hence, this model is supposed to be more reliable than the standard fixed-effects approach. It also yields a higher R^2 value.

Table 2 Variable Definitions

Variable Name	Variable Description
employment	log of employment
age	log of firm age
empl*age	interaction between log of employment and log of firm age
ltd_liability	limited liability or stock company
numb_pat	number of patent applications in current period
patent	indicator taking value 1 if firm applied for at least one patent in current period, 0 otherwise
attr_dead	leading selection indicator taking value 1 if firm leaves the panel due to firm closure in subsequent period, 0 otherwise
attr_perm	leading selection indicator taking value 1 if firm leaves the panel permanently for reasons other than death in subsequent period, 0 otherwise
att_temp	leading selection indicator taking value 1 if firm leaves the panel temporarily in subsequent period, 0 otherwise
mills 92-99	inverse Mills ratios estimated from probit regressions (equation 5)

than whether a firm has applied for any patents. Thus, it is rather the act of carrying out patenting activities itself than a firm's number of patent applications which enhances employment growth. The number of applications might be less meaningful due to the varying quality and economic significance of patents. Unfortunately, using patent grants and citations to construct a quality-weighted patent index is prevented by the nature of the underlying data set: the panel is too short to observe a sufficiently large portion of the time period over which the patents can be granted and cited.

An obvious weakness of the present model specification is the absence of any financial indicators serving as explanatory variables. Patenting activity might just be a signal of available internal financing, an important factor for growth. Unfortunately, there are no time-varying financial variables available for the present data set, only information on whether investment activities are being carried out by external firms. Such investments should provide an indication of a firm's financial situation. However, the corresponding variable proves to be insignificant in the estimations. Other important time-varying variables that are important for the growth performance of start-ups (e.g. business plan changes, networks with established companies) are not available for this study, either.

As the estimation results of the models containing selection indicators show, firms leaving the panel consistently show a relatively low employment growth rate in the precedent period. As expected and as ascertained by Almus (2002), attrition due to firm closure is preceded by bad growth performance. This also holds for drop-outs due to other reasons. The findings indicate the presence of attrition bias. In order to correct for the bias, the selection correction procedure is applied to the fixed-effects estimations with autocorrelated residuals (columns 6 and 7). The joint significance of the coefficients of the inverse Mills ratios again confirms the presence of attrition bias. Consequently, one would tend to have more confidence in the results of the regression when correcting for the bias. However, the estimated coefficients of the fixed-effects regressions with and without selection correction differ only slightly. This indicates that the leading selection indicators already correct for the bulk of the attrition bias.

Table 3 Fixed-effects and First-difference Employment Growth Regressions

	FE	FE	FE AR(1)	FE AR(1)	FE AR(1)	FE AR(1)	FD	FD
employment t-1	-0.4077***	-0.4083***	-0.6954***	-0.6961***	-0.7022***	-0.7030***	0.5377***	0.5353***
	[0.0171]	[0.0171]	[0.0270]	[0.0270]	[0.0273]	[0.0273]	[0.1260]	[0.1260]
(employment)^2 t-1	-0.003	-0.0029	-0.0078	-0.0076	-0.0058	-0.0056	–	–
	[0.0038]	[0.0038]	[0.0059]	[0.0059]	[0.0059]	[0.0059]		
age t-1	0.0042	0.0044	-0.0989***	-0.0993***	-0.0712**	-0.0714**	0.0442	0.0452
	[0.0042]	[0.0042]	[0.0277]	[0.0277]	[0.0302]	[0.0302]	[0.0364]	[0.0364]
(age)^2 t-1	-0.0048***	-0.0047***	0.0076*	0.0078**	0.006	0.0062	–	–
	[0.0018]	[0.0018]	[0.0039]	[0.0039]	[0.0042]	[0.0042]		
empl*age t-1	-0.0040**	-0.0039**	0.0319***	0.0321***	0.0283***	0.0284***	–	–
	[0.0019]	[0.0019]	[0.0087]	[0.0087]	[0.0088]	[0.0088]		
ltd_liability	0.1373***	0.1364***	0.3041***	0.3039***	0.3081***	0.3078***	0.2612***	0.2614***
	[0.0311]	[0.0311]	[0.0447]	[0.0447]	[0.0447]	[0.0447]	[0.0515]	[0.0515]
patent t	0.0747*	–	0.0466	–	0.0396	–	0.0452	–
	[0.0442]		[0.0456]		[0.0458]		[0.0520]	
patent t-1	0.0326	–	0.0479	–	0.0435	–	0.0695	–
	[0.0464]		[0.0475]		[0.0477]		[0.0570]	
patent t-2	0.0716	–	0.1283**	–	0.1183**	–	0.1144**	–
	[0.0504]		[0.0521]		[0.0524]		[0.0557]	
numb_pat t	–	0.0530**	–	0.0416	–	0.0384	–	0.0379
		[0.0211]		[0.0260]		[0.0261]		[0.0301]
numb_pat t-1	–	0.0082	–	0.0200	–	0.019	–	0.0249
		[0.0226]		[0.0216]		[0.0217]		[0.0282]
numb_pat t-2	–	0.0306	–	0.0452*	–	0.0418*	–	0.0438*
		[0.0242]		[0.0232]		[0.0233]		[0.0241]
attrition_dead	-0.1313***	-0.1316***	-0.1145***	-0.1147***	–	–	-0.0830***	-0.0832***
	[0.0174]	[0.0174]	[0.0190]	[0.0190]			[0.0215]	[0.0215]
attrition_perm	-0.2148**	-0.2150**	-0.5196***	-0.5198***	–	–	-0.3510***	-0.3511***
	[0.0986]	[0.0986]	[0.1353]	[0.1353]			[0.1207]	[0.1206]
attrition_temp	-0.0235	-0.0232	-0.3329***	-0.3323***	–	–	0.1575	0.1576
	[0.0587]	[0.0587]	[0.1012]	[0.1012]			[0.1336]	[0.1334]
mills 92-mills99	–	–	–	–	***	***	–	–
constant	0.7521***	0.7528***	1.2465***	1.2469***	1.2286***	1.2290***	-0.0206**	-0.0206**
	[0.0245]	[0.0245]	[0.0347]	[0.0347]	[0.0364]	[0.0365]	[0.0089]	[0.0089]
No. of observations	8233	8233	6648	6648	6648	6648	5547	5547
No. of firms	1387	1387	1241	1241	1241	1241	1175	1175
R^2 within	0.3381	0.3384	0.3761	0.3759	0.37	0.3698	0.3079	0.3079

*** (**,*) indicates a significance level of 1% (5%, 10%); standard errors in brackets.

8 Conclusion

This paper analyzes the post-entry growth performance of German start-up firms using panel data methods, controlling for time-constant, unobserved heterogeneity. The estimation results obtained can therefore be accepted as unbiased by firm-specific factors such as flexibility, entrepreneurial skills, and organizational abilities, which presumably exert considerable influence on firm growth. The econometric methods chosen also account for observed constant heterogeneity resulting, for example, from specific industries or regions or from cohort effects. Moreover, a procedure to correct for attrition bias is applied.

The analysis leads to a clear rejection of Gibrat´s law: Employment growth in the surveyed start-ups is negatively related to firm size in the previous year. It is also negatively related to firm age. The findings are consistent with the empirical literature on post-entry performance. The other important result is that involvement in patenting activities enhances a firm's employment growth performance. This is the overall picture arising from the use of different estimation methods. The positive effect of patenting activity appears to emerge two years after application. It seems that the very act of performing patenting activities is more important than the number of patent applications. This finding might be due to the varying quality and economic significance of patents.

The primary objective of the paper is to contribute to the scarce empirical evidence on the impact of innovative activity on post-entry employment growth. Since patents are mostly used to protect product innovations, the results seem to correspond to the empirical literature which mostly reveals a positive impact of these innovations on employment in established firms. However, the findings cannot be generalized because patents are only a partial indicator of innovativeness. Moreover, since no other innovation indicators are used in the analysis, the result may not only reflect the effect of patents per se but may relate to innovative activities in general; this could also include the ability to appropriate technical knowledge, which is presumably enhanced by patenting activities. It is clear, however, that the results do not just reflect time-constant, unobserved factors such as certain technical abilities or open-mindedness to change, which innovative firms are assumed to have – these are already captured by the firm-specific effects. Patenting firms do not generally exhibit higher growth rates than their non-patenting counterparts; instead, growth performance depends on their patenting behavior over time. Undertaking innovative activities of this kind is apparently beneficial for start-ups in order to become competitive and grow.

References

Almus M (2002) Wachstumsdeterminanten junger Unternehmen. Empirische Analysen für Ost- und Westdeutschland. Nomos, Baden-Baden

Almus M, Engel D, Prantl S (2000) The ZEW foundation panels and the Mannheim enterprise panel (MUP) of the Centre for European Economic Research (ZEW). Schmollers Jahr, 120, 2: 301–308

Almus M, Nerlinger E (1999) Wachstumsdeterminanten junger innovativer Unternehmen: Empirische Ergebnisse für West-Deutschland. Jahr Nationalökon Statist, 218, 3-4: 257–75

Almus M, Nerlinger E (2000) Testing "Gibrat's Law" for young firms – Empirical results for West Germany. Small Bus Econ, 15, 1: 01.Dez

Almus M, Prantl S, Brüderl J, Stahl K, Woywode M (2001) Die ZEW-Gründerstudie – Konzeption und Erhebung. ZEW-Dokumentation Nr. 01-01, Mannheim
Audretsch DB (1995) Innovation, growth and survival. Int J Ind Organ, 13, 4: 441–457
Audretsch DB, Klomp L, Thurik AR (1999) Do services differ from manufacturing? The post-entry performance of firms in Dutch services. In: Audretsch DB, Thurik AR (eds.) Innovation, Industry Evolution, and Employment, 230–252, Cambridge University Press
Audretsch DB, Mahmood T (1994) Firms selection and industry evolution: The post-entry performance of new firms. J Evol Econ, 4: 243–260
Becchetti L, Trovato G (2002) The determinants of growth for small and medium sized firms. The availability of external finance. Small Bus Econ, 19: 291–306
Blechinger D, Kleinknecht A, Licht G, Pfeiffer F (1998) The Impact of Innovation on Employment in Europe – An Analysis using CIS Data. ZEW-Dokumentation Nr. 98-02, Mannheim
Blechinger D, Pfeiffer F (1999) Qualifikation, Beschäftigung und technischer Fortschritt. Empirische Evidenz mit den Daten des Mannheimer Innovationspanels. Jahr Nationalökon Statist, 218, 1-2: 128–146
Bloom N, Van Reenen J (2002) Patents, real options and firm performance. Econ J, 112, 478: C97–C116
Brouwer E, Kleinknecht A, Reijnen JON (1993) Employment growth and innovation at the firm level. J Evol Econ, 3: 153–159
Cesher A (1979) Testing the law of proportionate effect. J Ind Econ, 27, 4: 403–411
Cohen WM, Goto A, Nagata A, Nelson RR, Walsh JP (2002) R&D spillovers, patents and the incentives to innovate in Japan and the United States. Res Pol, 31, 8-9: 1349–67
Das S (1995) Size, age and firm growth in an infant industry: The computer hardware industry in India. Int J Ind Organ, 13: 111–126
Doms M, Dunne T, Roberts MJ (1995) The role of technology use in the survival and growth of manufacturing plants. Int J Ind Organ, 13, 4: 523–542
Engel D (2002) The Impact of Venture Capital on Firm Growth: An Empirical Investigation. ZEW Discussion Paper 02-02
Evangelista R, Savona M (2003) Innovation, employment and skills in services: Firm and sectoral evidence. Struct Change Econ Dynam, 14, Special Issue Dec 03: 449–74
Evans DS (1987a) The relationship between firm growth, size, and age: Estimates for 100 manufacturing industries. J Ind Econ, 35: 567–581
Evans DS (1987b) Tests of alternatives theories of firms growth. J Polit Economy, 95, 4: 657–674
Freel MS (2000) Do small innovating firms outperform non-innovators? Small Bus Econ, 14, 3: 195–210
Geroski PA (1995) What do we know about entry? Int J Ind Organ, 13: 421–440
Gibrat R (1931) Les Inégalités Économiques; applications: aux inégalités des richesses, à la concentration des entreprises, aux populations des villes, aux statistiques des familles, etc., d´une loi nouvelle, la loi de l´effet proportionnel. Librairie du Recueil Sirey, Paris
Greenhalgh C, Longland M, Bosworth D (2001) Technological activity and employment in a panel of UK firms. Scot J Polit Economy, 48, 3: 260–82
Griliches Z (1990) Patent statistics as economic indicators: A survey. J Econ Lit, 18, 4: 1661–1707
Harhoff D, Stahl K, Woywode M (1998) Legal form, growth and exit of West German firms – Empirical results for manufacturing, construction, trade and service industries. J Ind Econ, 46, 4: 453–488
Heckman JJ (1979) Sample selection bias as a specification error. Econometrica, 47: 153–161
Heshmati A (2001) On the growth of micro and small firms: Evidence from Sweden. Small Bus Econ, 17: 213–228
Honjo Y (2004) Growth of new start-up firms: Evidence from the Japanese manufacturing industry. Appl Econ Letters, 11: 21–32
Hsueh L, Tu Y (2004) Innovation and the operational performance of newly established small and medium enterprises in Taiwan. Small Bus Econ, 23, 2: 99–113
Jovanovic B (1982) Selection and the evolution of industry. Econometrica, 50, 3: 649–670
Katsoulacos YS (1986) The Employment Effect of Technical Change. Wheatsheaf Books, Brighton
Kirchhoff BA, Phillips BD (1989) Innovation and growth among new firms in the US economy. In: Frontiers of Entrepreneurship Research, 173–188, Babson College, Wellesley, MA
Kline SJ, Rosenberg N (1986) An overview of innovation. In: Landau R, Rosenber N (eds.) The Positive Sum Strategy: Harnessing Technology for Economic Growth, National Academy

Press, Washington DC
Klomp L, Van Leeuwen G (2001) Linking innovation and firm performance: A new approach. Int J Econ Bus, 8, 3: 343–364
König H, Licht G (1995) Patents, R&D and innovation: Evidence from the Mannheim Innovation Panel. Ifo Studien, 41, 4: 21–43
Leo H, Steiner V (1995) Technological Innovation and Employment at the Firm Level. WIFO, Austrian Institute of Economic Research, Vienna
Licht G, Zoz K (1996) Patents and R&D. An Econometric Investigation Using Applications for German, European and US Patents by German Companies. ZEW Discussion Paper 96-19
Liu JT, Tsou MW, Hammitt J (1999) Do small plants grow faster? Evidence from Taiwan electronics industry. Econ Letters, 65: 121–129
Lööf H, Heshmati A (2004) Finance, R&D, and Productivity: Correlation or Causality? Preliminary Version presented at the Schumpeter Conference at 9-12 June 2004 in Milan, the 10th Meeting of the International Schumpeter Society (ISS)
Lotti F, Santarelli E, Vivarelli M (2001) The relationship between size and growth: The case of Italian newborn firms. Appl Econ Letters, 8, 7: 451–454
Lotti F, Santarelli E, Vivarelli M (2003) Does Gibrat´s law hold among young, small firms? J Evol Econ, 13: 213–235
Mansfield E (1962) Entry, Gibrat´s Law, and the growth of firms. Amer Econ Rev, 52, 5: 1023–1051
Mata J (1994) Firm growth during infancy. Small Bus Econ, 6, 1: 27–39
Peters B (2004) Employment Effects of Different Innovation Activities: Microeconometric Evidence. ZEW Discussion Paper 04-73
Rottmann H, Ruschinski M (1997) Beschäftigungswirkungen des technischen Fortschritts. Eine Paneldaten-Analyse für Unternehmen des verarbeitenden Gewerbes in Deutschland. ifo Studien, 43: 55–70
Tether BS (1997) Growth diversity amongst innovative and technology-based new and small firms: An interpretation. New Tech Work Employ, 12, 2: 91–107
Van Reenen J (1997) Employment and technological innovation: Evidence from U.K. manufacturing firms. J Lab Econ, 15, 2: 255–284
Wooldridge JM (2002) Econometric Analysis of Cross Section and Panel Data. MIT Press, Cambridge, MA

Part 7:

Public Policy – competition policy and the patent system

Jean-Luc Gaffard · Michel Quéré

What's the aim for competition policy: Optimizing market structure or encouraging innovative behaviors?

Abstract Innovation is usually thought of as a change in the fundamentals of an economy, which can require adjustments by policy-makers. The latter are usually thought as in regard to a dominant vision, which is to restore an optimal market structure, and leads to a competition policy mainly aimed at controlling for antitrust practices and limiting market power. In this paper, we favor another vision of innovation, as a discovery process that cannot allow *ex ante* a definition of best practices. Dealing with information issues in two different and alternative perspectives, we argue that antitrust authorities confront a market imperfection–market failure dilemma (MI–MF dilemma) which leads them to favor the existence of appreciative and discretionary policy rather than encouraging the existence of any market structure thought of as optimal as regards the current state of information. We conclude with policy implications, contrasting the EU with the US.

Keywords Competition · Information · Innovation

JEL Classification L2 · L4 · L5 · O3

This paper was presented at the 10th International J.A. Schumpeter Society Conference on Innovation, Industrial Dynamics and Structural Transformation: Schumpeterian Legacies, Milan 2004. It owes very much to the collaboration of J-L Gaffard with M. Amendola in the recent years. However he is not responsible for the way we have taken advantage of it.

J.-L. Gaffard (✉)
University of Nice Sophia Antipolis, Institut Universitaire de France and OFCE,
250 avenue Albert Einstein, 06560 Valbonne, France
E-mail: gaffard@idefi.cnrs.fr

M. Quéré
CNRS-GREDEG and OFCE, Valbonne, France
E-mail: quere@idefi.cnrs.fr

1 Introduction

In the perspective of understanding the current debate about the relation between competition and growth, it is worthwhile to contrast two conceptions of competition policy, derived from two contrasted analyses of innovation. The one is focused on the necessity for antitrust authorities to control damaging behaviors by firms, which try to increase their monopoly power thanks to collusive agreements, mergers and acquisitions, or concentration. In such a perspective, results of technological change are supposed to be known, as antitrust authorities intend to impose a specific (optimal) state of affairs. The other conception of competition policy is derived from an analysis of innovation as a discovery process, which implies that no optimal market structure can be a priori identified. Thus, the major issue for public authorities is to prevent excessive market disequilibria from occurring. In this context, coordination among the firms is not only "highly beneficial to the economy" (Baumol 2001, p. 727), but a necessary condition for innovative investment to be carried out. Practices usually perceived as anti-competitive can be tolerated by antitrust authorities, at least so long as they do not lead to actual market failures.

Within the traditional framework, "the best way to understand market failure is first to understand market success, the ability of a collection of idealized competitive markets to achieve an equilibrium allocation of resources which is Pareto optimal" (Ledyard 1989, p. 185). This particular definition of market success and hence of market failure is nothing but a reading of the first fundamental theorem of welfare, according to which: "if there are enough markets, if all consumers and producers behave competitively, and if an equilibrium exists, then the allocation of resources in that equilibrium will be Pareto optimal" (ibid.). Therefore, "market failure, the inefficient allocation of resources with markets, can occur if there are too few markets, non-competitive behavior, or non-existence problems" (ibid. p. 189). It is acknowledged that, for market imperfection to be corrected, we need devices for the creation of more markets. As we shall see, things can be much more complex. Once market imperfection appears as a means for making viable any innovative choice, it is no longer acceptable to reduce it to market failure or to oppose it to market success. Market failure as well as market success must receive another definition, which refers to the viability conditions of innovation process rather than to the optimality conditions of the allocation of given resources. Moreover, market imperfection remains a concept difficult to understand from a normative viewpoint insofar as the same behavior (read as non-competitive behavior in the standard analysis) may favor dynamic efficiency (in fact, innovative choice), while it is an obstacle to the achievement of static efficiency. Thus, antitrust authorities have to address a market imperfection–market failure dilemma (MI–MF dilemma hereafter). This dilemma is intrinsic to any innovation process, as coordination among firms is required (market imperfections) but should not lead to abusive market power that would block innovative choices (market failures).

The rest of the paper is as follows. Section 2 reviews the dominant paradigm for competition policy based on a traditional vision of innovation. The main result is that, while anything can happen, it is always possible to establish a Pareto ranking of the different outcomes. Perfect competition remains the benchmark. Nevertheless, such a Pareto-ranking holds only if information structure pre-exists to the carrying out of market transactions, as it is the case within the Industrial Organization framework. At the opposite end of the spectrum, monopolistic practices might be necessary when information has to be produced by the market process. Taking advantage of the latter statement, Section 3 proposes we consider innovation as a process of change that takes time, and so we must revisit competition policy, which, instead of consisting in applying rigid rules, must be active or appreciative. Section 4 concludes with some insights about the EU competition policy.

2 Competition or monopoly: the innovation challenge

Competition is usually thought of as a state of affairs that reflects a particular market structure. Whatever the kind of analysis in use, be it developed within the old SCP paradigm or within the new theory of industrial organization, full or perfect competition appears as the benchmark for competition policy. The latter has to maintain conditions of full competition as far as possible. Nevertheless, when innovative choice comes to the fore, things are much more complex, as underlined by Schumpeter (1942). The introduction of new methods of production and new commodities is hardly conceivable with perfect (and perfectly prompt) competition from the start. Thus, there is a real dilemma for the policy makers in charge of maintaining a competitive environment.

2.1 From the SCP paradigm to the new theory of industrial organization: anything can happen

The SCP (Structure–Conduct–Performance) paradigm, which has dominated industrial economics for a couple of decades, has led to clear-cut and basic guidance mechanisms. A major means to identify a competitive structure is the market share that is supposed to reveal the level of market power. Competition policy then focuses on the identification of the actual firms' market power by focusing on the identification of the relevant market. Even if perfect competition cannot be claimed as a device in modern industrial contexts, it is still thought to be "a useful measuring-rod against which certain aspects of market failures can be evaluated, most notably the welfare implications of monopoly conditions" (Rowley 1972, p. xix). As such, it gives a place for extensive discussions about the conditions by which welfare gains and losses can be obtained with regard to changes in market structures. Within this framework, economies of scale may be invoked in a defense against anti-trust plaintiffs when pointing out the trade-off existing between the loss of consumer benefits due to market power increase and the gain from the cost savings resulting from the larger scale of activity (Williamson 1968). On the other hand, due to the large scale, X-inefficiency can stifle the gains of welfare (Leibenstein 1966, Comanor and Leibenstein 1969). Thus, "cost savings derived from a reduction in X-inefficiency would provide additional reasons for an anti-trust policy which was weakened by Williamson's trade-off analysis" (Crew and Rowley 1972, p. 144). In short, a debate already exists about the real impact of the market structure on welfare, which does not have an indisputable solution.

The SCP paradigm was further challenged from the 1980s onwards both by the contestable market theory and by strategic interaction analysis. The one leads to a weakening of the importance of the 'S' in the SCP paradigm, as potential entry appears sufficient to induce actual competitive pressure and prevent any market power. The other gives rise to a huge amount of analytical apparatus from which any market structure can result from an optimizing behavior and be reproduced as the outcome of a specific game situation. The difficulty lies in the predictable character of those analytical specifications, as well as the estimation of the actual market context. "Theoretical findings and prescriptions are difficult to translate into workable and enforceable standards that in *actual market settings* would, without fail, promote conduct that enhances social welfare and would, without fail, promote conduct that harms welfare ... In the context of strategic interactions, it is difficult to distinguish between those actions, which are intended to harm actual (and potential) rivals, that stifles competition, and thereby reduce economic welfare, and those actions which harm present rivals and discourage future entry but which, nevertheless, promote economic welfare" (Ordover and Saloner 1989, pp. 538–539).

This is clearly pointed out when dealing with the influential role of R&D on the market structure. R&D spending appears as the major expression of an innovative choice. It results in lower unit costs, and determines both the profitability of firms and the organization of industry. As in other models based on game theory, a multiplicity of outcomes corresponds to various information structures. Thus, there is no a unique prevailing organization of industry that would correspond to the maximum of welfare (Dasgupta and Stiglitz 1980; Reinganum 1989). Nevertheless, even if information is conveyed through acting and the attempt to convey information leads to distorted behavior (Stiglitz 2003, p. 591), information structures, actions, and outcomes (market structures and profits) can be Pareto-ranked and policy (competition policy) can be thought to be Pareto-improving by changing the institutional or organizational frameworks. A potent illustration is given by the analysis of the impact of technological competition on the social advantages or disadvantages of R&D agreements (Petit and Tolwinski 1999). Symmetric firms are distinguished from asymmetric ones—the former use similar technologies over time, whereas the latter are unequal, differing with regard to the initial technology carried out or the innovation rate. In the case of symmetric firms, sharing R&D results (knowledge spillovers) thanks to technological agreements will induce higher individual profits, but the net outcome is lower R&D spending and higher prices than socially desired, due to the existence of a free rider effect. At the opposite end, coordination of investment decision in R&D (joint research ventures) by eliminating the free rider effect will result in an increase of social welfare in terms of accumulated knowledge. In that case, antitrust authorities should limit technological agreements, but encourage joint research ventures. In the case of asymmetric firms, technological agreements are beneficial for consumer welfare because they prevent the monopolization of the market that would lead to higher prices. Asymmetry is here a strong argument in favor of technological agreements that should be encouraged by antitrust authorities. Joint research ventures are also beneficial and should be encouraged, as some firms may lack incentives to cooperate.

These examples illustrate how competition policy is becoming quite a complex issue. Analytical refinements do not lead to clearly identify the estimation of expected losses and gains in terms of social welfare. But the real issue is less in the variety of market structures that results from optimizing behaviors than in the difficulty of identifying anti-competitive practices. As a matter of fact, in any antitrust case, the main issue is to know whether the particular conduct by the defendant contributes to an anti-competitive result or not. Unfortunately, game theory does not provide a definitive answer. "Modern theory by mainly showing that a variety of things *can* happen is likely to stimulate plaintiffs' imagination. It can certainly be suggestive; it will almost never be definitive" (Fisher 1991, p. 222). Thus, antitrust authorities have to decide whether departures from competition can be explained by collusion or whether the observed results have occurred through oligopolistic behaviors without explicit agreements. Here, the only guidance provided by the game literature is that "the outcome could be an equilibrium in a stylized non cooperative game. That is no help at all" (ibid.). What we need is "an examination of the detailed facts of the industry and the firm involved" (ibid.). Characterizing the competition environment cannot be the result of a pure analytical reasoning. It requires deep empirical and contextual studies. Nevertheless, these studies have to rely upon a clear understanding of the role of restrictive practices.

2.2 From the market structure to the market process

Within the SCP paradigm, as well as in the new theory of industrial organization, firms are assumed to obtain immediately gains associated with the new technology once they have decided to implement it. As a consequence, the only motive to innovate lies in the perspective of benefiting from a monopoly rent, which is at least transitory. But, on the other hand, antitrust policy necessarily focuses on the market power associated with this monopoly position. Thus, what is at stake is the existence of a technological leadership that can lead to excessive concentration on the product market. For example, as for mergers and acquisitions, they will be assessed with respect to the market structure involved rather than to the related decrease in production costs, as it is generally the case with European competition policy. In other words, there is a dilemma between competition and innovation.

In order to deal with this dilemma, it is worthwhile to come back to the meaning of competition and to consider the actual organization and information issues associated with any innovation process. A fundamental characteristic of perfect competition is the avoidance of any reference to law and regulatory rules. The existence of this state of affairs requires abstract conditions (a large number of independent producers supplying a homogenous good in a context where entry of firms is free) that have nothing to do with any particular institutional framework. But "should we define competition by its requisites or by its consequences" (Stigler 1957, p. 12)? Such an interrogation is relevant insofar as there is no a relation univocally determined between requisites and consequences. In fact, "it is possible to analyze the effects of a type of market organization, but it is not possible to enumerate all the types of market organization that lead to a given result" (Stigler ibid.). Moreover, the effects of a type of market organization are not those involved by the properties of equilibrium associated with it.

This is where Richardson's argument about the real nature of the information process comes to the rescue. This can be seen in the following quotation. "It is of the essence of the private enterprise that although its individual members are independent (in the sense that they are free from central direction) yet their activities are nevertheless interrelated" (Richardson 1960, p. 30). "Any single investment will in general be profitable only provided, first, that the volume of competitive investment does not exceed a critical limit set by the demand available, and, secondly, that the volume of complementary investment reaches some minimum level" (ibid. p.31). On the other hand, the investment decisions of entrepreneurs depends on expectations, the reliability of which is a function of adequate information or evidence. In this perspective, "the availability to entrepreneurs of the information on which to base investment decisions is a function of the structure of the model in which they are presumed to operate. Alternative market forms may be compared, in other words, according to the predictability of the environment that they afford entrepreneurs. Perfect competition represents an environment in which predictability (of the appropriate kind) is zero; only by postulating some restriction on the freedom of entrepreneurs can this predictability be increased. Once this admitted, practices in restraint of competition no longer appear as so much sand in the works" (Richardson 1965, pp. 435–436). The assumed market structure has therefore an important bearing on the way expectations are formed, and "some market imperfections may be essential to the process of successful economic adjustment" (Richardson 1960, p. 38).

Furthermore, focus must be on the process itself and not on the market structure, whatever it is. "Prices in every market equate supply and demand and thus render mutually compatible the equilibrium behavior of all the members of the economy... it is called an equilibrium theory because it shows what economic organization would be at the point of equilibrium. What happens when a market or the economic system is in disequilibrium is an entirely different problem" (Scitovsky 1961, p. 231). Agreeing with this statement, Hicks concludes in the form of a question. "Is it not time that economists abandoned the discussion of these problems in terms of *general* models, Perfect Competition, Free Competition, Restricted Competition, and the like?" That question is legitimate when "in a world of large-scale production and scientific technology (whatever may have been the case in a simpler world, but the same conclusion probably holds there too), there is no practicable market form which could be made to exist at all generally, and which does not have great disadvantages to match its advantages. The search for the ideal is therefore a will-of-the-wisp" (Hicks 1983, p. 162). In a context where innovation and change are becoming a continuous issue for firm practices, the question to ask is not with respect to the identification of economic organization and market form, i.e., which are the most appropriate in terms of welfare, but how firms are organized in order to match market requirements. "Firm behaviours can certainly distort competition either in terms of market selection processes or the generation of future innovation but the appropriate question to ask is their effects on the process not their effects on market structure" (Metcalfe and Ramlogan 2005, p. 232). This should lead us to abandon any reference to general models of optimal market forms under specific conditions. This can be interpreted as a call for elaborating more appreciative theories and, consequently, as a need for appreciative policy-making with regard to competition issues.

3 Innovation, competition, and growth: the 'MI–MF dilemma'

When departing from the dominant definition of innovation as an adoption of new available techniques, and considering a more broader view of innovation as a process modifying the existing context (with regard to technology characteristics, market structures, cost structures or organizational design of firms), we get quite a different perspective about the role and characteristics of competition and competition policy.

3.1 Innovation as a distributed phenomenon

In the largest sense, innovation necessarily implies the breaking up of the existing industrial structure and a modification in market conditions, followed by a gradual reshaping which reflects changes in cost conditions, profitability, relative prices, and modifications of the consumers' preference systems. Innovation thus is a process of research, learning, and selection, which results in the appearance of new productive options that bring about a modification of the environment itself. It generates sunk costs that cannot simply be assimilated to costs recovered later, but reveals the existence of a divorce between costs and revenues that call for specific adjustments along the way. Thus defined, innovation is a sequential process, which takes (and can change) form, content and direction at each successive step of its implementation. Firms do not know ex ante whether it is profitable to innovate. "Indeed the answer to this question for any single firm depends on the choices made by other firms, and reality does not contain any provisions for firms to test their policies before adopting them. Thus there is little reason to expect equilibrium policy configurations to arise. Only the course of events over time will determine and reveal what strategies are the better ones" (Nelson and Winter 1982, p. 286).

The above considerations point at innovation as a *distributed phenomenon*, even from the single firm's viewpoint. As a matter of fact, most innovations are the result of new forms of coordination among several firms and institutions rather than of the independent actions of single dominant innovating firms. In this light, particularly important is "how the innovating firms acquire, accumulate and develop knowledge other than scientific and technical knowledge which is material to innovation, (namely) knowledge about the specific characteristics of customers and markets, which in turn has wider connections to knowledge about economic, social and regulatory changes" (Metcalfe 2000, pp. 148–149). This is actually achieved "by means of several firms (or other institutions) contributing various technical, marketing or production resources, and coordinating the deployment of those resources in the innovating process" (ibid.). The traditional definition of the environment as something *within which* the activity of the firms is performed, and that exists in its own right in the sense that it is structurally unaffected by the firms' actions, fades away when the above distributed character of innovation is considered. Major innovations bring about "a significant change of both the innovating firm and its environment: in the particular sense that certain products, or productive phases, or relationships which existed before the innovation no longer exist or have become obsolete, and then are bound to disappear, or that entirely new products, or productive phases, or relationships, have come into existence" (Amendola and Bruno 1990, p. 423).

From this perspective, innovative choice consists not so much in the choice between given alternatives (whether based on complete or incomplete information) as in a search for coordination. What matters is no longer the 'rationality' of the choice between known alternatives. It is the 'viability' of the process through which a different alternative is brought about: a viability that depends on how co-ordination problems are dealt with step by step. Within such a sequential framework, several firms can coexist in the market, despite the existence of increasing returns, remaining differentiated not so much because they supply differentiated goods, but because they are each at a different step of the life cycle of the production process (Amendola and Gaffard 1998; Amendola et al. 2000, 2003, 2004). Interestingly, increasing returns raise only a transitory competitive advantage. Thus, they do not systematically involve market failures and hence they do not call systematically for regulatory interventions. But, this does not mean that innovation processes are necessarily viable and rivalry systematically maintained. Viability requires coordination among firms, which are aimed at bounding investment behaviors and hence avoiding excessive market disequilibria.

Competition plays a central role in the coordination process, as it determines the way in which the market information relevant for coordination is being made available, so that the required adjustments in productive capacity can actually take place. Thus it helps to make this process viable and to realize the productivity gains deriving from it. In this light, competition is not only aimed at equalizing supply and demand in a given market and technological environment, but "has also to adapt both structure and technology to the fresh opportunities created by expanding markets" (Richardson 1975, p. 353). Therefore, competition policy cannot be conducted in isolation without considering the distortions that are in the nature of the growth process. Instead of targeting any optimal market structure, it must be aimed at enforcing viability (and growth) conditions.

3.2 Revisiting competition policy

Innovation as a process makes the firm more fragile during the transition phases, i.e. during the period of search for and learning about new economic opportunities. As such, competition authorities have to take care of the construction period and of the capital structure of the firm. It can be the case that horizontal agreements concerning marketing and distribution are a means of compensating for a temporary increase in the production costs of firms due to the relative weight of innovative investment. Far from being a search for abusive market power, such agreements can be legitimated by high innovative costs that cannot be immediately recovered. It is essential to scrutinize the actual cost structure of the firm and to take account of its innovative characteristics in order to estimate the extent to which antitrust rules should apply or not.

In fact, innovation is necessarily a source of increasing uncertainty, due to the *ex ante* inability of firms to secure their investment properly. As a consequence, facing that uncertainty requires some coordination mechanisms among firms in order not to 'secure' but to maintain innovative investment within 'channels' or 'corridors' which complement that of other firms in the future (Quéré 2000). That blend of market information and of collusive R&D activity is something to be encouraged in order to promote firm investment. In such a situation, "the antitrust authorities' willingness to show forbearance in their toleration of research joint ventures and technology licensing is fully justified" (Baumol 2002, p. 119). Moreover, all stages of the production process may be concerned with such coordinating devices. Beyond the R&D issue, co-ordination of manufacturing, marketing and distribution among firms may also be critical in order to structure a proper demand for new technologies or products. They may also contribute to a sharing of financial risk as well as training costs induced by new production opportunities.

All these coordination issues put into light market imperfections (MI), usually equated with market failures (MF), which are always defined with respect to the conditions of perfect competition, and have to be corrected by means of specific incentives rules, without considerations of any process in time. At the opposite end, in our perspective, imperfect competition is a characteristic of any market process, which can neither be removed nor systematically corrected. It is a normal state, quite often associated with the persistence of rivalry (Quéré 2003). This rivalry, when it is maintained, is the means for productivity gains to be transformed in lower prices to the benefit of customers and normal profits to the benefit of entrepreneurs. The difficulty faced by anti-trust authorities is that market imperfections are, on the one hand, necessary to convince firms to launch innovative investment and, as such, they are not something to be systematically condemned but, on the other hand, reveal real market failures as they hamper the viability of the innovation process. The "MI–MF dilemma" is very close to the distinction developed by Metcalfe about good and bad market imperfections: "the imperfections identified in the market failure approach, therefore, can be viewed in a different perspective, as integral and necessary aspects of the production and the dissemination of knowledge in a market economy. In this perspective it is surely perverse to call them imperfections. This is, of course, not a new point; for those who have studied Schumpeter they are the natural features of an economic process driven by creative destruction" (Metcalfe 1998, p. 114).

The MI–MF dilemma has a very strong implication, which is in a direct line with Hicks' previous comments: it requires avoiding any general rule and substituting an analysis of the context and case under scrutiny. Properly solving for the MI–MF dilemma is a necessary requirement for encouraging firms' innovative behaviors. Consequently, assessing the MI–MF dilemma should be at the core of an appropriate competition policy.

One can revisit one of the most recent controversial cases of antitrust, the US against Microsoft, by stressing the importance of this MI–MF dilemma. That case shows interesting characteristics with regard to innovation. The PC (personal computer) industry can be thought of as very representative of the Schumpeterian competition framework, where monopoly power is challenged over time by the introduction of new technologies and new commodities, applications and services. The Microsoft controversy is fundamentally based on some actions by Microsoft to promote the use of its own browser (internet explorer (IE)). Fischer (2000, p. 183) listed a series of "made no business sense" actions that suggest that Microsoft was simply improving the protection of a monopoly power (for instance, pushing Apple into adopting IE, paying AOL to adopt its browser, and signing restrictive contracts with some internet content providers such as the Walt Disney Company). According to Fisher, all those actions were thought to be deliberate attempts to protect Microsoft's monopoly power because it was simply "removing the threat to the applications barrier to entry" (Fischer 2000, p. 183). In other words, those actions only showed how Microsoft aimed at mitigating the existence of alternative browsers (Netscape's Navigator, among others) and limiting the weight of application software driven from those alternative options. The latter was thought of as a source of anti-competitive behavior, as the "applications barrier to entry" argument applied (the more Windows users, the more new applications could be developed; the more new applications, the more new users). Then, Microsoft was accused of leveraging its monopoly power by using its initial Windows platform to suppress Netscape's threat. The perception of defendants (see Schmalensee 2000) was mainly to highlight the PC industry as a very innovative context in which major innovations occur repeatedly and displace established leaderships. In such context, the intensity of competition is obvious and "in a struggle for survival that will have only one winner, any firm must exclude rivals to survive" (Schmalensee 2000, p. 194). Then, the fragility of the players is a major reason to improve continuously their market bases and to add further characteristics and functionalities to their products. That fragility justifies a continuous improvement of product characteristics. Thus, integrating IE into Windows was a logical improvement and not a predatory behavior based on Windows's previous success.

From this rough depiction of the case, one can stress two major related remarks based on our previous analytical reflection. One is the difficulty in identifying the product market under consideration; the other is the importance of time in relation to innovation. Clearly, in the PC industry, there is a definitional problem in identifying the product market. "Plaintiffs defined the relevant markets as operating systems for Intel-compatible PC's and browsers. The defendant contended that the plaintiffs' case and the competition between Microsoft and Netscape were about *platforms*" (Schmalensee 2000, p. 194). This, on the one hand, shows the difficulty to which we previously referred for competition authorities to base their decisions on traditional tools in a dynamic market (i.e. innovative) environment; on the other hand, it illustrates the importance of the MI–MF dilemma we stressed as central for competition policy. Depending on the definition given to platforms (restricted or open) in that industry, and depending on the time horizon (short or large) considered, we get a different perspective about the case. Obviously, plaintiffs promoted a restricted and short-term vision of the market in order to highlight Microsoft's monopoly power and to apply the applications barrier-to-entry argument. Defendants promoted a larger and open conception of the platforms whereby it appears natural to improve continuously the latter as well as to add further functionalities.

Within our analytical framework, we clearly reject the restricted and short-term vision of the limits of product market in the industry because of its intrinsic dynamic character. In other words, there should be a place for market imperfections, as the latter guarantee the viability of the industry as well as its growth-enhancing effect over time. Part of Microsoft's reactions to the success of Netscape can be interpreted as a necessary requirement in order to ensure this objective. In some sense, it expresses the need for coordinating the rate and direction of innovation and change as well as for ensuring the success of future markets. But it should also be added that most of the previous actions listed by Fisher are more than market imperfections and fall into some kind of market failures defined as behaviors or modes of co-ordination that prevent innovative choices from being viable. Innovation in the industry did not require those actions to ensure Microsoft innovativeness. Indeed, if our interpretation stands in a sort of "grey" area, whereby we consider that the defendants were right by promoting a more open perspective about the market issue due to innovation requirements, we also consider that the plaintiffs were right in expressing the incompatibility of some Microsoft actions with the insurance that industry competitiveness can prevail. In other words, it offers a stimulating illustration of what we previously defined as the MI–MF dilemma.

4 What about the EU competition policy

Competition policy cannot be but a discretionary policy when innovation is at stake. The fundamental reason for this continuous adjustment is the nature of the articulation between innovation, competition and growth. As innovation is a disequilibrium process, it cannot be reduced to the adoption of (more) efficient technologies. It is essential to consider what innovation (associated in any change, be technology, organization, or market) has to do with adjustment costs (which are the costs of building new productive options and taking advantage of market opportunities). Mitigating the damaging consequences of these costs is a major issue and has significant consequences in terms of competition policy. In fact, it is certainly important that "market are open, that they facilitate and create incentives to challenge established positions and that they eliminate activities which are no longer viable in the prevailing environment" (Metcalfe and Ramlogan 2004, p. 230). But it is also important that rivalry among firms does not result in excessive turbulence that hampers the viability of the process of change. This requires appropriate (i.e. temporary and changing) market imperfections that prevent excessive and cumulative market disequilibria and hence sustain the growth process. From this viewpoint, it is interesting to contrast the EU with the US competition policy.

Articles 81 and 82 of the EC treaty guarantee a fairly uniform law across the EU countries. The first one restricts agreements that facilitate concentration or introduce discrimination, and hence create market power, while the second one outlaws the abuse of monopoly power. They clearly pay tribute to the standard analysis of competition that focuses on the market structure rather than on the process of innovation. However, article 81 (3) makes inapplicable article 81 (1) if an agreement contributes to promoting technical or economic progress. Thus, sunk costs are considered and dynamic efficiency is explicitly taken into account. The real issue with the EU legislation is much more related to the way policy makers interpret it than to its content. In fact, promoting fair market relationships is a goal sufficiently general and vague to be compatible with the two conceptions of competition policy previously mentioned. Thus, the main difficulty for EU competition policy is to systematically balance between 'a statement of facts' and 'a normative statement' (Neuman 2001, pp. 39–42). This results in various (and changing) policy decisions. The real change lies in an explicit reference to perfect competition as a benchmark, which appears in the Maastricht treaty and corresponds to the general statement consisting in privileging rules over discretion in the different compartments of economic policy, including competition policy.

Indeed, since the late 1980s the European Commission has been using competition policy as a tool for liberalizing markets. This policy plays an essential role in ensuring that market opening leads to the achievement of potential productivity and variety gains. It increasingly questions whether mergers and acquisitions are guided to promote innovation; exercises control over public aids to avoid distortions through subsidies to some companies; has been active in promoting a reorientation from aids to individual companies or sectors to less distorting horizontal measures addressing specific market failures; promotes rivalry, new entry, and hence turbulences at the micro level that are supposed to sustain macroeconomic stability, and goes hand-to-hand with deregulation and privatization. Thus, when examining particular agreements, authorities focus on the market power rather than on the effects on long-term production costs.

Interestingly, this gives a distinctive dimension to the EU competition policy. While the US competition policy only prohibits the illicit use of market power and does not condemn it per se when identified as a major incentive to innovate, EU competition policy has been used as a regulatory policy with the consequence of making unclear the distinction between an ex post and an ex ante intervention, as well as the distinction between MI and MF. Thus it mainly focuses on the properties of the market structure rather than on the conditions and outcomes of the process of change. Moreover, while US competition policy interacts with other dimensions of public intervention in a practical way that favors innovation and growth in an imperfect world, the EU competition policy is defined independently of the global context as if it were in itself a growth policy. In the former case, the guidance of competition policy is not defined in isolation and dedicated to the application of rules that an expected stage of perfect competition should require. This is reinforced by the fact that the common law courts are parts of a legislative tradition in which the aim is not systematically setting a world that would be determined from a priori hypotheses about technology efficiency or consumers' preferences. Decisions result from a more adjusting process based on conjectures about what suited coordination is needed in order to ensure innovation and growth. Summing-up, how to reconcile competition policy with innovation policy is still a challenging issue for the EU.

References

Amendola M, Bruno S (1990) The behaviour of the innovative firm: relations to the environment. Res Policy 19(5):419–433

Amendola M, Gaffard J-L (1998) Out of equilibrium. Clarendon Press, Oxford

Amendola M, Gaffard J-L, Musso P (2000) Competition, innovation and increasing returns. Econ Innov New Technol 9:149–181

Amendola M, Gaffard J-L, Musso P (2003) Co-ordinating the competition process: the role of finance constraints. Rev Austrian Econ 16(2–3):183–204

Amendola M, Gaffard J–L, Musso P (2004) Viability of innovation processes, emergence and stability of market structures. In: Gallegati M, Kirman A, Marsili M (eds) The complex dynamics of economic interaction. Springer, Berlin Heidelberg New York

Baumol WJ (2001) When is inter-firm coordination beneficial? The case of innovation. Int J Ind Organ 19:727–737

Baumol WJ (2002) The free-market innovation machine. Princeton University Press, Princeton

Comanor WS, Leibenstein H (1969) Allocative efficiency, x-efficiency and the measurement of welfare losses. Economica 36:304–309

Crew MA, Rowley CK (1972) Anti-trust policy: economics vs management science. In: Rowley CK (ed) Readings in industrial economics, vol 2, pp 136–149

Dasgupta P, Stiglitz J (1980) Industrial structure and the nature of innovative activity. Econ J 90:270–293
Fisher FM (1991) Organizing industrial organization : reflections on the handbook of industrial organization. Brookings Pap Econ Act Microecon:201–240
Fischer FM (2000) The IBM and Microsoft cases: what's the difference? Am Econ Rev 90 (2):180–183
Hicks J (1983) Classics and moderns, collected essays on economic theory, vol 3. Harvard University Press, Cambridge
Ledyard JO (1989) Market Failure. In: Earwell J, Milgate M, Newman P (eds) The new Palgrave, allocation, information, and markets. Macmillan, London
Leibenstein H (1966) Allocative efficiency vs x-efficiency. Am Econ Rev 15:341–392
Metcalfe JS (1998) Evolutionary economics and creative destruction. Routledge, London
Metcalfe JS (2000) CRIC (Centre for Research on Competition and Innovation) Mid-term review, Part 2. Research Report, University of Manchester, ESRC
Metcalfe JS, Ramlogan R (2005) Competition and the regulation of economic development. Q Rev Econ Finance 45:215–235
Nelson R, Winter S (1982) An evolutionary theory of economic change. Belknap Press of Harvard University Press, Cambridge, MA
Neuman M (2001) Competition policy, history, theory and practice. Edward Elgar, Cheltenham
Ordover JA, Saloner G (1989) Predation, monopolization, and antitrust. In: Schmalensee R, Willig RD (eds) Handbook of industrial organization, vol 1. North Holland, Amsterdam
Petit M-L, Tolwinski B (1999) R&D cooperation or competition? Eur Econ Rev 43:185–208
Quéré M (2000) Competition as a process : insights from the Marshallian perspective. In: Krafft J (ed) The process of competition. Edward Elgar, Cheltenham, pp 49–64
Quéré M (2003) Increasing returns and competition: learning from a Marshallian perspective. In: Arena R, Quéré M (eds) The economics of Alfred Marshall: revisiting Marshall's legacy. Palgrave, London, pp 182–201
Reinganum J (1989) The timing of innovation: research, development and diffusion. In: Schmalensee R, Willig R (eds) Handbook of industrial organization, vol 1. North Holland, Amsterdam
Richardson GB (1960) Information and investment. Clarendon Press, Oxford
Richardson GB (1965) The theory of restrictive trade practices. Oxf Econ Pap 17(3):432–449
Richardson GB (1975) Adam Smith on competition and increasing returns. In: Skinner A, Wilson T (eds) Essays on Adam Smith. Oxford University Press, Oxford. Reprinted in: Richardson GB (1998) The economics of imperfect knowledge. Edward Elgar, Cheltenham
Schmalensee R (2000) Antitrust issues in Schumpeterian industries? Am Econ Rev 90(2):192–196
Schumpeter JA (1942) Capitalism, socialism, and democracy. Allen & Unwin, London
Scitovsky T (1961) Welfare and competition. Allen and Unwin, London
Stigler GJ (1957) The theory of price, 5th edn. Macmillan, New York
Stiglitz JE (2003) Information and the change in the paradigm in economics In: Arnott R, Greenwald B, Kanbur R, Nalebuff B (eds) Economics for an imperfect world. Essays in honor of Joseph E. Stiglitz. MIT, Cambridge, MA
Williamson OE (1968) Economics as an anti-trust defense: the welfare tradeoffs. Am Econ Rev 58:18–36

Thomas Vallée · Murat Yıldızoglu

Social and technological efficiency of patent systems

Abstract This article develops an evolutionary model of industry dynamics in order to carry out a richer theoretical analysis of the consequences of a stronger patent system. The first results obtained in our article are rather consistent with the anti-patent arguments and do not favor the case for a stronger patent system: higher social welfare and technical progress are observed in our model in industries with milder patent systems (lower patent height and patent life).

Keywords Innovation · Technical progress · Patent system · Intellectual property rights (IPR) · Technology policy

JEL Classification O3 · O34 · L52

1 Introduction

The demand for a stronger patenting system has become in the recent period a major source of tension between the U.S. government and the E.U. (see the recent debate on software patents). The US demand is generally motivated by the conventional economic wisdom affirming that a strong patenting system yields convenient incentives for the private investment in Research and Development (R&D) and hence for technical progress in society. This rather mechanistic approach of technological dynamics and of the role of the patenting is mainly based

Murat Yıldızoglu gratefully acknowledges the support of the CCRRDT Program of the Aquitaine region.

T. Vallée
LEN, Université de Nantes, Chemin de la censive du Tetre,
44322 Nantes, France
E-mail: Thomas.Vallee@univ-nantes.fr

M. Yıldızoglu (✉)
E3i, IFReDE-GRES, Université Montesquieu Bordeaux IV,
Avenue Léon Duguit, 33608 Pessac, France
E-mail: yildi@u-bordeaux4.fr

on the neoclassical theory of technical progress that strongly focuses on the agents' incentives rather than on the dynamics of the existing technological systems.

1.1 What is a patent?

The US Patent and Trademarks Office (USPTO) gives the following definition on its web site:

> A patent for an invention is the grant of a *property right to the inventor*, issued by the United States Patent and Trademarks Office. Generally, *the term of a new patent is 20 years* from the date on which the application for the patent was filed in the United States or, in special cases, from the date an earlier related application was filed, subject to the *payment of maintenance fees*(...). The right conferred by the patent grant is, in the language of the statute and of the grant itself, *"the right to exclude others from making, using, offering for sale, or selling" the invention in the United States or "importing" the invention into the United States.* (...) Once a patent is issued, the patentee must enforce the patent without aid of the USPTO. [US Patent and Trademarks Office legal web site[1]]

This definition puts a particular stress on the role of a patent as a certificate of private ownership of an invention. This ownership gives to the inventor the right to exclude anyone from the provision of this invention in the US. From the point of view of the inventor, the benefits of this exclusivity (the monopoly rent) should be opposed to the costs of filing and maintaining a patent. These dimensions of the patent system directly follow from the theoretical arguments that have been used to defend the establishment of patent systems during the 19th century patent controversy.

1.2 Arguments for a patent system

Following the arguments that have been developed by the defendants of a patent system during the patent controversy of the 19th century (see van Dijk 1994), a patent is

- the natural property right in ideas;
- the just reward for the inventor;
- the best incentive to invent;
- the best incentive to disclose secret information.

These arguments are based on the assumption that the main motivation of an *intentional* innovation is the monopoly rent that can be obtained using a superior technology or product.

Modern theoretical arguments emphasize the particularities of new technical knowledge for justifying the necessity of protecting intellectual property rights.

---

[1] http://www.uspto.gov/

Following these arguments,

- knowledge has two important characteristics: nonrivalry and nonexcludability;
- technical knowledge is a source of externalities in R&D.

These two dimensions may induce an underprovision of this *public good*. As a consequence, the correct incentives must be established through a temporary monopoly position and the public disclosure of private information. These arguments have regularly been questioned by the opponents of a strong patent system. This is quite natural given that patent systems concern the reallocation of rents in society and possess, as such, a political dimension.

1.3 Some stylized facts about patenting

Following van Dijk (1994), Cohen et al. (2000), Gallini and Scotchmer (2002), Hall (2002), and Mansfield (1986), we can specify some stylized facts about patenting:

- most patentable innovations are patented (the exact proportion is industry-specific);
- inventing around a patent occurs (with an average cost advantage of 35%);
- most innovations combine elements from existing products;
- the effective lifetime of a patent is generally shorter than the legal lifetime (less than 8 years for the 50% of the patents in the UK and France);
- patents are useful to impede imitation (the supplementary imitation cost due to the existence of a patent is industry-specific, with weights from 7% to 30%);
- the propensity of patenting has heavily increased in the last decade. This propensity is industry-specific and it is higher for larger firms.

There are more than four million patents in force in the world today, and every year applications are filed for a further 700000 inventions. In 2002, the European Patent Office (EPO) received over 160000 patent applications. Figure 1 clearly shows the *explosion* of the number of patent applications and of effectively granted patents in Europe.

1.4 Behind the incentives

Is patenting the only tool for protecting an innovation? The results of the 1994 Carnegie Mellon Survey of the U.S. manufacturing sector are summarized in Cohen et al. (2000). Following these results, firms declare that they have at their disposal a variety of tools, and they do not necessarily prefer patents to other means for protecting their innovations. In fact, firms' declarations imply the following ordering of these mechanisms with decreasing protective effectiveness:

1. Secrecy;
2. Lead time;
3. Complementary sales/services;
4. Complementary manufacturing;
5. Patenting

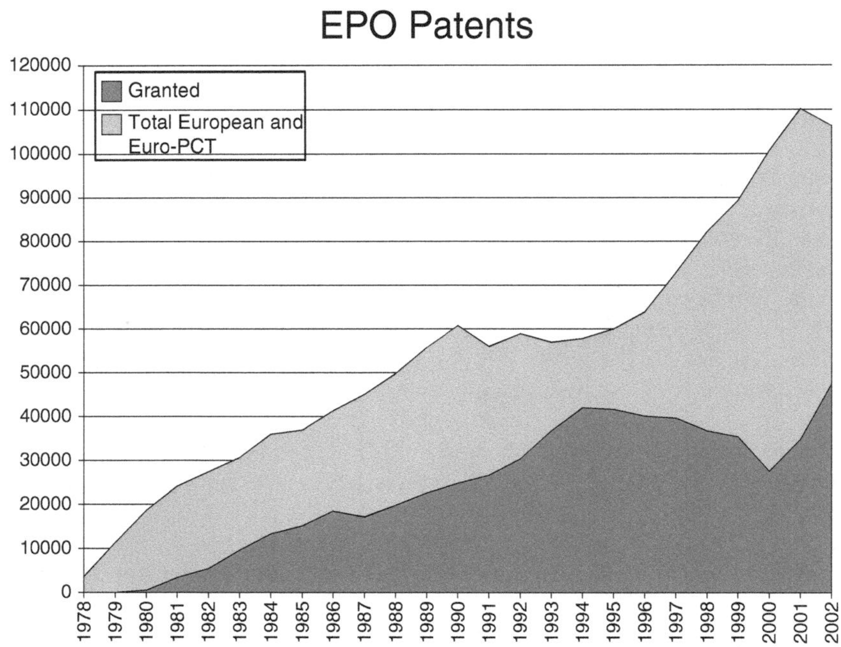

Fig. 1 Applications to the European Patent Office (EPO)

Secrecy is considered by the firms as the most effective mechanism for protecting process innovations, while lead time is considered to be slightly more effective for product innovations. This survey clearly shows that the main motivation for patenting does not correspond to the theoretical argument used in defense of a stronger patenting policy (better incentives for R&D). This observation, combined with the recent surge in patenting gives rise to what is called today the *patent paradox* (low effectiveness but high patenting).[2]

Patenting is mainly used by firms for strategic reasons: constructing patent fences around discrete inventions; building negotiation power through a patent portfolio in complex industries, especially for cross-licensing issues, etc. This strategic use of the patent system must be taken into account in the evaluation of its social costs and benefits. Patent systems are promoted on the basis that they are the least expensive means to provide incentives for innovation. The anti-patent movement argues that patents are inefficient and expensive: the costs of bureaucracy (strongly increased during the recent patent surge– see Fig. 1), court personnel and lawyers make the patent system very costly and unattractive. These costs add to the welfare lost due to monopoly granted by the patent system. Even worse, building *patent fences* around discrete innovations can constitute *patent thickets* implying strong dynamical inefficiencies in innovation systems.

The social cost of defensive patenting (Cohen et al. 2000) follows from the fact that, in these cases, patenting does not foster inventions. Merges and Nelson (1990) and Mazzoleni and Nelson (1998) emphasize the complex nature of the dynamics of technology in many industries. Merges and Nelson (1990) show that a stronger

[2] See Hall and Ziedonis (2001) for electronics firms.

patent system can have very different effects on different industries. They distinguish four classes of technologies in which the role of patents can be strongly contrasted: *discrete inventions* (new pharmaceuticals), *cumulative technologies* (aircraft), *chemical technologies* and *science-based technologies* (biotechnology).

A "one size fits all" system of intellectual property rights seems quite illusionary and can sometimes generate strong social dynamic costs by blocking the development of complementary innovations or of better substitutes. The diversity-reduction effect of broad patents on prospect opening inventions can generate high social costs. As a consequence, models of innovation and of patenting must take into account the complexity of different technological regimes while evaluating the global impact of a stronger patent system and of its dimensions (mainly scope and length). An agent based approach to the industrial system, composed of boundedly rational firms, can help us to locate main issues in this debate. Moreover, at the more specific level of the patent race models, the main results are generally *too* strongly sensitive to the rational expectations (or perfect foresight) assumption and to the assumed homogeneity of the firms.[3] The main results of this literature should be tested against more realistic assumptions before using them in the analysis of intellectual property regimes. This article develops an evolutionary model of industry dynamics, aiming to enhance our theoretical understanding of the consequences of a stronger patent system. The next section will briefly present the main characteristics of our model. The third section will be dedicated to the presentation of our simulation protocol and of the first results of the basic model. The last section will conclude the presentation.

2 The model

This model concerns an industry producing a homogenous good and facing a decreasing market demand. The only production factor is physical capital, and technology has constant returns to scale (it is linear). In each period, each firm shares its gross profits between different investment outlets: R&D, physical capital, patent budget, saving (equity) and distribution of dividends. R&D investment is necessary for the imitative and innovative activity of the firm, and these are the only sources of productivity gains in the model. Technical progress is disembodied and corresponds to the increase of the productivity of the firm's capital stock.

The industry is initially populated by firms with random characteristics (drawn following a normal distribution centered around common averages). The short-period market equilibrium fixes the price at which the consumers agree to buy this product, given their demand in each period. Market price determines the firm's gross profits, and these profits are used for investing in different assets (strategies): innovation and imitation follow from the R&D investment; physical capital increases as a consequence of the investment; the patent budget is used to finance new patents or to renew the patent portfolio of the firm; dividends are distributed to the consumers and they can increase demand; the equity is used as saving and can provide supplementary revenues for investing in future periods (see Fig. 2). In this model, we dedicate particular attention to the patenting strategies of the firms.

[3] See Silverberg and Yildizoglu (2002) for a discussion of this problem in the context of the Aghion and Howitt (1992) model.

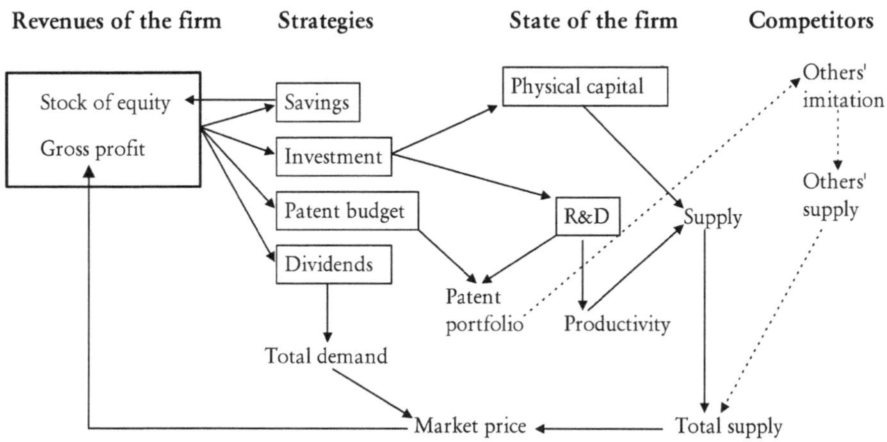

Fig. 2 The main connections in the model

In the rest of this section we will briefly present the main components of the model.

2.1 Strategies and learning processes of firms

At the beginning of each period, the firm must decide how to spend the gross profits and the savings from the previous periods. In our model, these revenues can be allocated between five alternative assets (see Fig. 2):

Investment in physical capital The firms expands its capital stock in order to increase its market share. *IKRATE* is the initial average value of this investment rate around which the strategies of the firms are created.

R&D investment R&D allows the firm to create new technologies, or to imitate the technology of a successful competitor. The obsolescence cost of the R&D stock, for a given firm, depends on its relative productivity, compared to the maximum level of productivity in the industry. Hence, if a firm's productivity is low compared to the maximum level of productivity in the industry, its R&D stock will be subject to a high degree of obsolescence. As a consequence, the firm with the highest productivity in the industry faces no obsolescence cost. Thus, even without any new specific investment, a leading firm will be able to keep constant its R&D stock. *IRDRATE* is the initial average value of this investment rate around which the strategies of the firms are created.

Patent budget In order to prevent other firms from benefiting from its own technological investments, a company can decide to protect an innovation. We assume that a technology may only be patented if it is sufficiently distinct from an already patented technology. The patent office can be more or less indulgent and this dimension of the patent system is measured in our model by the variable *PATENTHEIGHT*. A patented technology can be protected for a maximum of

PATENTLIFE periods. A new patent costs *NEWPATENTCOST*, and renewing a patent for a period necessitates the payment of *RENEWPATENCOST*. *PATENTRATE* is the initial average value of this investment rate around which the strategies of the firms are created.

Dividends Companies can redistribute a part of their profits to shareholders and thus to households. In this simplified model, this is the only way by which the total demand increases. The coefficient that transforms the distributed dividends into demand increases is γ. *DIVIDENDRATE* is the initial average value of this investment rate around which the strategies of the firms are created.

Savings Companies can save voluntarily and/or involuntary, part of their profits. Involuntary savings arise when one of the budget lines is not spent in its totality. This saving is precautionary, since it enables a company to offset certain consequences of unforeseen events (e.g. negative profits). In our model, if a company experiences negative profits and does not have any more saving, it quits the industry. *EQUITY RATE* is the initial average value of this investment rate around which the strategies of the firms are created.

In each period, the learning of the firms is represented through an evolutionary algorithm: firms learn through imitation of the strategies of others and through random experimenting (mutations). In our model, imitation is based on the market size of the opponents, rather than on their profits (as in Silverberg and Verspagen 1994). As a consequence, a bigger competitor will have a higher probability of being imitated. These two mechanisms are, respectively, commanded by the probabilities *PROBMIMITATE* and *PROBMUTATE*.

2.2 Technical progress and patenting

Technical progress is a result of the innovation or imitation processes of firms. The success of these processes is an increasing function of the R&D investment of the firms. Firms may file patents in order to protect their technologies from imitation by competitors.

2.2.1 Productivity gains: innovation and imitation

In our model, innovation is a two-stage stochastic process. A first draw determines if the firm has been successful to innovate. The probability of this success increases with the R&D investment. A second draw then gives the effective new productivity that results from the innovation.

It should be noted that a new technology may only be used and patented if it is not covered by an existing patent.

A firm can also benefit from imitating a successful competitor's technology. Imitation is rather rare and the probability of success again increases with the R&D investment of the firm. Only unpatented technologies can be imitated. When the imitation happens, each competitor has a probability of being imitated that increases with its market share.

2.2.2 Patenting

The management of the patent portfolio is very crucial in our model. In this version we do not allow any sleeping patents. Hence, when a new technology is found, the inventor can choose to protect it by filing a patent. If the firm does not protect it, the technology may be imitated or invented around by the competitors. A firm will only desire to patent a technology if (a) the technology is seen as sufficiently interesting to patent, and (b) the firms has a sufficient budget. More specifically, the probability of adopting (or keeping) a particular patent is given by a normal distribution that depends on the relative efficiency of the technology. Efficiency of a given technology is measured by the number of firms with a productivity lower than the productivity of this technology: the higher the number of such firms, the more efficient the patenting. We assume that firms cannot perfectly observe the efficiency of their innovation and they are prone to errors.

2.2.3 Management of the patent budget

In the beginning of each period, the firm will try to reserve a budget for patenting. This budget should cover two kinds of expenses: (a) the cost of maintaining previously filed patents, (b) the possible cost of filing a patent for a new innovation. This budget will result from the investment strategy of the firm on patenting.

2.3 Entry and exit

In this model, the size of the industry, in terms of active firms, is allowed to change at each period. Nevertheless, an upper bound is fixed, which is the initial size N.

Even with negative profits, a firm may stay in the industry as long as it holds some positive savings that offset the loss. When this is no longer the case, the firm exists the industry (the case of bankruptcy). If the number of active firms is lower than N at the beginning of a period, some new firms may enter the industry. For example, persistent high profits or increases of the level of profits in an industry may be an attractive signal for new entry. When entry occurs, the characteristics of the new firms are drawn from values around the industry-averages. If a potential new entrant is not profitable at the current market price, and/or if the technology found by it is patented, the potential firm will not be able to enter. The probability of entry is *ENTRY PROB*. The entry is also limited, in our model, by the inverse Herfindhal concentration index.

2.4 The pseudo-code of the model

We start with a population of N firms in the industry. We assume that each firm is initialized with random strategies that are drawn from the same normal distributions.

The algorithm of the model runs the following steps in each period t:

1. Populating the industry:
   - if t=1: creation of an industry composed of N firms
   - if t>1: death and birth process
2. For each period t, until t=T:
   (a) Computation of the production levels: Q_t^i and the total supply Q_t
   (b) if t>1 evolution of the demand D_t (depends on past dividend strategies)
   (c) Computation of the intra-period price (as a function of the inverse demand function): p_t
   (d) Randomize the order of play of firms in the current period t
   (e) Computation of the gross profits
   (f) Definition of the different budget levels for R&D, investment, patenting, savings and dividends
   (g) Computation of the list of all patented productivities in the industry
   (h) Imitation of technologies
   (i) Innovation of firms
   (j) Management of the patent portfolio and patenting
   (k) Diffusion of the best strategies in the industry (depends on the market shares of the firms)
   (l) Mutation of strategies: possible change of the individual set of strategy rates

3 Simulation protocol and first results

3.1 Simulation protocol

Given the complexity of the interactions that we model, we adopt a methodology that allows quite a systematic exploration of the parameter space of the model. This methodology is close to the Monte Carlo method. We run 1,000 series of 500 periods[4] each, where the results from each period have a probability of 2% of being saved. So, for each run we obtain an average number of ten randomly chosen observations for all the measured variables. The simulations are initialized with a randomly drawn vector of values for the main parameters of the model. As a result, we obtain a set of 10000 observations covering quite a diversified subset of the parameter space. The values from which different parameters are drawn can be read in Appendix A. We do not necessarily discuss in the text all the parameters that appear in this appendix, but only the most significant ones. We analyze the observations sampled from the last half of each run, for dates higher than the second quartile of the saved periods ($t \geq Q_2^t = 254$). We use for this analysis box plots (giving the four quartiles of the distributions of the variables), Wilcoxon–Mann–Whitney

[4] Running 1,000 simulations is sufficiently robust and secure in our case since $\sigma^2/\bar{x}$ becomes stable after 500 runs for any variable x in our model.

tests between subsets, and regression trees. The statistical analysis is conducted using R (see R Development Core Team 2003).

3.2 First results on patents and social welfare

Table 1 shows the influence of the dimensions of the patent system on some of our aggregate indicators (market price, productivities, number of firms...). We only present here the influences that are statistically significant at a unilateral test level higher than 5% in linear regressions computed between each of the dependent variables and the indicated four dimensions of the patent system.

Table 1 indicates that a stronger patent system (with a longer PATENTLIFE, and a more indulgent patent office corresponding to a stronger PATENTHEIGHT) would imply higher concentration, market prices and profits. These benefits for the firms in the economy would also have a social cost in terms of technical progress, since the average and maximal productivity would be lower under such a system. This phenomenon would also be concomitant with longer effective patent lives and fewer innovations. The influence of the cost dimensions is rather obvious and marginal.

Would these influences necessarily yield a lower social welfare? Answering this question using only the market price is not very straightforward in our model since, when distributed as dividends, the profits increase the revenues of the consumers and yield a higher consumers' surplus. As a consequence, a higher price does not automatically imply in our model a lower consumer' surplus. This is even more likely given the increasing relationship between the PATENTLIFE and the

Table 1 The role of characteristics of the patent system ($t \geq Q_2^t = 254$)

Variable	PL	PH	NPC	RNPC
price	+	+		
maxprod	–	–		+
averprod	–	–		+
activeN	–	–		–
invCI	–	–		
log (averprofit)	+	+		
nbinnov	–	–		
nbpat	+	–		–
cumnbpat	+	–		–
maxpatage	+	+	+	–
nbpatfirms	+	–	+	–
Behaviors				
avpatrate	+		–	–
avirdrate	–	–	+	+
avikrate			+	–
avequitrate		–	+	–
avdivrate	+		–	

PL PATENTLIFE; *PN* PATENTHEIGTH; *NPC* NEWPATENTCOST; *RNPC* RENEWPATENTCOST average dividend rate of the firms (see the last line of Table 1). So, we must take a better look at the social welfare for judging its evolution under a stronger patent system.

Figure 3 gives the distribution of the social welfare (consumers' surplus and total social surplus[5]) for different patent systems: for each of the dimensions (PATENTLIFE and PATENTHEIGHT) we call *low* the value of this dimension if it is lower than the second quantile of this variable and we call it *high* otherwise. The configuration *hl* corresponds, for example, to a situation where the *PATENTLIFE* is *high* and the *PATENTHEIGHT* is *low*. As is shown by the boxplot,[6] the highest social surplus is observed when both dimensions are low and, hence, the patent system is mild. Non-parametric Wilcoxon–Mann–Whitney tests[7] confirm this graphical result. As a consequence, the positive impact of a stronger patent system on the profits of the firms does not finally outweigh the negative impact on consumer welfare. This result casts a shadow on the admitted social efficiency of strong intellectual property rights. Moreover, the similarity between the configurations *hh* and *lh*, on the one hand, and between *hl* and *ll* on the other, indicates that PATENTHEIGHT dominates the impact of the patent system on social surplus. This point calls for a more detailed analysis of the determinants of social welfare in our model.

We analyze the role of different parameters of the model using regression trees. A regression tree establishes a hierarchy between independent variables using their contributions to the overall fit (R^2) of the regression. More exactly, it splits the set of observations in sub-classes characterized by their value in terms of their contribution to the overall fit and of their predictions for the dependent variables (all parameters that are modified by the Monte Carlo procedure are included as explanatory variables in each of the following regressions). This value is validated

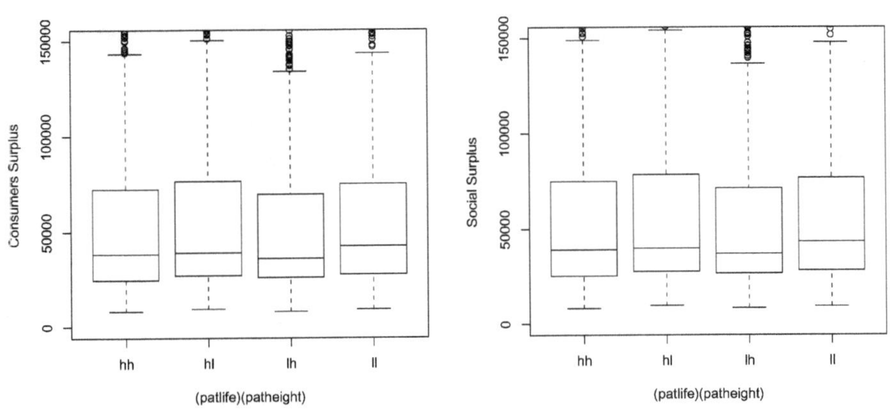

Fig. 3 Patent strength and social welfare

[5] Social suplus=consumers' surplus+total profits of the firms.

[6] These boxplots show four quartiles of the distributions of our indicators: the statistically significant minimum and the maximum correspond to the extreme end of the whiskers, while Q_1 and Q_3 correspond to the edges of the central box and the median corresponds to the horizontal line inside the box.

[7] The statistical appendix may be obtained from the following address: http://beagle.u-bordeaux4.fr/yildi/files/tvmy1appendix.pdf.

against a fraction (10%) of the sample that is not used during the estimation. Regression trees are very flexible and powerful in the clarification of the structure of the observations. The tree gives a hierarchical sequence of conditions on the variables of the model: the higher the role of a condition in the classification of the observed cases, the higher its status on the tree. For each condition, the left branch gives the cases for which the condition is true and the right branch gives the cases that are compatible with the complementary condition. We give now a step-by-step interpretation of the main elements of the regression tree exposed in Fig. 4.

Figure 4 shows that the main determinant of the social welfare in this model is the demand effect of the distributed dividends. γ is the coefficient through which dividends are transformed into supplementary demand by the consumers. The top first branching of Fig. 4 shows that the social surplus is the lowest ($\log(SS)= 10.47) \Rightarrow SS=35242.22$) when this coefficient is very small (the left branch: $\gamma< 0.008993$). In our model, 2703 observations correspond to this case. When γ is higher, the second component of this demand effect enters onto the scene: the initial average dividend rate (*DIVIDENDRATE*) around which the firms are initialized during the creation of the industry. The highest social welfare is obtained in the model when this value is very high (*DIVIDENDRATE*$\geq$0.8674 – the highest result corresponding to $\log(SS)=19.06$ can only be obtained in the right branch of this test). If the dividend rate is more reasonable (*DIVIDENDRATE*<0.8674 – the right branch), distinguishes again two sets of cases: on the left, we have the cases where $\gamma<0.01686$ and, on the right the ones with $\gamma\geq0.01686$. The first set of cases confirms again the preponderant role played by the demand effect in our model.

We should also note an interesting result that concerns the role played by the cost of new patents: the highest welfare in this case is obtained when firms distribute large dividends and when the *NEWPATENTCOST* is higher (so that they issue fewer patents given the lower resources dedicated to patenting and the higher cost of each patent). The second set of cases (corresponding to $\gamma\geq0.01686$ and *DIVIDENDRATE*<0.8674) is more interesting. Here the highest social welfare ($\log(SS)=17.31) \Rightarrow SS=32933469$) is observed when *(i)* initially firms invest sufficiently in physical capital (*IKRATE*$\geq$0.21), *(ii)* the height of the granted patents is not too high or too low (*PATENTHEIGHT*$\in$[1.138, 2.124]) and *(iii)* the depreciation of technological knowledge is not *too* strong ($\alpha<0.1207$). When

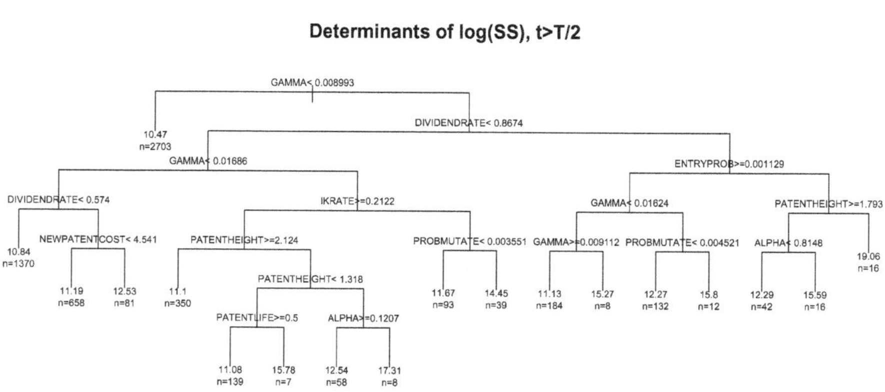

Fig. 4 Determinants of social surplus

PATENTHEIGHT<1.318, the highest welfare is obtained when the official patent life is nil and thus patents do not play any positive role in the profits of the firms (*PATENTLIFE*<0.5⇔*PATENTLIFE*=0). These results show that, under each set of conditions concerning other variables, welfare is higher when the patent system is milder.

We summarize in the following proposition our main conclusions concerning the welfare effects of a stronger patent system.

Proposition 1 (Welfare effects) *A stronger patent system implies a lower social welfare even if it implies higher profits for the firms. The indulgence of the patent office on the height of the accepted patents is the main determinant of this negative social effect. As a consequence, a more demanding and restrictive patent office that attributes very tightly defined protection with each patent would be beneficial to social welfare.*

What about the technological effects of the patent system? Figure 5 shows the distribution of the average productivity and of the maximal productivity, both conditioned by the dimensions of the patent system: the officially allowed patent life (*patlife*) and the patent height (*patheight*) that also corresponds to the patent breadth in our one-dimensional technological system. As before, the qualifications high (*h*) and low (*l*) result from the comparison with the second quantile (Q_2) of each variable. We again observe here the predominant role played by the *PATENTHEIGHT*: the technical progress is stronger when this dimension of the patent system is lower and hence, when the protection accorded to each patent is narrower in the technology space. The weakest technical progress is observed when both dimensions are high, and hence the patent system is strong. *A contrario*, a weak patent system yields the strongest technical progress.

We use a regression tree for a more detailed analysis of these results. Figure 7 exposes the determinants of technical progress measured by the average productivity in the industry during the second half of its life. This figure first shows that technical progress is the lowest when the *PATENTHEIGHT* is very high (>2.734),

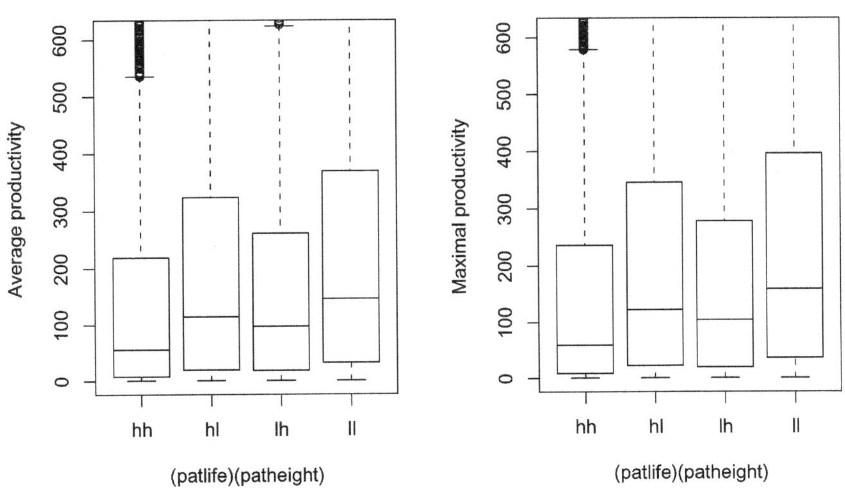

Fig. 5 Patent strength and technical progress

and hence the patent office is very indulgent. The predominant role played by this dimension of the patent system confirms the result of Fig. 3.

For lower heights, the role played by the *PATENTLIFE* variable is more ambiguous: its value (PATENTLIFE$\gtreqless$1.5) does not really have a consequence on technical progress itself but on the nature of the model, since two different dimensions of learning play an important and distinct role in each of these two sets of cases. When the patent life is very low, random exploration of strategies represented by the mutation operator must remain very weak (*PROBMUMATE*< 0.0285%) in order for the industry to attain the highest productivities. When the allowed maximal life of the patents is longer, some significant entry (*ENTRY PROB*>0.726%) and learning through imitation (*PROBIMITATE*≥0.45%) are necessary for transforming lower investments rates on physical capital (*IKRATE*< 0.1568) to the highest productivities for the industry. Moreover, Fig. 6 shows that the firms rarely maintain their patents during the full legal patent life, and thus confirms the stylized fact that we have underlined in the introduction. As a consequence, the variable *PATENTLIFE* only plays a role when its value is very low. *A contrario*, when the patent life allows a significant role for the patents, the learning of strategies, and the budget constraint that binds together all strategies, become crucial for technical progress.

We can also check that these results comply with the main result of the Fig. 3, that is, the highest average productivity (2877) is compatible with the configuration (*hl*) (high *PATENTLIFE* and low *PATENTHEIGHT*). Figure 3 shows that this configuration provides a higher consumer surplus than configuration (*ll*): the higher the average productivity, the lower the price, and, the higher the consumer surplus.

Figure 7 also demonstrates the role of learning and diversity in technical progress: in both cases (with low or high *PATENTLIFE*), a lower *ENTRY PROB*

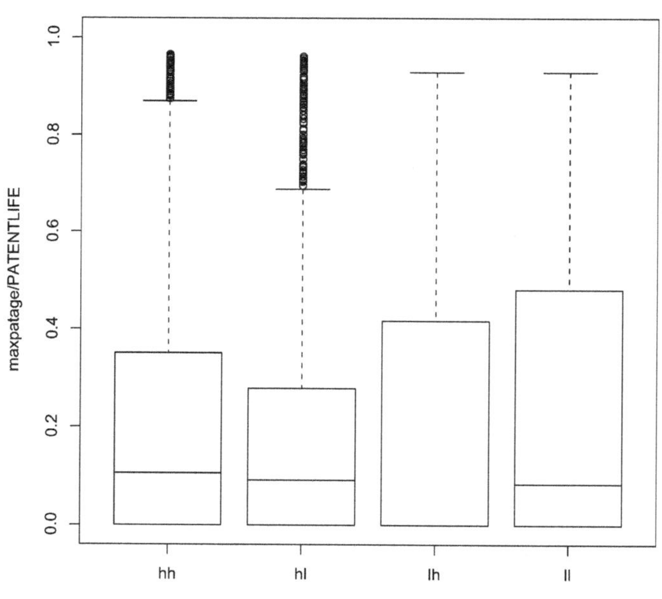

Fig. 6 Patent strength and effective patent life

Determinants of averProd, t>T/2

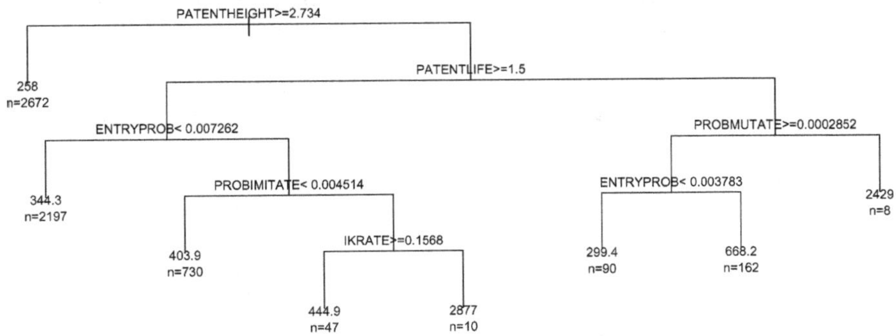

Fig. 7 Determinants of average productivity

decreases the average productivity. This result is quite normal, since a lower value of this variable implies a lower degree of diversity of the strategies and the technologies of the firms and, hence, a possibly higher degree of market concentration. The latter can, of course, lead to fewer innovations and, consequently, to a lower average productivity in the industry.

Proposition 2 (Technical progress) *A stronger patent system implies weaker technical progress even if the firms do not fully exploit the officially permitted patent life. The height/breadth of the granted patents play a major role in this result and the weakest technical progress is observed when this height is very strong. When the allowed maximal life of the patents is longer, some significant entry and some intraindustry communications that allow learning through imitation are necessary for transforming lower investments rates on physical capital to the highest productivities for the industry.*

4 Conclusion

> "If we did not have a patent system, it would be irresponsible, on the basis of our present knowledge of its economic consequences, to recommend instituting one. But since we have had a patent system for a long time, it would be irresponsible, on the basis of our present knowledge, to recommend abolishing it." (Machlup 1958 – cited by Hall 2002).

This article develops an evolutionary model of industry dynamics in order to carry out a richer theoretical analysis of the consequences of a stronger patent system. Even if we agree with Machlup that the question of abandoning the existing patent system cannot be easily posed, the necessity for a stronger patent system, as it is defended by the North-American authorities, must be questioned. The first results obtained in our article are rather consistent with the anti-patent arguments and they do not favor the case for a stronger patent system: higher social welfare and technical progress are observed in our model in industries with milder patent systems (lower patent height and patent life). Even if the firms' profits are higher with a stronger patent system, the global effects are negative, since they correspond to lower social welfare and weaker technical progress. If the actual patent system is to be reformed, the system should be limited in scope instead of strengthened as demanded by American authorities. We do not have any evidence in this model for advising the generalization of patent systems to areas where innovations are not actually protected through this type of IPR.

But we must now refine and improve this analysis in two directions. First, we must carry out a more detailed analysis of individual results in order to understand the mechanisms that are behind our results. Second, we must develop a richer technology space in order to include in the analysis different technological regimes as stressed by Merges and Nelson (1990). Moreover, by adopting a multidimensional technology space, we can distinguish the length and the breadth of the patents and have a more subtle and more realistic apprehension of the consequences of a stronger patent system. On a complementary level, we should also study the possible role of more forward-looking behavior on behalf of the firms.[8]

Appendix

1. Initialisation of the main parameters of the model

1.1 Exogenous variables

N=50: Number of firms
T=500: Number of periods
PROBIMITATE ∈ [10, 0.005]: Probability of imitation
PROBMUTATE ∈ [0, 0.005]: Probability of mutation
SIGMA_IN ∈ [0.1, 5]: Standard deviation of the innovative draws
DIVIDENDRATE ∈ [0, 1]: Initial average share of the distributed dividends in the gross profits
PATENTRATE ∈ [0, 1]: Initial average share of the patent budget in the gross profits
EQUITY RATE ∈ [0, 1]: Initial average share of the savings in the gross profits
IKRATE ∈ [0, 1]: Initial average share of the investment in physical capital in the gross profits
IRDRATE ∈ [0, 1]: Initial average share of the R&D budget in the gross profits
ENTRY PROB ∈ [0, 0.01]: Probability of new entry
ALPHA ∈ [0, 1]: Depreciation rate of the technological knowledge of the firm
GAMMA ∈ [0, 0.02]: Transformation rate of dividends into supplementary demand
NEWPATENTCOST ∈ [0, 5]: Cost of filing a new patent
RENEWPATENTCOST ∈ [0, 1]: Cost of renewing an existing patent
PATENTHEIGHT ∈ [0, 5]: The height of the granted patents. If the patent correspond to the productivity A_0, all productivities in [A_0−*PATENTHEIGHT*, A_0+*PATENTHEIGHT*] are protected from the competitors.
PATENTLIFE ∈ [0, 30]: Legal maximal life of patents

[8] See Yildizoglu (2001) for a possible modelling strategy.

EQUITY ∈ [10, 50]: Initial equity of the firms
CF ∈ [0, 2]: Fixed costs of the firms
K ∈ [10, 50]: Initial average capital stock of the firms
PROD ∈ [0.2, 1.2]: Initial average productivity of the firms
COST ∈ [0, 1]: Initial average unit using cost of the capital
DEM ∈ [300, 1000]: Initial demand coefficient
ETA ∈ 0.9: Elasticity of demand

1.2 Endogenous variables

price: Market price
max prod: Maximal productivity of the period
averprod: Average productivity of the period
activeN: Number of active firms in the industry
invCI: Inverse Herfindal index of the period
averprofit: Average profits
nbinnov: Number of innovations in the period
nbpat: Total number of active patents in the period
cumbpat: Cumulated number of the patents in the industry history
max patage: Age of the oldest active patent
nbpatfirms: Number of patenting firms in the period
avpatrate: Average percentage of the patent budget in the gross profits
avirdrate: Average percentage of the R&D budget in the gross profits
avikrate: Average percentage of the capital investment budget in the gross profits
avequitrate: Average percentage of the savings in the gross profits
avdivrate: Average percentage of the distributed dividends in the gross profits

References

Aghion P, Howitt P (1992) A model of growth through creative destruction. Econometrica 60:323–351

Cohen WM, Nelson R, Walsh J (2000) Protecting their intellectual assets: appropriability conditions and why U.S. manufacturing firms patent (or not). Working paper series 7552, NBER, Cambridge, MA

Gallini N, Scotchmer S (2002) Intellectual property: when is it the best incentive system? In: Jaffé A, Lerner J, Stern S (eds) Innovation policy and the economy, Vol. 2. MIT, Cambridge: MA, pp 51–78

Hall B (2002) Current issues and trends in the economics of patents. Lecture to the ESSID Summer School in Industrial Dynamics, http://emlab.berkeley.edu/users/bhhall/papers/BHH

Hall BH, Ziedonis RH (2001) The patent paradox revisited: an empirical study of patenting in the U.S. semiconductor industry, 1979–1995. Rand J Econ 32(1):101–128

Machlup F (1958) An economic review of the patent system. Study No. 15 of Commission on Judiciary, Sub comm. on Patents, Trademarks, and Copyrights, 85th Congress, 2nd Session

Mansfield E (1986) Patents and innovation: an empirical study. Manage Sci 32:173–181

Mazzoleni R, Nelson RR (1998) The benefits and costs of strong patent protection: a contribution to the current debate. Res Policy 27:273–284

Merges R, Nelson RR (1990) On the complex economics of patent scope. Columbia Law Rev 90:839–916

R Development Core Team (2003) R: A language and environment for statistical computing. Vol. http://www.r-project.org/, R Foundation for Statistical Computing, Vienna

Silverberg G, Verspagen B (1994) Learning, innovation and economic growth: a long run model of industrial dynamics. Industrial and Corporate Change 3

Silverberg G, Yildizoglu M (2002) An evolutionary interpretation of the Aghion & Howitt (1992) model. Working Paper 2202-3, IFReDE-E3i, Univ. Montesquieu Bordeaux IV

van Dijk TWP (1994) The limits of patent protection. Essays on the Economics of Intellectual Property Rights, Universitaire Pers Maastricht

Yildizoglu M (2001) Connecting adaptive behaviour and expectations in models of innovation: the potential role of artificial neural networks. Eur J Econ Soc Syst 15(3):203–220

Kolibri 360 GmbH
Ferdinand-Jühlke-Straße 7
99095 Erfurt, Deutschland
produktsicherheit@kolibri360.de